The Adaptive Landscape in Evolutionary Biology

EDITED BY

Erik I. Svensson

*Department of Biology,
Lund University, Sweden*

and

Ryan Calsbeek

*Department of Biological Sciences,
Dartmouth College, USA*

OXFORD
UNIVERSITY PRESS

OXFORD
UNIVERSITY PRESS

Great Clarendon Street, Oxford, OX2 6DP,
United Kingdom

Oxford University Press is a department of the University of Oxford.
If furthers the University's objective of excellence in research, scholarship,
and education by publishing worldwide. Oxford is a registered trade mark of
Oxford University Press in the UK and in certain other countries

© Oxford University Press 2012

The moral rights of the authors have been asserted

First Edition published 2012

Reprinted 2013

Impression: 2

All rights reserved. No part of this publication may be reproduced, stored in
a retrieval system, or transmitted, in any form or by any means, without the
prior permission in writing of Oxford University Press, or as expressly permitted
by law, by licence or under terms agreed with the appropriate reprographics
rights organization. Enquiries concerning reproduction outside the scope of the
above should be sent to the Rights Department, Oxford University Press, at the
address above

You must not circulate this work in any other form
and you must impose this same condition on any acquirer

British Library Cataloguing in Publication Data

Data available

Library of Congress Cataloging in Publication Data

Library of Congress Control Number: 2012931979

ISBN 978–0–19–959537–2 (Hbk.)
     978–0–19–959538–9 (Pbk.)

Printed and bound by
CPI Group (UK) Ltd, Croydon, CR0 4YY

Links to third party websites are provided by Oxford in good faith and
for information only. Oxford disclaims any responsibility for the materials
contained in any third party website referenced in this work.

Contents

List of Contributors x
Preface xii
Acknowledgments xv

Part I Historical Background and Philosophical Perspectives

1 A Shifting Terrain: A Brief History of the Adaptive Landscape 3
Michael R. Dietrich and Robert A. Skipper, Jr.

   1.1 Introduction 3
   1.2 The origins of the Adaptive Landscape 3
   1.3 The genetic landscape 5
   1.4 The phenotypic landscape 10
   1.5 The molecular landscape 12
   1.6 Conclusion 14

2 Sewall Wright's Adaptive Landscape: Philosophical Reflections on Heuristic Value 16
Robert A. Skipper, Jr. and Michael R. Dietrich

   2.1 Introduction 16
   2.2 Sewall Wright's Adaptive Landscape 16
   2.3 Models, metaphors, and diagrams 17
   2.4 Questioning value of the Adaptive Landscape diagram 18
   2.5 The case for heuristic value 20
   2.6 Conclusion 23

3 Landscapes, Surfaces, and Morphospaces: What Are They Good For? 26
Massimo Pigliucci

   3.1 Introduction 26
   3.2 Four types of landscapes 27
   3.3 What are landscapes for? 30
   3.4 What to do with the landscapes metaphor(s) 32

Part II Controversies: Fisher's Fundamental Theorem Versus Sewall Wright's Shifting Balance Theory

4 Wright's Adaptive Landscape versus Fisher's Fundamental Theorem   41
Steven A. Frank

    4.1 Introduction   41
    4.2 The disagreement about drift and dynamics   42
    4.3 Fisher's goal for the fundamental theorem   44
        4.3.1 Background   45
        4.3.2 Fisher's framing of the problem   45
        4.3.3 The fundamental theorem explained   46
        4.3.4 Fisher's conservation law for mean fitness   47
        4.3.5 Average excess, average effect, and genetic variance   47
        4.3.6 What is fitness?   48
        4.3.7 Deterioration of the environment   48
        4.3.8 Misunderstandings about additivity   49
    4.4 Fisher's laws versus Wright's dynamics   50
    4.5 Key points in the Fisher–Wright controversy   52
        4.5.1 Wright lacked interest in Fisher's general laws   52
        4.5.2 Long-term dynamics and the rate of adaptation   53
        4.5.3 Additivity versus genetic interactions   53
        4.5.4 Wright's expression of fitness surfaces in an Adaptive Landscape   54
        Acknowledgments   56

    Appendix   57

5 Wright's Adaptive Landscape: Testing the Predictions of his Shifting Balance Theory   58
Michael J. Wade

    5.1 Introduction   58
    5.2 Conceptual foundations of Wright's shifting balance theory   59
    5.3 Challenges of the premises of the shifting balance theory   62
    5.4 Empirical predictions of the shifting balance theory   63
    5.5 Empirical tests of the predictions of Wright's shifting balance theory   66
    5.6 Direct empirical tests of Wright's shifting balance theory   69
    5.7 Conclusion   70

6 Wright's Shifting Balance Theory and Factors Affecting the Probability of Peak Shifts   74
Charles J. Goodnight

    6.1 Introduction   74
    6.2 A modified view of the Adaptive Landscape   75
    6.3 Wright's shifting balance theory   76
    6.4 Controversies surrounding Wright's shifting balance theory   77
    6.5 Peak shifts in metapopulations   81
    6.6 The importance of peak shifts in evolution   82

Part III   Applications: Microevolutionary Dynamics, Quantitative Genetics, and Population Biology

7   Fluctuating Selection and Dynamic Adaptive Landscapes   89
Ryan Calsbeek, Thomas P. Gosden, Shawn R. Kuchta, and Erik I. Svensson

  7.1 Introduction   89
  7.2 Empirical support for shifting landscapes   90
       7.2.1 Establishing criteria for demonstrating variation in selection   92
  7.3 Frequency-dependent selection   93
  7.4 Density-dependence   95
  7.5 Competition, predation, or both?   96
  7.6 Fluctuations in sexual selection: intergenerational changes in male–male
       competition and mate choice   97
  7.7 Abiotic environmental factors, fluctuating selection, and the limits
       of ecological speciation   100
  7.8 Population variants: genetic morphs and phenotypic plasticity   101
  7.9 Conclusions and future directions   103

8   The Adaptive Landscape in Sexual Selection Research   110
Adam G. Jones, Nicholas L. Ratterman, and Kimberly A. Paczolt

  8.1 Introduction   110
  8.2 Sexual selection *is* selection   110
  8.3 The Adaptive Landscape and the individual selection surface   111
  8.4 Peculiarities of the Adaptive Landscape in sexual selection research   113
  8.5 Achievements of the Adaptive Landscape in sexual selection research   114
       8.5.1 The Bateman gradient as the cause of sexual selection   114
       8.5.2 The quantification of directional sexual selection on
              phenotypic traits   115
       8.5.3 Non-linear individual selection surfaces   117
       8.5.4 Correlational selection   119
       8.5.5 Selection on mating preferences   119
       8.5.6 Temporal and spatial variation in sexual selection   120
       8.5.7 Integrating precopulatory and postcopulatory sexual selection   120
  8.6 Conclusions and future directions   122

9   Analyzing and Comparing the Geometry of Individual Fitness Surfaces   126
Stephen F. Chenoweth, John Hunt, and Howard D. Rundle

  9.1 Introduction   126
  9.2 Analysis of the individual fitness surface within a single population   127
       9.2.1 Approximating the individual fitness surface using
              multiple regression   127
       9.2.2 Analyzing non-linear selection through canonical analysis   128
       9.2.3 Testing the significance of selection gradients and
              vectors of selection   129
       9.2.4 Issues and caveats   130
  9.3 Comparing fitness surfaces among groups   131
       9.3.1 Testing among-group variation   137

		9.3.2 Characterizing among-group differences	139
		9.3.3 Issues and caveats	140
	9.4	Integrating multivariate selection analyses with evolutionary genetics	141
		9.4.1 Model misspecification and the nature of genetic variance under selection	142
		9.4.2 The contribution of selection and genetic constraints to adaptive divergence	143
	9.5	Future directions	145

10 Adaptive Accuracy and Adaptive Landscapes 150
Christophe Pélabon, W. Scott Armbruster, Thomas F. Hansen, Geir H. Bolstad, and Rocío Pérez-Barrales

10.1	Introduction	150
10.2	Theory	152
	10.2.1 Adaptive accuracy of a population	152
	10.2.2 Adaptive accuracy of the genotype	153
	10.2.3 Maladaptation due to variation in the fitness landscape	154
	10.2.4 Inaccuracy of focal traits and marginal Adaptive Landscapes	154
10.3	An example from the pollination ecology of *Dalechampia*	156
	10.3.1 Phenotypic selection study in *Dalechampia scandens*	156
	10.3.2 The Adaptive Landscape	159
	10.3.3 Adaptive inaccuracy	159
10.4	Final remarks	164
Acknowledgments		165

11 Empirical Insights into Adaptive Landscapes from Bacterial Experimental Evolution 169
Tim F. Cooper

11.1	Introduction	169
11.2	The repeatability of evolution	169
	11.2.1 Assessing divergence of replicate populations	169
	11.2.2 The effect of environment on between-population divergence	172
11.3	Diversity within populations	173
11.4	Correlation of fitness across environments	174
11.5	Form of Adaptive Landscapes around adaptive peaks	175
11.6	Reversibility	176
11.7	Conclusion	176

12 How Humans Influence Evolution on Adaptive Landscapes 180
Andrew P. Hendry, Virginie Millien, Andrew Gonzalez, and Hans C. E. Larsson

12.1	Introduction	180
12.2	Our Adaptive Landscape	181
12.3	Altering evolution on Adaptive Landscapes	184
	12.3.1 Topography	184
	12.3.2 Dimensionality	186
	12.3.3 Phenotypic excursions	187

12.4 Human impacts — 188
    12.4.1 Invasive species — 188
    12.4.2 Climate change — 190
    12.4.3 Hunting and harvesting — 191
    12.4.4 Habitat loss and fragmentation — 192
    12.4.5 Habitat quality — 194
12.5 What have we gained? — 195
12.6 Summary and conclusions — 196
Acknowledgments — 197

Part IV Speciation and Macroevolution

13 Adaptive Landscapes and Macroevolutionary Dynamics — 205
Thomas F. Hansen

13.1 Introduction — 205
13.2 Patterns of evolution — 206
    13.2.1 Potential for evolution: evolvability and strengths of selection — 206
    13.2.2 The rate of evolution — 209
    13.2.3 The rate of adaptation — 211
13.3 Processes of evolution — 213
    13.3.1 Evolution on the Adaptive Landscape — 213
    13.3.2 Evolution of the Adaptive Landscape — 215
    13.3.3 A role for constraints? — 219
13.4 Final remarks — 220
Acknowledgments — 221

14 Adaptive Dynamics: A Framework for Modeling the Long-Term Evolutionary Dynamics of Quantitative Traits — 227
Michael Doebeli

14.1 Introduction — 227
14.2 Basics of adaptive dynamics: convergence stability — 229
14.3 Basics of adaptive dynamics: evolutionary stability and evolutionary branching — 234
14.4 The importance of ecological dynamics — 238
14.5 Conclusions — 239

15 Adaptive Landscapes, Evolution, and the Fossil Record — 243
Michael A. Bell

15.1 Evolutionary mechanism and paleontological pattern — 243
15.2 Sewall Wright's fitness (genotypic) landscape — 243
15.3 G. G. Simpson's Adaptive (phenotypic) Landscape — 244
15.4 Paleontology and Adaptive Landscapes — 246
15.5 Temporal resolution in paleontology and evolutionary rates in populations — 246
    15.5.1 The limits of temporal resolution in the stratigraphic and fossil records — 246

	15.5.2 Simulated rates of species-level morphological evolution	248
	15.5.3 Rates of phenotypic evolution in contemporary populations	248
15.6	Application of Adaptive Landscapes in paleontology	248
	15.6.1 Evolution of simulated individual lineages on Adaptive Landscapes	248
	15.6.2. Ascent of adaptive peaks by an individual fossil lineage	250
	15.6.3 Punctuated equilibria and natural selection in sets of fossil lineages	251
15.7	Adaptive Landscapes, evolution, and the fossil record	252
15.8	Conclusions	253
	Acknowledgments	254

Part V Development, Form, and Function

16 Mimicry, Saltational Evolution, and the Crossing of Fitness Valleys 259
Olof Leimar, Birgitta S. Tullberg, and James Mallet

16.1	Introduction	259
16.2	Transitions between adaptive peaks	261
16.3	Peak shift in multidimensional trait spaces	261
16.4	Feature-by-feature saltation	262
16.5	Fisherian peak shifting	263
16.6	Wrightian shifts in aposematic coloration	264
16.7	Well-studied cases	265
16.8	Concluding comments	267
	Acknowledgments	267

17 High-Dimensional Adaptive Landscapes Facilitate Evolutionary Innovation 271
Andreas Wagner

17.1	Introduction	271
17.2	Metabolic network space	272
17.3	From metabolic genotype to phenotype	273
17.4	Exploring metabolic genotype space	274
17.5	Regulatory circuits and novel gene expression patterns	274
17.6	Macromolecules	276
17.7	Genotype networks as a consequence of high dimensionality	277
17.8	Outlook	279
	Acknowledgments	280

18 Phenotype Landscapes, Adaptive Landscapes, and the Evolution of Development 283
Sean H. Rice

18.1	Introduction	283
18.2	Informal theory	283
18.3	Formal analysis	286

	18.4 Reconstructing phenotypic landscapes	290
	18.5 Covariance between traits and the evolution of heritability	292

Part VI Concluding Remarks

19 The Past, the Present, and the Future of the Adaptive Landscape — 299
Erik I. Svensson and Ryan Calsbeek

	19.1 Historical context	299
	19.2 Which Adaptive Landscape?	299
	19.3 Model or metaphor?	301
	19.4 Achievements	301
	19.5 Remaining conceptual problems: do we really need Adaptive Landscapes?	302
	19.6 Empirical challenges	304
	19.7 Alternatives to the Adaptive Landscape	305
	19.8 Conclusion and the next 80 years	306

Author Index — 309
Subject Index — 315

List of Contributors

W. Scott Armbruster, University of Portsmouth, School of Biological Sciences, King Henry Building, King Henry I Street, Portsmouth, PO1 2DY, UK
scott.armbruster@port.ac.uk

Michael A. Bell, Stony Brook University, Department of Ecology & Evolution, 650 Life Sciences Building, Stony Brook, NY 11794-5245, USA
mabell@life.bio.sunysb.edu

Geir H. Bolstad, Centre for Conservation Biology, Department of Biology, Norwegian University of Science and Technology, 7491 Trondheim, Norway
geir.bolstad@bio.ntnu.no

Ryan Calsbeek, Dartmouth College, Department of Biological Sciences, 338 Life Sciences Center, Hanover, NH 03755, USA
Ryan.calsbeek@dartmouth.edu

Stephen F. Chenoweth, The School of Biological Sciences, The University of Queensland, Brisbane QLD 4072, Australia
s.chenoweth@uq.edu.au

Tim F. Cooper, Department of Biology and Biochemistry, University of Houston, Houston, TX 77204, USA
tfcooper@uh.edu

Michael R. Dietrich, Dartmouth College, Department of Biological Sciences, Life Sciences Center, Hanover, NH 03755, USA
Michael.Dietrich@Dartmouth.EDU

Michael Doebeli, University of British Columbia, Departments of Zoology and Mathematics, 6270 University Boulevard, Vancouver, BC V6T 1Z4, Canada
doebeli@zoology.ubc.ca

Steven A. Frank, Department of Ecology and Evolutionary Biology, University of California, Irvine, CA 92697-2525, USA
safrank@uci.edu

Andrew Gonzalez, McGill University, 1205 Docteur Penfield Montreal, Quebec H3A 1B1, Canada
andrew.gonzalez@mcgill.ca

Charles J. Goodnight, The University of Vermont, Burlington, VT 05405, USA
charles.goodnight@uvm.edu

Thomas P. Gosden, The School of Biological Sciences, The University of Queensland, Brisbane QLD 4072, Australia
t.gosden@uq.edu.au

Thomas F. Hansen, Department of Biology, University of Oslo, P.O. Box 1066, Blindern, NO-0316, Oslo, Norway
t.f.hansen@bio.uio.no

Andrew P. Hendry, McGill University, Redpath Museum, 859 Sherbrooke St. West, Montreal, Quebec H3A OC4, Canada
andrew.hendry@mcgill.ca

John Hunt, Biosciences, University of Exeter, Cornwall Campus, Penryn, Cornwall, TR10 9EZ, UK
J.Hunt@exeter.ac.uk

Adam G. Jones, Texas A&M University, Biological Sciences Building, 3258 TAMU College Station, TX 77843-3258, USA
ajones@mail.bio.tamu.edu

Shawn R. Kuchta, Ohio University, Department of Biological Sciences, Life Science Building 233, Athens, OH 45701, USA
kuchta@ohio.edu

Hans C.E. Larsson, McGill University, Redpath Museum, 859 Sherbrooke St. West, Montreal, Quebec H3A OC4, Canada
hans.ce.larsson@mcgill.ca

LIST OF CONTRIBUTORS

Olof Leimar, Stockholm University, Department of Zoology, SE-106 91, Stockholm, Sweden
olof.leimar@zoologi.su.se

James Mallet, Department of Organismal and Evolutionary Biology, Harvard University, 16 Divinity Avenue, Cambridge MA, USA
jmallet@oeb.harvard.edu

Virginie Millien, McGill University, Redpath Museum, 859 Sherbrooke St. West, Montreal, Quebec H3A OC4, Canada
virginie.millien@mcgill.ca

Kimberly A. Paczolt, Texas A&M University, Biological Sciences Building, 3258 TAMU College Station, TX 77843-3258, USA
kpaczolt@bio.tamu.edu

Christophe Pélabon, Centre for Conservation Biology, Department of Biology, Norwegian University of Science and Technology, 7491 Trondheim, Norway
christophe.pelabon@bio.ntnu.no

Rocío Pérez-Barrales, University of Seville, Department of Plant Biology and Ecology, Reina Mercedes s/n 41012, Seville, Spain
ropeba@us.es

Massimo Pigliucci, City University of New York, The Graduate Center, 365 Fifth Ave, New York, NY 10016, USA
massimo@platofootnote.org

Nicholas L. Ratterman, Texas A&M University, Biological Sciences Building, 3258 TAMU College Station, TX 77843-3258, USA
nickratterman@tamu.edu

Sean H. Rice, Texas Tech University, Department of Biological Sciences, Lubbock, TX 79409, USA
sean.h.rice@ttu.edu

Howard D. Rundle, University of Ottawa, Department of Biology, Gendron Hall, Ottawa, ON K1N 6N5, Canada
hrundle@uOttawa.ca

Robert A. Skipper, Jr. University of Cincinnati, Department of Philosophy, Cincinnati, OH 45221-0374, USA
robert.skipper@uc.edu

Erik I. Svensson, Lund University, Department of Biology, Sölvegatan 37, SE-223 62 Lund, Sweden
Erik.Svensson@biol.lu.se

Birgitta S. Tullberg, Stockholm University, Department of Zoology, SE-106 91, Stockholm, Sweden
Birgitta.Tullberg@zoologi.su.se

Michael J. Wade, Indiana University, Department of Biology, Jordan Hall 131C, Bloomington, IN, 47405-3700, USA
mjwade@indiana.edu

Andreas Wagner, University of Zurich, Institute of Evolutionary Biology and Environmental Studies, Winterthurerstrasse 190, CH-8057, Zürich, Switzerland
andreas.wagner@ieu.uzh.ch

Preface

The Adaptive Landscape is one of those classic concepts in evolutionary biology that seems to emerge again and again despite having been declared dead many times. It was first formalized by the late population geneticist Sewall Wright in 1932 (Wright 1932), 80 years ago, and is the topic of this volume. Wright's idea immediately caught the attention of some leading evolutionary biologists, including Theodosius Dobzhansky, who popularized the idea in several books and articles. Eventually, the Adaptive Landscape became intimately connected to the Modern Synthesis in evolutionary biology, which started in the late 1930s with several influential volumes (Dobzhansky 1937; Mayr 1942; Simpson 1944), and which remains subject to much controversy and discussion even today (Pigliucci & Müller 2010). It was probably the highly appealing visual aesthetic of the Adaptive Landscape, as well as Dobzhansky's charismatic personality and strong influence that contributed to the popularity and success of the Adaptive Landscape concept.

The editors of this volume (Erik Svensson, ES; Ryan Calsbeek, RC) got our research training long after the Modern Synthesis was completed, and long after the death of Sewall Wright, yet the Adaptive Landscape has been with us since we first entered our PhD programs in Sweden and California, respectively. This influence of the Adaptive Landscape was part of our formal curriculae, but we also encountered it on various other occasions, such as at scientific meetings, but often in also quite unexpected contexts. For instance, ES went on field excursions with some other Swedish PhD students to the Galápagos Islands in 1993, and while standing watching giant tortoises from the slope of one of the five volcano peaks (Alcedo) on the large island of Isabela one of his colleagues (Anders Hedenström, currently a professor at Lund University) remarked: "These volcano peaks remind me about Sewall Wright's Adaptive Landscape." Anders was neither the first nor last person who has drawn the analogy between physical (geographic) landscapes and the landscape of fitness that Sewall Wright described in the powerful metaphor he formulated in 1932. A few years later, ES went to California to pursue postdoctoral research on side-blotched lizards (*Uta stansburiana*), where RC had just started as a graduate student. We both did our fieldwork on the hilly grass-covered slopes at Los Banos (Merced County) in Central California. These rolling hills also provided a natural visual connection to Sewall Wright's Adaptive Landscape. Interestingly, Sewall Wright actually spent at least one vacation in Central California during the critical time period when the Adaptive Landscape was formulated (Provine 1986), so it is actually possible that he was partly inspired by the very same hilly landscape that caught our attention.

After his postdoc, ES published a paper from his postdoctoral research that reveals the inspiration of the Adaptive Landscape in evolutionary ecological field studies: "Experimental excursions on Adaptive Landscapes: density-dependent selection on egg size" (Svensson and Sinervo 2000). Over the ensuing decade, we both published several more papers containing fitness surfaces and selection gradients as part of our research programs in field evolutionary ecology using insects, reptiles, and amphibians (see Chapter 7 for a review of this and many other selection studies).

The Adaptive Landscape has certainly also been popular outside of our research environments, otherwise we would probably not have been able to publish our studies. In fact, the Adaptive

Landscape seems to hold a central position in some fields, which are almost impossible to imagine without the concepts of adaptive fitness peaks and valleys. These include, for instance, the ecological theory of adaptive radiation and ecological speciation (Schluter 2000; Rundle and Nosil 2005). The study of local adaptation and fitness trade-offs between environments, for instance, postulates the existence of fitness valleys between fitness peaks that can be revealed by reciprocal transplant experiments (Schluter 2000). Moreover, in evolutionary genetics, fitness epistasis and multiple fitness peaks are central topics of research, and these topics are very difficult, if not impossible, to understand without referencing the Adaptive Landscape (Whitlock et al. 1995; Wolf et al. 2000; Phillips 2008). Finally, in comparative evolutionary biology and phylogenetic studies of adaptation there is often a strong link to the Adaptive Landscape in formal theoretical modeling (Hansen & Martins 1996; Hansen 1997).

We, along with many other evolutionary biologists, are obviously strongly influenced by "landscape thinking" in our day-to-day research. Still, we also became aware, quite early on, of the many criticisms and potential pitfalls of the Adaptive Landscape. These criticisms appear at regular intervals in philosophy journals and books (Gavrilets 2004; Kaplan 2008; Pigliucci 2008), and are important, as they challenge supporters of the Adaptive Landscape concept to clarify their points and sharpen their arguments about why it is important and useful. Science and the production of new knowledge is often stimulated as much by discussions and controversies, as by elegant ideas. Science is a process, rather than a product that will be finished with all questions resolved. These criticisms have therefore been as useful and stimulating to us as we prepared this book, as were the many examples and areas where the Adaptive Landscape appeared to have been useful and influential.

This volume aims to present different voices on the Adaptive Landscape, its past, present, and future position in evolutionary biology. We got the idea for this book a few years ago, when we realized that the 80-year anniversary of Sewall Wright's classic paper was approaching rapidly (2012), and we both felt that it was an excellent opportunity to summarize the state of the art of the Adaptive Landscape. Our hope is that this volume will be intellectually stimulating to both senior researchers, postdocs, as well as new-comers and fresh PhD students in evolutionary biology and related fields. We also hope that this volume does not mark the end of the scientific discussions about the Adaptive Landscape, but rather a new beginning. And finally, we hope that if the Adaptive Landscape will not survive another 80 years, it will hopefully be replaced by an even better concept or metaphor that will push evolutionary biology forward and increase our knowledge about adaptation, speciation, and the origins and preservation of biodiversity on this fragile planet.

Erik I. Svensson and Ryan Calsbeek

References

Dobzhansky, T. (1937). Genetics and the origin of species. Columbia University Press, New York.

Gavrilets, S. (2004). Fitness landscapes and the origin of species. Princeton University Press, Princeton, NJ.

Hansen, T. F. (1997). Stabilizing selection and the comparative analysis of adaptation. Evolution, 51, 1341–1351.

Hansen, T. F. and Martins, E. P. (1996). Translating between microevolutionary process and macroevolutionary pattern: The correlation structure of interspecific data. Evolution, 50, 1404–1417.

Kaplan, J. (2008). The end of the Adaptive Landscape metaphor? Biology & Philosophy, 23, 625–638.

Mayr, E. (1942). Systematics and the origin of species. Columbia University Press, New York.

Phillips, P. C. (2008). Epistasis—the essential role of gene interactions in the structure and evolution of genetic systems. Nature Reviews Genetics, 9, 855–867.

Pigliucci, M. (2008). Sewall Wright's Adaptive Landscape: 1932 vs. 1988. Biology & Philosophy, 23, 591–603.

Pigliucci, M. and Müller, G. B. (2010). Evolution—The Extended Synthesis. The MIT Press, Cambridge, MA.

Provine, W. B. (1986). Sewall Wright and evolutionary biology. University of Chicago Press, Chicago, IL.

Rundle, H. D. and, Nosil, P. (2005). Ecological speciation. Ecology Letters, 8, 336–352.

Schluter, D. (2000). The Ecology of Adaptive Radiation. Oxford University Press, Oxford.

Simpson, G. G. (1944). Tempo and mode in evolution. Columbia University Press, New York.

Svensson, E. and Sinervo, B. (2000). Experimental excursions on adaptive landscapes: density-dependent selection on egg size. Evolution, 54, 1396–1403.

Whitlock, M. C., Phillips, P. C., Moore, F. B. G., and Tonsor, S. J. (1995). Multiple fitness peaks and epistasis. Annual Review of Ecology, Evolution, and Systematics, 26, 601–629.

Wolf, J. B., Brodie III, E. D., and Wade, M. J. (2000). Epistasis and the evolutionary process. Oxford University Press, New York.

Wright, S. (1932). The roles of mutation, inbreeding, crossbreeding and selection in evolution. Proceedings of the Sixth Annual Congress of Genetics, 1, 356–366.

Acknowledgments

The idea for this volume was born at the European Evolutionary Biology Congress ("ESEB") in Turin (Italy) in August 2009, after we organized a symposium on phenotypic selection in natural populations. Shortly after this meeting, we presented the idea and general concept of this book to Ian Sherman at Oxford University Press (OUP), who enthusiastically endorsed it and encouraged us to move on with the project. We are most grateful to Ian and Helen Eaton at OUP who have been very supportive and helpful throughout the entire process, from the idea and initial contacts of authors until the final submissions.

We are also grateful to all the contributors to this volume for being so enthusiastic when invited to write chapters, for patiently responding to our reminders and emails and for their willingness to revise your chapters, even after critical reviews. This volume is very much a collective effort by all contributors, and we hope that readers will appreciate it as such, and find the different perspectives stimulating, even when conflicting with each other (indeed, *particularly* when perspectives conflict!).

To encourage cross-citations between chapters and the creation of a coherent volume, we have largely relied on the chapter authors themselves reviewing other chapters. However, we have also relied upon several external reviewers, who do not contribute to this volume, but who did an excellent job providing critical input for many of the chapters. These are (in alphabetical order): Jessica K. Abbott, Mats Björklund, Brittny Calsbeek, Robert Cox, Tom Van Dooren, Fabrice Eroukhmanoff, Duncan Greig, Rebecca Irwin, Kristina Karlsson, Joel McGlothlin, Borja Mila, Allen J. Moore, Jörgen Ripa, Adam Siepielski, Maja Tarka, Tobias Uller, and Machteld Verzijden.

Finally, we wish to thank our wives and families for encouragement, patience, and support in our daily work, both when working with this volume and in general.

Erik Svensson and Ryan Calsbeek

PART I
Historical Background and Philosophical Perspectives

CHAPTER 1

A Shifting Terrain: A Brief History of the Adaptive Landscape

Michael R. Dietrich and Robert A. Skipper, Jr.

1.1 Introduction

Sewall Wright's graphical metaphor of the Adaptive Landscape is touted as one of the most famous metaphors in the history of biology (e.g. Provine 1986; Ruse 1996; Coyne et al. 1997). In 1932, Wright analogized (the contours of) a physical landscape to a surface whose contours marked differences in fitness. Underlying this fitness surface were the many different possible gene combinations that might be realized. Borrowing representations familiar from topographical maps, Wright's fitness topographies were marked by peaks and valleys corresponding to high and low adaptive value. Populations or individuals could be located on these fitness surfaces and the action of selection pushing and pulling on their different features would help determine their trajectories over time.

Wright developed his Adaptive Landscape metaphor and its diagrams as a way to translate his shifting balance theory from the mathematics of population genetics to a more accessible idiom for the general biologist. The shifting balance theory assumed that genes produced their effects through complex interactions and that subdivided populations created an opportunity for selection to act efficiently as these subpopulations moved through phases of drift and selection (see Chapters 4, 5, and 6 for more detailed discussions of the shifting balance theory). The landscape diagrams also allowed Wright to set up his problem of shifting from one adaptive peak to another (see Chapter 6). While the influence of Wright's landscape analogy is undeniable, similar representations and related ideas had previously been presented. In this chapter, we will first consider the possibility that the Adaptive Landscape was independently co-discovered, before turning to the conceptual and representational lineages of different forms of the Adaptive Landscape originating from Wright's 1932 representation. We will distinguish between genetic, phenotypic, and molecular versions of the Adaptive Landscape as a way of marking significant moments of change in the history of the Adaptive Landscape concept. The history of these different conceptual lineages supports our claim that one of the chief reasons for the influence and persistence of Wright's Adaptive Landscape metaphor was its plasticity in the hands of different communities of evolutionary biologists and in the face of new forms of data.

1.2 The origins of the Adaptive Landscape

The Adaptive Landscape is commonly thought to have been introduced in Sewall Wright's 1932 paper presented at the Sixth International Congress of Genetics in Ithaca, New York. However, in 1979, J. Wynne McCoy pointed out that fitness curves and their representation as "peaks and valleys" were first introduced by Armand Janet in 1895 (McCoy 1979). Janet was a French engineer with an interest in entomology. Indeed, Janet served as president of the Société entomologique de France in 1911. McCoy makes a compelling case for parallels between Janet's representations of fitness surfaces and that of McCoy's contemporaries, but Janet's ideas have several differences from Wright's, and we argue that Wright's "discovery" of the Adaptive Landscape was independent of Janet's, and certainly more influential historically.

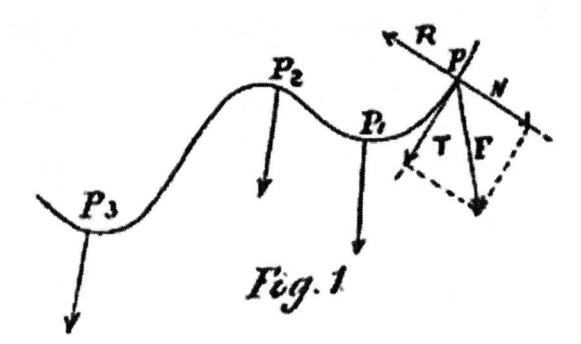

Figure 1.1 Armand Janet's selection surface. Species were represented as occupying different points of equilibrium (P) on the curve. Selection was represented as a vector at each point. The selection vector (F) could be decomposed into two vectors, N, which is normal to the surface, and T, which is tangential to the surface. (Janet 1895.)

At the Third International Congress of Zoology in Leyden, the Netherlands, Janet presented a paper describing the phenotypic change of a species in terms of a point moving along a selection surface (Janet 1895; see Fig. 1.1). The shape of the curve was determined by the external environment of the organism and the population's motion on the curve was directed by multiple forces modeled as a composite vector. In Janet's figure, selection acted like gravity to pull a population to the curves' minima or valleys. A population at a maxima or peak would tend to move toward a valley under the direction of selection.

Janet recognized that populations were variable and so added a measure of complexity to his representation by depicting populations as areas (see Fig. 1.2). The size of the area corresponded to the amount of variability in the population. Populations under strong selection would have narrow valleys and correspondingly smaller areas representing less variability, in contrast to the greater variability possible under weaker selection, depicted as a broader valley.

Moreover, Janet recognized that as environments change so would the shapes of the selective surfaces. So, what was once a valley might become a ridge or a peak. In so far as a phenotype was tracking this environmental change, its position would alter as a result of the changing environment and the force of selection in the new environment. For Janet such environmental transitions must have occurred in the past in order for new species to evolve. These transitional periods would then

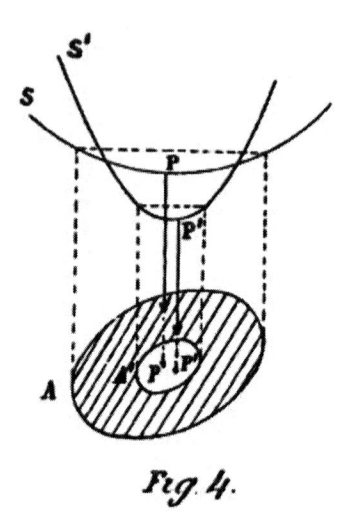

Figure 1.2 Armand Janet's depictions of population variability on a selection surface. Variability in a population was represented by the area (A) occupied by a species (P). The degree of concavity of the selection surface represented the strength of selection. S' represents stronger selection which reduces the variability that can be maintained in the population. Weaker selection represented in the selection surface S allows for greater phenotypic variability. (Janet 1895.)

correspond to transitional species, which would be relatively rare and short lived when compared to the species typical of periods of environmental stability (Janet 1895; McCoy 1979). In terms of his selective surface diagram, points of stability were likely to be found in the valleys and the steepness of the slopes leading to peaks and valleys represented areas of transition where selection would move a population quickly to an equilibrium point. The relative abundance of different species of fossils, especially the rarity of transitional forms, confirmed for Janet that environmental shifts underlying phenotypic change must be relatively swift.

McCoy's "rediscovery" of Janet was motivated by a paper in the *American Naturalist* published in 1977 by Maurice M. Dodson and Anthony Hallam where they too used a model of selective surfaces to explain phenotypic shifts in a changing environment (Dodson and Hallam 1977). McCoy was confident that Dodson and Hallam arrived at their ideas "in complete ignorance of Janet's paper." Indeed, Dodson and Hallam cite Wright and Simpson with regard to their representations, but not Janet. Regardless of its other virtues, McCoy could not help noting that Dodson and Hallam's paper may have been timely, but was also "out of date, for its argument is over 80 years old!" (McCoy 1979).

Implicit in McCoy's claims that Janet's construct was "recreated" by Wright, Simpson, and Dodson and Hallam is an appeal to recognize Janet as an originator of this diagrammatic form of representing selection. Any meaningful claim for Janet as the originator of the modern Adaptive Landscape concept, however, is undercut by important differences between Janet's and Wright's representations, as well as Janet's almost complete lack of historical influence. Janet's selective surface, coming before the rise of genetics, depicted phenotypic change, where Wright's represented genetic change, subsequently G. G. Simpson would adapt Wright's landscapes to represent phenotypes (see section 1.4). Janet's peaks and valleys represent the inverse of Wright's. Moreover, Janet's representations do not make an explicit analogy to topographic representation as Wright's will later. In fact, Janet's analogy was to forces in mechanics, where Wright's was to traversing hilly terrain. Janet's influence was mediated by his better-known brother, Charles Janet, who was a polymath best known for his reconfiguration of the periodic table (Stewart 2010). Charles referred to Armand's analysis of sudden or saltational changes when trying to explain the evolution of cocoons in ants (Janet 1896). This hypothesis and its references to Armand Janet's selective surface were discussed in William Morton Wheeler's 1904 and 1915 essays on cocoons in ants (Wheeler 1904, 1915). Even Wheeler though was more interested in Janet's ideas as a kind of mutationism, rather than as a consequence of thinking in terms of Janet's selective surface. In the end, Janet's unappreciated diagram stands in isolation from Wright's independent creation of what we now recognize as a diverse lineage of Adaptive Landscape diagrams. This is not to say that aspects of Wright's Adaptive Landscapes did not have important precursors, as we shall see later in this chapter, but that Janet was not one of those precursors.

1.3 The genetic landscape

Sewall Wright publicly debuted his Adaptive Landscape diagrams in Ithaca, New York at the Sixth International Congress of Genetics in 1932. Wright was an American biologist trained by E. W. Castle at Harvard University in physiological genetics. Prior to his groundbreaking research in evolutionary theory, which he carried out at the University of Chicago and University of Wisconsin, Madison, he worked as a staff scientist for the US Department of Agriculture (Provine 1986). In 1932, at the invitation of Harvard biologist, Edward Murray East, Wright joined R. A. Fisher and J. B. S. Haldane in a session on the newly emerging field of population genetics. Unlike their very mathematically challenging work from the previous 2 years (Fisher 1930; Wright 1931; Haldane 1932), the papers at the International Congress were intended to be accessible to general biologists interested in evolutionary theory.

Wright's principal contribution to evolutionary biology at the time was his 1931 paper "Evolution in Mendelian Populations" (Wright 1931). Wright condensed this long and technical article into his 1932 presentation for the Congress, published in the proceedings as "The Roles of Mutation, Inbreeding, Crossbreeding and Selection in Evolution" (Wright 1932). Rather than explain evolutionary dynamics with detailed mathematical models, Wright developed an analogy between fitness and altitude, between genetic combinations and the features of a hilly, physical landscape. In order to convey his metaphor, Wright offered two different kinds of diagrams. One represented the space of underlying genetic possibilities (see Fig. 1.3) and the

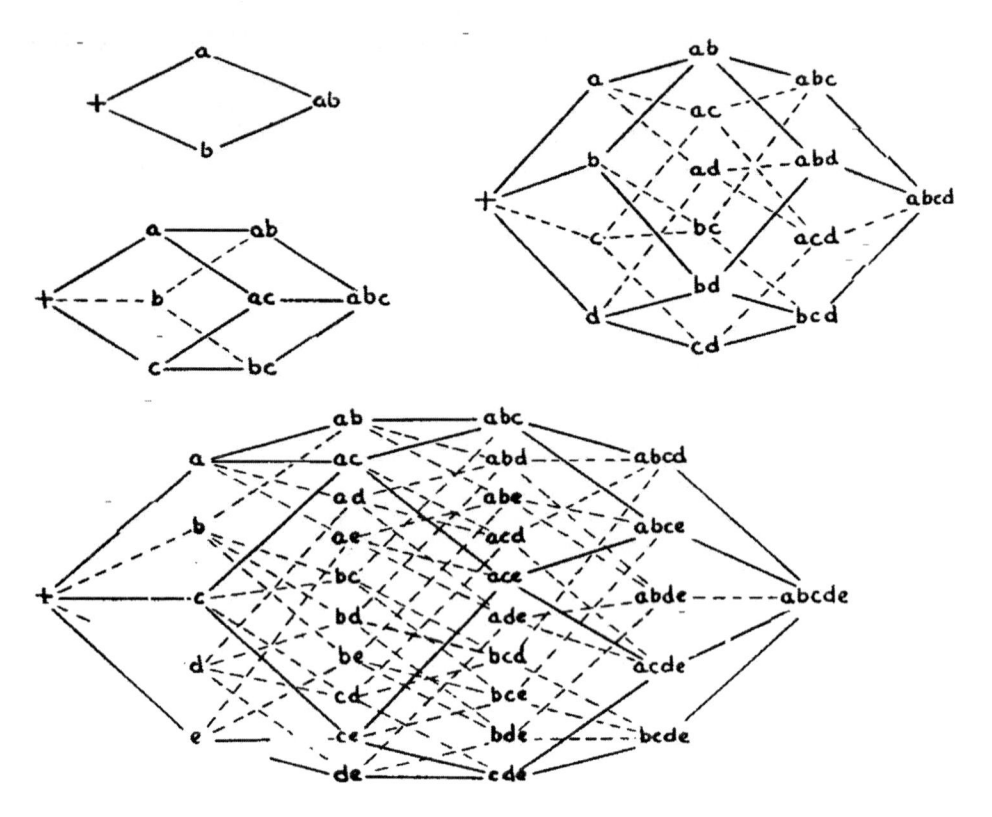

Figure 1.3 Sewall Wright's gene combination network diagrams. Networks of gene combinations or genotypes ranging from two to five loci represented the increasing complexity of all of the one-step transitions composing the space of genotypic possibilities. (Wright 1932.) Genetics by GENETICS SOCIETY OF AMERICA. Copyright 1931 Reproduced with permission of GENETICS SOCIETY OF AMERICA.

other analogized fitness differences to differences in altitude and used a topographic representation to depict those differences (see Fig. 1.4).

The foundation of the genetic version of the Adaptive Landscape is a network of relationships between different gene combinations or genotypes. Wright represented these as a network of single gene differences, where every node represented one possible set of gene combinations and every path between nodes represented one genetic change or difference between the two connected nodes (see Fig. 1.3). Wright offered a simplified image of this genetic space with diagrams showing the increasing complexity of the network as it expanded from two loci to five. Of course, Wright knew that the space of genetic possibilities for any organism in nature were much more complicated, and estimated that the field of gene combinations would number on the order of 10^{1000} (Wright 1932).

Wright used the two-dimensional graphical depiction of an Adaptive Landscape in Fig. 1.4 as a way of intuitively conveying what can only be realistically represented in thousands of dimensions. From its inception then, Wright's landscapes were multidimensional models represented on two dimensions in his figures. The axes of Wright's Adaptive Landscape diagrams are not labeled as such, because we believe that there was no way to provide a metric for the multidimensional network that had been idealized to just two dimensions. The surface of the landscape is typically understood as representing the adaptive value assigned to the underlying gene combinations. So, the vertical axis or height was assigned a measurable value in terms of fitness or relative adaptive value. Peaks on this adaptive topography represented areas of high adaptive value. Valleys represented areas of low adaptive value.

A. Increased Mutation or reduced Selection
4NU, 4NS very large

B. Increased Selection or reduced Mutation
4NU, 4NS very large

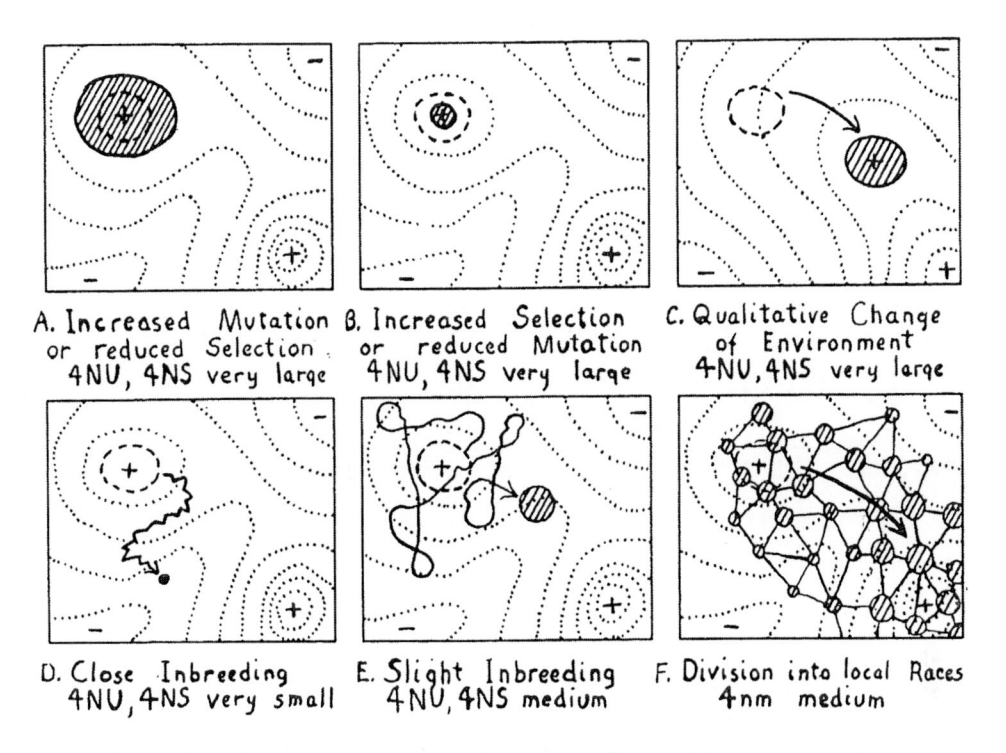

C. Qualitative Change of Environment
4NU, 4NS very large

D. Close Inbreeding
4NU, 4NS very small

E. Slight Inbreeding
4NU, 4NS medium

F. Division into local Races
4nm medium

Figure 1.4 Sewall Wright's six-frame Adaptive Landscape diagram. Each frame represents different evolutionary scenarios and their impact on the population in question. Frame C differs from the other five frames in that it represented a changing environment, which would create a dynamic landscape, so the population is shown tracking a moving landscape by the arrow. Frame F represents the dynamics of Wright's shifting balance theory. (Wright 1932.) Genetics by GENETICS SOCIETY OF AMERICA. Copyright 1931 Reproduced with permission of GENETICS SOCIETY OF AMERICA.

Wright had initially presented his idea for the Adaptive Landscape to Fisher in a letter in 1931, where he asked Fisher to:

Think of the field of visible joint frequencies of all genes as spread out in a multidimensional space. Add another dimension measuring degree of fitness. The field would be very humpy in relation to the latter because of epistatic relations, groups of mutations, which were deleterious individually producing a harmonious result in combination. (Wright to Fisher, February 3, 1931 in Provine 1986, p. 272.)

Each peak then would represent a point of harmonious interaction among the genes and the environment that would contribute to evolutionary stability. Fisher doubted that these points of stability could be common. According to Fisher, as the dimensionality of the field of gene combinations *increases*, the number of stable peaks on the surface of the landscape *decreases* (Fisher to Wright May 31, 1931 in Provine 1986, p. 274; Fisher 1941). Rather than a hilly landscape, Fisher argued that a landscape with a single peak with ridges along it was more likely. On Fisher's landscape, evolution does not require the complex of evolutionary factors of Wright's shifting balance process, but only selection and mutation (see Chapter 4).

In response to Fisher, Wright invoked J. B. S. Haldane's work on population genetics (Haldane 1932; Provine 1986). Haldane had also suggested that populations could be represented as a "multidimensional space" and had worked through the conditions to produce stable equilibria in a two-factor case (Wright to Fisher, June 5, 1931 in Provine 1986). Because Haldane argued that more than one apex in a hypercube or multidimensional space could be stable, Wright interpreted Haldane as holding a middle ground between his view and Fisher's (Wright 1935). Haldane's ideas and Wright's Adaptive Landscape were certainly very similar: both imagined a space of possibilities and a surface with stable equilibria. However, even though Haldane described his idea in print before Wright's 1932 essay, Wright's correspondence indicates that he had thought of his Adaptive Landscape before reading Haldane's essay (Provine 1986). Moreover, the landscape metaphor itself was absent from Haldane's essay where he opted for a much more complex mathematical argument rather than an analogy that made his theory generally accessible.

In his 1931 and 1932 essays, Wright was trying to describe the ideal conditions for evolution to occur, given specific assumptions about the relationship between Mendelian heredity and the adaptive value of gene complexes. The ideal conditions would produce the fastest rate of evolution to the highest "adaptive peak." Wright believed that these conditions required that populations be subdivided and semi-isolated, and that selection along with random genetic drift and migration operated in a "shifting balance" of phases. Wright tried to capture these conditions in his six-frame Adaptive Landscape diagram (Fig. 1.4). Each frame represented a different evolutionary scenario. When he first published this image in 1932, Wright was predisposed to frame F, which represented shifting balance dynamics mixing drift, selection, and migration among subdivided, semi-isolated populations, as the most efficient and so the most favorable (Wright 1932; Provine 1986). Fisher's work had impressed upon him that the large population in a changing environment represented in frame C was also an important scenario. It is worth emphasizing that frame C represented a dynamic landscape whose peaks and valleys would rise and fall over time like waves on the ocean (Wright 1932, see also Chapter 7). In 1944, as part of the Bulletin of the Committee on Common Problems of Genetics, Paleontology and Systematics, Wright offered the following appraisal of the evolutionary possibilities in his six-frame figure (Fig. 1.4). He wrote,

> The cases in which an indefinitely large species is subject to qualitative change of environment (C) or is subdivided into partially isolated local races (F) are stressed as enormously more favorable for evolution than the case of a random breeding population of intermediate size (E) even though this is more favorable than either a small population (D) or an indefinitely large one under the highly improbable condition of complete panmixia and no change of conditions for selection (or an increase (B) or reduction (A) in severity of selection not associated with any change in the direction of selection). (Wright 1944, p. 34.)

The favorability of the scenario with the large population changing with the environment was a concession to Fisher's views. However, Wright was not convinced that multiple peaks would not occur and that the dynamics of shifting from one peak to another was not a significant problem for evolutionary biology (see Chapter 4).

Despite Fisher's reservations, Wrights landscape was embraced by two key, but very different, proponents of the Neo-Darwinian synthesis, Theodosius Dobzhansky and G. G. Simpson. Simpson would transform the Adaptive Landscape by shifting its basis from genes to phenotypes, discussed in the next section. Dobzhansky would promote Wright's original genetic landscape as a means of understanding the basic processes of evolution.

Dobzhansky was trained in a Soviet tradition of evolutionary biology that emphasized the variability of natural populations. As a student of Iurii Filipchenko, he studied populations of *Coccinellidae* before getting a Rockefeller Foundation fellowship to join T. H. Morgan's famous Fly Group in 1927 (Provine 1986). At Columbia and later Cal Tech, Dobzhansky quickly became a leading figure in *Drosophila* genetics. In collaboration with A. H. Sturtevant and later Sewall Wright, Dobzhansky returned to questions of evolution—taking *Drosophila* genetics from the laboratory to the field. Dobzhansky's 1937 book, *Genetics and the Origin of Species,* articulated a foundational program of research for evolutionary genetics. The theoretical underpinning of Dobzhansky's program was deliberately based upon Wright's shifting balance theory, as Dobzhansky understood it. Because Dobzhansky was not a mathematically expert, his understanding of Wright rested heavily on Wright's non-mathematical 1932 presentation. Accordingly, Dobzhansky's explanation of the shifting balance theory rested heavily on the Adaptive Landscape metaphor (Provine 1986). *Genetics and the Origin of Species* translated and popularized one of the dominant general theories of evolution into a research program for evolutionary genetics, and served as a foundational publication for the emerging Neo-Darwinian synthesis (Levine 1995; Smocovitis 1996).

The influence of Dobzhansky as a biologist and teacher on twentieth-century biology has been tremendous. Versions of Wright's genetic landscape have appeared in almost every major textbook on evolutionary biology since 1937. Dobzhansky reprinted Wright's diagrams in each of the three editions of *Genetics and the Origin of Species*, as well as in the evolution textbook that he co-authored with Francisco Ayala, G. Ledyard Stebbins, and James Valentine (Dobzhansky 1942, 1951; Dobzhansky et al. 1977). He used the Adaptive Landscape as Wright had—to convey the range of evolutionary possibilities for a range of genotypes and population structures under the influence of selection, drift, mutation, and migration. Even Wright used *Genetics and the Origin of Species* as the textbook for his evolution courses at the University of Chicago from 1937–1954 (Provine 1986). Newer diagrams of the genetic version of the Adaptive Landscape continue to appear in evolution textbooks to this day (Ridley 2003; Barton et al. 2007; Freeman and Herron 2007; Futuyma 2009).

In 1960, one of Dobzhansky's students, Richard Lewontin, transformed the genetic landscape from a depiction of evolutionary possibilities to a graph of measured genetic frequencies and real populations. Lewontin brought an interest in both evolution and statistics to Dobzhansky's laboratory as a graduate student in the early 1950s. This expertise in statistical modeling allowed Lewontin to approach population genetic problems from a more theoretical perspective than Dobzhansky and many others at the time. In Lewontin's words:

> While many people like me start with observations in nature and develop theoretical tools to deal with them, which then may become part of the general theoretical apparatus of the field, my way of working has always been the reverse. I start by thinking about some general phenomenon (frequency-dependent selection, multilocus selection, selection in age distributed continuously breeding life histories, etc.), and explore the theoretical dynamics of such systems. (Lewontin to Michael R. Dietrich, October 1, 2010, personal correspondence.)

When he later came across relevant observations, he applied them. According to Lewontin, this pattern of thinking describes his work on Adaptive Landscapes in the late 1950s and early 1960s. At his first job at North Carolina State in the 1950s, Lewontin and his graduate student, Ken-ichi Kojima, worked on two-locus, two-allele models of selection (Lewontin and Kojima 1960). In the late 1950s, M. J. D. White visited North Carolina from Australia and told Lewontin about the chromosomal inversion data that he had been collecting from populations of the grasshopper, *Moraba scurra*. Lewontin realized that White's inversion data fitted the two-locus models that he and Kojima had been developing. Lewontin then applied the *Moraba* data to the two-locus models in papers with both White and Kojima (Lewontin and Kojima 1960; Lewontin and White 1960).

Lewontin's use of Adaptive Landscapes differed from Wright's 1932 presentation in so far as he did not take the Adaptive Landscape to be based on all possible gene combinations and their assumed adaptive values. Instead, Lewontin created a surface based on the frequencies of two different inversions and their fitness values (see Fig. 1.5). In doing so, Lewontin made the Adaptive Landscape into an empirical representation. And with that, Lewontin returned to a second form of Wright's Adaptive Landscape based on gene frequencies, not gene combinations.

When Wright first discussed the idea of the Adaptive Landscape in private correspondence with Fisher in 1931, he proposed that the landscape represent ranges of gene frequencies such that a point or area on the surface would correspond to a population with genes at those frequencies, and the height of the surface would represent the average fitness of the population (Provine 1986). In the version presented in 1932, the surface corresponds to all of the possible gene combinations that might be in individual organisms, and the height of the surface represents the fitness of the individual with that type of genetic combination. He would offer both versions in print in 1939 in a French collection on statistical biology that does not seem to have been widely read (Wright 1939). Wright seemed to shift between these two views, creating ambiguities that have vexed critics ever since (Provine 1986;

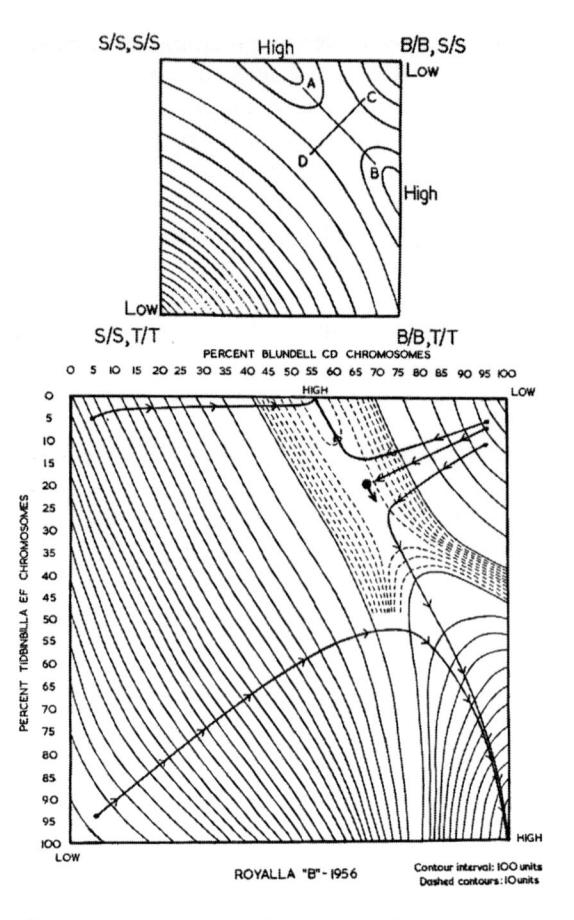

Figure 1.5 Lewontin and White's representation of the Adaptive Landscape and population trajectories for chromosomal inversions in populations of *Moraba scurra*. (a) A generalized version of the Adaptive Landscape with main features identified. (b) A plot of the Adaptive Landscape with current population location and trajectories. Reproduced from Lewontin, R. C. and White, M.J. D. (1960), "Interaction between inversion polymorphism of two chromosome pairs in the grasshopper, Moraba scurra," Evolution 14: 116–129.

Skipper 2004; Kaplan 2008). That said, Wright did not do what Lewontin and White did in 1960: he did not actually empirically construct an Adaptive Landscape.

Using White's data on chromosomal inversions from different populations, Lewontin calculated an Adaptive Landscape. The x- and y-axes represented the frequency of the *Tidbinbilla* EF chromosomes and the Blundell CD chromosomes in a certain population, such as *Royalla B* in Fig. 1.5b. White had measured the chromosomal composition of this population in 1958. The viability of each of the chromosomal genotypes was then estimated as a ratio of the actual to the expected number of individuals (Lewontin and White 1960). The graph of the topography was created by allowing the frequency of the *Tidbinbilla* EF chromosomes and the Blundell CD chromosomes to range from 0–1 at intervals of 0.05. With 21 values on each axis, 441 points on the landscape were calculated. An average fitness value was then calculated for each of these 441 points. Topographic lines were then drawn connecting the points with equal mean adaptive values (Lewontin and White called these isodapts, instead of isoclines). All of the landscapes calculated in this way showed a ridge with a saddle. The different populations from different years were always located in the saddle of their respective topography. The problem is that saddles are points of instability. Using the Adaptive Landscape as an empirical representation generated a new problem of explaining why the *Moraba* populations would all be found at similar points of instability.

Lewontin and White's genetic landscape lent the representation more reality than it ever had before. But it also opened it to more questioning. Turner and others questioned Lewontin's method for calculating the landscape surface (Turner 1972). Turner's recalculations put the *Moraba* populations on peaks, not saddles. Moreover, as Lewontin and White note, their results raised a number of questions about the adequacy of their model. In 1974, Lewontin would generalize these concerns in terms of the dynamic sufficiency of any two-locus system to describe the evolution of much more complex genetic systems (Lewontin 1974). This problem of representational adequacy, however, did not stop the genetic landscape from being widely used in evolutionary biology, although it does suggest that the Adaptive Landscape was often understood as a heuristic even when it represented measured empirical values (see Chapter 2).

1.4 The phenotypic landscape

Where Dobzhansky brought contemporary genetics to evolutionary thinking in 1937, G. G. Simpson brought paleontology to the mix in his *Tempo and Mode in Evolution*, published in 1944. Having earned his PhD at Yale in 1923, Simpson had rapidly become one of the most prominent paleontologists in the United States. From his position at the

American Museum of Natural History, Simpson's expertise on the evolution and classification of fossil mammals allowed him to articulate a paleontological understanding of evolution compatible with Dobzhansky's genetic approach. *Tempo and Mode* was drafted in the late 1930s and completed after Simpson's service in the Second World War. Where Dobzhansky brought modern genetics to the evolutionary synthesis, Simpson brought paleontological evidence of evolution in deep time and an appreciation for how processes of adaptive change had created patterns of phenotypic change across species.

Like Dobzhansky, Simpson was taken with Wright's Adaptive Landscapes, but as a paleontologist without access to genes, Simpson redrew Wright's diagrams and fundamentally altered their interpretation by casting them in terms of phenotypes (see Chapter 18). Where Wright's diagrams generally represented a selective landscape, Simpson refined the representational vocabulary to denote different forms of selection as topographic patterns. He began by depicting the range of variation in a population as a shaded area. Selection was represented as arrows, which indicated both strength and direction of selection with regard to the population. Selection decreasing population variation was labeled centripetal selection, while centrifugal selection was that which allowed variation to increase (see Fig. 1.6). As in Wright's topographies, the distance between the lines represented the intensity of selection (intensity increases with the slope of gradient).

Simpson then put his phenotypic landscape to work to explain equine evolution (see Fig. 1.7). Simpson divided the *Equidae* lineage into browsers and grazers based on tooth morphology and several other features. Grazing horses evolved from browsing ancestors from the Eocene to the late Miocene. Simpson postulates that in the Eocene the Adaptive Landscape was marked by two distinct adaptive peaks, one for browsing and one for grazing, but only the browsing peak was occupied. As the horses grew bigger, the adaptive peaks moved closer together, because with size came larger tooth crowns, which moved then toward a grazer morphology (Simpson 1944). Asymmetry in the strength of selection on the browsing peak meant that more variants on the grazing side were allowed to persist. As the two peaks moved closer, this asymmetry resulted in individuals located in the saddle between the two peaks. From the saddle, individuals were under selection to move higher on one of either of the two peaks. The grazing peak was steeper and so individuals climbed it fairly quickly resulting in a relatively sudden appearance of the grazing morphology (Simpson 1944). Later in *Tempo and Mode*, Simpson will use this as a case of what he called quantum evolution, but in doing so will impose his idea of adaptive zones on the Adaptive Landscape (Simpson 1944). Adaptive zones were partitions of the environment that were particularly favorable for certain forms of adaptations. Adaptive zones changed over time as the environment changed, and organisms had to track these changes over time or risk extinction. Simpson's interest in using the Adaptive Landscape to represent changing environments over time follows directly from Wright's depiction of the Adaptive Landscape in frame C, although Simpson was not obligated to large population sizes (see Fig. 1.4).

Wright reviewed Simpson's *Tempo and Mode*, but did not criticize this reinterpretation beyond urging Simpson to consider lower levels of biological organization, such as local populations. The shift to phenotypes was acknowledged as simply "different in point of view" (Wright 1945; Provine 1986).

Simpson's phenotypic landscape was perpetuated in its own lineage of textbooks, especially those written by paleontologists. For instance, Simpson's images were duplicated in G. S Carter's *Animal Evolution* in 1951, Terrell Hamilton's *Process and Pattern in Evolution* in 1967, and Niles Eldredge's *Unfinished Synthesis* in 1985 and *Macroevolutionary Dynamics* in 1987 (Carter 1951; Hamilton 1967; Eldredge 1985, 1987).

In the 1970s, Russell Lande gave Simpson's phenotypic landscape a mathematical model (Lande 1976, 1979) that broadened its use beyond paleontology (see Chapter 19). In doing so, Lande drew on a tradition of biometrical modeling of phenotypic evolution going back to Karl Pearson. In 1903, Pearson had proposed a mathematical model for the evolutionary selection of two traits in terms of what he called a selection surface or surface of survivals (Pearson 1903). The surface was a way

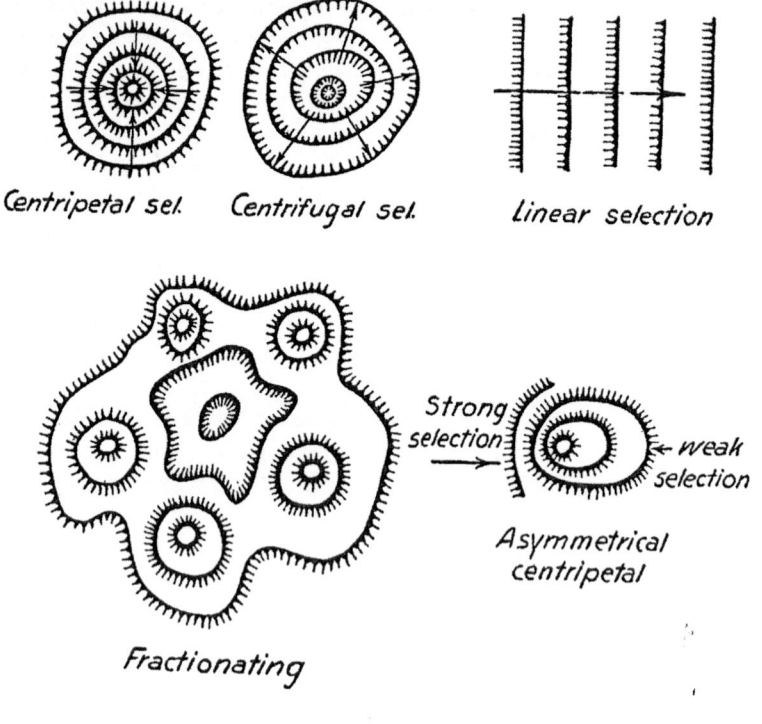

Figure 1.6 G. G. Simpson's reinterpretation of the Adaptive Landscape. Different forms of selection represented as topographic contours. The hash-mark side of a contour line represents a lower area. Tempo and mode in evolution by SIMPSON, GEORGE G. Copyright 1944 Reproduced with permission of COLUMBIA UNIVERSITY PRESS.

of representing the proportions of the population selected as either fit or unfit under different selection regimes. While Wright was familiar with Pearson's mathematically complex papers, there is no evidence that Simpson was. Lande certainly built on the biometrical tradition from Pearson as he developed a mathematical theory for phenotypic evolution and applied it to cases of micro- and macroevolution.

Simpson created a separate lineage of Adaptive Landscape diagrams. Where Wright originated a lineage of genetic landscapes, Simpson reinterpreted the foundation of the Adaptive Landscape in terms of a space of phenotypic possibilities. The foundation of the Adaptive Landscape would be reinterpreted once again with the rise of molecular evolution.

1.5 The molecular landscape

As biology grew increasingly molecular in the 1950s and 1960s, evolutionary biologists began to consider how these newly discovered sequences of DNA, RNA, and proteins themselves evolved (Anfinsen 1959; Jukes 1966). Reframing evolutionary change in terms of molecular sequences instead of alleles may not seem like a radical reconsideration, but considering networks of proteins and nucleic acids had several important consequences for Adaptive Landscapes.

John Maynard Smith was the first evolutionary biologist to imagine evolution in a space of possible protein sequences (Maynard Smith 1962, 1970). Maynard Smith suggested that in order to imagine how protein evolution proceeds we must imagine a network of proteins, each one mutational step away from the other (Maynard Smith 1970). The question then is how it is possible to evolve from one functional protein to another by natural selection. If natural selection is the only operative means of evolution, then the network must have continuous paths between functional proteins, because a mutation to a non-functional protein would not be favored by natural selection. This would require

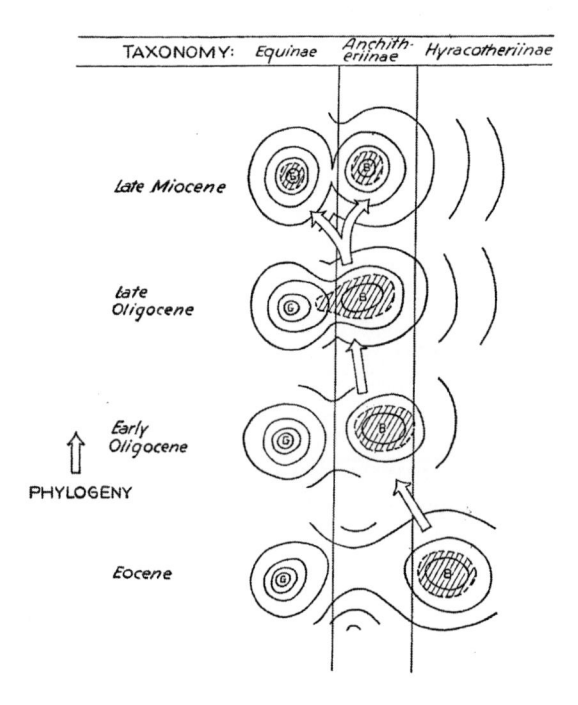

Figure 1.7 G. G. Simpson's changing Adaptive Landscapes for the case of equine evolution. Movement of both a population and the landscape over geological time was used by Simpson to explain patterns of equine evolution in the fossil record. Tempo and mode in evolution by SIMPSON, GEORGE G. Copyright 1944 Reproduced with permission of COLUMBIA UNIVERSITY PRESS.

that a certain proportion of proteins that are one step away be at least as functional as the original protein.

Even as Maynard Smith was imagining protein evolution, molecular evolution was beginning to be understood in very different terms from organismal evolution. In the late 1960s, Motoo Kimura, Jack King, and Thomas Jukes articulated their reasons for thinking that most detected molecular substitutions were not subject to selection, but were instead neutral (Kimura 1968; King and Jukes 1969). The neutral theory sparked an intense controversy over the relative power of natural selection at the molecular level (Dietrich 1994, 1998). This controversy was not an all-or-nothing contest pitting selection against drift. Instead, it was a relative significance controversy, where each side admitted that selection and drift occurred, but the big question was how often did each occur, or, put another way, how much of the genome was evolving neutrally and how much selectively.

In terms of Maynard Smith's model of evolution in protein space, the neutral theory changed the nature of the problem. Maynard Smith asked, "How often, if ever, has evolution passed through a non-functional sequence?" (Maynard Smith 1970). The possibility that many substitutions are neutral, which Maynard Smith acknowledged by citing King and Jukes, allowed him to realize that the "functional" paths may not always lead to higher functionality. The paths could follow equal functionality or neutrality to form a random walk through protein space that circumvented non-functional paths. In terms of Adaptive Landscapes, an idiom that Maynard Smith did not use, every path in protein evolution need not lead uphill.

Molecular evolution and the Adaptive Landscape were more explicitly connected in 1984, when John Gillespie introduced the metaphor of the mutational landscape (Gillespie 1984). Gillespie was a mathematical geneticist and a selectionist partisan in the neutralist–selectionist controversies raging at the time (Dietrich and Skipper 2007). In the mid 1980s, Gillespie was challenging the neutralist mechanism for the molecular clock by proposing a selectionist alternative that depended on episodic bursts of mutations. In order to model the mutational process in DNA sequences, Gillespie proposed a space of nucleotide sequences, each one nucleotide-substitution away from the other. The distance that can be traveled through this molecular landscape depends on the mutational paths from the original sequence that are selectively tractable. So, on the one hand, a set of substitutions may occur in rapid succession if they follow a path of mutations with selective advantage. On the other hand, a much more fit sequence could exist in the sequence space, but if it was two or more mutational steps away and those intermediate steps are through less fit alleles, then that more fit sequence would never be reached (Gillespie 1984). In Gillespie's words, "the mutational structure, in effect, creates innumerable selective peaks in the adaptive topography" (Gillespie 1984). Using this model, Gillespie calculated the time to cross a valley between two adaptive peaks on a selective mutational landscape and drew important implications for the rate of molecular evolution as understood from a selectionist perspective.

Gillespie's model of the mutational landscape did not include neutral mutations. Gillespie did not deny that a model of neutral and selected mutations might be useful and interesting; he was simply involved in a polemic that led him to explore the selectionist alternative to neutrality. If the selectionist molecular landscape was very hilly, the opposite neutral molecular landscape was flat. A molecular landscape that blended selected and neutral changes could still retain its hills and valleys, but would have many new plateaus constituting a neutral space for molecular change. Shifting the foundation of the Adaptive Landscape to a space of nucleotide sequences, thus, has profound effects for the possible topographies (Gavrilets 2004).

1.6 Conclusion

Adaptive Landscape diagrams are often unlabelled giving the impression that the same biological entities underlie every representation. In this brief overview of the history of the Adaptive Landscape in evolutionary biology, we divide representations of the Adaptive Landscape into three historical lineages based on what kind of biological entities were thought to ground the landscape. Although Armand Janet articulated a phenotypic selective surface in 1895, we mark the origin of the Adaptive Landscape with Sewall Wright's 1932 description of a genetic landscape. Phenotypic landscapes quickly re-emerged in 1944 when Simpson reimagined Wright's diagram in terms of organismal phenotypic traits. The molecular revolution inspired a second bifurcation from the genetic landscape when the space of genetic possibilities was reconsidered in terms of the space of possible proteins and nucleic acids. Each of these lineages reaches to the present and is represented by contributions to this volume. Taken together these three different foundations and histories of research and representation have given the Adaptive Landscape a much greater range of application than was imagined by Sewall Wright in 1932. These three different historical lineages reveal the adaptability of the Adaptive Landscape itself and offer one explanation for why the Adaptive Landscape has persisted so long in evolutionary biology.

References

Anfinsen, C. (1959). The Molecular Basis of Evolution. John Wiley and Sons, Inc., New York.

Barton, N. H., Briggs, D. E. G., Eisen, J. A., Goldstein, D. B., and Patel, N. (2007). Evolution. Cold Spring Harbor Laboratory Press, New York.

Carter, G. S. (1951). Animal Evolution. Sidgwick and Jackson Limited, London.

Coyne, J. A., Barton, N. H., and Turelli, M. (1997). Perspective: A critique of Sewall Wright's shifting balance theory of evolution. Evolution, 51, 643–671.

Dietrich, M. R. (1994). The origins of the neutral theory of molecular evolution. Journal of the History of Biology, 27, 21–59.

Dietrich, M. R. (1998). Paradox and persuasion: Negotiating the place of molecular evolution within evolutionary biology. Journal of the History of Biology, 31, 85–111.

Dietrich, M. R. and Skipper, Jr., R. A. (2007). Manipulating underdetermination in scientific controversy: The case of the molecular clock. Perspectives on Science, 15, 295–326.

Dobzhansky, T. (1937). Genetics and the Origin of Species, 1st edn. Columbia University Press, New York.

Dobzhansky, T. (1942). Genetics and the Origin of Species, 2nd edn. Columbia University Press, New York.

Dobzhansky, T. (1951). Genetics and the Origin of Species, 3rd edn. Columbia University Press, New York.

Dobzhansky, T., Ayala, F., Stebbins, G. L., and Valentine, J. (1977). Evolution. W. H. Freeman, New York.

Dodson, M. M. and Hallam, A. (1977). Allopatric speciation and the fold catastrophe. American Naturalist, 111, 415–433.

Eldredge, N. (1985). Unfinished Synthesis. Oxford University Press, New York.

Eldredge, N. (1987). Macroevolutionary Dynamics: Species, Niches, and Adaptive Peaks. McGraw-Hill, New York.

Fisher, R. A. (1930). The Genetical Theory of Natural Selection. Oxford University Press, Oxford.

Fisher, R. A. (1941). Average excess and average effect of a gene substitution. Annals of Eugenics, 11, 53–63.

Freeman, S. and Herron, J. (2007). Evolutionary Analysis, 4th edn. Benjamin Cummins, New York.

Futuyma, D. (2009). Evolution, 2nd edn. Sinauer Associates, Sunderland, MA.

Gavrilets, S. (2004). Fitness Landscapes and the Origin of Species. Princeton University Press, Princeton, NJ.

Gillespie, J. (1984). Molecular evolution over the mutational landscape. Evolution, 38, 1116–1129.

Haldane, J. B. S. (1932). The Causes of Evolution. Longmans Green, London.

Hamilton, T. (1967). Process and Pattern in Evolution. Macmillan, New York.

Janet, A. (1895). Considerations mechaniques sur l'evolution et le probleme des especes. Comptes-rendus de séances du Triosieme Congres International de Zoologie, Leyden, 16–21 Septembre, pp. 136–145.

Janet, C. (1896). Les Fourmis. [Conference February 28, 1896, pp. 3–4.] Societe Zoologique de France, Paris.

Jukes, T. (1966). Molecules and Evolution. Columbia University Press, New York.

Kaplan, J. (2008). The end of the Adaptive Landscape metaphor? Biology and Philosophy, 23, 625–638.

Kimura, M. (1968). Evolutionary rate at the molecular level. Nature, 217, 624–626.

King, J. and Jukes, T. (1969). Non-Darwinian evolution. Science, 164, 788–798.

Lande, R. (1976). Natural selection and random genetic drift in phenotypic evolution. Evolution, 30, 314–334.

Lande, R. (1979). Effective deme sizes during long-term evolution estimated from rates of chromosomal inversion. Evolution, 33, 314–334.

Levine, L. (Ed.) (1995). Genetics of Natural Populations: The Continuing Importance of Theodosius Dobzhansky. Columbia University Press, New York.

Lewontin, R. C. (1974). The Genetic Basis of Evolutionary Change. Columbia University Press, New York.

Lewontin, R. C. and Kojima, K. (1960). The evolutionary dynamics of complex polymorphisms. Evolution, 14, 458–472.

Lewontin, R. C. and White, M. J. D. (1960). Interaction between inversion polymorphism of two chromosome pairs in the grasshopper, *Moraba scurra*. Evolution, 14, 116–129.

Maynard Smith, J. (1962). The limits of molecular evolution. In I. J. Good (ed.) The Scientist Speculates. Capricorn Books, New York, pp. 252–255.

Maynard Smith, J. (1970). Natural selection and the concept of protein space. Nature, 225, 563–564.

McCoy, J. W. (1979). The origin of the 'Adaptive Landscape' concept. American Naturalist, 113, 610–613.

Pearson, K. (1903). Mathematical contributions to the theory of evolution. XI. On the influence of natural selection on the variability and correlation of organs. Philosophical Transactions of the Royal Society of London. Series A, 200, 1–66.

Provine, W. B. (1986). Sewall Wright and Evolutionary Biology. University of Chicago Press, Chicago, IL.

Ridley, M. (2003). Evolution. Wiley-Blackwell Publishing, New York.

Ruse, M. (1996). Are pictures really necessary? The case of Sewall Wright's 'adaptive landscapes'. In B. Baigrie (ed.) Picturing Knowledge: Historical and Philosophical Problems Concerning the Use of Art in Science. University of Toronto Press, Toronto, pp. 303–337.

Simpson, G. G. (1944). Tempo and Mode in Evolution. Columbia University Press, New York.

Skipper, Jr., R. A. (2004). The heuristic role of Sewall Wright's 1932 Adaptive Landscape diagram. Philosophy of Science, 71, 1176–1188.

Smocovitis, V. B. (1996). Unifying Biology: The Evolutionary Synthesis and Evolutionary Biology. Princeton University Press, Princeton, NJ.

Stewart, P. (2010). Charles Janet: Unrecognized genius of the period system. Foundations of Chemistry, 12, 5–15.

Turner, J. G. (1972). Selection and stability in the complex polymorphism of Moraba scurra. Evolution, 26, 334–343.

Wheeler, W. M. (1904). On the pupation of ants and the feasibility of establishing the Guatamalan kelep or cotton-weevil ant in the United States. Science, 20, 437–440.

Wheeler, W. M. (1915). On the presence and absence of cocoons among ants, the nest-spinning habits of the larvae and the significance of the black cocoons among certain Australian species. Annals of the Entomological Society of America, 8, 323–342.

Wright, S. (1931). Evolution in Mendelian populations. Genetics, 16, 97–159. Reprinted in W. B. Provine (1986). Sewall Wright: Evolution: Selected Papers. University of Chicago Press, Chicago, IL, pp. 98–160.

Wright, S. (1932). The roles of mutation, inbreeding, crossbreeding and selection in evolution. Proceedings of the Sixth Annual Congress of Genetics, 1, 356–366. Reprinted in W. B. Provine (1986). Sewall Wright: Evolution: Selected Papers. University of Chicago Press, Chicago, IL, pp. 161–177.

Wright, S. (1935). Evolution in populations in approximate equilibrium. Journal of Heredity, 30, 257–266.

Wright, S. (1939). Statistical Genetics in Relation to Evolution. [Actualités scientifiques et industrielles, 802. Exposés de Biométrie et de la statistique biologique XIII.] Paris: Hermann & Cie. Reprinted in W. B. Provine (1986). Sewall Wright: Evolution: Selected Papers. University of Chicago Press, Chicago, IL, pp. 283–341.

Wright, S. (1944). Letter to Ernst Mayr, 1 June 1944. In Bulletin of the Committee on Common Problems of Genetics, Paleontology and Systematics, 2, 31–35. H. J. Muller Papers. Lilly Library, Indiana University, Bloomington, IN.

Wright, S. (1945). Tempo and mode: A critical review. Ecology, 26, 415–419.

CHAPTER 2

Sewall Wright's Adaptive Landscape: Philosophical Reflections on Heuristic Value

Robert A. Skipper, Jr. and Michael R. Dietrich

2.1 Introduction

Sewall Wright's 1932 Adaptive Landscape diagram is arguably the most influential visual heuristic in evolutionary biology, yet the diagram has met with criticism from biologists and philosophers since its origination. In our view, the diagram is a valuable evaluation heuristic for assessing the dynamical behavior of population genetics models (Skipper 2004). Although Wright's particular use of it is of dubious value, other biologists have established the diagram's positive heuristic value for evaluating dynamical behavior. In what follows, we will survey some of the most influential biological and philosophical work considering the role of the Adaptive Landscape in evolutionary biology. We will build on a distinction between models, metaphors, and diagrams to make a case for why Adaptive Landscapes as diagrams have heuristic value for evolutionary biologists.

2.2 Sewall Wright's Adaptive Landscape

E. M. East invited the architects of theoretical population genetics, R. A. Fisher, J. B. S. Haldane, and Sewall Wright, to present their work at the 1932 Sixth International Congress of Genetics. They were to present compact and accessible forms of their seminal but mathematically intimidating work on evolutionary theory. Wright's principal evolutionary paper was his 1931, "Evolution in Mendelian Populations" (Wright 1931). The paper Wright delivered at the congress in 1932 was, basically, a distillation of the 1931 paper, and was published in the proceedings as "The Roles of Mutation, Inbreeding, Crossbreeding and Selection in Evolution" (Wright 1932). The Adaptive Landscape was first publicly presented in the 1932 paper.

Wright's aim in the 1931/1932 papers was to determine the ideal conditions for evolution to occur given specific assumptions about the relationship between Mendelian heredity and the adaptive value of gene complexes (Wright 1931, 1932). In Wright's 1932 paper, he used the Adaptive Landscape diagram to demonstrate his solution (see Chapter 5).

According to Wright (1932), accurately representing the population genetics of the evolutionary process requires thousands of dimensions. This is because the field of possible gene combinations of a population is vast (approximately 10^{1000}). Wright used the two-dimensional graphical depiction of an Adaptive Landscape in Fig. 2.1a as a way of intuitively conveying what can only be accurately represented in thousands of dimensions. Wright's interpretation of his diagrams is confusing, as we will discuss later in this chapter. The surface of the landscape is typically understood as representing populations with each point on the landscape representing a unique combined set of allele frequencies. Each point or population is then graded for adaptive value. Presumably, populations with very similar sets of alleles at similar frequencies will have similar adaptive values and so the adaptive surface will show relatively gradual transitions from low to high adaptive value, although this is not a necessary condition. The surface of the landscape is very "hilly," says Wright, because of epistatic

Figure 2.1 Wright's (1932, pp. 161–163) key figures. (a) Wright's Adaptive Landscape diagram. (b) Diagrams depict evolution occurring on the Adaptive Landscape under alternative assumptions. Genetics by GENETICS SOCIETY OF AMERICA. Copyright 1932. Reproduced with permission of GENETICS SOCIETY OF AMERICA.

relations between genes, the consequences of which (for Wright) are that genes adaptive in one combination are likely to be maladaptive in another. Given Wright's view of epistasis and the vastness of the field of gene combinations in a field of gene frequencies, Wright estimates the number of adaptive "peaks" separated by adaptive "valleys" at 10^{800}. Peaks are represented by "+"; valleys are represented by "–."

The Adaptive Landscape diagram sets up Wright's signature problem, viz., the problem of peak shifts (see Chapter 6). That is, given that the Adaptive Landscape is hilly, the ideal conditions for evolution to occur must allow a population to shift from peak to peak to find the highest peak. Otherwise, a population would remain fixed at the nearest local peak regardless of its adaptive value. In Wright's 1931 paper, he demonstrated mathematically the statistical distributions of genes under alternative assumptions of population size, mutation rate, migration rate, selection intensity, etc. In the 1932 paper, the graphs displaying the results appear, and he uses them in combination with the landscape diagram to argue for his three-phase shifting balance model of the evolutionary process (window F in Fig. 2.1b) as the solution to his problem of peak shifts via assessments of alternative models of the process (windows A–E in Fig. 2.1b). Wright's view was that his "shifting balance" process of evolution satisfied the ideal conditions for evolution to occur. Evolution in the shifting balance process occurs in three phases: Phase I—random genetic drift causes subpopulations semi-isolated within the global population to lose fitness; Phase II—selection on complex genetic interaction systems raises the fitness of those subpopulations; Phase III—interdemic selection then raises the fitness of the large or global population.

2.3 Models, metaphors, and diagrams

Wright's 1931 exposition of his shifting balance theory relied on a series of mathematical models. When he presented his work for a general biological audience in 1932, he chose to describe the elements, behaviors, and consequences of these models by developing a metaphor and interpretations of a series of diagrams. These different entities, i.e. models, metaphors, and diagrams, are recognized by philosophers of science and some biologists as having importantly different features and significantly different roles in the manufacture of scientific knowledge.

Diagrams are visual representations that use spatial configurations within an image to convey information (Perini 2004, 2005). Understanding how the form of a representation assigns meaning or content is a major topic among scholars interested in visual representation (Lynch and Woolgar 1990). Extracting the meaning of some images may involve interpreting an image symbolically in terms of conventions that associate a particular form and a particular meaning. For other images, their meaning can be interpreted in terms of the resemblance of the form to the object represented. For propositionally-oriented philosophers though, images are one of many forms for conveying information and so seem like they ought to be replaceable by linguistic descriptions (Ruse 1996; Perini 2005). Other philosophers eager to explain why images are so prevalent in science, see them as a particularly concise and effective way of communicating complex information, such as the functional relationships between different molecular structures to form the active site of an enzyme (Perini 2005).

Models, like many diagrams, are extralinguistic representations. However, the relationships between the variables and parameters of a model are articulated linguistically and mathematically (Lloyd 1988). Philosophers advocating the semantic approach to theories have developed a sophisticated understanding of the features of biological models in terms of rules of coexistence, interaction, and temporal succession for the variables and parameters in any particular type of model. In other words, the semantic approach views theories as models describing the many different relationships between a set of variables and parameters. An important feature of the semantic approach is its natural understanding of models in contemporary biology (Lloyd 1988). Richard Lewontin, for instance, defines a scientific model as a set of entities with corresponding quantities that are connected to each other by rules of transformation (Lewontin 1963, 1974).

Two of the central virtues of scientific models are that they allow scientists to make very precise claims about relationships between variables in the models and that they allow scientists to make precise claims about which empirical entities are represented by which features of that model. Metaphors by contrast depend on an analogy between two objects or two systems (Hesse 1966). A metaphor is an accurate analogy in so far as its components and relationships correspond to the object and relationships with which it is meant to be compared. A metaphor is didactically useful, however, in so far as it allows us to understand those elements and relationships better by way of analogy (Lewontin 1963). The didactic function of the Adaptive Landscape as a metaphor depends on our familiarity with actual hilly landscapes and their representation as topographical maps.

We are convinced that the metaphor of the Adaptive Landscape has been didactically useful in generating novel theoretical and empirical paths of inquiry. In the remainder of this essay, we justify our claim by evaluating influential criticisms of the usefulness of the metaphor, analyzing a series of criticisms of Wright's version of the Adaptive Landscape, and describing the way in which we understand the didactic role of the metaphor.

2.4 Questioning value of the Adaptive Landscape diagram

There is considerable agreement among biologists and philosophers that Wright's particular version of the Adaptive Landscape is flawed. Roughly, this agreement revolves around the extent to which Wright's evolutionary assumptions in fact underwrite a hilly landscape of the sort depicted, and therefore the problem of peak shifts, that he thought is so central to evolutionary biology. Among these same biologists and philosophers, however, is sharp disagreement about the heuristic value of the Adaptive Landscape diagram. Some, notably William Provine (1986) and Jonathan Kaplan (2008), argue that Wright's Adaptive Landscape ought to be abandoned as a visual representation of the evolutionary process because confusions surrounding its interpretation have led scientists down fruitless paths of inquiry.[1] Others, such as Michael Ruse (1996), Robert Skipper (2004), and Anya Plutynski (2008), argue that in spite of the technical problems with Wright's diagram and the fact that as a heuristic it may lead scientists astray, the Adaptive Landscape metaphor has led and will continue to lead scientists down fruitful paths of inquiry. We endorse this latter position in the following critical analysis of the Adaptive Landscape metaphor. We begin with those who adopt the former view.

---

[1] Pigliucci and Kaplan (2006) and Pigliucci (2008) come very close to this view.

Provine (1986), Wright's biographer, harshly criticized Wright's view of the Adaptive Landscape. Provine argues that Wright interpreted the diagram in two main ways. First, Wright (1932, 1977) interpreted the diagram as the multidimensional field of *all possible gene combinations* graded for their adaptive value. Call this the *genotype interpretation*. Second, Wright (correspondence to Fisher, 3 February 1931 in Provine 1986; Wright 1939, 1978) interpreted Fig. 2.1a as the multidimensional field of *joint frequencies of all genes* in a population graded for their adaptive value Call this the *population interpretation*.

According to Provine (1986), Wright's genotype interpretation of the diagram is mathematically incoherent and the two interpretations are incommensurable. Provine claims that on the genotype interpretation each axis of the graphic is a gene combination. But, Provine argues, there are no gradations along the axes, no indications of what the units along the axes are, and no point along them to indicate where a gene combination is to be placed. Given this, Provine concludes that there is no way of generating the continuous surface represented in Fig. 2.1a. On the population interpretation, Provine argues, each point on the surface represents a population, and the entire surface is of mean population fitness rather than genotype fitness. Each axis is now graded between 0 and 1 for gene frequency; the result is a continuous surface. However, Provine claims, there is no way to plot genotype fitness values on the surface of gene frequencies; one is attempting to plot individual haplotypes onto a surface of which the points are populations. The result is a surface that collapses into a single point because the axes are incompatible. For these reasons, Provine concludes that Wright's Adaptive Landscape diagram does not successfully illustrate the shifting balance process and, so, must be abandoned.

Wright (1988) responded to Provine's criticism by uncritically claiming that Provine confused a metaphor for a mathematical model. Ruse (1996) develops the criticism and rightly disagrees with Provine: *even if* Wright's interpretation of the diagram *is* incoherent (meaning that it does not accurately represent Wright's mathematical model), it may *still* be a valuable heuristic. Ruse thinks that Provine's assessment of the heuristic value of Wright's diagram is too conservative. Heuristics are devices that are used to generate paths of inquiry whether those paths are fruitful or not. The fact that Wright's interpretation of the diagram does not cohere with his mathematical models is beside the point. Rather, because uses of the diagram have a track record of generating fruitful paths of inquiry, it is a valuable heuristic. Ruse cites T. Dobzhansky (1951), G. G. Simpson (1953), G. Ledyard Stebbins (1969), and C. H. Waddington (1956) as having used the Adaptive Landscape to produce apparently positive results: Dobzhansky used the landscape to illustrate the shifting balance process, Simpson used the landscape to illustrate species and speciation, and Waddington used the diagram for his own illustrative purposes in population genetics. Moreover, the biological work we discuss later in this essay includes instances of the heuristic used to produce positive results, as we will see (briefly) (see Chapter 1). Provine's doubts about the heuristic value of the Adaptive Landscape diagram are misplaced.

Kaplan (2008) argues that the Adaptive Landscape metaphor in general ought to be abandoned. Kaplan lists a number of complaints about the diagram, e.g. it is not explanatory in the way a mathematical model is, the diagram does not map onto a relevant mathematical model, it is imprecise, it lacks a univocal heuristic role, and it has done more to confuse than to enlighten. But the main point of Kaplan's argument is that the diagram is no longer useful given the continued growth in computational power that we are experiencing with respect to scientific modeling. That is, there is no need for a simplified landscape diagram when biologists have access to computational power that will allow them to build models that will more precisely do the work that the landscape diagram is meant to do. Kaplan then recommends biologists abandon the diagram in favor of computational models.

Many of Kaplan's complaints are familiar. Indeed, the Adaptive Landscape diagram has led biologists astray, it is not explanatory in the way that a mathematical model is, it is imprecise, and so on. We think these criticisms are misplaced. One certainly expects mathematical models to be explanatory, precise, etc. But, in the vein of Wright's and Ruse's arguments against Provine, these demands are misdirected when aimed at the landscape diagram. Wright developed the Adaptive Landscape in an effort to defend his shifting balance process in an intuitive, informal way using a simple, unlabeled topographical map as a metaphor for the mean fitness of populations. It is wrongheaded to expect such a heuristic to meet the standards of a mathematical model. Rather, as we described above, we expect positive analogies from metaphors, but we also expect negative analogies (Hesse 1966). Nothing Kaplan has said in criticizing the landscape diagram addresses it as a metaphor. After all, the diagram may not meet the expectations we have for mathematical models, but that does not mean the diagram does not meet expectations properly construed. The review of biological work in section 2.5 we think demonstrates that the diagram is heuristically valuable.

Finally, it is not at all clear that Kaplan's suggestion to abandon the landscape diagram in favor of complex mathematical models is reasonable. Certainly the suggestion is not descriptive of the way many if not most population geneticists practice, though there are exceptions. And, at any rate, we do not think there is good reason to think that the computational power that is available to biologists is sufficient to yield a biologically realistic population genetics model. Idealizations and simplifications are the norm and will no doubt continue to be the norm.

We think Provine's and Kaplan's reasons to reject the Adaptive Landscape diagram are poor. Both seem to think that the diagram must play the role of an explanatory model, and because it cannot do that, it is a defective scientific tool. For Provine, the diagram is simply a failure. For Kaplan, the diagram has led scientists down fruitless paths of inquiry. On our view, not all aspects of scientific models are explanatory; some are metaphorical (Lewontin 1963). These metaphorical elements play heuristic roles in scientific theorizing. And heuristics are rules of thumb, i.e. strategies that work well sometimes but not always. We now consider a patchwork of biological work that suggest that the Adaptive Landscape metaphor is a valuable heuristic in spite of the flaws in Wright's particular view of it (cf. Plutynski 2008).

2.5 The case for heuristic value

Fisher immediately pointed out the central problem for Wright's version of the diagram. Fisher and Wright discussed the landscape metaphor in correspondence prior to the 1932 presentation (Wright correspondence to Fisher, 3 February 1931 in Provine 1986, pp. 271–273; Fisher correspondence to Wright, 31 May, 1931 in Provine 1986, p. 274). According to Fisher, the Adaptive Landscape diagram is flawed because, in fact, as the dimensionality of the field of gene combinations in the field of gene frequencies *increases* the number of stable peaks on the surface of the landscape *decreases* (see also Fisher 1941). Thus, claims Fisher, representation of the mean fitness of populations in multiple dimensions will not result in a *hilly* landscape, but one that is a single peak with ridges along it. As a consequence, evolution on the landscape does not require the complex of evolutionary factors of Wright's shifting balance process, but only selection and mutation. Fisher's informal critique of Wright was taken up perhaps most influentially by P. A. P. Moran (1964), A. W. F. Edwards (1994), and Jerry Coyne et al. (1997). Mark Ridley (1996) developed an Adaptive Landscape diagram based on Fisher's critique, reproduced in Fig. 2.2a (see Chapter 4).

Sergey Gavrilets (1997, 1999, 2004) has criticized Wright's version of the landscape. On Gavrilets's view, the Adaptive Landscape in multiple dimensions will not have adaptive peaks and valleys. Instead, the landscape will be the holey one depicted in Fig. 2.2b. That is, the higher the number of possible gene combinations in a field of gene frequencies, the higher the number of incompatible combinations in that field. The incompatible gene combinations cause reproductive isolation within populations, which cause genetically-driven speciation events. The holes represent locations of incompatible combinations of genes and replace the peaks. Gavrilets's argument, as he recognizes, is based on a set of specific assumptions that must be relaxed in fundamental ways if his theoretical intuitions are to be tested empirically. Gavrilets assumes that (1) fitnesses of gene

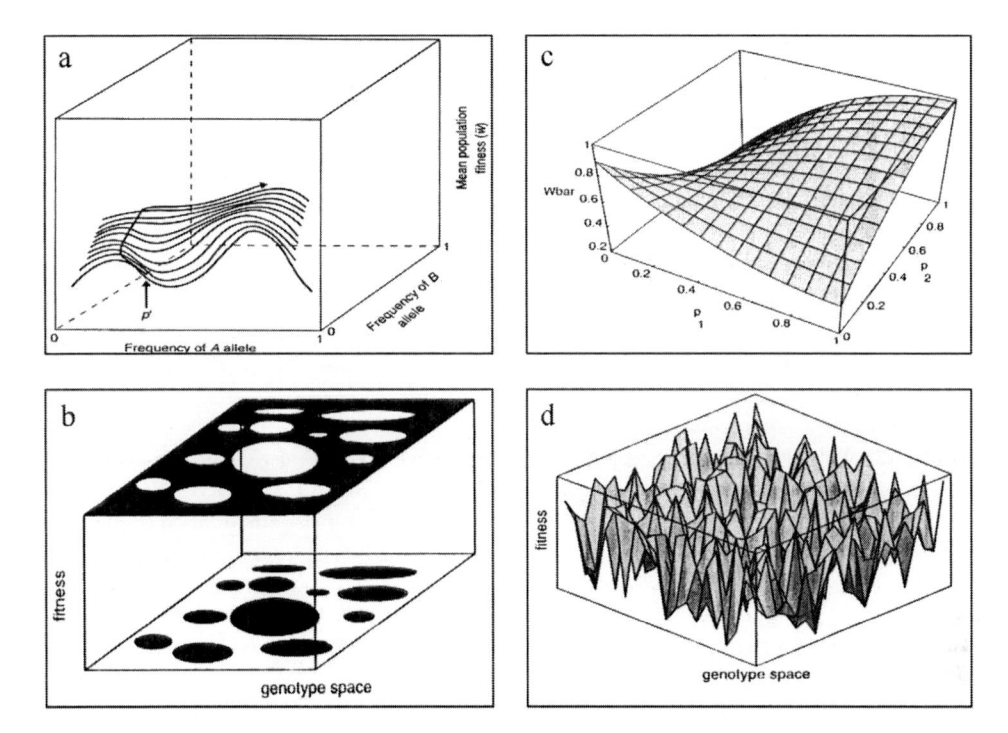

Figure 2.2 Four Adaptive Landscape diagrams, resulting from critiques of Wright's 1932 diagram. (a) Ridley's (1996, p. 219) depiction of a Fisherian landscape. Evolution. 2nd edition. Copyright 1997. Reproduced with permission of Wiley-Blackwell. (b) Gavrilets's (1997, p. 309) depiction of the holey landscape. Trends in Ecology and Evolution. Copyright 1997. Reproduced with permission of Elsevier. (c) Coyne et al.'s (1997, p. 647) simplified Wrightian landscape. Reproduced from Coyne, J., Barton, N., and Turelli, M. (1997), "Perspective: A Critique of Sewall Wright's Shifting Balance Theory of Evolution", Evolution 51: 643–671 (d) The Kauffman and Levin (1987, p. 33) rugged Adaptive Landscape. The Journal of Theoretical Biology. Copyright 1987. Reproduced with permission of Elsevier.

complexes are generated randomly; (2) fitnesses are generated independently; and (3) fitness values are either 0 or 1. Nevertheless, Gavrilets (1999) has discussed the evolutionary dynamics of speciation on holey landscapes as driven by random genetic drift, mutation, recombination, and migration. Indeed, Gavrilets points out that Wright's apparently restrictive combination of drift, mass selection, interdemic selection, and migration is not necessary to traverse the Adaptive Landscape if Gavrilets is correct. There are, after all, no adaptive valleys between peaks to pass through.

A series of papers by collaborators working on the genotype–phenotype map problem have raised an issue against Wright's version of the landscape diagram that is related to that of Gavrilets (e.g. Fontana and Schuster 1998; B. Stadler et al. 2001; P. Stadler 2002). The basic claim here is that a discontinuous landscape surface is more likely than a continuous one. The argument, however, is different. Roughly, the argument is that taking seriously the developmental processes involved in going from genotype to phenotype in evolution, one will discover that there are many phenotypes inaccessible from genotypes, resulting in discontinuities on the surface of the landscape. The argument is rooted in computational work on the biophysical genotype–phenotype model defined by the folding of RNA sequences into secondary structures. The RNA sequences are considered to be genotypes, and the role of the phenotype is played by the structure of the molecule. Based on this model, a general, mathematical theory of landscapes has resulted, reaching far beyond the informal use of topographical map-making that Wright used in 1932. Despite the implications for Wright's landscape, apparently a landscape on which to depict evolutionary trajectories in population genetics based on the general landscape theory is not immediately forthcoming due to constraints on computational power.

Although there are serious problems for Wright's view that he can transform a hilly landscape in two dimensions into a hilly landscape in thousands of dimensions, the notion that there are simple cases of hilly landscapes persists (e.g. Coyne et al. 1997, p. 647). Indeed, in simple and rather restrictive cases, i.e. two loci cases assuming complete dominance at each locus, Wright's landscape has been given some plausibility (e.g. Lande 1976, 1979; Coyne et al. 1997) (Fig. 2.2c). Further, Kauffman and Levin (1987) responded to problems for Wright's view of the landscape by developing "rugged Adaptive Landscapes" as a way of understanding the fitness of gene combinations given simple Wrightian epistatic gene interaction (Fig. 2.2d). In Kauffman and Levin's *NK* model, the fitness contribution of each of *N* loci depends in a random way on *K* other loci. The parameter *K* describes the degree of epistasis. If $K = 0$, then an Adaptive Landscape with one peak results. But as *K* increases, the number of peaks on the landscape increases and the mean fitness of the nearest peak decreases toward that of an entirely random genotype. Typically, the result is a rugged Adaptive Landscape. Kauffman and Levin's work has been applied in biochemistry (e.g. Fontana et al. 1989, 1991, 1993), and interestingly that work is a principal ancestor of the genotype–phenotype mapping work discussed earlier. Kauffman (1993) extends the use of rugged Adaptive Landscapes in his work on complexity and on artificial life modeling.

Notice that these biologists do not view the landscape diagram as a failed heuristic. Rather, each critically evaluates the foundational assumptions of the other out of which a new Adaptive Landscape metaphor is created, along with new, arguably fruitful paths of inquiry. This pattern of contrasts can be seen across all of the cases discussed, but consider the contrast between Fisher's single-peak landscape and Wright's hilly landscape. The single-peak landscape reflects Fisher's view that there is one optimal gene combination, which can be found by the combination of mutation and selection. Wright's hilly landscape reflects his emphasis on the evolutionary effects of epistatic gene interaction.

For Wright, genes adaptive in one combination will be maladaptive in another. The consequence of such epistasis was, for Wright, a hilly Adaptive Landscape. More generally, depending on the basic genetic assumptions in a model, whether it is Fisher's, Wright's, Coyne's, Gavrilets's, etc. the corresponding surface of selective value will take different forms. Indeed, it is important to note that it is the set of core genetic assumptions embedded in a population genetics model rather than assumptions about the balance of evolutionary causes that go toward shaping the surface of an Adaptive Landscape. Given that, it seems plausible that while, for example, Wright's shifting balance process is dependent on his version of the Adaptive Landscape, the landscape is in no way dependent upon the shifting balance process (cf. Ruse 1996; Coyne et al. 1997).

Ultimately, we think the cases we have discussed demonstrate that these biologists find the heuristic useful. Next, we describe the way in which we think the Adaptive Landscape metaphor has been a valuable heuristic.

The diagram is used importantly as a model evaluation heuristic (Skipper 2004). Indeed, we think it is clear that Wright used the Adaptive Landscape diagram as a visual heuristic to evaluate the *dynamical behavior* of population genetics models of evolutionary processes constructed with alternative assumptions to demonstrate his own. The dynamical behavior of a mathematical model refers to the way(s) in which some system being described by the model change(s) according to changes in the model's state(s).[2] The dynamical behavior of such a model in population genetics includes, for example, the changes in the mean fitness of a population against the parameters that hold the measured intensity of specific evolutionary factors such as population size, migration rate, selection, mutation, etc. and describes the ways that the states of the model change. The Adaptive Landscape diagram, as a simplified visualization of common, core assumption of all the models, is where the

---

[2] Our "dynamical behavior" should not be confused with Lewontin's (1974) "dynamical sufficiency." Lewontin's dynamical sufficiency refers to a model's empirically being demonstrated to contain all of the relevant parameters, etc. required to describe evolutionary change. And Lewontin outlines a specific and probably unattainable view of "sufficiency." "Dynamical behavior" is a practical specification of one way to assess a much more modest notion of sufficiency.

evaluation of the behaviors of the models takes place; the diagram is the heuristic with which the evaluation is being made. A model is positively evaluated in case a model system described by it can traverse the landscape, shifting from one adaptive peak to the highest adaptive peak.

Wright assesses the dynamical behavior of evolutionary systems described by six alternative models, i.e. windows A–F in Fig. 2.1b. Each window is a piece of the larger Adaptive Landscape in Fig. 2.1b. What Wright does is to visually depict the dynamical behavior of a model system on the landscape. Consider window A in Fig. 2.1b. Here, Wright sets up a model with the following assumptions: populations are very large and panmictic, mutation rate is high, selection intensity and mutation rate are low. Depicting that model by way of Wright's landscape, the model system will not be able to get to the highest adaptive peak because it will not be able to traverse the hilly surface—it will not be able to move from its initial position to a higher peak. Wright repeats this process in windows B–F, demonstrating how the dynamical behavior of the various mathematical models changes as the assumptions change. Only the model sketched in window F succeeds in traversing the heuristic landscape; that is the model Wright interprets as describing the shifting balance process.

Ultimately, Wright's evaluative strategy, using the landscape diagram, led him to his view that evolution is a process that includes a constellation of factors. That is, out of his evaluation of alternative evolutionary hypotheses, driven by the landscape diagram, Wright was led to his shifting balance process: the evolutionary factors delineated in the shifting balance process are necessary for traversing the Adaptive Landscape. Now, Wright's own heuristic use of the diagram is problematic because of a flaw in his interpretation of it. However, critics who have pointed out Wright's flaw have then gone on to create landscape diagrams of their own for the same sorts of assessments. And they further use the landscape to assess the dynamical behavior of alternative population genetics models. Indeed, Ridley (1996), using his version of the Fisherian landscape, shows that Wright's shifting balance process is unnecessary for traversing the landscape:

The models that describe the shifting balance process overdetermine the process that is required given the surface of the landscape. Gavrilets (1997, 2004) shows, using his holey landscape, that neither Wright's nor Fisher's mechanisms are necessary to traverse the landscape. In Gavrilets's case, Wright's shifting balance process is unnecessary because there are no peaks. And Coyne et al. (1997) show that, in simple cases, Wright's three-phase evolutionary process is one among many possible mechanisms for traversing a hilly landscape. The Adaptive Landscape diagram is valuable, in spades, as a heuristic for evaluating the dynamical behavior of evolutionary models (Skipper 2004).

2.6 Conclusion

In spite of the fact that Wright's Adaptive Landscape diagram is one of the most influential visual heuristic in evolutionary biology, there has been considerable confusion about whether it is actually valuable in that role. We have argued that the confusion about the heuristic value of the diagram is rooted in misunderstandings about models and metaphors. The landscape diagram is not a mathematical model; it is a visual metaphor intended to represent the ways in which a system described by a mathematical model may behave. Broadly, metaphors play a didactic and not explanatory role in scientific theorizing. The Adaptive Landscape plays that didactic role as a model assessment heuristic. Understanding the diagram in this way makes it clear why the Adaptive Landscape has been so influential.

References

Coyne, J. A., Barton, N. H., and Turelli M. (1997). Perspective: A critique of Sewall Wright's shifting balance theory of evolution. Evolution, 51, 643–671.

Dobzhansky, T. (1951). Genetics and the Origin of Species, 3rd edn. Columbia University Press, New York.

Edwards, A. W. F. (1994). The fundamental theorem of natural selection. Biological Reviews of the Cambridge Philosophical Society, 69, 443–474.

Fisher, R. A. (1941). Average excess and average effect of a gene substitution. Annals of Eugenics, 11, 53–63.

Fontana, W., Greismacher, T. Schnabl, P. Stadler, F., and Schuster P. (1991). Statistics of landscapes based on free energies, replication and degradation rate constants of RNA secondary structures. Monatshefte für Chemie, 122, 795–819.

Fontana, W., Schnabl, P., and Schuster, P. (1989). Physical aspects of evolutionary optimization and adaptation. Physical Review A, 40, 3301–3321.

Fontana, W. and Schuster, P. (1998). Continuity in evolution: On the nature of transitions. Science, 280, 1541–1455.

Fontana, W., Stadler, F., Tarazona, P. Weinberger, E., and Schuster, P. (1993). RNA folding and combinatory landscapes. Physical Review, E 47, 2083–2099.

Gavrilets, S. (1997). Evolution and speciation on holey adaptive landscapes. Trends in Ecology & Evolution, 12, 307–312.

Gavrilets, S. (1999). A dynamical theory of speciation on holey adaptive landscapes. American Naturalist, 154, 1–22.

Gavrilets, S. (2004). Fitness Landscapes and the Origin of Species. Princeton University Press, Princeton, NJ.

Hesse, M. (1966). Models and Analogies in Science. Notre Dame University Press, Notre Dame, IN.

Kaplan, J. (2008). The end of the Adaptive Landscape metaphor? Biology and Philosophy, 23, 625–638.

Kauffman, S. (1993). The Origins of Order: Self-Organization and Selection in Evolution. Oxford University Press, New York.

Kauffman, S. and Levin, S. (1987). Towards a general theory of adaptive walks on rugged landscapes. Journal of Theoretical Biology, 128, 11–45.

Lande, R. (1976). Natural selection and random genetic drift in phenotypic evolution. Evolution, 30, 314–334.

Lande, R. (1979). Effective deme sizes during long-term evolution estimated from rates of chromosomal inversion. Evolution, 33, 314–334.

Lewontin, R. (1963). Models, mathematics, and metaphors. Synthese, 15, 222–244.

Lewontin, R. (1974). The Genetic Basis of Evolutionary Change. Columbia University Press, New York.

Lloyd, E. A. (1988). The Structure and Confirmation of Evolutionary Theory. Greenwood Press, New York.

Lynch, M. and Woolgar, S. (Eds.) (1990) Representation in Scientific Practice. MIT Press, Cambridge, MA.

Moran, P. A. P. (1964). On the non-existence of adaptive topographies. Annals of Human Genetics, 27, 383–393.

Perini, L. (2004). Convention, resemblance and isomorphism: understanding scientific visual representations. In G. Malcolm (ed.) Multidisciplinary Approaches to Visual Representations and Interpretations. Elsevier, Amsterdam, pp. 39–42.

Perini, L. (2005). Explanation in two dimensions: Diagrams and biological explanation. Biology and Philosophy, 20, 257–269.

Pigliucci, M. (2008). Sewall Wright's Adaptive Landscapes: 1932 vs. 1988. Biology and Philosophy, 23, 591–603.

Pigliucci, M. and Kaplan, J. (2006). Making Sense of Evolution: The Conceptual Foundations of Evolutionary Biology. University of Chicago Press, Chicago, IL.

Plutynski, A. (2008). The rise and fall of the Adaptive Landscape? Biology and Philosophy, 23, 605–623.

Provine, W. B. (1986). Sewall Wright and Evolutionary Biology. University of Chicago Press, Chicago, IL.

Ridley, M. (1996). Evolution, 2nd edn. Blackwell Science, Inc., Cambridge, MA.

Ruse, M. (1996). Are pictures really necessary? The case of Sewall Wright's 'Adaptive Landscapes'. In B. Baigrie (Ed.) Picturing Knowledge: Historical and Philosophical Problems Concerning the Use of Art in Science. University of Toronto Press, Toronto, pp. 303–337.

Simpson, G. G. (1953). The Major Features of Evolution. Columbia University Press, New York.

Skipper, Jr., R. A. (2004). The heuristic role of Sewall Wright's 1932 Adaptive Landscape diagram. Philosophy of Science, 71, 1176–1188.

Stadler, B. M. R., Stadler, P. Wagner, G., and Fontana, W. (2001). The topology of the possible: Formal spaces underlying patterns of evolutionary change. Journal of Theoretical Biology, 213, 241–274.

Stadler, P. (2002). Fitness landscapes. In M. Lässig and A. Valleriani (eds.) Biological Evolution and Statistical Physics. Springer-Verlag, Berlin, pp. 187–207.

Stebbins, G. L. (1969). The Basis of Progressive Evolution. University of North Carolina Press, Chapel Hill, NC.

Waddington, C. H. (1956). Principles of Embryology. Macmillan, New York.

Wright, S. (1931). Evolution in Mendelian populations. Genetics, 16, 97–159. Reprinted in W. B. Provine (1986). Sewall Wright: Evolution: Selected Papers. University of Chicago Press, Chicago, IL, pp. 98–160.

Wright, S. (1932). The roles of mutation, inbreeding, crossbreeding and selection in evolution. Proceedings of the Sixth Annual Congress of Genetics, 1, 356–366. Reprinted in W. B. Provine (1986). Sewall Wright: Evolution: Selected Papers. University of Chicago Press, Chicago, IL, pp. 161–177.

Wright, S. (1939). Statistical Genetics in Relation to Evolution. [Actualités scientifiques et industrielles, 802. Exposés de Biométrie et de la statistique biologique XIII.] Hermann & Cie, Paris. Reprinted in W. B. Provine (1986). Sewall Wright: Evolution: Selected Papers. University of Chicago Press, Chicago, IL, pp. 283–341.

Wright, S. (1969). Evolution and the Genetics of Populations, Vol. 2: The Theory of Gene Frequencies. University of Chicago Press, Chicago, IL.

Wright, S. (1977). Evolution and the Genetics of Populations, Vol. 3: Experimental Results and Evolutionary Deductions. University of Chicago Press, Chicago, IL.

Wright, S. (1978). The relation of livestock breeding to theories of evolution. Journal of Animal Science, 46, 1192–1200. Reprinted in W. B. Provine (1986). Sewall Wright: Evolution: Selected Papers. University of Chicago Press, Chicago, IL, pp. 1–11.

Wright, S. (1988). Surfaces of selective value revisited. American Naturalist, 131, 115–123.

CHAPTER 3

Landscapes, Surfaces, and Morphospaces: What Are They Good For?

Massimo Pigliucci

3.1 Introduction

Few metaphors in biology are more enduring than the idea of Adaptive Landscapes, originally proposed by Sewall Wright (1932) as a way to visually present to an audience of typically non-mathematically savvy biologists his ideas about the relative role of natural selection and genetic drift in the course of evolution. The metaphor, however, was born troubled, not the least reason for which is the fact that Wright presented different diagrams in his original paper that simply cannot refer to the same concept and are therefore hard to reconcile with each other (Pigliucci 2008). For instance, in some usages, the landscape's non-fitness axes represent combinations of individual genotypes (which cannot sensibly be aligned on a linear axis, and accordingly were drawn by Wright as polyhedrons of increasing dimensionality). In other usages, however, the points on the diagram represent allele or genotypic *frequencies*, and so are actually populations, not individuals (and these can indeed be coherently represented along continuous axes).

Things got even more confusing after the landscape metaphor began to play an extended role within the Modern Synthesis in evolutionary biology and was appropriated by G. G. Simpson (1944) to further his project of reconciling macro- and microevolution, i.e. to reduce paleontology to population genetics (some may object to the characterization of this program as a reductive one, but if the questions raised by one discipline can be reframed within the conceptual framework of another, that is precisely what in philosophy of science is meant by reduction; see Brigandt 2008). This time the non-fitness axes of the landscape were phenotypic traits, not genetic measures at all. Even more recently, Lande and Arnold (1983) proposed a mathematical formalism aimed at estimating actual (as opposed to Simpson's hypothetical) fitness surfaces, making use of standard multiple regression analyses. But while Simpson was talking about macroevolutionary change involving speciation, Lande and Arnold were concerned with microevolutionary analyses within individual populations of a single species.

In principle, it is relatively easy to see how one can go from individual-genotype landscapes to genotypic-frequency landscapes (the two Wright versions of the metaphor). However, the (implied) further transition from either of these to phenotypes (either in the Lande–Arnold or in the Simpson version) is anything but straightforward because of the notorious complexity and non-linearity of the so called genotype–phenotype mapping function (Alberch 1991; Pigliucci 2010). This is a serious issue if—as I assume is the case—we wish to use the landscape metaphor as a unified key to an integrated treatment of genotypic and phenotypic evolution (as well as of micro- and macroevolution). Without such unification evolutionary biology would be left in the awkward position of having two separate theories, one about genetic change, the other about phenotypic change, and no bridge principles to connect them.

One more complication has arisen in more recent years, this one concerning Wright-style fitness landscapes. Work by Gavrilets (2003; see also Chapter 17 this volume) and collaborators, made possible by the availability of computing power exceeding by several orders of magnitudes what was achievable throughout the twentieth century, has explored the features of truly highly dimensional landscapes—as opposed to the standard two- or three-dimensional ones explicitly considered by Wright and by most previous authors. As it turns out, evolution on so-called "holey" landscapes is characterized by qualitatively different dynamics from those suggested by the standard low-dimensional version of the metaphor—a conclusion that has led some authors to suggest abandoning the metaphor altogether, in favor of embracing directly the results of formal modeling (Kaplan 2008; though see Plutynski 2008 and Chapter 2 for a somewhat different take).

In this essay I wish to discuss the implications of four versions of the metaphor, often referred to as fitness landscapes, Adaptive Landscapes, fitness surfaces, and morphospaces. I will argue that moving from one to any of the others is significantly more difficult than might at first be surmised, and that work with morphospaces has been unduly neglected by the research community.

3.2 Four types of landscapes

As I mentioned earlier, there are several versions of the "landscape" metaphor that have proliferated in the literature since Wright's original paper. Indeed, Wright himself was referring—in that very same paper—to at least two conceptions of landscapes, one with individual genotypes as the points on the (necessarily non-continuous) landscape, the other with populations identified by gene or genotype frequencies (in a continuous space). I will ignore the distinction between the two types of Wright landscapes for two reasons: first, they are connected in a conceptually straightforward manner, since it is obviously possible to go from individual genotypes to populations of genotypes without any theoretical difficulty. Second, most of the post-Wright literature, including the presentation of the metaphor in textbooks and the more recent work on "holey" landscapes, is framed in terms of gene/genotype frequencies, not individual genotypes. This is understandable within the broader context of the Modern Synthesis (Mayr and Provine 1980) as a theory essentially rooted in (though certainly not limited to) population genetics, where evolution is often simply defined as change in allelic frequencies (Futuyma 2006).

Despite some confusion due to the often interchangeable use of terms like "fitness" and "Adaptive" Landscape, I will use and build here on the more rigorous terminology followed by authors such as McGhee (2007) and distinguish the following four types of landscapes:

Fitness landscapes. These are the sort of entities originally introduced by Wright and studied in a high-dimensional context by Gavrilets and collaborators. The non-fitness dimensions are measures of genotypic diversity. The points on the landscape are population means, and the mathematical approach is rooted in population genetics.

Adaptive Landscapes. These are the non-straightforward "generalizations" of fitness landscapes introduced by Simpson, where the non-fitness dimensions now are phenotypic traits. The points on the landscape are populations speciating in response to ecological pressures or even above-species level lineages (i.e. this is about macroevolution).

Fitness surfaces. These are the Lande–Arnold type of landscapes, where phenotypic traits are plotted against a surrogate measure of fitness. They are statistical estimates used in quantitative genetic modeling, and the points on the landscape can be either individuals within a population or population means, in both cases belonging to a single species (i.e. this is about microevolution).

Morphospaces. These were arguably first introduced by Raup (1966), and differ dramatically from the other types for two reasons: (a) they do not have a fitness axis; (b) their dimensions, while representing phenotypic ("morphological") traits, are generated via a priori geometrical or mathematical models, i.e. they are not the result of observational measurements. They typically refer to across-species (macroevolutionary) differences, though they can be used for within-species work.

Let us take a look in a bit more detail at each type of landscape, to familiarize ourselves with their respective similarities and differences. Throughout

I will use both the terminology just summarized and the names of the relevant authors, to reduce confusion at the cost of some redundancy. The first thing we notice from even a cursory examination of the literature is that there are few actual biological examples of fitness landscapes (Wright-style) or Adaptive Landscapes (Simpson-style) available, while there is a good number of well understood examples of morphospaces (Raup-style) and particularly of adaptive surfaces (Lande–Arnold style). These differences are highly significant for my discussion of the metaphor.

Beginning with Wright-type, fitness landscapes, many of the examples in the literature are entirely conceptual, i.e. presented by various authors only for heuristic purposes. It should be obvious why it is so: to actually draw a "real" fitness landscape we need a reasonably complete description of the genotype → fitness mapping function, i.e. we need data about the fitness value of each relevant combination of genotypes that we happen to be interested in. For most purposes this is next to impossible. Dobzhansky (1970, p. 25) famously put it this way:

> Suppose there are only 1000 kinds of genes in the world, each gene existing in 10 different variants or alleles. Both figures are patent underestimates. Even so, the number of gametes with different combinations of genes potentially possible with these alleles would be 10^{1000}. This is fantastic, since the number of subatomic particles in the universe is estimated as a mere 10^{78}.

Indeed, things are complicated even further by the fact that the genotype → fitness function can be thought of as the combination of two subfunctions: genotype → phenotype and phenotype → fitness. The second function requires an understanding of fitness (Lande–Arnold) surfaces, which can then be translated into fitness landscapes through the first function. I know of very few instances in which anything like that has even been attempted, given the so far (and possibly in principle) enormous difficulties, both empirical and computational.

The major class of exceptions to the paucity of actual fitness (Wright) landscapes is in itself highly illuminating of both the potential and limitations of the metaphor: studies of the evolution of RNA and protein structures. Consider for instance the work of Cowperthwaite and Meyers (2007) on 30-nucleotide long binary RNA molecules. This is arguably one of the simpler models of genotype–phenotype–fitness relations, and it is computationally tractable and empirically approachable. Still, 30-nucleotide binary molecules correspond to a bewildering one billion unique sequences! These in turn generate "only" about 220,000 unique folding shapes in a G/U landscape and a "mere" 1000 shapes in the A/U landscape, both of which these authors have tackled. In other words, a genotypic space of a billion sequences corresponds (obviously in a many-to-many manner) to a phenotypic space of thousands to hundreds of thousands of possibilities, each characterized by its own (environment-dependent, of course) fitness value.

Things get even more complex when we move from RNA to proteins: Wroe et al. (2007) have explored evolution in protein structure-defined phenotypic space, focusing in particular on so-called "promiscuity," the ability of a given protein to perform two functions because of the alternation between multiple thermodynamically stable configurations. Proteins are, of course, much more complicated biochemical objects than RNA molecules. They are made of more types of building blocks, and their three-dimensional structure is more difficult to predict from their linear sequence. Still, studies like those of Wroe et al. show some interesting similarities between RNA and protein landscapes, perhaps one of the most significant being that—as predicted by the "holey" landscape models studied by Gavrilets—large portions of phenotypic hyperspace are actually neutral in terms of the fitness of the forms that define that space, which means that much of the time evolution is a matter of sliding around a fitness landscape via genetic drift, with occasional "punctuations" of selective episodes.

Let us turn now to Adaptive (Simpson) Landscapes. The original idea was to use the metaphor to show how macroevolutionary events like ecological speciation can be understood in terms of the then nascent theory of population genetics, developed to directly address microevolutionary processes. Simpson, for instance, presented a (hypothetical) analysis of speciation in the *Equidae* (horse) lineage from the Cenozoic (Simpson 1953). The subfamily (i.e. significantly above species level) *Hyracotheriinae* is represented as occupying an adaptive peak characterized by their browsing-suited teeth morphology during the Eocene. Simpson also imagined a contemporary, empty, adaptive peak for species whose teeth are suitable for grazing. The diagram then shows how gradually, during the Oligocene, the subfamily *Anchitheriinae* evolved anagenetically (i.e. by replacement) from their *Hyracotheriinae* ancestors. The new subfamily, as a result of evolving greater body size, also acquired a teeth morphology that corresponded to a gradual movement *of the peak itself* closer to the empty grazing zone. Finally, in the late Miocene, the *Anchitheriinae* gave origin, cladistically (i.e. by splitting) to two lineages, one of which definitely occupied the grazing peak and gave rise to the *Equinae*. Simpson's reconstruction of the events is, of course, compatible with what we knew then of the phylogeny and functional ecology of the horse family, but the landscape's contour and movements are by necessity entirely hypothetical. This is a truly heuristic device to make a conceptual point, not the result of any quantitative analysis of what actually happened.

The situation is quite different for fitness (Lande–Arnold) surfaces, which are supposed to be closely related to Simpson's landscapes, but in fact work significantly differently in a variety of respects. For one thing, the literature is full of Lande–Arnold type selection analyses, because they are relatively straightforward to carry out empirically and in terms of statistical treatment. As is well known, phenotypic selection studies of this type are based on the quantification of selection vectors by means of statistical regression of a number of measured traits on a given fitness proxy, appropriate or available for the particular organism under study. There are, however, several well-known problems of implementation (Mitchell-Olds and Shaw 1987). Among other things, surveys of Lande–Arnold type studies show that many actual estimates of selection coefficients are unreliable because they are based on too small sample sizes (Kingsolver et al. 2001; Siepielski et al. 2009). Moreover, they tend to have very low replication either spatially (from one location to the other) or temporally (from one season to another), thereby undermining any claim to the generality or reproducibility of the results (unlike the much more experimentally confined cases of RNA and protein structures mentioned earlier). Most crucially, of course, there is essentially no connection between fitness surfaces and either fitness (Wright) landscapes or adaptive (Simpson) landscapes. The reason for the former lack of connection is that we do not have any idea of the genotype → phenotype mapping function underlying most traits studied in selection analyses, so that we cannot articulate any transition at all from Wright landscapes to Lande–Arnold surfaces. In terms of a bridge from Lande–Arnold surfaces to Simpson landscapes, this would seem to be easier because they are both expressed in terms of phenotypic measurements versus fitness estimates. But there is where the similarity ends: Lande–Arnold selection coefficients cannot be compared across species, and they do not incorporate any of the functional ecological analyses of the type envisioned by Simpson. Indeed, a major problem with the literature on selection coefficients is that they simply ignore functional ecology altogether: we can measure selection vectors, but we usually have no idea of what causes them, and no such idea can come from multiple regression analyses (this is simply a case of the general truth that correlation does not imply causation, let alone a specific type of causation). All of this means that—despite claims to the contrary (Arnold et al. 2001; Estes and Arnold 2007; see also Chapter 13) adaptive surfaces are *not* a well worked out bridge between micro- and macroevolution (see Kaplan 2009 for a detailed explanation based on the specific claims made by Estes and Arnold 2007).

The situation is again different when we move to morphospaces. Arguably the first example of these was proposed by Raup (1966), who generated a theoretical space of all possible forms of bivalved shells based on a simple growth equation, with the parameters of the equation defining the axes of the morphospace. As I have already noted, there is no fitness axis in morphospaces, though as we shall see later, fitness/adaptive considerations do enter into how morphospaces are *used*. Moreover, morphospaces do not depend on any actual measurement at all: they are not constructed empirically, by measuring gene frequencies or phenotypic traits, they are drawn from a priori—geometrical or mathematical—considerations of what generates biological forms. As documented in detail by McGhee (2007), morphospaces have been generated for a variety of organisms, from invertebrates to plants, and they have been put to use by comparing actually existing forms with theoretically possible ones that are either extinct or that for some reason have never evolved. Which brings us to the question of what biologists actually do with the various types of landscapes.

3.3 What are landscapes for?

When it comes to asking what the metaphor of landscapes in biology is for we need to begin by distinguishing between the visual metaphor, which is necessarily low-dimensional, and the general *idea* that evolution takes place in some sort of hyperdimensional space. Remember that Wright introduced the metaphor because his advisor suggested that a biological audience at a conference would be more receptive toward diagrams than toward a series of equations. But of course the diagrams are simply not necessary for the equations to do their work. More to the point, the recent papers by Gavrilets and his collaborators, mentioned previously, have shown in a rather dramatic fashion that the original (mathematical) models were too simple and that the accompanying visual metaphor is therefore not just incomplete, but highly misleading. Gavrilets keeps talking about landscapes of sorts, but nothing hinges on the choice of that particular metaphor as far as the results of his calculations are concerned—indeed, arguably we should be using the more imagery-neutral concept of hyperdimensional spaces, so not to deceive ourselves into conjuring up "peaks" and "valleys" that do not actually exist.

In a very important sense Wright's metaphor of what we have been calling fitness landscapes was meant to have purely heuristic value, to aid biologists to think in general terms about how evolution takes place, not to actually provide a rigorous analysis of or predictions about the evolutionary process (it was for the math to do that work). Seen from this perspective, fitness landscapes have been problematic for decades, generating research aimed at solving problems—like the "peak shift" one (Whitlock et al. 1995)—that do not actually exist as formulated, or that at the very least take a dramatically different form, in more realistic hyperdimensional "landscapes." Even when (relatively) low-dimensional scenarios actually apply, as in the cases of RNA and protein functions briefly discussed earlier, the work is done by intensive computer modeling, not by the metaphor, visual or otherwise.

The peak shift problem started captivating researchers' imagination soon after Wright's original paper, and consists in explaining how natural selection could move a population off a local fitness peak. The landscape metaphor seems to make it obvious that there is a problem, and that it is a significant one, because of course natural selection could not force a population down a peak to cross an adaptive valley in order to then climb up a nearby (presumably higher) peak. That would amount to thinking of natural selection as a teleological process, something that would gratify creationists of all stripes, but is clearly not a viable solution within the naturalistic framework of science (that said, there are situations described in population genetics theory where natural selection does not always increase fitness, as in several scenarios involving frequency dependent selection (Hartl and Clark 2006)).

Wright famously proposed his shifting balance theory of interdemic selection as an alternative mechanism to explain peak shifts (Wright 1982): genetic drift would move small populations off-peak, and natural selection would then push some of them up a different peak. This particular solution to the problem has proven theoretically unlikely, and it is of course very difficult to test empirically (Coyne et al. 1997). Several other answers to the peak shift problem have been explored, including the (rather obvious) observation that peaks are not stable in time (i.e. they themselves move), or that phenotypic plasticity and learning may help populations make a local "jump" from one peak to another. It is also interesting to observe that of course there is no reason to think that natural selection *has* to provide a way to shift between peaks, since it is a satisficing, not an optimizing, process. Getting stuck on a local peak and eventually going extinct is the fate shared by an overwhelming majority of populations and species (van Valen 1973). The current status of this particular problem, as I said, is that Gavrilets's work has shown that there simply aren't any such things as peaks and valleys in hyperdimensional genotypic spaces, but rather large areas of quasi-neutrality (where therefore there is ample room for drift, in that sense at least indirectly vindicating Wright's original intuition of the importance of stochastic events), punctuated by occasional fitness holes and "multidimensional bypasses" (Gavrilets 1997), i.e. connections between distant areas of the hyperspace that can be exploited by natural selection to "jump," though the term now means nothing like what the classical literature on peak shift refers to.

Similar considerations to the ones made in the case of fitness landscapes apply to Adaptive Landscapes, the phenotype-based version of the metaphor introduced by Simpson—albeit with some caveats. Again, the few actual visual examples of such landscapes to be found in the literature have heuristic value only, though at least they are potentially less misleading than Wright-type fitness landscapes simply because low dimensionality is a more realistic situation when we are considering specific aspects of the phenotype. (The phenotype *tout court* of course is a high-dimensionality object, but biologists are rarely interested in that sort of phenotypic analyses, focusing instead either on a small sample of characters, or on a particular aspect of the phenotype—such as skull shape or leaf traits—that can be studied via a small number of variables.) As we have seen, Simpson's classic example concerned the evolution of equids, the horse family, and the corresponding landscape first appeared in his *The Major Features of Evolution* (Simpson 1953). That diagram makes for a great story, which could even be true (and is certainly consistent with the paleontological data), but the whole point of the diagram is to capture the reader's imagination, not to present empirical data or provide testable hypotheses about the observed morphological shifts and alleged ecological context.

Things are somewhat different for (Lande–Arnold; see Chapters 7 and 9) fitness surfaces, because studies quantifying selection coefficients in natural populations are common and can fairly be thought of as statistical analyses of multidimensional fitness (but strictly speaking not adaptive) surfaces. Tellingly, though, what does the work is not the occasional graph of a partial surface (or its often hard to interpret multivariate rendition) but, instead, a tabular output from Lande–Arnold style multiple regression analyses. What understanding we do have of fitness surfaces comes from the actual statistics and our ability to make sense of them (and of their limitations (Mitchell-Olds and Shaw 1987; Pigliucci 2006)), not from visualizations of biologically interpretable surfaces.

The situation is significantly better, I suggest, in the case of the fourth type of landscape: Raup-style morphospaces. McGhee (2007) discusses several fascinating examples, but I will focus here on work done by Raup himself, with crucial follow-up by one of his graduate students, John Chamberlain. It is a study of potential ammonoid forms that puts the actual (i.e. not just heuristic) usefulness of morphospaces in stark contrast with the cases of fitness and Adaptive Landscapes/surfaces discussed so far.

Raup (1967) explored a mathematical-geometrical space of ammonoid forms defined by two variables: W, the rate of expansion of the whorl of the shell; and D, the distance between the aperture of the shell and the coiling axis. As McGhee shows in his detailed discussion of this example, Raup arrived at two simple equations that can be used to generate pretty much any shell morphology that could potentially count as "ammonoid-like," including shells that—as far as we know—have never actually evolved in any ammonoid lineage. Raup then moved from theory to empirical data by plotting the frequency distribution of 405 actual ammonoid species in W/D space and immediately discovered two interesting things: first, the distribution had an obvious peak around $0.3 < D < 0.4$ and $W \sim 2$. Remember that this kind of peak is not a direct measure of fitness or adaptation, it is simply a reflection of the actual occurrence of certain forms rather than others. Second, the entire distribution of ammonoid forms was bounded by the $W = 1/D$ hyperbola, meaning that few if any species crossed that boundary on the morphospace. The reason for this was immediately obvious: the 1/D line represents the limit in morphospace where whorls still overlap with one another. This means that for some reason very few ammonites ever evolved shells in which the whorls did not touch or overlap.

Raup's initial findings were intriguing, but they were lacking a sustained functional analysis that would account for the actual distribution of forms in W/D space. Why one peak, and why located around those particular coordinates? Here is where things become interesting and the morphospace metaphor delivers much more than just heuristic value. John Chamberlain, a student of Raup, carried out experimental work to estimate the drag coefficient of the different types of ammonoid shells. His first result (Chamberlain 1976) clarified why most actual species of ammonoids are found below the $W = 1/D$ hyperbola: shells with whorl overlap have a significantly lower drag coefficient, resulting in more efficiently swimming animals.

However, Chamberlain also found something more intriguing: the experimental data suggested that there should be *two* regions of the W/D morphospace corresponding to shells with maximum swimming efficiency, while Raup's original frequency morphospace detected only one peak. It seemed that for some reason natural selection found one peak, but not the other (Fig. 3.1). Four decades had to pass from Raup's paper for the mystery of the second peak to be cleared up: the addition of 597 new species of ammonoids to the original database showed that indeed the second peak had also been occupied (Fig. 3.2)! Notice that this is a rather spectacular case of confirmed prediction in evolutionary biology, not exactly a common occurrence, particularly in paleontology.

Let me briefly go over a second example—from the same line of inquiry—of how practically (as opposed to simply heuristically) useful morphospaces can be. Fig. 3.3 again shows a W/D space, this time occupied by two different groups of animals with similar morphology and ecology. The top diagram plots the frequency distributions of Cretaceous ammonoids (the large area on the right of the figure) and of nautilids of the same period (the narrow area on the top-left section of the morphospace). Ammonoids, but not nautilids, went extinct at the end of the Cretaceous, a fact reflected by the lower graph, plotting the distribution of Cenozoic nautilids in the same morphospace. Two things need be noted about this second graph: first, the nautilids shifted their major peak in a position previously occupied by ammonoids. This may represent a nice example of competitive exclusion that got released by extinction. Second, why have the nautilids—which are structurally and developmentally similar to ammonoids—not expanded to occupy the full morphospace left empty by the demise of the ammonoids? We do not currently have a satisfactory answer to that question. It seems unlikely that natural selection hasn't had the time to explore the empty morphospace (after all, the ammonoids went extinct 65 million years ago). Given the similarity in architecture and development between the two types of organisms, it also seems unlikely—though certainly not impossible—that a developmental constraint played a role, and we know that the empty space can be colonized, since that's where the extinct ammonoids were to begin with. Finally, considering that there doesn't seem to be competition by any other ammonoid-like group, we are left with the possibility of some sort of genetic constraint, a hypothesis that is however difficult to test given the current status of nautilid genomics.

3.4 What to do with the landscapes metaphor(s)

It is time to put together our thoughts about what the various landscape metaphors are supposed to accomplish on one hand, and what they have so far actually accomplished on the other hand, so to arrive at some conclusion concerning whether any of the four variants of the metaphor examined in this essay is actually useful to the biological scientific community.

Wright-style fitness landscapes are supposed to help biologists think through how the genetic makeup of populations changes in response to evolutionary mechanisms, chiefly natural selection (one can also visualize drift, but not the other classical mechanisms, such as mutation, assortative mating, and migration). The metaphor was confused from the beginning, sometimes referring to individual genotypes and at other times to populations; it has historically been used in the low-dimensional version (typically with two "genetic" dimensions

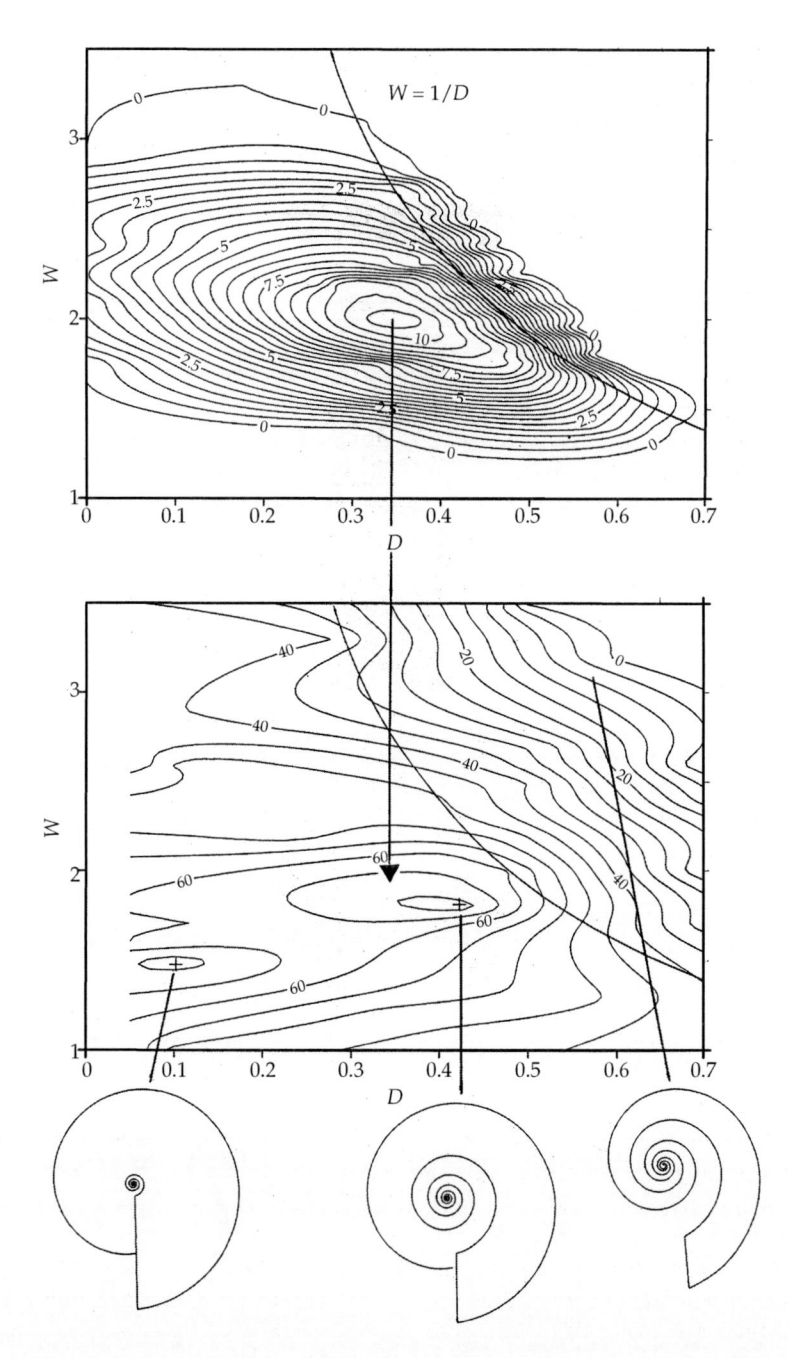

Figure 3.1 A frequency plot and functional analysis of the W/D morphospace of ammonoid shells, following work by Raup and Chamberlain. The upper graph is based on Raup's (1967) original paper, and shows that of the 405 species of ammonoids known at that time all of them could be found below the $W = 1/D$ hyperbole—in agreement with the fact that shells found within that parameter space have higher swimming efficiency. Notice the one frequency peak around the $0.3 < D < 0.4$ and $W \sim 2$ coordinates. The lower graph is based on Chamberlain's (1981) data, and shows two adaptive peaks in terms of swimming efficiency. One peak corresponds to the actual ammonoid peak shown in the top graph, but the second one seemed to indicate that natural selection had somehow "missed" a second W/D combination that maximizes swimming efficiency. (Graphs from McGhee (2007), reproduced with permission from Cambridge University Press.)

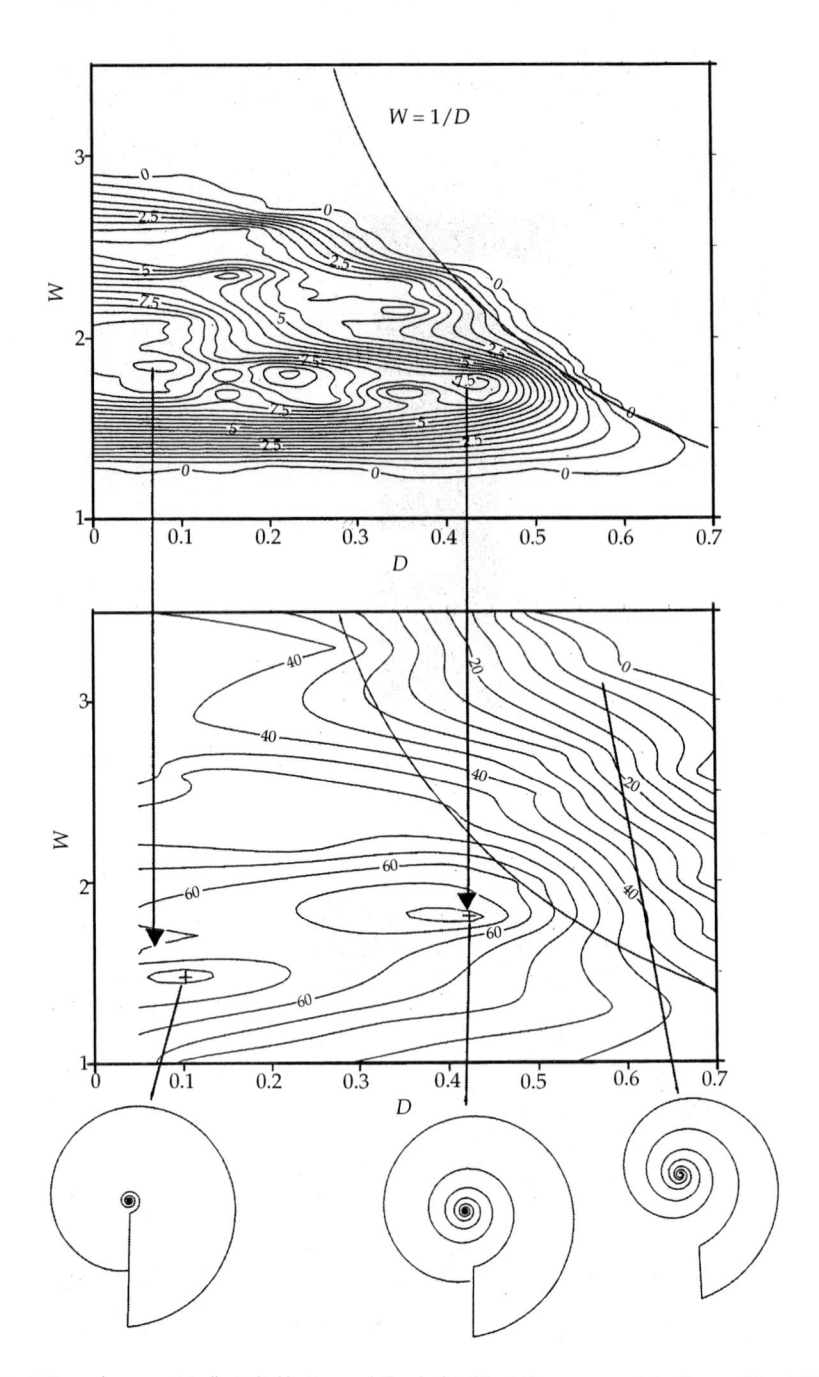

Figure 3.2 The same W/D morphospace originally studied by Raup and Chamberlain (Fig. 3.1), now augmented with an additional 597 newly discovered species of ammonoids. The frequency distribution of actual forms (top), showing two peaks, now nicely corresponds with the two predicted adaptive peaks based on drag coefficients (or its reverse, swimming efficiency, bottom). (Graphs from McGhee (2007), reproduced with permission from Cambridge University Press.)

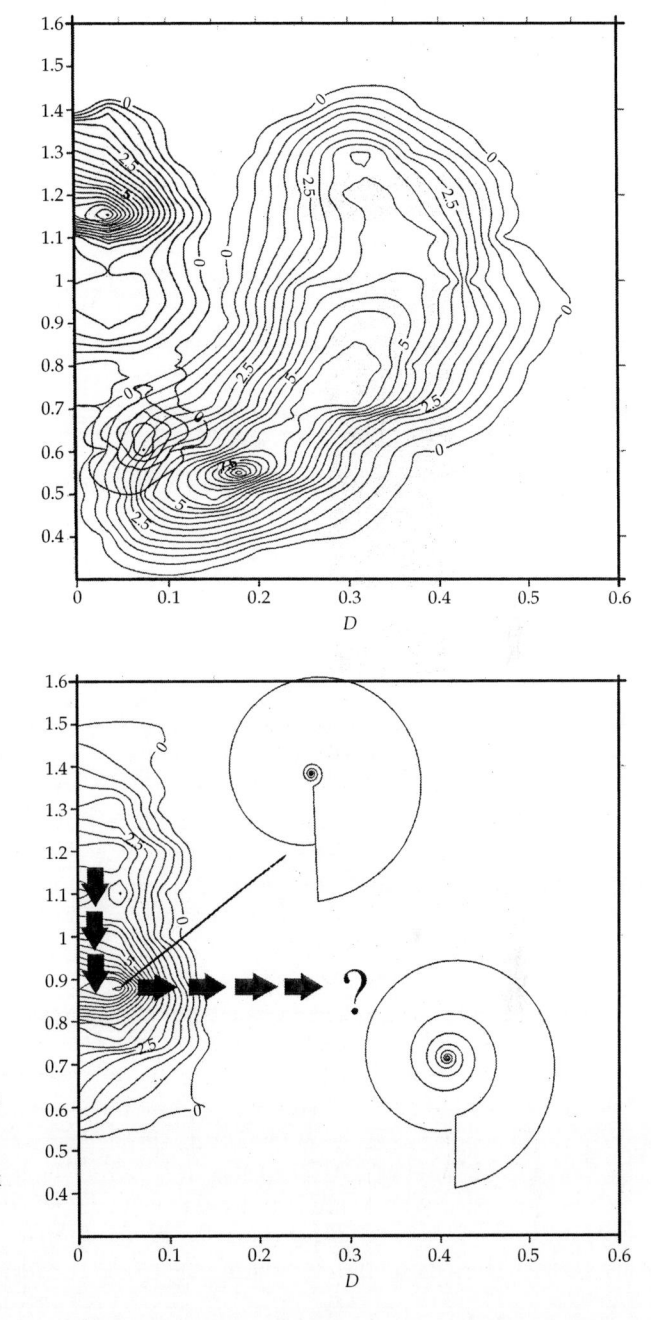

Figure 3.3 Two more morphospaces defined by W/D: the top graph shows the frequency distribution of Cretaceous ammonoids (large area on the center-right) and nautilids (smaller area on the upper left). The bottom graph shows what happened after the extinction of the ammonoids: the nautilids moved their main peak down, but failed to expand toward the right of the morphospace. Possible reasons are discussed in the main text. (Graphs from McGhee (2007), reproduced with permission from Cambridge University Press.)

and one fitness dimension), while recent work clearly shows that whatever intuitions one derives from low-dimensionality fitness landscapes they are likely to prove profoundly misleading. Indeed, an argument can be made that entire research programs, such as the search for mechanisms causing "peak shifts," have been informed by a faulty assumption, since more realistic hyperdimensional genotypic spaces simply do not have anything that resembles peaks and valleys. It seems like the rational thing to do in this case would in fact be to follow Kaplan's (2008) advice, abandon the metaphor altogether and simply embrace directly the results of formal modeling—as both the cases of Gavrilets' "holey" spaces and the research on the evolution of RNA and protein function elegantly illustrate. Wright may have needed to soften his math with pictures in the 1930s, but surely modern biologists ought to be able to take on the full force of the mathematical theory of evolution.

Simpson-style Adaptive Landscapes also aimed from the beginning at a heuristic value, as demonstrated by the fact that there are few examples in the literature that are not hypothetical. However, it is arguable that these landscapes are in fact less misleading than fitness landscapes, since their low dimensionality reflects the real fact that often biologists are interested in selected aspects of an organism's phenotype, and rarely consider simultaneously hundreds or thousands of characters (again, unlike the genetic scenarios). Still, the point of Adaptive Landscapes is to study adaptation and its macroevolutionary consequences in terms of speciation and lineage divergence, which means that far more than Simpson-style suggestive pictures are necessary. Serious studies of Adaptive Landscapes need to integrate historical records (via paleontology and/or cladistics), functional ecology, as well as morphology—a splendid area for fertile interdisciplinary work, much of which remains to be done.

Lande–Arnold style multiple regression analyses of natural selection—and the graphical rendition of the underlying fitness surfaces—has been well underway, and arguably represents the bulk of the empirical work inspired by landscape metaphors. Yet Lande–Arnold selection analyses themselves are characterized by many well-known issues, such as problems with multicollinearity of different traits, the problem of the effect of "missing" (i.e. unmeasured) traits, the assumption of linearity of the statistical models, the dearth of spatially and temporally replicated studies with sufficient sample sizes, etc. (Mitchell-Olds and Shaw 1987; Kingsolver et al. 2001; Pigliucci 2006; Siepielski et al. 2009). None of this affects the conclusion that fitness surfaces can be rigorously quantified, though their visualization must be left largely to ineffective multivariate compound variables.

Where both Simpson-type Adaptive Landscapes and Lande–Arnold type fitness surfaces do not deliver is when we assume that they are phenotypic versions of Wright-type landscapes, as Simpson himself surely did. This simply cannot be the case because of the high complexity and non-linearity of the genotype → phenotype mapping function, as argued elsewhere (Alberch 1991; Pigliucci 2010). This is a problem insofar as we are interested in an evolutionary theory that provides us not just with an account of genetic change (as given by population and quantitative genetics), but also with accounts of phenotypic change and of how, precisely, the two are connected.

We finally get to Raup-style morphospaces, which I think are not just heuristically useful (to visualize the range of possible organismal forms within particular aspects of the phenotype), but actually have a nice if small record of generating new understanding, as well as testable hypotheses, in biological research, despite still being somewhat of a backwater topic in need of further attention. As philosopher of science James Maclaurin (2003) aptly put it: "Theoretical morphology might allow us to sort life into the actual, the non-actual and the impossible," a research agenda splendidly illustrated by the examples of ammonoids and nautilids shell shapes discussed above and that can be traced back to general attempts at theorizing about biological form that even predate an explicitly evolutionary approach (Thompson 1917). However, a survey of the literature on morphospaces (McGhee 2007) shows that most of the available examples are restricted to a small range of animal taxa and limited aspects of their phenotype (with some exceptions concerning plants (Niklas 2004)). This could be due to the fact that comparatively few scientists (mostly drawn from paleontology) have even thought about organizing their research using the framework of morphospaces, because until recently the study of phenotypic evolution had taken a backseat in biology (Schlichting and Pigliucci 1998), or because there are actual conceptual issues to be dealt with that may limit the general applicability of morphospaces across living organisms and types of characters. Only further research will be able to tell.

References

Alberch, P. (1991). From genes to phenotypes: dynamical systems and evolvability. Genetica, 84, 5–11.

Arnold, S.J., Pfrender, M.E., and Jones, A.G. (2001). The Adaptive Landscape as a conceptual bridge between micro- and macroevolution. Genetics, 112–113, 9–32.

Brigandt, I. (2008). Reductionism in biology. Stanford Encyclopedia of Philosophy, http://plato.stanford.edu/entries/reduction-biology/

Chamberlain, J.A. (1976). Flow patterns and drag coefficients of cephalopod shells. Paleontology, 19, 539–563.

Chamberlain, J.A. (1981). Hydromechanical design of fossil cephalopods. Systematics Association, Special Volume 18, 289–336.

Cowperthwaite, M.C. and Meyers, L.A. (2007). How mutational networks shape evolution: lessons from RNA models. Annual Review of Ecology Evolution and Systematics, 38, 203–230.

Coyne, J.A., Barton, N.H., and Turelli, M. (1997). A critique of Sewall Wright's shifting balance theory of evolution. Evolution, 51, 643–671.

Dobzhansky, T. (1970). Genetics of the Evolutionary Process. Columbia University Press, New York.

Estes, S. and Arnold, S.J. (2007). Resolving the paradox of stasis: models with stabilizing selection explain evolutionary divergence on all timescales. American Naturalist, 169, 227–244.

Futuyma, D. (2006). Evolutionary Biology. Sinauer, Sunderland, MA.

Gavrilets, S. (1997). Evolution and speciation on holey adaptive landscapes. Trends in Ecology & Evolution, 12, 307–312.

Gavrilets, S. (2003). Evolution and speciation in a hyperspace: the roles of neutrality, selection, mutation, and random drift. In J.P. Crutchfield and P. Schuster (eds.) Evolutionary Dynamics: Exploring the Interplay of Selection, Accident, Neutrality, and Function. Oxford University Press, Oxford, pp. 135–162.

Hartl, D.L. and Clark, A.G. (2006). Principles of Population Genetics. Sinauer, Sunderland, MA.

Kaplan, J. (2008). The end of the Adaptive Landscape metaphor? Biology and Philosophy, 23, 625–638.

Kaplan, J. (2009). The paradox of stasis and the nature of explanations in evolutionary biology. Philosophy of Science, 76, 797–808.

Kingsolver J.G., Hoekstra, H.E., Hoekstra, J.M., Berrigan, D., Vignieri, S.N., Hill, C.E., et al. (2001). The strength of phenotypic selection in natural populations. American Naturalist, 157, 245–261.

Lande, R. and Arnold, S.J. (1983). The measurement of selection on correlated characters. Evolution, 37, 1210–1226.

Maclaurin, J. (2003). The good, the bad and the impossible. Biology and Philosophy, 18, 463–476.

Mayr, E. and Provine, W. (1980). The Evolutionary Synthesis: Perspectives on the Unification of Biology. Harvard University Press, Cambridge, MA.

McGhee, G.R. (2007). The Geometry of Evolution: Adaptive Landscapes and Theoretical Morphospaces. Cambridge University Press, Cambridge.

Mitchell-Olds, T. and Shaw, R.G. (1987). Regression analyses of natural selection: statistical inference and biological interpretation. Evolution, 41, 1149–1161.

Niklas, K.J. (2004). Computer models of early land plant evolution. Annual Review of Earth and Planetary Sciences, 32, 47–66.

Pigliucci, M. (2006). Genetic variance covariance matrices: a critique of the evolutionary quantitative genetics research program. Biology and Philosophy, 21, 1–23.

Pigliucci, M. (2008). Sewall Wright's adaptive landscapes: 1932 vs. 1988. Biology and Philosophy, 23, 591–603.

Pigliucci, M. (2010). Genotype-phenotype mapping and the end of the "genes as blueprint" metaphor. Philosophical Transactions of the Royal Society, B365, 557–566.

Plutynski, A. (2008). The rise and fall of the Adaptive Landscape? Biology and Philosophy, 23, 605–623.

Raup, D.M. (1966). Geometric analysis of shell coiling: general problems. Journal of Paleontology, 40, 1178–1190.

Raup, D.M. (1967). Geometric analysis of shell coiling: coiling in ammonoids. Journal of Paleontology, 41, 43–65.

Schlichting, C.D. and Pigliucci, M. (1998). Phenotypic Evolution: A Reaction Norm Perspective. Sinauer, Sunderland, MA.

Siepielski, A.M., DiBattista, J.D., and Carlson, S.M. (2009). It's about time: the temporal dynamics of phenotypic selection in the wild. Ecology Letters, 12, 1261–1276.

Simpson, G.G. (1944). Tempo and Mode in Evolution. Columbia University Press, New York.

Simpson, G.G. (1953). The Major Features of Evolution. Columbia University Press, New York.

Thompson, d'A. W. (1917). On Growth and Form. Cambridge University Press, Cambridge.

van Valen, L. (1973). A new evolutionary law. Evolutionary Theory, 1, 1–30.

Whitlock, M.C., Phillips, P.C., Moore, F.B.G., and Tonsor, J. (1995). Multiple fitness peaks and epistasis. Annual Review of Ecology and Systematics, 26, 601–629.

Wright, S. (1931). Evolution in Mendelian populations. Genetics, 16, 97–159.

Wright, S. (1982). The shifting balance theory and macroevolution. Annual Review of Genetics, 16, 1–20.

Wroe, R., Chan, H.S., and Bornberg-Bauer, E. (2007). A structural model of latent evolutionary potentials underlying neutral networks in proteins. HFSP Journal, 1, 79–87.

PART II

Controversies: Fisher's Fundamental Theorem Versus Sewall Wright's Shifting Balance Theory

CHAPTER 4

Wright's Adaptive Landscape Versus Fisher's Fundamental Theorem

Steven A. Frank

Fisher is frequently portrayed in the contemporary literature as believing in a strictly additive basis for the inheritance of quantitative characters, and as dismissing any evolutionary importance for epistatic interactions in fitness effects. This is accompanied by a sub-text that this is in some way less virtuous than embracing a less 'reductionist' view, which assigns a prominent role to epistasis, as in Wright's 'shifting-balance' theory. This is, in fact, a travesty of Fisher's views.

—Charlesworth (2000)

4.1 Introduction

Wright developed the Adaptive Landscape to support his shifting balance theory of evolution. The shifting balance theory emphasized that progressive improvement by natural selection is too slow by itself to account for biological diversity and the rate of adaptive change. Wright suggested that random perturbations of gene frequencies in small partially isolated populations may act synergistically with natural selection to explain rapid adaptation and diversity. One may visualize the synergy of random perturbations and deterministic natural selection by imagining the dynamics of populations on an Adaptive Landscape.

In an Adaptive Landscape, natural selection corresponds to climbing local hills of increasing fitness. A local peak traps a population to a narrow range of phenotypes that limits opportunities for major improvement in fitness. However, a small population may, by stochastic sampling and drift, change its common non-additive (epistatic) combinations of interacting genes. Such perturbations of epistatic gene combinations can move a population down a hill and across a valley of lower fitness to the base of a nearby and potentially higher fitness peak. Natural selection then pushes the population up that higher peak, causing a major improvement in fitness relative to the recently abandoned lower peak. Such peak shifts lead to major diversification of phenotype—the shifting balance process (Chapter 5).

Wright often presented the local hill climbing aspect of natural selection in terms of Fisher's fundamental theorem. That theorem describes the rate of improvement in fitness caused by natural selection. Wright also ascribed to Fisher the view that local hill climbing by natural selection was the primary force of evolutionary change over long periods of time. Such local hill climbing does not require a key role for non-additive gene combinations, so Wright also ascribed to Fisher the view of natural selection acting locally on additive gene effects (Chapter 6).

Fisher strongly rejected Wright's characterization of the fundamental theorem and, in turn, severely criticized the Adaptive Landscape. At first glance, it may seem that the Wright–Fisher controversy ultimately comes down to the opposing views given by each combatant's primary slogan: the Adaptive Landscape on Wright's side versus the fundamental theorem on Fisher's side.

I clarify two points. First, Wright and Fisher did disagree about whether random perturbations of drift were essential to explain the long-term processes of adaptation and diversification. That disagreement was in fact their primary battle. Much of their sometimes acrimonious mischaracterizations of each other's work on various topics often derived from this single and often unspoken rift with regard to the importance of stochastic fluctuations in small populations. They did not disagree about whether such fluctuations occurred, only the relative importance of those fluctuations in adaptive evolution (Provine 1986).

My second point concerns the very different goals of Wright and Fisher with respect to the Adaptive Landscape and the fundamental theorem. Wright spent decades of intensive work refining the Adaptive Landscape theory. He made that effort to provide support for the shifting balance theory as the prime mover of evolutionary change and biological diversity.

Fisher presented the fundamental theorem in his 1930 book and rarely commented on it again except to criticize Wright or provide a few minor corrections. Wright actually produced more commentary on the fundamental theorem than Fisher. However, Wright's commentary almost always misrepresents both Fisher's particular results and Fisher's deeper goals for the fundamental theorem. When discussing Fisher's work, Wright promoted his own views as defining the long-term consequences of non-additive gene interactions over a global multipeaked landscape and Fisher's views as defining the short-term consequences of additive gene action within a local and narrowly confined fitness peak.

Why did Fisher let Wright's misrepresentations go mostly unanswered? In my opinion, Fisher did not see the fundamental theorem as having anything to do with their primary disagreement over the roles of geographic isolation and random perturbations in long-term adaptive evolution. The fundamental theorem is about the logical nature of selection as a universal law of biology. That law expresses an invariant rate of change caused by natural selection when considered alone as an isolated force, as distinct from the total change to a population caused by a variety of processes including mutation, recombination, competition, and so on. Fisher also presented a quasi-conservation principle: the amount of adaptive improvement by natural selection is typically balanced by an equal and opposite decline in fitness caused by increased competition from simultaneously improved competitors. The total fitness of a population must typically change hardly at all, because population growth rates must typically be close to zero. A continuously growing population would overrun the world; a continuously declining population would soon be extinct.

Fisher had training in mathematical physics. For him, the fundamental theorem had the same power in biology that the great laws of invariance and conservation had in physics. Those physical laws do not predict how a real, complex, heterogeneous, and open physical system will evolve over time. Such predictions of complex dynamics in real systems are often impossible and at best not reducible to a brief and simple expression. In the same way, Fisher never suggested that the fundamental theorem predicted how real populations would evolve over time. Rather, he intended only to express how natural selection as a force necessarily acted within a complex evolutionary system subject to many distinct types of forces. When Fisher and Wright argued, the issues primarily concerned how real populations evolve over long periods of time. Fisher was very interested in the problem of long-term evolution, but he also realized that the fundamental theorem had little to say on that topic.

Wright's published commentary continues to define the dominant view of Fisher's outlook on evolutionary dynamics and on the fundamental theorem. Here, I describe what Fisher actually wrote about these topics, which differs greatly from the picture painted by Wright.

4.2 The disagreement about drift and dynamics

In Fisher and Ford (1950), Fisher clearly expressed his deepest disagreement with Wright:

> The widest disparity...which has so far developed in the field of Population Genetics is that which separates those who accept from those who reject the theory of "drift" or "non-adaptive radiation," as it has been called by its author, Professor Sewall Wright of Chicago.
>
> [T]his theory of Sewall Wright...claims that the subdivision of a population into small isolated or semi-isolated colonies has had important evolutionary effects; and this through the agency of random fluctuation of gene ratios, due to random reproduction in a small population.

We have long felt that there are grave objections to this view...[O]ne, however, is completely fatal to the theory in question, namely that it is not only small isolated populations, but also large populations, that experience fluctuations in gene ratio. If this is the case, whatever other results isolation into small communities may have, any effects which flow from fluctuating variability in the gene ratios will not be confined to such subdivided species, but will be experienced also by species having continuous populations.

This fact, fatal to "The Sewall Wright Effect," appeared in our own researches from the discovery that the year-to-year changes in the gene ratio in a wild population were considerably greater than could be reasonably ascribed to random sampling, in a population of the size in question.

Fisher and Ford agreed that random fluctuations by sampling and drift will always occur. But they argued that the fluctuations they observed were too great to be explained by sampling. Instead, fluctuating selection caused by a varying environment appeared to be the cause. They noted that others, such as Dobzhansky, had also presented data on fluctuating gene ratios (frequencies) most likely explained by selection. Fisher and Ford conclude:

> Sub-division into small isolated or semi-isolated populations is clearly favourable to evolutionary progress through the variety of environmental conditions to which the colonies are exposed. Moreover, so long as it could be believed that large fluctuations in gene ratios occur only in small isolated colonies by reason of fluctuations of random survival, then it might have been true that such fluctuations themselves favoured evolutionary change in a way that would not be allowed in a continuous distribution of the species. If now it is admitted that large populations with continuous distributions also show year-to-year fluctuations of comparable or greater magnitude in their gene ratios, due to variable selection, the situation is entirely altered. In these circumstances, the claim for ascribing a special evolutionary advantage to small isolated communities due to fluctuations in gene ratios, had better be dropped.

Fisher and Ford are not saying that major adaptive changes occur only in large, panmictic populations. Rather, they argue that subdivision into small populations and drift are not necessary conditions for significant adaptive change by natural selection. The fluctuating gene frequencies in large populations caused by fluctuating selection may be sufficient to allow shifts in favored gene combinations and the equivalent of a Wrightian peak shift. To repeat, Fisher's primary argument about major adaptive change is against a necessary role of subdivision, small population size, and drift. Those factors may occur, but major adaptive shifts by altered epistatic gene combinations can arise in other ways.

Provine (1986, pp. 301–302) clearly traces the origin of this disagreement between Fisher and Wright to the early 1930s. At that time, the evolution of dominance formed the particular subject of debate, rather than the dynamics of gene frequencies under selection as in the Fisher and Ford paper (1950). But, as Provine emphasizes, the real argument in the early 1930s that led to the original rift between Fisher and Wright also turned on the alternative views of selection and drift. Provine makes his case by quoting from Wright (1934, pp. 50–51):

> From the standpoint of the theory of dominance it may seem of little importance which mechanism is accepted if it be granted that selection has been an important factor. This is not at all the case, however, with the implication of Fisher's and Plunkett's selection theories, for the theory of evolution. Fisher used the observed frequency of dominance as evidence for his conception of evolution as a process under complete control of selection pressure, however small the magnitude of the latter.
>
> My interest in his theory of dominance was based in part on the fact that I had reached a very different conception of evolution (1931) and one to which his theory of dominance seemed fatal if correct. As I saw it, selection could exercise only a loose control over the momentary evolutionary trend of populations. A large part of the differentiation of local races and even of species was held to be due to the cumulative effects of accidents of sampling in populations of limited size. Adaptive advance was attributed more to intergroup than intragroup selection.

Provine (1986, p. 302) nicely summarizes the key point:

> I think Wright is correct in saying that what really was at stake in the argument with Fisher over the evolution of dominance was not the particular problem of dominance but their differing conceptions of evolution. If either was correct on the evolution of

dominance, it was perceived by the other as fatal to his entire conception of evolution.

Fisher's criticism of the Adaptive Landscape focused on the claim that drift by random sampling is not a necessary condition for significant evolutionary change. Wright's early work did specify random sampling as the key perturbation in small, local populations. However, Wright expanded his framing of drift in later work, probably in response to Fisher and Ford's argument that fluctuating selection could explain how populations may be perturbed from a fixed, local peak on an Adaptive Landscape. In Wright's (1977, p. 455) grand synthesis, he describes the first phase of the shifting balance process as:

> *Phase of Random Drift.* In each deme, the set of gene frequencies drifts at random in a multidimensional stochastic distribution about the equilibrium set characteristic of a particular fitness peak or goal. The set of equilibrium values is the resultant of three sorts of pressures on the gene frequencies: those due to recurrent mutation, to recurrent immigration from other demes, and to selection. The fluctuations in the gene frequencies responsible for the stochastic distribution (or random drift) may be due to accidents of sampling or to fluctuations in the coefficients measuring the various pressures [e.g. mutation, immigration, and selection].

Here, Wright clearly allows that fluctuating selection may be the cause of perturbations to local populations. Fisher was long dead by this time. Fisher might have replied that fluctuating selection works just as well in large populations, so there would in this case be no need to invoke small separated populations as essential to the process. Wright, in turn, may have answered that many small separated populations allow the many parallel independent lines an opportunity to initiate a peak shift, greatly increasing the chance that one local population makes the jump to another peak and then exports its enhanced adaptive combinations through the population. In this view, Wright's primary idea is subdivision of the population into local populations, allowing multiple parallel exploration of the Adaptive Landscape and thereby greatly accelerating the pace of evolutionary change. Fisher probably would have accepted that subdivision might under some conditions have an effect on evolutionary rate, but that such subdivision is neither necessary nor likely to be a commonly important factor.

4.3 Fisher's goal for the fundamental theorem

Fisher (1958a) stated the fundamental theorem as: "The rate of increase in fitness of any organism at any time is equal to its genetic variance in fitness at that time" (p. 37) and "The rate of increase of fitness of any species is equal to its genetic variance in fitness" (p. 50). At first glance, these expressions seem closely related to Wright's study of the Adaptive Landscape, which is usually described as a surface of population mean fitness. Fisher's result would then describe how fast natural selection can push a population up a surface of mean fitness.

Wright frequently quoted the fundamental theorem in support of his gradient formulation of the Adaptive Landscape, in which gene frequencies change at a rate proportional to the slope of the fitness surface, $d\overline{W}/dq$, where $\overline{W}$ is the mean fitness of the population, and q is the frequency of a gene. In what I believe to be Wright's (1988, p. 118) last publication, he said: "The effects [on gene frequencies in an Adaptive Landscape] may be calculated using Fisher's fundamental theorem."

These quotes from both Fisher and Wright seem to say that Fisher's theorem is about the rate of change in the mean fitness of a population. That interpretation was adopted by essentially everyone who subsequently commented on the theorem until papers by Price (1972) and Ewens (1989) that I will come to later. But we can see clearly from other statements by Fisher that something is wrong: "In regard to selection theory, objection should be taken to Wright's equation [the expression $d\overline{W}/dq$] principally because it represents natural selection, which in reality acts upon individuals, as though it were governed by the average condition of the species or inter-breeding group" (Fisher 1941, p. 58) and "I have never, indeed, written about $\overline{w}$ and its relationships...the existence of such a potential function [i.e. a function nondecreasing in time]...is not a general property of natural populations, but arises only in the special and restricted cases which Wright has chosen to consider" (Fisher 1958b, p. 290).

Fisher and Wright never suggested the other's equations were incorrect. Once again the disagreement is about how to interpret evolutionary process. Wright's goal remains easy to follow. He wanted to understand the various forces that change gene frequency in order to argue for his shifting balance theory of evolution. In developing his theory, he needed expressions for how natural selection changes gene frequencies. Wright repeatedly invoked Fisher's fundamental theorem to describe how natural selection moves populations up a hill of increasing fitness. By contrast, Fisher's goal for the fundamental theorem seems obscure at first glance.

When Fisher argued against Wright's shifting balance theory, he clearly focused on the key issues of population subdivision and the role of drift in perturbing gene frequencies in small, isolated populations. Thus, Fisher's complaint about Wright's use of the fundamental theorem does not have to do with shifting balance and the controversy over long-term evolutionary dynamics. If not about shifting balance and evolutionary dynamics, what was Fisher ultimately arguing by saying the fundamental theorem expressed "the rate of increase in fitness of any species" and at the same time sharply criticizing Wright by saying "In regard to selection theory, objection should be taken to Wright's equation...principally because it represents natural selection...as though it were governed by the average condition of the species" and "I have never, indeed, written about $\bar{w}$ [mean fitness] and its relationships"?

4.3.1 Background

Essentially everyone interpreted Fisher's theorem in relation to the long-term dynamics of populations. The theorem seemed to say, at the very least, that the average fitness of a population never decreased. More strongly, the theorem described the dynamical path of mean fitness in relation to genetic variance.

Fisher (1930, 1958a) emphasized strongly that his theorem is exact. Yet essentially every commentator in the 40 years following the 1930 announcement qualified the theorem by the wide variety of special assumptions required: random mating, large populations, pure additivity of genic interactions (no epistasis), free recombination with no linkage disequilibrium, and no frequency or density-dependent interactions. Several analyses showed that mean fitness could decrease under a variety of conditions. Other analyses quantified how closely mean fitness tracked additive genetic variance and thus the extent to which the fundamental theorem was a good approximate result under certain special conditions.

Price (1972) provided the first clues about the theorem as Fisher meant it. Ewens (1989) followed with a full, clear proof and exposition. The Price–Ewens exposition showed that Fisher never meant to discuss the long-term dynamics of populations. Thus, Wright's use of the theorem and all of the prior commentary about evolutionary dynamics had nothing to do with Fisher's view of the theorem.

I do not give the mathematical details here. Interested readers should consult the extensive literature that has developed, which can be found by tracing citations to Price (1972) and Ewens (1989). My own more technical interpretations are in Frank (1997, 2009). Here, I give a simplified expression of the key ideas based on Frank and Slatkin (1992).

4.3.2 Fisher's framing of the problem

Fisher realized that one cannot make a complete model of evolutionary dynamics. Too many factors come into play: changes in the physical environment, changes in competitive intensity within and between species, and changes in the complex non-additive interactions between genes that fluctuate in frequency. Given the complexity of "open" systems in which forces flow from a variety of unknown sources, Fisher sought a way to define a "closed" subset in which one could completely and exactly study the process of natural selection. Indeed, the first sentences of *The Genetical Theory* are (Fisher 1958a):

> Natural Selection is not Evolution. Yet, ever since the two words have been in common use, the theory of Natural Selection has been employed as a convenient abbreviation for the theory of Evolution by means of

Natural Selection, put forward by Darwin and Wallace. This has had the unfortunate consequence that the theory of Natural Selection itself has scarcely ever, if ever, received separate consideration. To draw a physical analogy, the laws of conduction of heat in solids might be deduced from the principles of statistical mechanics, yet it would have been an unfortunate limitation, involving probably a great deal of confusion, if statistical mechanics had only received consideration in connexion with the conduction of heat. In this case it is clear that the particular physical phenomena examined are of little theoretical interest compared to the principle by which they can be elucidated. The overwhelming importance of evolution to the biological sciences partly explains why the theory of Natural Selection should have been so fully identified with its role as an evolutionary agency, as to have suffered neglect as an independent principle worthy of scientific study.

The expression of intent seems clear. Fisher wishes to isolate natural selection as a process from the context of particular aspects of evolutionary dynamics as they occur in particular instances. Put another way, to study evolutionary dynamics, one must make many particular assumptions that confine the analysis to a particular kind of problem, and so obscure any general principles that may hold for natural selection across all assumptions and particular instances of evolutionary dynamics.

4.3.3 The fundamental theorem explained

Fisher started his argument by first isolating the general aspects of natural selection from those aspects of evolutionary dynamics that are particular to each system. To do this, Fisher set the standard for measurement of fitness as the full conditions of the population and environment at a particular instant in time.

Those conditions, together called "environment," include all of the gene frequencies that set the genetic environment in which each gene lives, all of the biotic interactions within and between species, and all aspects of the physical environment. By fixing those environmental conditions at a particular instant, Fisher obtained a fixed standard against which he could measure the exact contribution of natural selection to changes in the adaptation of populations. Fisher fully recognized that the actual evolutionary change in adaptation and mean fitness would then include two components: one component caused by natural selection in relation to the original fixed environmental standard of measurement, and one component caused by the changes in the environmental standard of measurement.

Perhaps the most confusing aspect arises because natural selection itself changes the environmental standard of measurement by changing gene frequencies (genetic environment), by changing competitive intensity, and perhaps by changing the physical environment. Those effects of natural selection on the standard of measurement are not, in Fisher's system, direct components ascribed to natural selection, but rather indirect components that Fisher lumped into the term for changes in the environment. Although such a partitioning of total evolutionary change may seem arbitrary with regard to defining the consequences of natural selection, there is no other way to isolate the role of natural selection, because natural selection is a force that acts instantaneously in relation to the conditions that hold at that instant. Once one sees this point of view, all else is detail.

Frank and Slatkin (1992) expressed Fisher's partition as follows. The total change in fitness over time, $\Delta \overline{W}$, in the context of the environment, E, can be defined as

$$\Delta \overline{W} = \overline{W}'|E' - \overline{W}|E, \qquad (4.1)$$

where $\overline{W}|E$ is mean fitness in the context of a particular environmental state, primes denote one time step or instant into the future, and $\Delta \overline{W}$ is the total change in fitness which nearly everyone had assumed was the object of Fisher's analysis. Fisher's theorem, however, was not concerned with the total evolutionary change, which depends at least as much on changes in the environment as it does on natural selection. Instead, Fisher partitioned the total change into

$$\Delta \overline{W} = \left(\overline{W}'|E - \overline{W}|E\right) + \left(\overline{W}'|E' - \overline{W}'|E\right)$$
$$= \Delta \overline{W}_{NS} + \Delta \overline{W}_E. \qquad (4.2)$$

Fisher called the first term the change in fitness caused by natural selection because there is a constant frame of reference, the initial environmental state, E. The fundamental theorem proves that the change in fitness caused by natural selection is equal to the genetic variance in fitness, where genetic variance is defined in a particular way (see Section 4.3.5). Fisher (1958a, pp. 45–46) referred to the second term as the change caused by the environment, or as the change caused by the deterioration of the environment to stress that this term is often negative, because natural selection increases fitness but the total change in fitness is usually close to zero:

> Against the action of Natural Selection in constantly increasing the fitness of every organism, at a rate equal to the genetic variance in fitness which that population maintains, is to be set off the very considerable item of the deterioration of its inorganic and organic environment. This at least is the conclusion which follows from the view that organisms are very highly adapted. Alternatively, we may infer that the organic world in general must tend to acquire just that level of adaptation at which the deterioration of the environment is in some species greater, though in some less, than the rate of improvement by Natural Selection, so as to maintain the general level of adaptation nearly constant...
>
> An increase in numbers of any organism will impair its environment in a manner analogous to, and more surely than, an increase in the numbers or efficiency of its competitors. It is a patent oversimplification to assert that the environment determines the numbers of each sort of organism which it will support. The numbers must indeed be determined by the elastic quality of the resistance offered to increase in numbers, so that life is made somewhat harder to each individual when the population is larger, and easier when the population is smaller. The balance left over when from the rate of increase in the mean value of m [fitness] produced by Natural Selection, is deducted the rate of decrease due to deterioration in environment, results not in an increase in the average value of m, for this average value cannot greatly exceed zero, but principally in a steady increase in population.

4.3.4 Fisher's conservation law for mean fitness

Fisher's argument that mean population growth rate (fitness) must always remain close to zero leads to an approximate conservation law: any increase in the mean fitness of a population caused by natural selection must usually be balanced by an equal and opposite decrease in mean fitness caused by "deterioration of the environment." Here, deterioration would most often arise from increased competition by members of the same or different species, as those competitors also increase their own level of adaptedness by natural selection (Chapter 7).

Fisher supported this approximate conservation law of mean fitness by arguing that total population growth cannot be continually above zero, otherwise the population would grow without bound. Similarly, total population growth cannot be continually less than zero, otherwise the population would soon disappear. Fisher recognized that one species can increase at the expense of other species, so the total mean growth rate applies to all species potentially in competition with each other.

Fisher clearly emphasized this balance between improvement by natural selection and deterioration by enhanced competition. However, this broad context of the theorem has been almost entirely ignored. Instead, the focus has been on the natural selection component of increase, as in the quote "The rate of increase in fitness of any organism at any time is equal to its genetic variance in fitness at that time" (Fisher 1958a, p. 50). Wright's use and commentary of the theorem concerned only this first component. So it is useful to look explicitly at Fisher's expression for the natural selection component of evolutionary change in mean fitness.

4.3.5 Average excess, average effect, and genetic variance

The fundamental theorem's logic and its relations to Wright's Adaptive Landscape depend on two key definitions. Each definition quantifies the contribution of a particular allele to a character, in this case fitness. Here, I give rough verbal descriptions to emphasize the main ideas. Details can be found in Ewens (1989) and Frank (1997). Note that minor variants of the definitions exist in the literature, but all forms have the same essential meaning.

It is easiest to think of a single diploid genetic locus with two alleles, B and b, and three genotypes, BB, Bb, and bb. The *average excess* measures the excess reproduction of B relative to an average individual. To calculate the excess, we start with the fitness of individuals with the BB genotype and one-half of the fitness of individuals with the Bb genotype, the half arising because the heterozygote carries half as many copies of B as the homozygote. From the average fitness for B we subtract the average fitness of all individuals, leaving the excess reproduction of the B allele compared with the population as a whole.

The average excess is a direct measure of the change in gene frequency, because it simply counts up the number of newly made alleles of a particular type compared with the average number of newly made alleles in the population. It is helpful to show this change in gene frequency in symbols. Suppose that each allele over all loci is associated with an index label j, with frequency q_j and average excess a_j in a population with average fitness $\overline{W}$. Then the change in the frequency of each allele after a round of reproduction is

$$\Delta q_j = q_j a_j / \overline{W}. \qquad (4.3)$$

Fisher (1958a, p. 31) emphasized that the average excess is not a good measure of the direct contribution of an allele to fitness, but rather is defined simply to describe the change in gene frequency that arises from the distribution of fitnesses among genotypes:

> The [average] excess in a factor will usually be influenced by the actual frequency...of the alternative genes, and may also be influenced, by way of departures from random mating, by the varying reactions of the factor in question with other factors.

The *average effect* is a more subtle measure of the contribution of a particular allele to fitness. Take a population in its current form, fully accounting for non-random mating, linkage associations between loci, non-additive epistatic interactions between genes, and so on. Measure the average effect of the allele B by taking each individual in the population and changing, one at a time, each copy of B to b, and measuring the effect of that change on fitness. The average of each of those changes is the average effect of a gene substitution. The advantage of this definition is that all aspects of mating pattern and interactions between genes are automatically accounted, because the change is made in each actual genetic combination that exists in the population. The average effect of an allele is the partial regression coefficient of the presence of that allele on fitness.

We use the symbol b_j for the average effect of the jth allele on fitness using the notation and definitions of Frank (1997). Then we can write the total change in fitness caused by natural selection as

$$\Delta \overline{W}_{NS} = 2 \sum_j (\Delta q_j) b_j, \qquad (4.4)$$

where the two arises because we assume two alleles at each locus in a diploid genetic system. This equation says that we can calculate the total change in fitness by natural selection by summing up each change in allele frequency, Δq_j, and weighting that change by the average effect of that allele on fitness, b_j. This form provides the clearest expression of Fisher's fundamental theorem. One can also show that this expression is equivalent to the variance in the average effects, which Fisher called the genetic variance in fitness (Ewens 1989). Thus, the change in fitness caused by natural selection is equal to the genetic variance in fitness.

4.3.6 What is fitness?

A key problem concerns the definition of fitness itself. Fisher referred to fitness as a rate of increase, but he was vague about the precise definition of what is actually measured. Fisher's vagueness in the conception of fitness caused confusion over the status of the fundamental theorem as a universally true mathematical theorem. However, the technical details of how one might define fitness are not needed to understand the history and the main conceptual points about Fisher's theorem. For those readers interested in this issue, I have added a brief Appendix.

4.3.7 Deterioration of the environment

What about the change in fitness caused by the "change in the environment" as expressed by $\Delta \overline{W}_E$ in equation 4.2? We account for environmental changes by the changes in the average effects. To obtain the changes in average effects, we recalculate the average effect of each allele in the changed population, including any changes in interactions between genes, changes in the array of genotypes caused by mating pattern, changes in competition between individuals, and changes in the physical environment. Frank (1997) wrote the total change in the environment as the total change in average effects

$$\Delta \overline{W}_E = 2 \sum_j q'_j (\Delta b_j), \qquad (4.5)$$

where the prime on q shows that we use the frequencies in the changed populations to weight the changes in average effects for each allele. Putting the pieces together and using the definitions in equation (1) of Frank (1997) yield the full partition

$$\Delta \overline{W} = \Delta \overline{W}_{NS} + \Delta \overline{W}_E$$
$$= 2 \sum_j (\Delta q_j) b_j + 2 \sum_j q'_j (\Delta b_j). \qquad (4.6)$$

The natural selection term is equivalent to the genetic variance in fitness. Conservation of total fitness implies that the deterioration of the environment term is typically close to the negative of the first term.

Recently, I have shown that the genetic variance in fitness can also be thought of as a distance between the population before natural selection and after natural selection (Frank 2009). The distance measures the information the population acquires about the environment through the changes in gene frequencies caused by natural selection. Quantifying the consequences of natural selection by an informational measure is conceptually more profound than quantifying the change in fitness by the genetic variance, although the descriptions are mathematically equivalent. I will not pursue here my own informational interpretation, although that interpretation may be necessary to understand the full significance of the fundamental theorem as a law.

4.3.8 Misunderstandings about additivity

Fisher's genetic variance is calculated by adding the contribution of each individual allele independently, leading to its common description as the additive genetic variance. This description suggests that the additive genetic variance ignores dominance and genic interactions, instead assuming that each allele has a fixed contribution that can be taken independently and additively with respect to other alleles. For example, Wright (1930, p. 353) noted in his review of Fisher (1930):

> One's first impression is that the genetic variance in fitness must in general be large and that hence if the theorem is correct the rate of advance must be rapid. As Dr. Fisher insists, however, the statement must be considered in connection with the precise definition which he gives of the terms. He uses "genetic variance" in a special sense. It does not include all variability due to differences in genetic constitution of individuals. He assumes that each gene is assigned a constant value, measuring its contribution to the character of the individual (here fitness) in such a way that the sums of the contributions of all genes will equal as closely as possible the actual measures of the character in the individuals of the population. Obviously there could be exact agreement in all cases only if dominance and epistatic relationships were completely lacking. Actually, dominance is very common and with respect to such a character as fitness, it may safely be assumed that there are always important epistatic effects. Genes favorable in one combination, are, for example, extremely likely to be unfavorable in another. Thus allelomorphs which are held in equilibrium by a balance of opposing selection tendencies...may contribute a great deal to the total genetically determined variance but not at all to the genetic variance in Fisher's special sense, since at equilibrium there is no difference in their contributions.

This quote is the sort of commentary from Wright that led many people to regard Fisher's view as one of genes acting additively and ignoring Wright's own emphasis on the importance of dominance and epistatic genetic interactions. However, one must parse this quote with care to understand what Wright is truly emphasizing.

The quote begins by framing the problem with respect to the rate of adaptive change. Wright characterizes Fisher's argument as inevitably leading to the conclusion that the rate of adaptive change by natural selection must in fact be slow, because Fisher's analysis strips away the most important contributions to variance that come from nonadditive genetic interactions. Wright continues by stressing the great importance of gene interactions, implying that a true theory of adaptive change must be based primarily on such interactions. The quote is not really about Fisher's theorem, but rather about Wright's characterization of his difference with Fisher.

According to Wright, the Wright view fully accounts for genetic interactions as the primary source of genetic variation and thus can fully account for the processes that may lead to rapid adaptive change. By contrast, Wright has Fisher limited to the small component of genetic variance associated with the purely additive effects of genes acting in isolation, and thus with a theory that must be associated with a very limited rate of adaptive evolution.

It is never quite true that Wright misunderstands Fisher's mathematics and arguments. Wright understood mathematical genetics far too well for that. But Wright's insistence on emphasizing his view of the Wright–Fisher contrast makes it very hard to get a fair characterization of Fisher's views from Wright. Of course, Fisher did no better in return. So, to understand their theories, we cannot read Wright on Fisher or read Fisher on Wright.

Fisher did not actually ignore dominance or genetic interactions. Instead, he fully and completely accounted for those interactions. The heritable contribution of each allele in the context of all of the genetic interactions in the population at any moment in time is exactly the average effect of the allele. Fisher was trying to quantify the evolutionary change in fitness caused by natural selection, which means that the only important quantity with respect to each allele is its heritable contribution to fitness. Heritable effects are the only effects that are passed to offspring, so they are the only effects that one must account in the calculation of change by natural selection.

The average effect of each allele is chosen statistically to be the effect one has to add to an individual carrying the allele to get the best prediction of the individual's phenotype or fitness. The average effect is a statistical form of additivity that accounts for all forms of non-additive gene interactions. The average effect is not a physiological statement about the presence or absence of dominance or genetic interactions. Hidden in Wright's statements is his own primary interest in how processes other than natural selection might rearrange the patterns of genetic interactions, thereby providing a different subsequent evolutionary path by natural selection. That sort of rearrangement of genetic interactions is a very interesting problem, but it has nothing to do with the fundamental theorem or with Fisher's accounting for additive and non-additive genetic effects with respect to natural selection.

4.4 Fisher's laws versus Wright's dynamics

The fundamental theorem expresses two laws. First, the rate of increase in fitness caused by natural selection is an invariant quantity equal to the genetic variance. This quantity is invariant in the sense that the many complexities of mating, environment, and genetic interactions are subsumed into a single value that does not depend on the large number of details that can differ. Invariant quantities tell us what does not matter; what is left is all that matters. Many of the deepest insights in science have arisen from a clear understanding of what matters and what does not matter—from a clear expression of invariance.

The second component of the theorem is an approximate conservation law. The total change in fitness tends to remain close to zero, so the deterioration of the environment tends to be equal and opposite to the rate of increase in fitness caused by natural selection. This conservation law captures the ever improving adaptation of individuals offset by the increasing pressure of competition from the improved adaptation of other individuals. Although other factors also contribute to the deterioration of the environment, Fisher emphasized this balance between individual improvement and enhanced competitive pressure.

The theorem is clearly designed to express laws rather than to calculate long-term dynamics. Laws play a key role in understanding natural phenomena. Laws also set boundaries that must be satisfied by all systems—necessary but not sufficient conditions by which we may calculate the dynamics of systems. To the extent that one wishes to calculate dynamics, the theorem is limited by its description of laws rather than dynamics.

Fisher (1958a, p. 39) used the principles of statistical mechanics to obtain simple laws:

> The regularity of [natural selection] is in fact guaranteed by the same circumstance which makes a statistical assemblage of particles, such as a bubble of gas obey, without appreciable deviation, the laws of gases.

To understand statistical mechanics, think of each allele in a population as an individual particle, like a particular atom of an element. The whole population is a collection of particles divided into discrete sets, each set forming the genotype of an individual. The population of a large number of alleles divided into many distinct genotypes is like a large collection of atoms divided into many distinct molecules.

One can study the dynamics of a collection in two distinct ways: particle dynamics or statistical mechanics. In particle-based dynamics, one analyzes the dynamics of the aggregate population by following the dynamics of each particle. In genetics, that would mean studying the dynamics of the population and its fitness by analyzing the dynamics of all alleles with respect to their assortment into genotypes. Particle-based dynamics is the most complete description possible. It is also hopelessly complex for all but the most unrealistically reduced of systems. Thus, any physical study of large aggregates applies statistical mechanics.

In statistical mechanics, one reduces all of the complex dynamics of the individual particles to a simple statistical summary. For example, the movement of each particle can be thought of as fluctuation, and each fluctuation is typically influenced by interactions with many other particles. To study explicit dynamics, each fluctuation of each particle must be analyzed with respect to all of the interactions between particles. In genetics, we can think of a fluctuation as a change in fitness caused by a particular gene, each fitness fluctuation ascribed to a gene depending on the interaction of that gene with many other genes.

To study statistical mechanics, we may use the variance of the individual fluctuations—a single aggregate measure that summarizes the overall intensity of fluctuation in the whole population. Thus, the variance in fitness of the individual genes is the single aggregate measure of genetic variance in the population. Fisher's fundamental theorem shows that a single aggregate measure of variance is sufficient to fix the total change in fitness caused by natural selection. The reduction in complexity is almost magical. Much of our understanding of the natural world arises from being able to reduce the overwhelming complexity of the dynamics of many particles to simple aggregate measures that capture essential features of system behavior.

Fisher (1958a, p. 39) felt very strongly about the deep power of the statistical laws of nature and of what he accomplished with his theorem:

> It will be noticed that the fundamental theorem proved above bears some remarkable resemblances to the second law of thermodynamics. Both are properties of populations, or aggregates, true irrespective of the nature of the units which compose them; both are statistical laws; each requires the constant increase of a measurable quantity, in the one case the entropy of a physical system and in the other the fitness...of a biological population...Professor Eddington has recently remarked that 'The law that entropy always increases—the second law of thermodynamics—holds, I think, the supreme position among the laws of nature'. It is not a little instructive that so similar a law should hold the supreme position among the biological sciences.

One of Fisher's main goals for his book was to demonstrate the law-like character of natural selection in shaping the biological world. He wanted to put to rest many of the groundless criticisms of natural selection that we continue to hear today. Fisher (1958a, p. 40) continued:

> The statement of the principle of Natural Selection in the form of a theorem determining the rate of progress of a species in fitness...puts us in a position to judge of the validity of the objection which has been made, that the principle of Natural Selection depends on a succession of favourable chances. The objection is more in the nature of an innuendo than of a criticism, for it depends for its force upon the ambiguity of the word chance, in its popular uses. The income derived from a Casino by its proprietor may, in one sense, be said to depend upon a suggestion of improbability more appropriate to the hopes of the patrons of his establishment. It is easy without any very profound logical analysis to perceive the difference between a succession of favourable deviations from the laws of chance, and on the other hand, the

continuous and cumulative action of these laws. It is on the latter that the principle of Natural Selection relies.

These quotes help to understand Fisher's motivation with regard to the fundamental theorem and to analyze his various arguments with Wright about the fundamental theorem and the Adaptive Landscape. Fisher viewed the fundamental theorem as an invariance law about natural selection rather than an expression of evolutionary dynamics. He fully acknowledged that other evolutionary processes affected dynamics. The fundamental theorem is not a complete statement of evolutionary change, only a statement about natural selection: "Natural Selection is not Evolution" is the first sentence of Fisher's book (Fisher 1958a, p. vii).

By contrast, Wright's mathematical theories analyzed gene frequency dynamics, that is, the full dynamics of the individual particles that make up the system. Wright needed to study particle dynamics because he wanted to characterize those situations in which evolutionary systems change from being dominated by particular particle interactions through a transition to which alternative particle interactions dominate. Put another way, Wright was concerned with epistatic gene interactions that bound a population to a local fitness peak, and the change in gene frequencies that would alter the gene combinations to shift a population to a different fitness peak. To study mathematically that sort of peak shift, Wright did not study full dynamics of real systems, which is not possible, but instead reduced system size to a small number of genes (particles) to capture the particular interactions in an explicit way.

4.5 Key points in the Fisher–Wright controversy

My main argument is that Fisher and Wright talked past each other with regard to the fundamental theorem and the Adaptive Landscape. They did so because each was usually arguing about some other issue, although in a way that often left the subtext obscure.

There are four main ways in which Fisher and Wright talked past each other. These four items help to parse the Fisher–Wright controversy and to understand more deeply the history and key concepts of evolutionary theory.

4.5.1 Wright lacked interest in Fisher's general laws

Wright always tried to parse the fundamental theorem in relation to its consequences for long-term evolutionary dynamics. I think almost everyone analyzed the fundamental theorem in this way. The reason is that mathematical theory in science is often regarded as simple dynamical expressions: start with initial conditions and hypothesized rules of change and calculate the predicted outcome. The predicted outcome is the dynamical expression of the future state of the system given the initial conditions and rules of change.

Certainly all of Wright's mathematical theory is cast in this standard dynamical framework. His mathematics may be technically dense at times, but the framing and goals are usually very clear in regard to the standard view of dynamical theory in science.

Reading Fisher's exposition of the fundamental theorem in the context of his book, I find it hard to understand how everyone could have tried to force the theorem into this standard dynamical context. As I showed earlier, Fisher gave an invariance law and an approximate conservation law. The invariance is that the rate of change in fitness caused by natural selection is always the genetic variance in fitness. He made clear that another component of evolutionary change in total fitness must always be ascribed to the deterioration of the environment. In the approximate conservation law, mean fitness remains nearly constant, thus the deterioration of the environment must usually be nearly equal and opposite to the increase by natural selection in order to maintain nearly constant total fitness.

In Wright's (1977) magnum opus, the index entry for "Fundamental theorem of natural selection" says "*See* Evolutionary transformation (panmictic species)." It is hard to think of Wright as being ironical. But the irony is certainly there: Fisher's (1941) scathing criticism of Wright's Adaptive Landscape focused most strongly on Wright's assumption of random mating (panmixia) in his early formulation, as opposed to Fisher's own carefully chosen definitions to distinguish the effects of genes under non-random mating. Wright certainly felt the sting of Fisher's criticism to which he replied in Wright (1964). But Wright took many opportunities after 1941 to label Fisher's theorem as a statement about randomly breeding, panmictic populations. For example, Wright (1977, p. 425):

> As noted, Fisher's theorem holds strictly only under the assumption of random combination of loci. It applies in equilibrium populations with respect to genes with wholly independent effects, in spite of linkage.

4.5.2 Long-term dynamics and the rate of adaptation

Wright labeled Fisher's theorem as one of random mating for two reasons. First, Wright ignored Fisher's development of laws about natural selection and instead interpreted Fisher with respect to a theory of long-term evolutionary dynamics. Technically, this means that Wright ignored Fisher's partition of total evolutionary change into two components, natural selection and change of the environment. In Fisher's theory, some of the total evolutionary change under non-random mating falls into the change of the environment, in the sense that changing genotype frequencies alter the genetic environment of each gene (Frank and Slatkin 1992).

Second, Wright wanted to create a contrast between their views on the rate of adaptation. For example, in Wright (1988, p. 122):

> Fisher's "fundamental theorem of natural selection" was concerned with the total combined effects of alleles at multiple loci under the assumption of panmixia in the species as a whole. He recognized that it was an exceedingly slow process.

Fisher was interested in this debate about long-term dynamics and the rate of adaptation. He realized that these issues did not concern his fundamental theorem about laws. So, when arguing with Wright about the rate of adaptation, he never answered directly with respect to the fundamental theorem.

Wright repeatedly stated that Fisher's fundamental theorem leads to a very slow rate of long-term adaptation. I have not found any statement by Fisher about the slowness of adaptive evolution following from either his theorem or his view of adaptation. I believe Wright emphasized the slowness of Fisher's view, because Wright believed that in a large, mixed population, the only source of new variation for adaptation must come from new favorable mutations. The rate of adaptation by new favorable mutations would, in Wright's view, be slow. Wright ascribed that view of slowness to Fisher, even though Fisher rejected such a conclusion.

In the quotes given previously from Fisher and Ford (1950), Fisher made clear that he believed fluctuating selection pressures are common in nature. Under fluctuating selection, gene frequencies may be perturbed in ways that change the combinations of interacting genes favored by natural selection. Once such changes occur, rapid adaptive change may follow by the process Wright, but not Fisher, ascribed to the fundamental theorem.

In spite of the clear comments in Fisher and Ford (1950), Wright continued to claim that Fisher believed adaptive evolution to be an exceedingly slow process in large, mixed populations. Wright contrasted this Fisherian strawman with his own view of rapid adaptive change driven by population subdivision, random perturbation of gene frequencies by small population size, followed by rapid adaptive evolution when new gene combinations are favored by natural selection. Fisher and Ford (1950) agreed that Wright's theory would lead to rapid adaptive evolution, but they regarded that theory as neither necessary nor likely for the explanation of rapid adaptive evolution.

4.5.3 Additivity versus genetic interactions

Wright's whole program turned on the novel variation generated by changes in the favored combinations of interacting genes. Those changes in favored combinations do not depend on new mutations, but rather on fluctuations in gene frequencies. For example, a particular gene cannot increase in frequency if it works well only with another gene that is rare and works poorly with the common alternative gene. A fluctuation that makes the rare gene become common changes the situation, allowing the beneficial combination to be favored.

Wright repeatedly characterized Fisher's theorem as incapable of dealing with such genetic interactions. This characterization must have puzzled Fisher, who devoted much of his famous 1918 paper on quantitative genetics to the complexities of genetic interactions. That paper describes the explicit partitioning of genetic variance into components that arise from the direct effects of each gene—the average effect—and the interactions that arise from dominance and epistasis. In the partitioning of total genetic variance, one adds the direct effects of each gene, that additive component is often called the additive variance. But that component does not arise physiologically from constant additive contributions of separate genes. The direct effect of a gene depends on the frequencies of all other genes with which it interacts, and the direct effect changes with the changing frequencies of those interacting loci.

It is true that inheritance is controlled by the sum of the direct effects calculated for each independent gene, because in the short term, it is only those direct effects that get transmitted from parent to offspring after the sexual mixture of parental genomes. The brilliance of Fisher's analysis was to find a simple expression for the heritable component within the complex system of genetic interactions that he assumed was universal. The fundamental theorem was a direct descendant of the statistical approach to genetic interactions originated in 1918.

Why did Wright ascribe to Fisher the assumption of constant additive effects of separate genes? To understand the shifting standard of the average effect in a theory that predicts the rate of change in mean fitness, one has to understand that the fundamental theorem is not about evolutionary dynamics but instead about the invariant quantity of genetic variance with respect to the natural selection component of evolutionary change. Wright was not interested in that invariant law, but rather in his own world view with respect to long-term evolutionary dynamics. Thus, he misrepresented the theorem, because he only discussed it within his own frame of reference.

In addition, I think that Wright favored sharp distinctions between Fisher's work and his own: Fisher associated with additive, independent gene action in large, randomly breeding populations versus Wright associated with complex genetic interactions in small, subdivided populations. By this characterization, Wright linked Fisher to slow adaptive change limited by the flow of rare beneficial mutations, in contrast with Wright's own claim for rapid evolutionary change by random fluctuations of gene frequencies creating newly favored beneficial combinations. Wright did not care about Fisher's laws or his statistical partitioning of genetic effects. He did care deeply about his own shifting balance theory based on newly favored beneficial combinations of genes.

Wright originally formulated the shifting balance theory during the early 1930s. He spent the following 50 years refining that theory primarily through the mathematical exploration of gene frequency dynamics over his metaphor of the Adaptive Landscape, which describes a surface of mean fitness.

4.5.4 Wright's expression of fitness surfaces in an Adaptive Landscape

Wright (1988, p. 118) stated in his final paper:

> The...diagrams...represent cases in which the population is assumed to be so large and its individuals so mobile that there can be no significant effects of accidents of sampling, giving rise to the panmixia assumed by Fisher (1930) to be characteristic of species in nature under similar environmental conditions throughout the range. This assumption was basic to the derivation of his "fundamental theorem of natural selection" (1930, p. 35)...The effects of these four processes [by which populations climb local peaks of fitness] may be calculated by means of Fisher's "fundamental theorem of natural selection."

When arguing with Wright about the rate of adaptation and long-term evolutionary dynamics, Fisher rarely answered directly with respect to Wright's misrepresentation of the fundamental theorem. On the few occasions that Fisher commented on Wright's use of the theorem, Fisher emphasized various isolated issues. But Fisher never tried to explain the distinction between his goal to formulate laws and Wright's goal to understand the rate of adaptation. Perhaps the reason Fisher did not defend his view was expressed by Haldane (1964): "Fisher...preferred attack to defense."

Fisher (1941) first noted the failure of Wright's formulation to handle non-random mating compared with the clear way in which non-random mating is handled by the fundamental theorem. Fisher (p. 377) then stated:

> It is, I think, clear from Sewall Wright's allusions to the subject that he has never clearly grasped the difficulties of interpretation of such expressions as
>
> $$\frac{d\overline{W}}{dp}$$
>
> in which the numerator involves the average of W for a number of different genotypes greatly exceeding the number of gene frequencies p on which their frequencies are taken to depend. It is likely, therefore, that he does not share my reasons for putting a particular and well defined meaning upon the phrase '*average effect of a gene substitution*'.

Fisher's point is that, for two different alleles at a locus, there are three different genotypes. Thus, for 1000 different loci, there are 3^{1000} different genotypes, which far exceeds the size of any population. Thus, the notion of dp for a change in gene frequency may not have much meaning with respect to mean fitness, because a discrete additional copy of an allele for a change dp in frequency must often be added in a way that creates a novel genotype, the composition of which would be hard to predict even in a very large population. The discreteness of genotypes and the rarity of many genotypes means that average fitness cannot reliably change in a smooth and regular way with smooth changes in gene frequencies.

Wright (1964, p. 219) acknowledged that the mathematical formulation of the Adaptive Landscape could be used only when there are a small number of loci or one makes very regular assumptions about genetic effects:

> The summation in the formula for $\overline{W}$ has, however, as many terms as there are kinds of genotypes, 3^{1000} for 1000 pairs of alleles. This, of course, points to a practical difficulty in calculating Δq for more than two or three pairs of interacting factors, unless a regular model is postulated.

I suggested earlier that Fisher was reluctant to argue about the fundamental theorem as a law rather than a statement of dynamics. In Fisher's exchanges with Wright, Fisher usually kept to the issue of long-term evolutionary dynamics and the rate of adaptation, topics that from his point of view were not directly related to the theorem. One clear exception shows the point, from Fisher (1958b, p. 290):

> I have never, indeed, written about $\overline{w}$ and its relationships... the existence of such a potential function [i.e. a function non-decreasing in time]... is not a general property of natural populations, but arises only in the special and restricted cases which Wright has chosen to consider.
>
> I should not have alluded to this storm in a teacup, but for the circumstance that I mean to put forward some ideas on... the possible adaptive value of polymorphisms, and, incidentally, to express my personal opinion that Dobzhansky was right in regarding polymorphism as very often properly described as an adaptation to the conditions of life in which a species finds itself, but for reasons quite distinct from the direct action of Natural Selection, by which the polymorphism is maintained, or indeed from Natural Selection as it acts among the individuals of any one interbreeding population.

Another brief mention by Fisher also points to the way in which Wright's formulation of change in mean fitness differed from Fisher's own view of natural selection:

> In regard to selection theory, objection should be taken to Wright's equation [the expression $d\overline{W}/dq$] principally because it represents natural selection, which in reality acts upon individuals, as though it were governed by the average condition of the species or inter-breeding group (Fisher 1941, p. 58).

Wright (1964, p. 219) responded to that criticism many years later:

> As I understood it, Fisher [was]... trying to arrive at a theorem on the rate of increase of "fitness" under natural selection that applies to a species as a whole. My purpose was to obtain a formula for change of gene frequency in a random breeding deme in cases that involve factor interaction.

I think it is generally true that Wright was interested in gene frequency change rather than mean fitness. But his Adaptive Landscape metaphor of climbing fitness peaks by natural selection was a prominent part of his view. Indeed, in Wright (1942, p. 241), he makes clear that he considered how climbing adaptive peaks does directly affect the mean fitness of populations:

> These [gene frequency] changes will be such that the mean selective value of the populations changes approximately by the amount
>
> $$\Delta \overline{W} = \sum \left(\Delta q \, \partial \overline{W} / \partial q \right)$$
>
> the species moving up the steepest gradient in the surface $\overline{W}$ except as affected by mutation pressures.

This statement is clearly about the rate of change in the mean fitness of populations, contradicting Wright's later comment. Perhaps the Fisher–Wright controversy remains alive long after the combatants have passed because of the odd dissonance between what these two seemed to be saying at any point, what they had said previously, and the underlying and often hidden basis of their disagreements.

Acknowledgments

My research is supported by National Science Foundation grant EF-0822399, National Institute of General Medical Sciences MIDAS Program grant U01-GM-76499, and a grant from the James S. McDonnell Foundation.

References

Charlesworth, B. (2000). Book review of 'The genetical theory of natural selection: a complete variorum edition' by R. A. Fisher. Genetical Research, 75, 369–370.

Ewens, W. J. (1989). An interpretation and proof of the fundamental theorem of natural selection. Theoretical Population Biology, 36, 167–180.

Fisher, R. A. (1930). The Genetical Theory of Natural Selection. Clarendon, Oxford.

Fisher, R. A. (1941). Average excess and average effect of a gene substitution. Annals of Eugenics, **11**: 53–63.

Fisher, R. A. (1958a). *The Genetical Theory of Natural Selection*, 2nd edn Dover, New York.

Fisher, R. A. (1958b). Polymorphism and natural selection. Journal of Ecology, 289–293.

Fisher, R. A. and Ford, E. B. (1950). The "Sewall Wright Effect". Heredity **4**: 117–119.

Frank, S. A. (1995). George Price's contributions to evolutionary genetics. Journal of Theoretical Biology, **175**: 373–388.

Frank, S. A. (1997). The Price equation, Fisher's fundamental theorem, kin selection, and causal analysis. Evolution, 51: 1712–1729.

Frank, S. A. (2009). Natural selection maximizes Fisher information. Journal of Evolutionary Biology, **22**: 231–244.

Frank, S. A. and Slatkin, M. (1992). Fisher's fundamental theorem of natural selection. Trends in Ecology & Evolution, 7: 92–95.

Haldane, J. B. S. (1964). A defense of beanbag genetics. Perspectives in Biology and Medicine. **7**: 343–359.

Price, G. R. (1972). Fisher's 'fundamental theorem' made clear. Annals of Human Genetics, **36**: 129–140.

Price, G. R. (1995). The nature of selection. Journal of Theoretical Biology, **175**: 389–396.

Provine, W. (1986). *Sewall Wright and Evolutionary Biology*. University of Chicago Press, Chicago, IL.

Wright, S. (1930). The genetical theory of natural selection: a review. Journal of Heredity, **21**: 349.

Wright, S. (1934). Physiological and evolutionary theories of dominance. American Naturalist, **68**: 24–53.

Wright, S. (1942). Statistical genetics and evolution. Bulletin of the American Mathematical Society, **48**: 223–246.

Wright, S. (1964). Stochastic processes in evolution. In J. Gurland (ed.), *Stochastic Models in Medicine and Biology*, Vol. 25, University of Wisconsin Press, Madison, WI, pp. 199–241.

Wright, S. (1977). *Evolution and the Genetics of Populations: Volume 3, Experimental Results and Evolutionary Deductions*. University of Chicago Press, Chicago IL.

Wright, S. (1988). Surfaces of selective value revisited. American Naturalist, **131**: 115–123.

Appendix

Because Fisher never gave a precise definition of fitness, there is no historical basis for ascribing any particular expression to Fisher himself. Fisher's vagueness about fitness also made it difficult to understand what might be meant by a deterioration of the environment in relation to an exact mathematical theorem.

My own view is that Price's (1995) definition of fitness is the only one that provides both mathematical consistency of the theorem and logical unity to Fisher's vision (Frank 1995, 1997, 2009). Here is the particular definition of fitness that provides a simple mathematical basis for the fundamental theorem and related topics. The following is taken from Frank (2009).

The fitness of a type defines the frequency of that type after evolutionary change. Thus, we write $q'_j = q_j(w_j/\overline{w})$, where w_j is the fitness of the jth type, and $\overline{w} = \sum q_j w_j$ is the average fitness. We may use j to classify by any kind of type, such as allele, genotype, or any other predictor of fitness.

Here, w_j is proportional to the fraction of the second population that derives from (maps to) type j in the first population. One often thinks of the second population as the descendants and the first population as the ancestors, but any pair of populations can be used, separated by an instant in time, by discrete generations, or by some other scale of divergence that is not related at all to time. The scale of divergence can be set by describing the point of measurement of the first population as θ, and the point of measurement of the second population as θ'. Thus, q'_j does not mean the fraction of the population at θ' of type j, but rather the fraction of the population at θ' that derives from type j at θ.

The fitness measure, w, can be thought of in terms of the number of progeny derived from each type. In particular, let the number of individuals of type j at θ be $N_j = Nq_j$, where N is the total size of the population. Similarly, at θ', let $N'_j = N'q'_j$. Then $\overline{w} = N'/N$, and $w_j = N'_j/N_j$.

Fitness can alternatively be measured by the rate of change in numbers, sometimes called the Malthusian rate of increase, m. This is the measure that Fisher typically used. To obtain the Malthusian rate of increase with respect to an infinitesimal change in scale, $\Delta \theta \to d\theta$, define the overdot as the differential $d/d\theta$, and write

$$\begin{aligned}\frac{\dot{q}_j}{q_j} &= \dot{\log}(q_j) \\ &= \dot{\log}(N_j/N) \\ &= \dot{\log}(N_j) - \dot{\log}(N) \\ &= \dot{N}_j/N_j - \dot{N}/N \\ &= m_j - \bar{m} \\ &= a_j, \end{aligned} \quad (4.7)$$

where a_j is the average excess in fitness. Because the changes here are infinitesimal, corresponding to continuous time and in equation 4.3 to $\overline{W} \to 1$, the expression here is equivalent to the expression in equation 4.3 for the average excess in fitness.

CHAPTER 5

Wright's Adaptive Landscape: Testing the Predictions of his Shifting Balance Theory

Michael J. Wade

5.1 Introduction

Sewall Wright proposed his shifting balance theory (SBT) in the 1930s (Wright 1931, 1932) to describe adaptive evolution occurring by a combination of mutation, random genetic drift, and natural selection. Wright used his famous "Adaptive Landscape" to depict the movement of a population toward adaptation under the direction of the constellation of evolutionary forces in his SBT theory. His Adaptive Landscape has many fitness peaks, separated by fitness valleys, rather than a single fitness optimum, because of the complexity of the map Wright envisioned between gene interactions, individual phenotypes, and the mean fitnesses of demes across a meta-population (see Chapter 2). Simpler maps, such as purely additive genetic maps without genotype-by-environment or gene–gene interactions, can have (but do not necessarily have) much simpler fitness landscapes. It is the interactions with respect to fitness that create the multiplicity of fitness peaks, so that selection for an intermediate optimum value of an additively determined character creates fitness epistasis:

> Thus selective value as a character, usually superimposes interaction effects of the most extreme sort, upon whatever interaction effects there may be among genes with respect to underlying characters. Mathematical treatment of gene interaction must obviously be a primary concern in the theory of population genetics (Wright 1968, p. 425).

Whether the fitness landscape has a single peak or a multiplicity of peaks separated by valleys, the metaphor of "hill climbing," wherein natural selection gradually moves the mean of a population toward higher values of mean fitness, has broad intuitive appeal. Indeed, the Adaptive Landscape has proven to be a very useful metaphor in phenotypic selection studies and would remain so even if Wright's SBT were proven to be wrong. Furthermore, it is a more or less adequate and traditional representation of the fitness gradient (regression) which quantitatively expresses the movement of a population mean in response to natural variations in the fitnesses of its members. The force of natural selection acting within each deme results in it climbing to the top of the nearest adaptive fitness peak. However, to navigate a landscape with a multiplicity of possible fitness peaks and arrive at the highest peak, a local population must pass through a region of lower mean fitness, a valley in the landscape, by the combination of random genetic drift and interdemic selection. Drift alone in Wright's view could permit the crossing of a valley and the capture of a deme by the domain of attraction of another fitness peak, but it could not guarantee that the new peak, found by trial and error, would be one of higher mean fitness than the original peak.

Wright's theory of evolution was founded on several general empirical observations and his deductions about the process of adaptation given those observations. Because the Adaptive Landscape itself is a simple picture of a much more complex underlying process, it would be possible to reject the landscape metaphor but nevertheless retain Wright's theory. Indeed, the metaphor of the Adaptive Landscape was widely adopted shortly after he proposed it, even though the underlying evolutionary processes and their actions in relation to complex genetics were poorly understood, even by some of Wright's strongest proponents (Provine 1986), perhaps because the "hill-climbing" image also characterizes much simpler adaptive processes than those imagined by Wright.

In order to understand whether or not the Adaptive Landscape is an adequate pictorial representation of Wright's SBT, it is necessary to understand both the origins of and the predictions of his theory. To further determine the respective utilities of the landscape and the SBT for guiding research strategies, we also need to understand results of empirical tests of the assumptions of the SBT as well as its unique predictions. Although others in this volume have also addressed Wright's theory in relation to the Adaptive Landscape, I begin with the conceptual foundations of the SBT in order to more clearly illustrate its predictions and how they might be tested. Many who use the Adaptive Landscape as a guiding interpretive metaphor have little appreciation of the theoretical and genetic underpinnings of the theory it depicts; conversely, many who study the predictions of the SBT do not explicitly reference the Adaptive Landscape.

5.2 Conceptual foundations of Wright's shifting balance theory

Early in his career, Wright spent a decade at the United States Department of Agriculture (USDA), and these years spent studying agriculture guided his thinking about evolution in natural populations. Wright's theory was founded on these several general empirical observations and deductions: (1) gene interactions (epistasis) are "ubiquitous" and underlie all adaptations; (2) there are an astronomical number of genotypes possible with genomes consisting of 1000 or more genes with several alleles at each locus; (3) populations in nature are genetically subdivided, consisting of many more or less isolated demes of finite size, such that random genetic drift plays a significant role locally in each; (4) random genetic drift causes the phenotypic and genetic diversification among inbred lineages of guinea pigs for a wide variety of traits; and, (5) the histories of the improvement of different breeds of domesticated animals can be reconstructed from pedigrees. The latter reconstruction was the empirical basis for including interdemic selection or Phase III (see later in this section) in his SBT. Just as Darwin used artificial selection by breeders as a model for natural selection, Wright reconstructed the genetic transformation of domesticated breeds and used that process as a model for his theory of adaptive evolutionary change. Breed histories showed Wright that the phenotypic transformation of a breed proceeded when chance and local selection produced an important combination of genes on a particular farm and subsequent demand for that combination by other breeders led to its dispersion out from the originating farm and into other farms. Phase III of Wright's SBT corresponded to this stage in the process of artificial breed transformation: interdemic selection occurred by differential dispersion of good gene combinations out from high fitness demes and into demes of lower mean absolute fitness.

Wright (1931, 1932) introduced his concept of the "Adaptive Landscape" or "surface of selective value" with the goal of elucidating how the simultaneous action of several evolutionary forces (mutations, migration, random genetic drift and natural selection) results in adaptation and speciation. In Wright's words (1931, p. 101):

> Selection, whether in mortality, mating or fecundity, applies to the organism as a whole and thus to the effects of the entire gene system rather than to single genes. A gene which is more favorable than its allelomorph in one combination may be less favorable in another. Even in the case of cumulative effects, there is generally an optimum grade of development of the character and a given plus gene will be favorably selected in combinations below the optimum but selected against in combinations above the optimum. Again the greater the number of unfixed genes in a population, the smaller must be the average effectiveness of selection for each one of them. The more intense the selection in one respect, the less effective it can be in others. The selection coefficient for a gene is thus in general a function of the entire system of gene frequencies.

In later writings, Wright (1969, p. 419–420) elaborated on the "inadequacy of the simple additive concept of gene effect" stating that, "all genes that approach additivity in their effects on varying characters will be favorable in some combinations and unfavorable in others in terms of natural selective value (fitness) and, thus, exhibit interaction effects of the most extreme sort in the latter respect." From these and other writings, it is clear that the focal trait of Wright's theory is *fitness* and it is his explicit premise that the fitness of an individual genotype is not the sum of the properties of its composite single genes but rather an indivisible property of the whole genetic system (*sensu* interactions in statistics, which cannot be assigned as main effects to the interacting factors). This foundational premise of Wright's SBT is different from the standard assumptions of evolutionary genetic theory in which constant selective effects are assigned to new alleles, when they arise *de novo* by mutation. It is further imagined in almost all cases that these effects are invariant with respect to genotypic background at other loci, to spatial and temporal environmental variations, and to other genotypes in the population (Williams 1966; Crow 2010; Hill et al. 2008). These assumptions of the constancy of the fitness effects of alleles have carried over into the empirical study of the contribution of mutations to heritable variation (Loewe and Hill 2010).

Interaction effects, whether between genes or between genes and environments (biotic or social), describe conditions under which the relationship between an allele and the fitness of the individual in which that allele resides changes in magnitude or in sign. Changes in magnitude affect the rate of evolution because they affect the additive variance in fitness and thereby the rate of change of an allele's frequency. Changes in sign affect the direction of evolution, i.e. whether the change of an allele's frequency is positive (favored by selection) or negative (opposed by selection). The existence of interaction effects for fitness limits the application of the standard theoretical approach by placing boundaries, in the form of genotypic and environmental contexts, on the applicable domain of the theory's central assumption, namely, the constancy of the selective effects of alleles. In his SBT, Wright assumed that interaction effects were ubiquitous and "of the most extreme sort" with respect to fitness.

This difference in viewpoint between Wright and standard theory can be understood using the two card games of poker and war as metaphors. Standard evolutionary theory, if applied to hands in a game, would predict that a card game can be understood in its essence by assigning each card a constant value and decomposing any hand dealt into its component cards. This is a strategy that works well for understanding the popular children's card game, war, where outcomes are determined by the rank order of the value of the cards played. The rank value is a constant property of each card and does not change with the identity of other cards in a hand or the other players in the game. Now consider the game of poker. Here, a winning (or losing hand) depends upon its combination of cards, the spatial order of in which the hands are played, the abilities of the other players in the game, and even the past history of the "population" of players. Here the value of a card is determined by both the local context of other cards in the same hand as well as by the "population context" of the other players in the game, and even the order of play of the hands. Although standard theory could provide a partial understanding of the broader features of poker, Wright's theory, with its emphasis on "good card combinations," might well give a deeper, more mechanistic understanding of poker's outcomes than a theory based on "additive" card values. In my metaphor, Wright's SBT is a better description of Nature if the process of adaptation is like a game of poker, where interactions predominate, while the standard theory is a better description if the adaptive process in nature is like the game of war, where interactions matter little if at all. It is likely, however, that gene–gene interactions are the key to biodiversity, since additive systems allow for fewer genetic architectures.

Wright's SBT of adaptive evolution was proposed as a three-stage process whereby a population could move toward more perfect adaptation by a combination of evolutionary forces (reviewed in Johnson 2008). In Phase I of Wright's process, random genetic drift operates to genetically differentiate more or less isolated local demes. In Phase II, individual (or mass) selection acting within populations on the basis of local "additive" effects moves the system of genotype frequencies toward local adaptive peaks. During this process, populations attracted to different local fitness peaks become even more genetically different from one another. Differentiation can be especially extreme when there are interactions between genes that change the sign of allelic effects, because an allele favored in one population may be selected against in another. By a combination of Phases I and II, a population could move from its local fitness optimum to another, not necessarily higher, fitness peak by random genetic drift and subsequent mass selection. This was the state of Wright's inbred lines of guinea pigs. But, Wright (1932, p. 358–359) considered evolution to this point incomplete:

> The problem of evolution as I see it is that of a mechanism by which the species may continually find its way from lower to higher peaks ... In order that this may occur, there must be some trial and error mechanism on a grand scale by which the species may explore the regions surrounding the small portion of the field which it occupies. To evolve, the species must not be under strict control of natural selection.

Phases I and II of the SBT provide the "trial and error mechanism" but do not guarantee transition to higher peaks. Wright's solution to this problem, was found in the breeding records of the USDA, and is his reason for including Phase III in his SBT. In Phase III, a population could be moved to a higher fitness peak by gene flow into it from a population(s) at a higher peak(s). By analogy with semi-conservative recombination, Wright's process permits populations to hold onto selective gains, while casting about in neighboring genotype frequency space by drift and selection for other fitness peaks, and includes as well a mechanism for biasing this exploration in favor of the higher peaks. A similar process mimicking the SBT more recently is the use of genetic algorithms to solve complex problems with multiple solutions (Holland 1975). Studies of social learning (Rendell et al. 2010) have also shown that a combination of trial-and-error learning (analogous to Phases I and II) and social copying of the better outcomes (Phase III) is an efficient strategy for acquiring adaptive behavior in a complex environment. While finding utility in other disciplines, the value of Wright's theory in his own field remains contentious (e.g. Coyne et al. 1997).

The SBT and its Adaptive Landscape represent a comprehensive synthesis of Wright's empirical observations and theoretical deductions that provide a useful perspective for organizing observations from nature. The Adaptive Landscape has been used as a conceptual metaphor since the time of Wright (see Chapter 2) but it has been used much less extensively as a research strategy in the sense of a guide to the design and planning of experiments for testing the SBT, which it portrays. Of course, Wright's statistical genetic measures of population genetic structure (F_{ST} and H_{ST}) and of mating systems have found wide acceptance and are an integral part of the standard theory of evolutionary genetics, used in studies at all levels, from field populations to genes to genomes. However, attempts to use the SBT as a research strategy have been restricted primarily to animal and plant breeders (cf. review by Hill and Caballero 1992; Schamber and Muir 2001; Hill and Kirkpatrick 2010) and to a handful of laboratory and field studies, many testing specific assumptions of the SBT, rather than the SBT directly. Breeders have been especially interested in the claim that artificial selection might proceed faster in genetically subdivided populations than in large panmictic populations. In fact, in the literature on the domestication of plants and animals, there are many discussion of the conditions under which family selection can be more effective than individual selection. These conditions tended to involve individual traits of low heritability, owing to large amounts of environmental variation, low additive genetic variation as a result of long-term artificial selection, large amounts of epistatic genetic variation, or a combination of these. In such cases, the family-mean value of a trait can be a better indicator of an individual's genotype than the individual's own phenotype. Thus, there are conditions under which it is more efficient to impose selection on the basis of the family mean. Among the traits typically subjected to family selection are threshold characters (Moorad and Linksvayer 2008) and "emergent" characters like sex ratio, density-dependent competitive ability, and family composition for, at least in the statistical sense, individuals do not have a "variance" or a "sex ratio." Of more relevance to the SBT are traits like milk yield in cattle, where cows are selected at the individual level but bulls, which do not express the trait under selection but do pass genes affecting its value to their daughters, are chosen using family selection (or other female kin information), based on the average milk yield of their sisters: the combination of individual and family selection, like individual and interdemic selection, being a more efficient way to increase milk yield than individual selection acting alone. Indeed, in the absence of among-family selection, alleles enhancing competitive ability increase over time, lowering yield to an extent comparable to inbreeding depression (Wade et al. 2010.)

The SBT itself has been the subject of only two direct experimental tests (Wade and Goodnight 1991; Schamber and Muir 2001). Overall, very few of the SBTs empirical predictions have been directly tested, although several of its essential components (population genetic subdivision, variation in mean fitness among populations, random genetic drift, and differential dispersion) have been shown to characterize natural populations of a very wide range of taxa. Over the past two decades, theoretical research on the premises of the SBT has added both refinements and additional predictions to Wright's grand theory; however, for the most part, these more recent hypotheses have not been extensively tested either.

In the next section, I critically review research directed toward testing the predictions of the SBT and indicate where recent theoretical work has extended the predictions of Wright's theory, adding additional opportunities for testing. Before reviewing the experimental studies, I will first address the apparent lack of interest in empirically testing the conditions for and the predictions of Wright's SBT.

5.3 Challenges of the premises of the shifting balance theory

In the years immediately following Wright's explication of the SBT and during the early phase of the Modern Synthesis, his theory was uncritically adopted. In microevolutionary and ecological genetics, in his very influential textbook, Dobzhansky (1937, p. 190) reproduced Wright's Adaptive Landscape and asserted that "differentiation into numerous semi-isolated colonies is the most favorable one, for a progressive evolution." In describing Wright's process, Dobzhansky (1937, p. 190) emphasized Phase III, stating,

> Natural selection will deal here not only with individuals of the same populations (intra-group selection) but also, and perhaps to a greater extent, with colonies as units (inter-group selection)... Here, then, conditions are given both for a differentiations of a single species into derived ones, and for a movement of the species as a whole to a new status.

In speciation and biogeography, in another classic work, Mayr (1942, p. 263) asserted that there was clear empirical support for Wright's SBT, "... evolution should proceed more rapidly in small populations than in large ones, and this is exactly what we find." Like Wright, Mayr (1954) employed random genetic drift as a creative mechanism in his theory of speciation by founder effects. In systematics, paleontology, and phylogeny, in his classic book, Simpson (1944) used Wright's SBT and his Adaptive Landscape to support his controversial concept of "quantum evolution," wherein rapid change occurs in the adaptive zone of a species, giving rise to new taxa in a relatively short period of evolutionary time.

However, random genetic drift and its interaction with epistasis, two of the central premises of Wright's theory, were incorporated into evolutionary theory differently by R. A. Fisher and other theoreticians. Fisher did not view gene interactions as important because he believed that populations in nature were so large that the effect of an allele could be defined as the average over all genotypes and over all temporal and spatial variations in the environment. He specifies that the population used to determine the average effect of a gene on fitness:

> Comprises, not merely the whole of a species in any one generation attaining maturity, but is conceived to contain all the genetic combinations possible, with frequencies appropriate to their actual probabilities of occurrence and survival, whatever these may be, and if the average is based upon the statures attained by these genotypes in all possible environmental circumstances, with frequencies appropriate to the actual probabilities of encountering these circumstances (Fisher 1958, pp. 30–31).

Fisher's use of this global average in his theory, caused theoretician and Fisher scholar, W. Ewens

(2000, p. 33), to remark that "Fisher never paid much attention to the concept [effective population size] as he should have...and used extremely high population sizes (up to 10^{12}) in his analyses, surely far too large in general." By assuming such large population sizes and such a thorough mixing of genotypes and environments, Fisher concluded that, though possibly ubiquitous, both gene–gene interactions and gene–environment interactions had little or no impact on the evolutionary process. From this perspective, random genetic drift induced fluctuations in gene frequency that only made natural selection less efficient. In fact, Fisher showed in his famous fundamental theorem of natural selection that selection acts only on the additive component of the genetic variance. Differently put, for gene interactions to be important as Wright asserted, local populations had to differ from one another in genetic background. Given Fisher's assumptions about population sizes, the effects of random drift (proportional to $[1/2N_e]$ in standard theory) are necessarily much smaller than those of selection (s), and, as a result, Fisher believed that such interactions could be safely overlooked when modeling the adaptive process, a viewpoint still common today (Hill et al. 2008; Crow 2010).

Other theoreticians (e.g. Haldane 1959; Williams 1966) questioned the necessity of or the feasibility of interdemic or group selection, Phase III of the SBT. Haldane (1959) argued that co-adapted gene combinations would be broken up by recombination when introduced by migration into demes of lower mean fitness and, thus, adaptive gains that depended upon such gene complexes could not be efficiently exported in the manner proposed by Wright. Williams (1966) invoked many of the same arguments as Fisher in discarding among-group selection as an evolutionary force, at best, much weaker than individual selection. Similarly, Barton (1992), though conceding that random genetic drift might lead to some genetic diversification among populations, dismisses interdemic selection as a viable force for adaptive evolution.

Lastly, the origin and spread of superior adaptive gene combinations was viewed as a rare event even by Wright himself. Wright (pers. comm.) believed that the origin of advantageous gene combinations was such a rare event that an experimental study of his SBT would require a "minimum of 10,000 populations." Crow (1991, p. 973) shared a similar insight into Wright's views:

> Wright never showed much interest in experimental tests of his theory; his arguments were based on plausibility and analogy. He thought that much of evolution, the steady improvement of adaptation, could happen by mass selection acting on the additive component of the genetic variance, as Fisher said. But he thought that evolutionary creativity demanded something more. This might not happen often, and hence would be difficult to test in nature, but would be important when it did happen...Fisher was interested in the steady improvement of fitness; Wright, in the occasional incorporation of novel gene complexes.

It is difficult to measure the characteristics of rare events, just as the distribution of mutational effects and rates of rare new mutations are difficult to quantify (Loewe and Hill 2010). A similar view of Wright's process was held by Alan Robertson and Douglas Falconer (pers. comm. 1985) who referred to the study of such gene combinations as "rather like studying a lottery" and therefore empirically infeasible.

In summary, Wright's SBT was challenged not only for its premises of epistasis, adaptive gene combinations, and population genetic subdivision but also for its reliance on the processes of random genetic drift and interdemic selection. This constellation of opposing arguments presents a serious challenge to Wright's SBT and it explains, in large part, the apparent lack of interest in empirically testing the conditions for and the predictions of Wright's theory.

5.4 Empirical predictions of the shifting balance theory

Wright's SBT makes several specific empirical predictions. Arguably the most general and important among them is the prediction drawn directly from Wright's own research: (1) *the rate of evolution in a metapopulation consisting of more or less isolated demes and with epistasis for fitness will be faster than that in a large, homogeneous population.* It is this prediction that has received the most experimental attention from animal breeders interested in enhancing the response to selection (see section 5.5). This prediction is qualitative in that it does not describe how much epistatic variation is required; it does not prescribe the optimal number of demes or their effective sizes; it does not specify the level of migration among demes nor the rate of differential migration in Phase III; and, it does not specify the optimal combination of within and among-deme selection. Differently put, the optimum balance among the evolutionary forces for achieving the maximum acceleration of the rate of evolution is not clear from Wright's discussions, leaving the experimenter with a very large number of decisions concerning parameters values and the regions of parameter space worth exploring.

A second prediction is directly related to the role of the among-deme genetic variance and selection among demes. Just as the response to individual selection is proportional to the additive genetic variance, the response to interdemic selection is proportional to the total genetic variation among populations: the greater the genetic variance among populations, the greater is among-deme heritability and the response to interdemic selection. It follows directly from Wright's theory of F_{ST} that: (2) *the response to interdemic selection will be stronger in a metapopulation with small effective deme sizes and little or no migration than it will be in a metapopulation with larger deme sizes and higher rates of migration.* The primary caveat here is that Wright considered one-gene measures of population genetic differentiation, like his F_{ST}, to be "wholly inadequate" (Wright 1978) for describing the effects of population subdivision on interacting gene combinations.

Additional and more specific predictions have been derived from the ongoing theoretical endeavor to incorporate the gene–gene and gene–environment interactions into the theory of evolution in genetically subdivided populations. The most important of these is the prediction based on the theoretical research of C. Goodnight (1985, 1987, 1988) into the interaction of epistasis and random genetic drift: (3) *when there is additive-by-additive epistasis, V_{AA}, the average additive genetic variance within population, V_A, will increase whenever the ratio, $\{V_{AA}/V_A\}$, exceeds 0.25 in the ancestral population.* This result is perhaps easiest to see from the formula of Whitlock, Phillips, and Wade (1993), who showed that the average amount of additive genetic variance within demes in a metapopulation equals $\{(1 - F_{ST}) V_A + 4F_{ST}(1 - F_{ST})V_{AA}\}$, where V_A and V_{AA} are defined with respect to the ancestral population. The first term is the standard expectation from the purely additive model: random genetic drift reduces the amount of additive genetic variance within demes by a factor of $(1 - F_{ST})$ at each generation. However, the second term, which is approximately $4F_{ST}V_{AA}$ (ignoring terms of order F^2), can replenish some of the additive variance lost to random genetic drift. In fact, whenever $4F_{ST}V_{AA} > F_{ST}V_A$ or, more simply, when $\{V_{AA}/V_A\} > 0.25$, additive genetic variance is not lost but rather *increased* by drift. Some consider the condition, $\{V_{AA}/V_A\} > 0.25$, "restrictive" (e.g. Lopez-Fanjul et al. 1999; Coyne et al. 2000), even though, for fitness, it is expected that V_{AA} will greatly exceed V_A. Thus, this is clearly a prediction in need of quantitative empirical testing. One complication with this prediction is that the non-additive, dominance variance (V_D) also contributes to the additive genetic variance upon inbreeding in proportion to $2F_{ST}V_D$ (Willis and Orr 1993). Thus, observing an increase in V_A by drift does not guarantee that the underlying genetic mechanism is the between-locus epistasis discussed by Wright.

At the level of gene frequencies, an increase in V_A does not change the fact that random drift always reduces the genetic variance within populations. What it changes is the *effect size* of segregating alleles as described in Goodnight (1987, 1988) and Wade (1992, 2002). The increase in effect size is similar to the increase in main effects often observed in factorial designs when one or more factors are held constant in an analysis of variance (Wade 1992). Neiman and Linksvayer (2006) describe this as the tendency of an allelic effect to become more predictable (i.e. more additive) as the genetic background with which it interacts trends toward constancy through fixation or loss of alleles at other loci by drift or selection.

A gene's contribution to the additive genetic variance is the square of its effect size multiplied by its variance. Thus, increasing the effect size, increases its contribution to the additive genetic variance.

In one-locus, two-allele models, this is the familiar term s^2pq; it is the s that is not constant and can increase owing to epistasis. With dominance, the effect size of a recessive allele is $s\{q + h[p - q]\}$, where q is the frequency of the allele and h is the degree of dominance. Clearly, when allele frequencies change by drift this effect size also changes, so that it contributes to V_A at least to the order $[q]^2$. For this reason, recessive alleles at high frequency are more effectively purged by selection within demes than those at low frequency. Cheverud and Routman (1996) have shown that other types of epistasis also contribute to an increase in additive variance by drift, although the additive-by-additive contribution is the largest as Goodnight (1985, 1987, 1988) established. Wade and Goodnight (2006) extended these effects to interactions between genes in different genomes and showed that cytonuclear epistasis (interactions between genes in the mitochondria and genes in the nucleus) also interact with random drift to increase additive genetic variation. Here, the effect is asymmetric owing to the relatively smaller effective population size of the mitochondrion: more additive variance is created for the nuclear genes than for the cytoplasmic genes, because the former fix more slowly in response to drift than the latter. Furthermore, these effects of drift on the local additive genetic variance may be more important when drift acts for several generations as opposed to immediately after a bottleneck (Goodnight 1987). In simulations, it has been shown that the effect extends for a larger number of generations within large populations than it does with small populations (Cheverud and Routman 1996). Lastly, the process is not dependent solely on random genetic drift. A similar *increase* in additive genetic variance may occur whenever selection changes gene frequencies at one of two (or more) interacting loci (Wade and Goodnight 1998; Goodnight 2004; also see Chapter 6).

An additional prediction, relevant to the SBT, was discovered empirically by Wade (1985) and given quantitative conceptual expression by Goodnight (1987). Goodnight found that additive-by-additive epistasis affects the local breeding value (LBV) of an individual in a manner that is likely to augment Wright's SBT. The breeding value for fitness of an individual is the average performance of its offspring relative to the mean number of offspring per individual produced by the population. In practice, the breeding value of a sire is determined by mating that sire to a number of randomly chosen dams and comparing the mean of his offspring to the mean of the entire population of offspring. With purely additive genetic variation, the difference in breeding values of two individuals when crossed to one set of dams remains constant if they are crossed to a second group of dams. In a metapopulation, this constancy of relative local breeding values means that the covariance of breeding values for a group of sires when tested using dams from two different populations should be nearly equal to1. Differently put, the sire with the highest fitness in one deme should remain the sire with the highest fitness in another and the sire with the poorest fitness in one deme should be the sire with the poorest fitness in any other deme as well. With additive-by-additive epistasis, Goodnight (1987) showed that this expectation changes in a way that permits our fourth prediction: (4) *As the amount of epistasis in the ancestral population increases, the covariance between LBVs decreases and the LBV of an individual becomes independent of the LBVs of other individuals.* This means that the fittest sire in one deme is not likely to be the fittest sire in another, not because of genotype-by-environment interactions, but because of the variation from deme to deme in genetic background. This phenomenon, if found to be widespread, is important in two ways for Wright's SBT. First, it potentially enhances Phase II of the SBT by allowing local selection in combination with random drift to augment the genetic divergence of traits, and especially fitness, between populations. Even if the direction of selection was the same in two demes, differences in genetic background between them would result in a gene being favored in one deme but selected against in the other. Second, it limits our ability to generalize our understanding of the genetic basis for adaptation from studies conducted in a single population to other, unstudied populations with different genetic backgrounds. In this regard, it is important to note that the quantitative genetic approach that Goodnight used to derive his theoretical findings not only quantifies a claim of

Wright (see earlier in this chapter) but also suggests the type of experimental design that could be used to test it.

A fifth prediction of Wright's SBT is that (5) *owing to epistasis, the order of fixation of alleles at different loci will matter to adaptive evolution*. In a purely additive model, the effect of a mutation on fitness is independent of the genetic background on which it occurs, while in the SBT, the effect may change in sign or magnitude with background. Thus, in an additive framework, the degree of reproducibility of adaptive evolution is higher than it would be in a non-additive framework. Experimental evolution, especially studies using populations of bacteria, where the molecular history of adaptation can be reconstructed and manipulated, can be used to test this prediction.

5.5 Empirical tests of the predictions of Wright's shifting balance theory

In this section, I will discuss the experimental tests of the five predictions of the SBT in the order in which they were presented in the preceding section. Each prediction will be repeated in italics and the experimental tests of it will follow.

(1) *The rate of evolution in a metapopulation consisting of more or less isolated demes and with epistasis for fitness will be faster than that in a large, homogeneous population*. Breeders were naturally interested in the possibility that they could achieve an increased rate of response to artificial selection if they practiced a combination of selection within groups and periodic selection between groups as outlined in Wright's SBT and as discussed by Dobzhansky, Mayr, and Simpson (see section 5.4). Hill and Caballero (1992) reviewed artificial selection experiments which incorporated population genetic subdivision as a means of increasing response to selection. Because Wright did not specify the degree of population subdivision necessary to accelerate the response to selection, many different population structures have been examined. The theoretical work by Goodnight (1985, 1987, 1988; Carson 1990; Cheverud and Routman 1996; Cheverud 2000) suggests that not only interdemic selection but also the increase in additive genetic variance within selection lines could accelerate the rate of response to selection. Most experimental selection studies have not distinguished between these two ways that genetic subdivision might enhance the response to artificial selection.

Using the *Tribolium* model system, Katz and Enfield (1977) compared within and among line selection for increased pupal weight, a primarily additive trait. In addition to a panmictic control line, they used experimental metapopulations with six lines and imposed within-line selection every generation and among-line selection either once every four or once every eight generations. For the first 12 generations they observed an equivalent response to selection in the control and the experimental populations, but over 42 generations the response in the panmictic control was superior. They concluded that epistasis was not important to this trait, even though earlier crosses between replicate selection lines showed clear evidence of epistasis (Enfield 1977).

Like Katz and Enfield (1977), most of the experimental investigations of the effect of population subdivision and selection among demes or lineages have targeted additively determined traits (Hill and Caballero 1992), where the predicted response in a subdivided population is expected to be lower not higher, except in the case of purging recessive deleterious alleles. These experiments are necessarily inconclusive regarding Wright's prediction because they study traits whose genetic basis was not the focus of Wright's theory. Lopez-Fanjul (1989) stated it well: "it is desirable that experiments of this kind be carried out for epistatic traits in order to test Wright's hypothesis."

(2) *The response to interdemic selection will be stronger in a metapopulation with small effective deme sizes and little or no migration than in a metapopulation with larger deme sizes and higher rates of migration*. There has been little empirical investigation of this prediction, but the series of experimental investigations by Wade and McCauley (e.g. 1980, 1984) using laboratory populations of flour beetles has produced surprising results that contradict the prediction. It is clear that a response to interdemic selection (Phase III) depends upon the prior existence of genetic variation among demes (Wade 1978) and the fraction of the total genetic variation that is among demes (F_{ST}), depends upon

the effective size of local demes and the amount of migration among them. The experimental work of Wade and McCauley (1980, 1984) did not confirm this prediction: they found that the fraction of the variation in mean fitness among-demes was as large for small demes ($N_e \sim 6, 12, 16$) as it was for demes severalfold larger ($N_e \sim 24, 48, 96$) and measurable variation among demes was found even in the presence of high levels of island model migration (m ~0.06, 0.12). These findings motivated a model of quantitative genetic trait in a subdivided population by Slatkin (1981, p. 869), who found that "The potential for response to group selection...[does] not depend strongly on deme size" and that "even for relatively large values of K (K = 24, 48), there is a substantial variance [among demes] in the populational characteristic." Furthermore, Slatkin found that "The model also predicts that there can be significant local genetic differentiation in the presence of migration among the demes." The data indicate that very small deme sizes and an absence of migration are not prerequisites for the operation of Phase III of the SBT, contrary to expectations based on F_{ST}. These results call into question the standard interpretation that small deme size is a pre-requisite for Wright's SBT. This issue has further been confounded with the issues of deme size raised in the group selection debates, where selection within and selection between groups are pitted against one another. This too is not relevant to Wright's SBT, where both levels of selection act in concert to produce individually advantageous traits, as opposed to the "group advantageous" traits of the group selection debate.

(3) *When there is additive-by-additive epistasis, V_{AA}, the average additive genetic variance within population, V_A, will increase whenever the ratio, $\{V_{AA}/V_A\}$, exceeds 0.25 in the ancestral population.* This prediction has been investigated in a number of experimental systems at least qualitatively, if not quantitatively, and these were reviewed by Buskirk and Willi (2006). They surveyed 22 studies comparing estimates of V_A from outbred, control populations with estimates for V_A of inbred lines (inbred to degree $F_{ST} \sim 0.40$) derived from them. For life-history traits, i.e. components of fitness believed to have large components of non-additive genetic variation, they found an average increase in V_A of 244% in the inbred lineages relative to that of the respective outbred control. However, for primarily additive traits, they found no such effect. This is a remarkably strong confirmation of this prediction of the SBT. Of these studies, Wade, Shuster, and Stevens (1996) is typical in its findings and used a very large number of selection lines with different prior histories of inbreeding, each with its unselected control, favoring increased pupal weight in T. castaneum. They found that the response to selection for increased pupal weight diminished as the degree of inbreeding increased in quantitative accord with additive theory acting on an additive trait. With respect to fitness, measured as total offspring numbers, however, they observed a different outcome:

> For some fitness components, the heritable variance was greatest in the most highly inbred and selected lines although not statistically so. In our study, life history parameters such as productivity and longevity appeared to respond differently than the morphological trait (pupal weight) to inbreeding and selection. (Wade et al. 1996, p. 732).

Thus, a study directly comparing an additive trait and a non-additive trait in the effects of population subdivision found results wholly consonant with the predictions of the SBT. This is not conclusive evidence for epistasis for fitness, however, because the experiment could not distinguish dominance variance from between-locus epistasis.

In general, for fitness traits, an increase in the additive genetic variance and/or an increased response to selection following an experimentally induced reduction in N_e has been shown in a wide array of taxa, including houseflies, *Musca domestica* (Bryant and Meffert 1995, 1996; Bryant et al. 1986), fruit flies, *D. melanogaster* (Lopez-Fanjul and Villaverde 1989; Garcia et al. 1994), mice (Cheverud et al. 1999), flour beetles, *T. castaneum* (Shuster et al. 1996), and the plant, *Brassica rapa* (Briggs and Goldman 2006). Only the study by Cheverud et al. (1999) specifically identifies epistasis as the cause of the increase in additive genetic variance and several studies (e.g. Lopez-Fanjul and Villaverde 1989; Willis and Orr 1993) attribute the increase in additive variance for viability to change in the frequency of rare deleterious recessives alleles, i.e. to the dominance variance.

(4) *As the amount of epistasis in the ancestral population increases, the covariance between LBVs decreases and the LBV of an individual becomes independent of the LBVs of other individuals.* In a strictly additive genetic system, the breeding value of a sire's phenotype predicts the mean phenotype of that sire's progeny. If one sire is significantly "better" than another sire with respect to breeding value, that difference between sires will be realized as a similar difference in means between their progeny mean phenotypes. This is true for any genetic background of the dams to which the sires are crossed, since in strictly additive systems only the "general combining ability" of the sires exists and there is no element of "special combining ability," i.e. background-dependent breeding value, because there is no epistasis. For this reason, the relative ranking of sires by their breeding values is fixed under the additive model. Consequently, the correlation in sire breeding values across genetic backgrounds should be 1.0 (the same ranking on every background). Wright argued differently, emphasizing what the breeders call "special combining ability," as seen in the quotes repeated earlier. Thus, empirically estimating the correlation in breeding values across backgrounds is a way to test whether or not Wright's genetic premise regarding epistasis for fitness and the variation in genetic backgrounds among demes holds. The lower that correlation, the more the world corresponds to Wright's tenets, while the higher the correlation, the more correct are the assumptions of the additive model. Goodnight (1988, 2000) went further and derived for two-gene epistatic models the expected magnitude of change in the correlation of sire breeding values across genetic backgrounds differentiated to some degree (some level of F_{ST}).

Several different experimental comparisons of the variance and covariance of local breeding values of sires when mated to groups of dames from different genetic backgrounds have been conducted. The genetic backgrounds tested have included different breeds of cattle (Damon et al. 1961), different natural populations of flour beetles (Drury and Wade 2011) and populations derived by random genetic drift from a common laboratory stock population of flour beetles (Wade 1985, 2002). For testing Wright's theory, the latter two backgrounds are probably the most relevant because they compare sire effects across backgrounds created in nature and backgrounds created only by drift, respectively. Importantly, when a half-sib design (randomly chosen sires from one population crossed to groups of dams from the same and other populations) is used, the dominance variance cannot contribute to the among-sire variation in half-sib family means even when it accounts for differences in the overall mean of crosses between two populations. So, unlike the previous predictions, the effects of the epistatic and dominance variance are not confounded in the relative differences among sires in breeding value when tested on different backgrounds.

Breeders have long been concerned with exploiting heterosis and have investigated whether or not the breeding values of sires from selected purebreds tend to increase when out-crossed to other breeds (reviewed in Ibanez-Escriche et al. (2009)). A general finding from these studies is that the covariance of breeding values is very low, especially for fitness traits, such as litter size and viability, in long selected lines. These data confirm a significant role for gene interactions in the manner suggested by Wright and given quantitative expression by Goodnight (1988, 1995, 2000; also see Chapter 6). Nevertheless, differences among domestic breeds may not be the most interesting type of difference in genetic background against which to test this prediction.

Drury and Wade (2011) used wild populations of *T. castaneum* collected in Bhopal, India, and Dar es Salaam, Tanzania, and an out-bred laboratory population to investigate the covariance of local breeding value for two fitness components, fertility, and adult offspring numbers. They crossed 40 sires from each of the three populations to groups of five dams from their own and the other two populations and estimated the genetic variances, co-variances, and the sire × dam-population interactions. There was significant variation among sires in offspring numbers on all population backgrounds, with the phenotypic variation being greater in the interpopulation crosses than in the intrapopulation crosses, despite all progeny being reared under identical environmental conditions. Such an *increase* in among-sire variance with change in background is consistent with genetic differences between the populations involving epistatic interactions that increase the additive variance. In all cases, the observed genetic correlations across backgrounds were significantly less than +1.0; indeed, *none* could be bounded away from zero despite significant among-sire variations on own-background. These observations are wholly consistent with the prediction drawn from Goodnight's research (Chapter 6) and are an important prediction, unique to the SBT.

Prior to Goodnight's theoretical research, I conducted a study, using a design similar to that of Drury and Wade (2011), but with ten different population backgrounds, all descended by random genetic drift for 13 generations from the same laboratory stock population (Wade 1985, 2000). The trait measured was the number of adult offspring produced in a generation, i.e. total fitness. The goal was to determine whether or not the among-deme variance in sire half-sib family mean fitness was quantitatively proportional to the average within-deme V_A for fitness in the manner predicted by the additive model. I also estimated the correlation of sire breeding values across the different genetic backgrounds (Wade 1985) and illustrated the correlation using plots similar to G × E plots. I found that the among-deme variance in mean total fitness was 54 times larger than expected from the additive model using the empirical estimates of F_{ST} and the average additive variance in fitness for this population. In addition, the average correlation of sire breeding values across backgrounds was only +0.23, consistent with a significant role for epistasis but not consistent with the expectation of 1.0 from the purely additive model. These observations are wholly consistent with the prediction drawn from Goodnight's research.

(5) *Owing to epistasis, the order of fixation of alleles at different loci will matter to adaptive evolution.* In the literature on experimental evolution using bacteria populations there is growing evidence that the order of mutations matters in adaptation. These systems have the tremendous advantages (Colegrave and Buckling 2005) that: (1) samples of populations can be collected and stored frozen throughout the adaptive process; (2) whole-genome sequencing of each such temporal collection can be used to identify the genes responsible for the adaptive advance; and, (3) each novel mutations identified in (2) can be transformed into the base population, alone or in combinations, to determine the effect not only of each genetic change but also the order of such changes. Several experimental studies of adaptation in replicates selection lines have established that even very large populations do evolve to different fitness peaks, the genetic bases of the underlying adaptations can be very different even between populations with the same mean fitness, and that the order of the mutational changes determines adaptive progress in a highly contingent manner (reviewed in Colegrave and Buckling (2005) and in Elena and Lenski (2003)). Each of these features is consonant with the genetic architecture on which the SBT is premised, even though none of the studies addresses the most controversial part of Wright's process, Phase III. This is why, even if Wright's SBT was wrong, his Adaptive Landscape would be useful for discussing this type of genetic complexity underlying adaptation.

5.6 Direct empirical tests of Wright's shifting balance theory

There have been two direct empirical tests of Wright's SBT, both using experimental populations of flour beetles (Wade and Goodnight 1991; Schamber and Muir 2001). The experiment by Wade and Goodnight (1991) directly modeled Wright's SBT, with Phase III interdemic selection imposed on mean absolute fitness (adult offspring numbers) in replicate 50-deme metapopulations (20 beetles per deme or 1,000 beetles total per metapopulation) at three different rates: every generation, every two generations, and every third generation. Each experimental metapopulation had a paired "control" metapopulation with a level of random (island model) migration among demes exactly equal to the level of differential migration imposed in its experimental companion. The strength of this design is that it directly examined the effect of Wright's Phase III process of differential dispersion out from demes of high mean fitness and into demes of lower mean fitness in contrast to random dispersal of migrants among demes. It has been asserted in theoretical studies that, random migration alone should be sufficient to export advantageous gene combinations out from the demes in which they arise by chance and into other demes, wherein the subsequent work of adaptive transformation of the deme is carried out by local natural selection.

The study by Schamber and Muir (2001) was significantly different in that it contrasted the response to selection on fitness (offspring numbers) in subdivided and non-subdivided populations, controlling exactly for the intensity of selection. Thus, where Wade and Goodnight (1991) contrasted metapopulation evolution with and without Phase III of the SBT, Schamber and Muir (2001) directly examined the efficiency of response to the same intensity of selection in subdivided and non-subdivided populations.

In all three experimental Phase III metapopulations, Wade and Goodnight (1991) found a large and significant increase in mean fitness relative to the paired control metapopulations. The "realized heritability" at the demic level associated with the Phase III process was also high. One of their most striking findings was that imposing Phase III only once every three generations produced a response equivalent to imposing it every generation. Furthermore, the absolute magnitude of the response (and its realized heritability) was greatest when Phase III was imposed every other generation. In strictly additive systems reducing the frequency and strength of artificial selection always reduces the short-term response to selection. Thus, the observed Phase III responses lie outside the predictions of additive theory. Clearly, this need not be the case in non-additive systems, like the experimental *Tribolium* metapopulation model investigated by Wade and Goodnight (1991).

Holding the intensities of selection constant, Schamber and Muir (2001) investigated the response to selection for fitness in smaller metapopulations, where there were ten demes (lines), each consisting of 15 pairs of beetles (150 beetles total per metapopulation). At generations four, eight, and 12, they eliminated the poorer performing demes and merged the better performing demes, a manipulation of differential extinction and colonization, like the experiments of Wade and McCauley discussed earlier. Overall, Schamber and Muir (2001) found no differences in mean fitness between subdivided and non-subdivided populations. Nevertheless, at *every* generation, they found demes within the metapopulation with a mean fitness significantly higher than those observed in any of the panmictic controls. They concluded that, if these high performing lines were preserved, rather than merged with the other high performing lines, that breeders could access the genes underlying the increased production.

5.7 Conclusion

In summary, our understanding of Wright's SBT has deepened and its empirical predictions have been enriched by the theoretical, field, and laboratory experimental research discussed in this chapter. Whereas models with strictly additive genetic effects provide the basis for much of evolutionary genetic theory, studies of the molecular genetic basis of adaptations find that gene interaction is the norm. I agree with the remarks in Chapter 6, that the impact of Wright's SBT will not be fully understood until we have a better understanding of how physiological interactions affect patterns of phenotypic variation. And, until we understand the theory, it will not be evident whether or not the Adaptive Landscape is an adequate representation of the adaptive process that Wright envisioned.

References

Barton, N. H. (1992). On the spread of new gene combinations in the third phase of Wright's shifting balance. Evolution, 46, 551–557.

Briggs, W. H., and Goldman, I. L. (2006). Genetic variation and selection response in model breeding populations of *Brassica rapa* following a diversity bottleneck. Genetics, 172, 457–465.

Bryant, E. H and Meffert, L. M. (1995). Analysis of selectional response in relation to a population bottleneck. Evolution, 49, 626–634.

Bryant, E. H and Meffert, L. M. (1996). Nonadditive genetic structuring of morphometric variation in relation to a population bottleneck. Heredity, 77, 168–176.

Bryant, E. H., McCommas, S. A., and Combs, L. M. (1986). The effect of an experimental bottleneck upon quantitative genetic variation in the housefly. Genetics, 114, 1191–1211.

Carson, H. L. (1990). Increased genetic variance after a population bottleneck. Trends in Ecology & Evolution, 5, 228–230.

Cheverud, J. M. (2000). Detecting epistasis among quantitative trait loci. In J. B. Wolf, E. D. Brodie, and M. J. Wade (eds.) Epistasis and the evolutionary process. Oxford University Press, New York, pp. 58–91.

Cheverud, J. M. and Routman, E. (1996). Epistasis as a source of increased additive genetic variance at population bottlenecks. Evolution, 50, 1042–1051.

Cheverud, J. M., Vaughn, T. T., Pletscher, L. S., King-Ellison, K. Bailiff, J., Adams, E., et al. (1999). Epistasis and the evolution of additive genetic variance in populations that pass through a bottleneck. Evolution, 53, 1009–1018.

Colegrave, N. and Buckling, A. (2005). Microbial experiments on adaptive landscapes. Bioessays, 27, 1167–1173.

Coyne, J. A., Barton, N. H., and Turelli, M. (1997). Perspective: A critique of Sewall Wright's shifting balance theory of evolution. Evolution, 57, 643–671.

Coyne, J. A., Barton, N. H., and Turelli, M. (2000). Is Wright's shifting balance process important in evolution? Evolution, 54, 306–317.

Crow, J. F. (1991). Was Wright right? Science, 30, 973.

Crow, J. F. (2010). On epistasis: why it is unimportant in polygenic directional selection. Philosophical Transactions of the Royal Society B: Biological Sciences, 365, 1241–1244.

Damon, R. A., Crown, R. M., Mccraine, S. E., Singletary, C. B., and Harvey, W. R. (1961). Genetic analysis of crossbreeding beef cattle. Journal of Animal Science, 20, 849–857.

Dobzhansky, T. (1937). Genetics and the origin of species, 1st edn. Columbia University Press, New York.

Drury, D. W. and Wade, M. J. (2011). Genetic variation and co-variation for fitness between intra-population and inter-population backgrounds in the red flour beetle, *Tribolium castaneum*. Journal of Evolutionary Biology, 24, 168–176.

Elena, S. F. and Lenski, R. E. (2003). Microbial genetics: Evolution experiments with microorganisms: the dynamics and genetic bases of adaptation. Nature Reviews Genetics, 4, 457–469.

Enfield, F. D. (1977). Selection experiments in *Tribolium* designed to look at gene action issues. In E. Pollak, O. Kempthorne, and T.B. Bailey Jr. (eds.) Proceedings of an International Conference on Quantitative Genetics. Iowa State University Press, Ames, IO, pp. 177–190.

Ewens, W. C. (2000). The mathematical foundations of population genetics. In R. Singh and C. Krimbas (eds.) Evolutionary genetics, Cambridge University Press, New York, pp. 24–40.

Fisher, R. A. (1958). The Genetical Theory of Natural Selection, 2nd edn. Dover Press, New York.

Garcia, N., Lopez-Fanjul, C., and Garcia-Dorado, A. (1994). The genetics of viability in *Drosophila melanogaster*: effects of inbreeding and artificial selection. Evolution, 48, 1277–1285.

Goodnight, C. J. (1985). The influence of environmental variation on group and individual selection in a cress. Evolution, 39, 545–558.

Goodnight, C. J. (1987). On the effect of founder events on the epistatic genetic variance. Evolution, 41, 80–91.

Goodnight, C. J. (1988). Epistatic genetic variance and the effect of founder events on the additive genetic variance. Evolution, 42, 441–454.

Goodnight, C. J. (1995). Epistasis and the increase in additive genetic variance: Implications for phase 1 of Wright's shifting balance process. Evolution, 49, 502–511.

Goodnight, C. J. (2000). Quantitative trait loci and gene interaction: the quantitative genetics of metapopulations. Heredity, 84, 587–598.

Goodnight, C. J. (2004). Genetics and evolution in structured populations. In C. Fox and J. Wolf (eds.) Evolutionary Genetics: Concepts and case studies. Oxford University Press, New York, pp. 80–102.

Haldane, J. B. S. (1959). The theory of natural selection today. Nature, 183, 710–713.

Hill, W. G. and Caballero, A. (1992). *Artificial selection experiments*. Annual Review of Ecology and Systematics, 23, 287–310.

Hill, W. G. and Kirkpatrick, M. (2010). What animal breeding has taught us about evolution. Annual Review of Ecology and Systematics, 41, 1–19

Hill, W. G., Goddard, M. E., and Visscher, P. M. (2008). Data and theory point to mainly additive genetic variance for complex traits. PLOS Genetics, 4, e1000008.

Holland, J. H. (1975). *Adaptation in Natural and Artificial Systems*. University of Michigan Press, Ann Arbor, MI.

Ibanez-Escriche, N., Fernando, R. L., Toosi, A., and Dekkers, J. C. M. (2009). Genomic selection of purebreds for crossbred performance. Genetics Selection Evolution, 41, 12–22.

Johnson, N. (2008). Sewall Wright and the development of shifting balance theory. Nature Education, 1(1).

Katz, A. J. and Enfield, F. D. (1977. Response to selection for increased pupa weight in *Tribolium castaneum* as related to population structure. Genetical Research, 30, 237–246.

Loewe, L. and Hill, W. G. (2010). The population genetics of mutations: good, bad and indifferent. Philosophical Transactions of the Royal Society B: Biological Sciences, 365, 1153–1167.

Lopez-Fanjul, C. (1989). Tests of theory by selection experiments. In W. G. Hill and T. F. C. Mackay (eds.) Evolution and animal breeding. CAB International, Wallingford, pp. 129–133.

Lopez-Fanjul, C., Fernandez, A., and Toro, M. A. (1999). The role of epistasis in the increase in the additive genetic variance after population bottlenecks. Genetical Research, 73, 45–59.

Lopez-Fanjul, C. and Villaverde, A, (1989). Inbreeding increases genetic variance for viability in *Drosophila melanogaster*. Evolution, 43, 1800–1804.

Mayr, E. (1942). Systematics and the origin of species. Columbia University Press, New York.

Mayr, E. (1954). Change of genetic environment and evolution. In J. Huxley, A. C. Hardy, and E. B. Ford (eds.) Evolution as a process. Allen and Unwin, London, pp. 157–180.

Moorad, J. A. and T. A. Linksvayer, T. A. (2008). Levels of selection on threshold characters. Genetics, 179, 899–905.

Neiman, M. and Linksvayer, T. A. (2006). The conversion of variance and the evolutionary potential of restricted recombination. Heredity, 96, 111–121.

Provine, W. (1986). *Sewall Wright and Evolutionary Biology*. University of Chicago Press, Chicago, IL.

Rendell, L., Boyd, R. Cownden, D. Enquist, M. Eriksson, K., Feldman, M. W., et al. (2010). Why copy others? Insights from the social learning strategies tournament. Science, 328, 208–213.

Schamber E. M. and Muir, W. M. (2001). Wright's shifting balance theory of evolution in artificial breeding programmes: empirical testing using the model organism, *Tribolium castaneum*. Journal of Animal Breeding and Genetics, 118, 181–188.

Shuster, S. M. and Stevens, L. (1996). Inbreeding: its effect on response to selection for pupal weight and the heritable variance in fitness in the flour beetle, *Tribolium castaneum*. Evolution, 50, 723–733.

Simpson, G. G. (1944). Tempo and Mode in Evolution. Columbia University Press, New York.

Slatkin, M. (1981). Populational heritability. Evolution, 35, 859–871.

Van Buskirk, J. and Willi, Y. (2006). The change in quantitative genetic variation with inbreeding Evolution, 60, 2428–2434.

Wade, M. J. (1978). A critical review of the models of group selection. Quarterly Review of Biology, 53, 101–114.

Wade, M. J. (1985). The effects of genotypic interactions on evolution in structured populations. In Genetics: New Frontiers, Proceedings of the XV International Congress on Genetics. Oxford University Press and IBH, UK, pp. 283–290.

Wade, M. J. (1992). Sewall Wright: Gene Interaction in the Shifting Balance Theory. In J. Antonovics and D. Futuyma (eds.) Oxford Surveys of Evolutionary Biology VI. Oxford University Press, New York, pp. 35–62.

Wade, M. J. (2000). Epistasis: genetic constraint within populations and accelerant of divergence among them. In J. Wolf, E. Brodie III, M. J. Wade (eds.) Epistasis and the evolutionary process. Oxford University Press, Oxford, pp. 213–231.

Wade, M. J. (2002). A gene's eye view of epistasis, selection, and speciation. Journal of Evolutionary Biology, 15, 337–346.

Wade, M. J., and Goodnight, C. J. (1991). Wright's shifting balance theory: An experimental study. Science, 253, 1015–1018.

Wade, M. J., and Goodnight, C. J. (1998). Perspective: The theories of Fisher and Wright in the context of metapopulations: When nature does many small experiments. Evolution, 52, 1537–1553.

Wade, M. J., and Goodnight, C. J. (2006). Cytonuclear epistasis: Two-locus random genetic drift in hermaphroditic and dioecious species. Evolution, 60, 643–659.

Wade, M. J., and McCauley, D. E. (1980). Group selection: the genetic and demographic basis for the phenotypic differentiation of small populations of *Tribolium castaneum*. Evolution, 34, 813–821.

Wade, M. J., and McCauley, D. E. (1984). Group selection: the interaction of local deme size and migration on the differentiation of small populations. Evolution, 38, 1047–1058.

Wade, M. J., Bijima, P., Ellen, E. D., and Muir, W. (2010). Group selection and social evolution in domesticated animals. Evolutionary Applications, 3, 453–465.

Whitlock, M. C., Phillips, P. C., and Wade, M. J. (1993). Gene interaction affects the additive genetic variance in subdivided populations with migration and extinction. Evolution, 72, 1758–1769.

Williams, G. C. (1966). Adaptation and natural selection. Princeton University Press, Princeton, NJ.

Willis, J. H. and Orr, H. A. (1993). Increased heritable variation following population bottlenecks: The role of dominance. Evolution, 47, 949–957.

Wright, S. (1931). Evolution in Mendelian populations. Genetics, 16, 97–159.

Wright, S. (1932). The roles of mutation, inbreeding, crossbreeding and selection in evolution. Proceedings of the Sixth International Congress on Genetics, 1, 356–366.

Wright, S. (1968). Evolution and the Genetics of Populations. Vol. I: Genetic and Biometric Foundations. University of Chicago Press, Chicago, IL.

Wright, S. (1969). Evolution and the Genetics of Populations. Vol. II: The Theory of Gene Frequencies. University of Chicago Press, Chicago, IL.

Wright, S. (1977). Evolution and the Genetics of Populations. Vol. III: Experimental Results and Evolutionary Deductions. University of Chicago Press, Chicago, IL.

Wright, S. (1978). Evolution and the Genetics of Populations. Vol. IV: Variability Within and Among Natural Populations: Variability Within and Among Natural Populations. University of Chicago Press, Chicago, IL.

CHAPTER 6

Wright's Shifting Balance Theory and Factors Affecting the Probability of Peak Shifts

Charles J. Goodnight

6.1 Introduction

Perhaps the most controversial feature of Sewall Wright's body of literature is his "shifting balance theory" (SBT) of evolution metaphor. It should not be surprising that the SBT was not developed beyond the level of a metaphor. It is intrinsically a "complex systems" model, which is a discipline that has only recently developed (Bar-Yam 1997). Prior to the advent of modern computers most models, including models of evolution, were necessarily purely analytic. The general approach was to use the "mean field" approximation, that is to assume that the specific environment an individual experiences is equal to the average environment that all individuals experience. Examples of the mean field assumptions in population genetics include assumptions of random mating, the assumption that populations are unstructured and that interactions are random, and for some applications, Fisher's concept of average effects (but see Frank Chapter 3). The mean field approximation has led to stunning advances in a large number of fields, and indeed, the field of classical quantitative genetics is primarily developed using the mean field approximation. However, with the advent of modern computers the study of complex systems has developed as it became possible to explore models that were not analytically tractable. The field of complexity theory holds promise for developing the SBT from the attractive metaphor that Wright first described; however, before this can be done the details of the model need to be developed adequately for formal modeling. It is apparent that when Wright developed his SBT and his concept of adaptive topographies he developed it more to illustrate his ideas about the complexity of evolution rather than to develop a formal model. I will discuss some of the controversies surrounding this theory, and some of the ways that it can be formalized.

Chief among the issues that need to be addressed is what is meant by the "Adaptive Landscape." As discussed by Provine (2001), there has been considerable controversy over to what the axes of the Adaptive Landscape refer. Indeed, this controversy apparently existed in Wright's own interpretation since he originally considered the adaptive topography to refer to characteristics of individuals, and later appears to consider the adaptive topography to refer to populations. Wright's original metaphorical figure (Wright 1931) described a system with two "horizontal" axes representing aspects of an individual's or population's genotype or phenotype, and one "vertical" axis representing fitness. It is frequently convenient to follow Wright and use the three-dimensional metaphor, but it is important to recognize that the adaptive topography necessarily has high dimensionality and there are a large number of orthogonal "horizontal" axes.

From Wright's description it is not clear whether it is individuals or populations that are plotted on the adaptive topography, and indeed both are seen in the literature (Provine 2001). In this paper I will assume that the axes represent aspects of the individual organism, or more correctly, the lowest level at which selection is acting. In this view each organism represents a point on the fitness landscape, and a population of interbreeding individuals, because they share a common gene pool and shared interactions, would be a cloud of points that are clustered in the same area of the Adaptive Landscape.

More problematical is the question of what the axes represent. Again, there is considerable confusion, and Provine (2001) argues that Wright himself was never clear on this. It is clear that Wright imagined that the horizontal axis represented some measure of the effect of the genotype on phenotype; however, his representation of the adaptive topography is not consistent with the discrete nature of allele frequencies within an individual, that is, an individual can only have zero, one, or two copies of any given allele. Although the true axes take on the discrete values that are a necessary feature of particulate inheritance, this may not be apparent when only a portion of the many axes are being examined. If only a subset of the axes are under consideration, this is the equivalent of a projection of the very highly dimensional adaptive topography on to the subset being examined. If there are interactions among loci then the axes that have been collapsed in the projection will shift the values of the genotypes being examined. Since this will potentially be a unique shift for each individual this will have the effect of making the examined axes essentially continuous in nature. Thus, it is possible to both assume that the horizontal axes represent the effects of a locus on an individual's phenotype, and that the axes are continuous because they are the results of three-dimensional projections of landscapes of very high dimensionality.

6.2 A modified view of the Adaptive Landscape

Several important developments in evolutionary thinking are not readily incorporated into Wright's original conception of Adaptive Landscapes. Although there are undoubtedly many developments deserving attention, I will discuss only one: social interactions and its potential corollary, multilevel selection. By social interactions I mean any manner in which other members of the population or multispecies community impinge on the fitness of the individual.

As mentioned in the previous section, the "true" Adaptive Landscape is an extraordinarily highly multidimensional concept. For convenience we inevitably deal with only a subset of this landscape, or effectively a projection of the true landscape onto the smaller subset we are interested in studying. As already pointed out, this has the effect of rendering the otherwise discrete allele differences at a single locus within an individual into a continuous variable. It also has the effect that we can project the landscape along any dimensions we choose, thus, the more recent uses of Adaptive Landscapes in which the "horizontal" axis are phenotypic trait values can be seen as projections of the underlying genetic landscape onto phenotypic axes rather than on the genetic locus axes that are more consistent with Wright's original metaphor.

Social interactions can be incorporated by first expanding the dimensionality of the landscape to include environmental components. This expansion could include both physical and social aspects of the environment. There is no conceptual difficulty with incorporating environmental axes into the adaptive topography; however, since physical aspects of the environment do not contribute to heritability in nearly all instances it will be appropriate to do a projection of the landscape that removes these dimensions. The one exception is when the "environment" is other members of the population or community. In structured populations the social environment can contribute to heritable variation, and thus, it may be appropriate to include these axes in the analysis.

Contextual analysis provides a particularly useful framework in this regard. Contextual analysis is a statistical methodology for studying multilevel selection in the field (Boyd and Iversen 1979; Heisler and Damuth 1987; Goodnight et al. 1992; Okasha 2006). The basic approach follows the Arnold and Wade methodology for studying selection in natural populations (Arnold and Wade 1984). That is, a "fitness trait" such as reproductive success or survival is identified, and then a multiple regression of phenotypic traits of individuals on their relative fitness is performed. A significant slope of the partial regression of a phenotypic trait on relative fitness indicates that selection is acting on that trait. In contextual analysis this basic approach is extended to include "contextual" or group level traits in the regression. Contextual traits are traits measured on the population as a whole, such as population size or density, or summary traits, such as group mean phenotype. These contextual traits are included in the multiple regression as if they were traits of the individual. Thus, in a classic metapopulation model all of the individuals in the same subpopulation would have the same value for a contextual trait such as population size (Heisler and Damuth 1987). As an example, Goodnight et al. (1992) theoretically examined a multiple regression in which the value of a trait in each individual, and the group mean value of that trait for each subpopulation were included in the analysis. The details of this method are not necessary for this discussion, but there are two conceptual changes that this approach uses that are very useful. First, in contextual analysis fitnesses are only assigned to the individual. Group selection occurs when the fitness of an individual is changed by group membership. For example, Eldakar et al. (2010) used contextual analysis to study mating behavior in water striders, and found that the mating success of males decreased as the average level of aggression in a group increased, independent of the aggressiveness of the focal individual. Second, contextual traits are effectively being treated as traits of the individual. Several studies have demonstrated that population density influences seed output in plants (Stevens et al. 1995; Aspi et al. 2003; Weinig et al. 2007). In these studies "population density" is a contextual trait that, while it cannot be measured on the individual, is nevertheless used in the analysis in the same manner as individual level traits.

Taken together this allows social interactions and multilevel selection to be relatively easily incorporated into the conceptual Adaptive Landscape. We need only to imagine that some of the axes in the multidimensional "horizontal" landscapes are contextual traits, that is, traits of the population or community that individuals experience. Examples might include the population gene frequency, which might be important in frequency-dependent selection, or population density, which might be important in density-dependent selection. Indirect genetic effects, such as maternal effects, may also be important and could be included as axes in the adaptive topography. In this case we can imagine that the one or several of the horizontal axes include measures of the parental genotypes, or the genotypes of interacting partners. Of course, even without adding these additional axes the Adaptive Landscape is a conceptual construct of impossibly high dimension. In most cases these axes would be ignored, in effect we would do a projection that collapsed those dimensions. The important point, however, is that should such interactions be of interest, there is no conceptual problem with including these interactions in an adaptive topography.

From a perspective informed by contextual analysis multilevel fitness effects are also easily incorporated into the adaptive topography framework. Because fitness is assigned only to individuals, there is no difficulty in incorporating group effects that impact an individual's fitness. Multilevel fitness effects would be occurring if there was significant variation in fitness along one of the contextual axes. The fitness landscape is simply a description of the relationship between the horizontal axes (genotype etc.) and absolute fitness. Thus, no statements can be made about whether multilevel selection is occurring. Instead, it could be used to describe whether there were fitness differences that could lead to evolutionary change if multilevel selection were occurring.

6.3 Wright's shifting balance theory

In examining problems associated with peak shifts it is worth revisiting Wright's conception of the SBT. Wright's SBT grew out of his study of the genetics of coat color, and the development of a set of seven generalizations (Wright 1968), the three most important of which are: (1) traits are polygenic; all traits, even apparently single locus Mendelian traits, are influenced by a large number of loci, (2) there is "universal pleiotropy"; all genes affect multiple traits, and (3) there is "universal epistasis"; all genes interact with a large number of other genes, and as a result the effect of a gene on the phenotype will change as the genetic background changes. From these observations he concluded that an adaptive topography would have multiple fitness peaks, and developed his SBT as a theory of how a population could move from one fitness peak to a second peak.

Wright believed that peak transitions were most likely to occur in what are today called metapopulations (Levins 1969; Hanski and Gaggiotti 2004), that is, a set of relatively small subpopulations that are tied together by a low level of migration. Wright envisioned a three-phase process by which peak shifts could occur; Phase I, the phase of genetic drift; Phase II, the phase of mass selection; and Phase III, the phase of interdemic selection (Fig. 6.1). These phases are described in detail by Wright (1968).

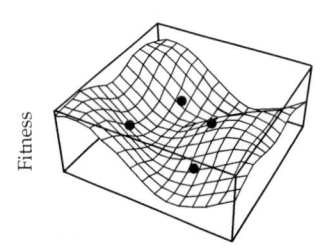

In phase one (Fig. 6.1a), the phase of genetic drift, the individual subpopulations explore the adaptive topography through a process of random drift. The thought is that the subpopulations are small enough that individual selection will be relatively ineffective, and because migration is low, the subpopulations will randomly move about the Adaptive Landscape. As this exploration of the Adaptive Landscape due to random genetic drift occurs, the individual subpopulations will come under the selective influence of different adaptive peaks. The result of Phase I is that the subpopulations will randomly distribute across the landscape without regard to fitness. Wright thought that this was the only way that a population could pass through an "adaptive valley." In Phase II (Fig. 6.1b), the phase of mass selection, individual selection drives the population up the nearest adaptive peak. In this phase each subpopulation climbs the nearest peak, resulting in a metapopulation consisting of subpopulations clustered around two or more peaks. In Phase III (Fig. 6.1c), the phase of interdemic selection the populations on the highest fitness peaks produce more migrants, and thus export their "adaptive gene complexes" to other populations on lower peaks. Wright envisioned these as distinct phases, but nevertheless thought that they would all be going on to some extent at all times and that the balance of fitness would be constantly shifting among peaks as new higher peaks were discovered. Thus, he appears to have thought of the shifting balance process as a fluid constantly changing structure.

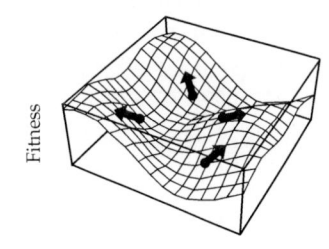

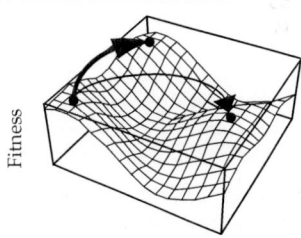

Figure 6.1 The three phases of Wright's shifting balance theory. (a) Phase I, the phase of genetic drift. During this phase demes (represented by dots) within a metapopulation explore the Adaptive Landscape at random with respect to fitness. (b) Phase II, the phase of mass selection. During this phase demes are driven up hill to the nearest adaptive peak by mass (individual) selection. (c) Phase III, the phase of interdemic selection. During this phase the demes on the highest peaks are net exporters of migrants, whereas those on lower peaks are net importers of migrants. This differential migration will tend to cause the all of the demes to move towards the peak of highest fitness.

6.4 Controversies surrounding Wright's shifting balance theory

Wright's vision of the shifting balance process has been controversial almost since its inception. This controversy resurfaced, and was well argued by a series of papers (Coyne et al. 1997, 2000; Wade and Goodnight 1998; Goodnight and Wade 2000), and only a summary will be offered here. Few argue that

Wright's SBT cannot occur, however many argue that it is rare. For example, in their review of the SBT Coyne et al. (1997) conclude: "Given the multifarious nature of evolution, almost every conceivable scenario must occur. Nevertheless, we have found no compelling evidence that Wright's SBT accounts for the evolution of a single adaptation, much less a significant proportion of adaptations in nature." Concerns raised about the shifting balance theory either revolve around concern that the conditions for this process to work are so restrictive that it will be very rare in nature, or the concern that the metaphor is not an appropriate representation of nature. There are concerns with each of the phases considered independently, some of which are exacerbated when it is recognized that the three phases must all operate simultaneously for the SBT to occur.

The first phase, the phase of genetic drift, is not particularly controversial in the sense that the theory of random change in gene frequencies is well worked out (e.g. Wright 1931; Hedrick 2005). However, it has been pointed out (e.g. Barton and Charlesworth 1984) that even very small bottlenecks will not cause major changes in inbreeding coefficients. Furthermore, Barton and Charlesworth (1984), in discussing founder event speciation, point out that bottlenecks reduce genetic variance, and thus are unlikely to cause a genetic revolution. Finally, using a conservation perspective, Lande (1988) has argued that in small populations extinction due to demographic stochasticity is of greater concern than inbreeding and genetic drift. Taken together these ideas can be used to argue that, while there certainly are examples of evolution due to genetic drift, in most circumstances genetic drift is unlikely to be a major evolutionary force (Coyne et al. 1997).

This line of reasoning is based on the assumption of additive genetic effects, there is no gene interaction. This assumption of additivity traces to Fisher (1918, 1930), who correctly reasoned that in a very large population with random interactions gene interaction would not contribute in a lasting response to selection, and thus could be ignored by including it in the residual or environmental variance (see Frank (Chapter 3) for a more nuanced discussion of this issue). This is an example of the "mean field approximation" (Bar-Yam 1997) in which for analytical tractability it is assumed that the actual interactions experienced by an individual are equal to the average interactions among all individuals. However, it has become clear that in structured populations the mean field approximation does not hold, and gene interactions cannot be ignored. Several authors have shown that when there are gene interactions in the form of dominance (Willis and Orr 1993) epistasis (Goodnight 1987, 1988, 2000a; Cheverud and Routman 1995, 1996), or cytonuclear interactions (Wade and Goodnight 2006) that the additive genetic variance can increase as a result of genetic drift (Wade Chapter 5).

Goodnight (1995, 2000a) examined the cause of this increase in additive genetic variance and found that it was due to a shifting of the local average effects of alleles. The local average effect of an allele is the effect that an allele in a particular deme has on the individual phenotype taken as a deviation from the metapopulation mean. Once corrected for the deme mean effect, there is residual variation in the local average effect that is due to the allele by deme interaction. The effect of this is that an allele that confers high fitness in one deme may be a low fitness allele in a different deme with a different genetic makeup. As a result, when there is gene interaction genetic drift does not cause a simple reduction in genetic diversity. Instead, it can cause shifts the local average effects of alleles can result in a shift in which alleles are favored by selection. In effect, these genetic shifts may underlie the process of a population coming under the domain of influence of a new peak in the SBT metaphor (Goodnight 2000a; Wade Chapter 5).

The second phase, the phase of mass selection, has never been controversial, since Wright envisioned this to be a phase in which individual selection caused a population to "climb" the nearest peak. This process is well understood both for single loci and for quantitative genetic variation (Falconer and Mackay 1996). One point that is worth elaborating on, however, is that in most models of selection only additive effects are considered, again, an outgrowth of Fisher's (1930) infinitesimal model. In these models uniform directional selection is a homogenizing force that drives a population towards the global optimal phenotype and, in many models, the single optimum genotype. Again, when there is gene interaction the model is not as simple. In a simulation study Goodnight (2004) showed that directional selection could drive a conversion of dominance and additive-by-additive epistasis into additive genetic in a manner that was very similar to the effects of genetic drift. One interesting result is that rather than depleting additive genetic variance, selection can potentially generate "new" additive genetic variance as adaptation proceeds. As with drift, however, this additive genetic variance is generated by the shifting of average effects of alleles. As a result, when there is gene interaction directional selection can be a force of diversification, even if, at the phenotypic level, it is acting in a uniform manner in all demes (Wade and Goodnight 1998).

Finally, Phase III, the phase of interdeme selection has always been controversial. Originally there was doubt as to whether interdeme selection, or similarly, group selection, could be effective (e.g. Williams 1966), however, it has long been recognized that group selection is indeed highly effective, and that the early concerns were based on faulty assumptions incorporated into the early models (e.g. Wade 1978). There has also been concern that group selection by differential migration would not be sufficient to cause major evolutionary change. However, this concern has been addressed both experimentally (Wade and Goodnight 1991), and theoretically (Crow et al. 1990) and there is little doubt that group selection by differential migration is a powerful enough force to cause evolutionary change. Indeed, a model developed by Barton (Barton 1992; Phillips 1993) can be interpreted to indicate that under many circumstances group selection by differential migration is powerful enough to override individual selection, and can lead to the invasion and fixation of gene combinations regardless of their fitness relative to the deme into which they are migrating.

As can be seen there is some controversy surrounding each of the individual phases, however, the larger controversy is whether the conditions required for all three phases can simultaneously be met. A major issue in this regard is that Phase I, the phase of genetic drift, requires small population sizes both to generate random changes in gene frequency, and because individual selection is less effective in small populations, however, Phase II, the phase of mass selection, requires large population sizes for precisely the opposite reasons, that is selection is more effective and drift less effective in large populations. Similarly, Phases I and II will be most effective if migration is very limited. However, in Wright's model Phase III depends on differential migration rates. Thus, it is on the basis of issues such as this that Coyne et al. (1997) argued that, while it is plausible that the SBT may occur, it is unlikely to be a common mode of evolution. In contrast, Wade and Goodnight (1998) argued, among other points, that experiments and models incorporating non-additive gene action suggest that the conditions under which the three phases can coexist are much broader than additive models would suggest. Which of these perspectives is correct will have to await a better understanding of genetic interactions and their influence on phenotypic variation.

From this discussion, it should be clear that an important issue is how important is gene interaction? Nobody denies that what has been called "physiological epistasis" (Cheverud and Routman 1995) is common. Physiological epistasis can be defined as the interaction among loci causing a phenotype that is unique compared to the phenotypic effects of the loci considered separately. Importantly, physiological epistasis concerns only the interactions among genes, and does not consider the effects of gene frequencies. From an evolutionary perspective far more important is the amount of "statistical epistasis" (Cheverud and Routman 1995). The statistical epistasis is the genetic variance component that can be attributed to interactions among genes. Unlike physiological epistasis, statistical epistasis is a function of both gene interaction and gene frequencies. The general consensus is that the amount of epistatic variance in most populations is relatively small (Coyne et al. 1997). The reason for this is that while gene interaction is common, most loci are presumably dominated by one or a few alleles at high frequency, and a larger number of alleles at low frequencies. This is exactly the type of situation that will maximize the conversion of epistasis to additive effects and minimize the amount of epistatic genetic variance

(Goodnight 2000b). Other factors will also reduce the epistatic genetic variance within populations. In general, epistatic genetic variance will be highest when all alleles at the interacting loci are equally frequent. Both genetic drift and selection will, in general, cause gene frequencies to deviate from equal representation of all alleles, and as a result cause a conversion of epistasis to additivity. The net result is that populations that have been under long-term selection, or have had a moderate population size for long periods of time are not expected to have high levels of epistatic genetic variance, and indeed, efforts to measure epistasis within populations have been largely unsuccessful, although it is not clear whether this is due to lack of epistatic variance, or the difficulty in detecting it (Fenster et al. 1997).

The difficulty of measuring the epistatic variance within populations may not be such a great problem, however, since it is the epistatic variance in the metapopulation as a whole that is actually of most importance. Measuring, or even defining, this component of variance is problematic and has not yet been done. It is suggestive however, that quantitative trait locus (QTL) studies in wide crosses often find substantial epistasis (Carlborg and Haley 2004). One possibility for measuring the role of epistasis in a metapopulation is suggested by the models developed by Goodnight (Goodnight 1995, 2000a; Wade Chapter 5). These models use the variance in the local breeding values (Goodnight 1995), or local average effects (Goodnight 2000a) as a measure of population differentiation, and show that differentiation for these measures only occurs when there is gene interaction (epistasis or dominance). The variance in local breeding values or local average effects can be measured using a diallel cross, or a modification of standard breeding designs. Drury and Wade (Drury and Wade 2011; Wade Chapter 5) have measured the variance in local breeding values among *Tribolium* strains, and found that there was a significant variance in local breeding values.

Another issue with the SBT was brought up by Gavrilets (1997, 1999). In his "holey landscapes model" Gavrilets argues that a highly dimensional landscape is not adequately depicted by a three-dimensional plot, or by the intuition of investigators used to visualizing relatively small numbers of dimensions. Instead he argues that in highly multidimensional space different points are typically connected by complex multidimensional ridges of constant fitness. In his words, "viable genotypes form 'clusters' in the genotype space. Members of a cluster can be connected by a chain of viable single-gene substitutions." (Gavrilets 1997) He further argues that steps to lower fitness would be sufficiently opposed by selection that they can be considered "holes" that populations can not enter, and that steps to higher fitness are sufficiently rare that they can be ignored for the purposes of his model (Fig. 6.2). Thus, he argues that an adaptive topography is best visualized as a plane of constant fitness that is perforated by numerous "holes" of reduced fitness. He acknowledges that occasionally a shift to a higher fitness plane can be achieved, and argues that when this is achieved it is the equivalent of shifting to a new adaptive plane with a new set of low fitness holes. The important feature is that all of the points on the adaptive plane are of equal fitness values. Thus, populations will not be affected by selection and will move around the plane by genetic drift alone. An adaptive advance will occur when a population drifts into the vicinity of a step up in the adaptive topography. At this point selection would rapidly move the population to the new adaptive plane, and the process of movement by random genetic drift would resume on this new plane.

This model is related to Wright's SBT in that it invokes an adaptive topography, albeit a topography that is quite different than the one that Wright proposed. It also can be seen that evolution could occur on this topography following a somewhat modified form of the three phases described by Wright. Wright's first phase becomes a phase of exploration of Gavrilets's adaptive plane. Gavrilets's model differs from Wright's model in that because populations are on an adaptive plane there is no necessity for drift to oppose selection as there is in Wright's model. Wright's second phase would presumably be a brief phase in Gavrilets's model that occurs when drift brings a population close to a step up in fitness. Interestingly, the issues associated with population size in Wright's model are much less important in Gavrilets's model

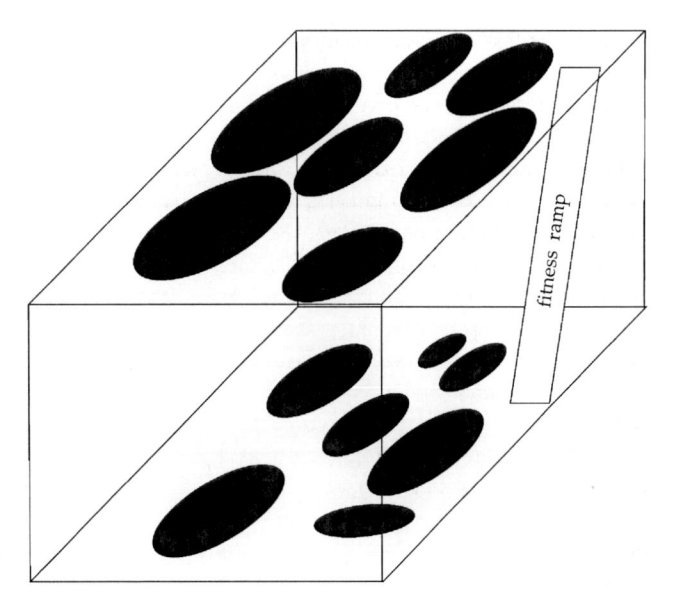

Figure 6.2 Gavrilets's (1997) holey landscape model. Gavrilets argues that in highly multidimensional landscapes different "peaks" will be connected by ridges of neutral fitness. This can be modeled by considering the adaptive surface to be a neutral plane on which demes move about by random drift. Selection will prevent demes from entering areas of low fitness, thus, it is adequate to model them as holes in the adaptive plane. Occasionally there may be the possibility for an adaptive advance (indicated by the ramp in the figure). A deme will drift at random until it encounters the ramp, at which point selection will drive it to a new adaptive plane with a new architecture.

since changes in population size will affect the rate of exploration of the adaptive plane, but since it is a plane of constant fitness the problem of drift opposing selection does not apply. Wright's second phase occurs in Gavrilets's model when a step to high fitness is encountered. Again, the issues in Wright's model are avoided since there is much less conflict between drift and selection in the holey landscape model then in Wright's adaptive topography. Finally, as with Wright's SBT, it is easy to imagine that differential migration could work in the same manner, that is introducing favorable gene combinations into populations, and causing them to migrate to a higher adaptive plane.

One issue with the Gavrilets's model that deserves modeling consideration is that it is reasonable to conjecture that as adaptation proceeds the effective dimensionality of successive adaptive planes will steadily be reduced. Based on this it may be reasonable to conclude that Gavrilets's holey landscape model may be more appropriate for the early stages of adaptation in which the adaptive plane dimensionality will be very high, and less appropriate for later stages of adaptation in which the dimensionality is reduced, and Wright's low dimensionality metaphor may be more appropriate.

6.5 Peak shifts in metapopulations

From the discussion of the controversies surrounding Wright's metaphor it should be clear that the SBT as originally proposed in the 1930s needs modification. First, it is apparent that there is considerable confusion surrounding the exact meaning of the axes in the adaptive topography metaphor. Regardless of the final interpretation that we accept for the axes it is clear that any realistic view of the Adaptive Landscape will be highly multidimensional. Accepting this highly multidimensional view of the Adaptive Landscape, suggests that, at least at the early stages, Gavrilets's holey landscape model may be a more realistic metaphor than Wright's low dimensionality metaphor. If this is the case then this would imply that much of the evolution of novel adaptations does not involve the crossing of adaptive valleys, but instead involves travel along multidimensional ridges. It is important to recognize that we cannot rule out the possibility that the evolution of adaptive novelty will require crossing adaptive valleys, at least in some cases. This will be especially true if, as I conjecture, effective dimensionality decreases as adaptations are refined.

If it is indeed the case that in many situations evolution can proceed along ridges of neutral fitness then the restrictions on a modified version of the SBT become much less stringent. As pointed out earlier, the conflict between the small population size required for drift and the large population size required for selection is resolved, since by definition all points on the adaptive plane are of equal fitness, thus the populations can move around on this plane by drift with population size only affecting the rate of drift. Similarly, because population sizes can be larger, when an adaptive step up is encountered on the adaptive plane, selection can be more effective. Finally, Wright's phase of interdemic selection becomes less controversial, since this becomes differential migration that introduces new genetic variants into a population causing a shift along the adaptive plane towards the adaptive step up discovered by the high fitness deme in the metapopulation.

Regardless of whether or not a population needs to cross an adaptive valley as envisioned in Wright's original formulation of the SBT, it is still necessary to update the how the processes of genetic drift, selection, and interdemic selection interact to cause evolutionary change. Our intuition is based on models that assume only additive effects; however, the SBT is only meaningful if there is gene interaction. As a result a full development of a modified SBT will have to await a better empirical and theoretical understanding of the effects of gene interaction on phenotypic variation. Nevertheless, currently available models give some indication of how our conception of the three phases of the SBT should be modified to incorporate gene interaction.

Phase I, the phase of random drift becomes much more interesting when there is gene interaction. Instead of simply causing a change in gene frequencies and a decline in the genetic variance as predicted by additive theory, genetic drift can cause an increase in the additive genetic variance, but more importantly, it can cause a shifting of the average effects of alleles. This phase of random drift is expected to reduce the amount of epistatic variance in a population within populations, and to have this decreased epistatic variance re-expressed as increased additive variance, variance among demes, and especially variance among demes in allelic effects. Thus, the phase of random drift can be viewed as a period of reshuffling gene associations causing random changes in the effects of alleles on the phenotype, and therefore causing random changes in the evolutionary trajectory that a deme can follow. Phase II, the phase of mass selection, remains a period of refinement of adaptations; however, because of the shifting in local average effects that occur as a result of gene interaction, this phase can be seen as a period of forming adaptive gene complexes. Thus, genes are favored by selection in the context of the genetic background of the deme. Even though selection is acting in a uniform manner it will still lead to differentiation of populations if the different populations start out at different allele frequencies at interacting loci. Phase III, the phase of interdeme selection remains as a process of dissemination of favorable gene complexes through differential migration. To the extent that Gavrilets's holey landscape model holds, this phase becomes less controversial because the interdemic selection is not opposing individual selection so much as pulling other demes across a neutral adaptive plane.

6.6 The importance of peak shifts in evolution

The final question to be addressed, then, is what is the role of peak shifts in evolution. As might be expected, the short answer to this is that at present we don't have the data or models to answer this question. As discussed earlier, one of the critical questions is assessing the contribution of gene interactions to the differentiation among populations. Unfortunately standard methods of measuring variance components within populations do not answer this question, both because of the lack of power in most designs for detecting epistatic genetic variance, and because gene interactions contributing to the differentiation of populations are not segregating within populations. Similarly it is important to recognize that differentiation for average effects of alleles need not be related to the variance in population means. Goodnight (2000a) shows that additive genetic effects contribute to the variance in population means, but not to the variance in local average effects. Conversely, dominance and dominance related epistasis, such as additive by dominance epistasis, contribute to the variance in local average effects, but have little or no effect on the differentiation of population means. Thus, new experimental methods need to be developed to measure the differentiation of local breeding values and local average effects. It is likely that such a method will follow the broad outline followed by Drury and colleagues (Drury and Wade 2011; Goodnight 1995, 2000a).

A second critical question is how will variance components change as a population moves across the Adaptive Landscape. This issue is discussed from a macroevolutionary perspective by Hansen (Chapter 13). One of the features of epistatic systems is that as gene frequencies change, the partitioning between variance components also changes. Looking at a simple additive by additive epistatic two-locus two-allele system (Fig. 6.3) it can be seen that the additive genetic variance changes across this simple landscape. Importantly, the additive genetic variance is lowest at the adaptive peaks and at the saddle separating the two peaks. If this is a general phenomenon it suggests that crossing adaptive valleys may be less difficult than typically modeled. The reason for this is that if a population, for what ever reason, manages to traverse the slope of the adaptive peak where the additive genetic variance is highest, as it gets to the lower slopes of the peak the additive genetic variance will tend to drop off, which will decrease the efficacy of selection relative to drift. Additive genetic variances and covariances can also change in unpredictable manners as the environment changes (Sgró and Hoffman 2004), with the additive genetic variance often increasing in stressful environments (e.g. Sgró and Hoffman 1998). Thus, the balance between Phase I and Phase II may be adjusted as much by changes in variance as it is by changes in population size. Current quantitative genetic models typically assume that additive genetic variance is approximately constant, or obeys the simple rules of additivity. Models that allow the additive genetic variance to change need to be developed if this conjecture is to be adequately explored.

In addition to changes in genetic variance components, changes in phenotypic variance can also contribute to the possibility of peak shifts. Whitlock (1995) developed a model in which he demonstrated that increases in phenotypic variance that

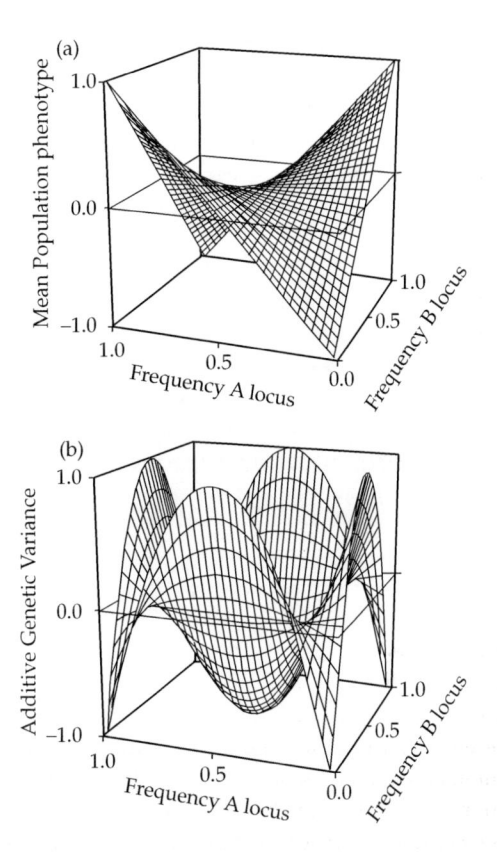

Figure 6.3 The relationship between additive genetic variance and mean population fitness for an example of additive-by-additive epistasis with two alleles. The additive genetic variance is highest when one locus if fixed and the second is segregating, but declines to zero at adaptive peaks, valleys, and saddle points. (a) The mean fitness of a population as a function of gene frequency at the interacting loci. (b) The additive genetic variance as a function of gene frequency at the interacting loci.

are often associated with bottlenecks may dramatically increase the potential for peak shifts to occur. He suggested that an increase in phenotypic variance that results in a decline in heritability will in many circumstances cause fitness valleys to disappear, and allow populations to move deterministically from a lower peak to a higher peak. Several experimental studies have confirmed this increase in phenotypic variance associated with population bottlenecks (Lopez-Fanjul and Villaverde 1989; Whitlock and Fowler 1996; Pray and Goodnight 1997).

Related to the idea of increased phenotypic variance is the idea that environmental variability could facilitate peak shifts (Frank Chapter 3). Wright

(1968) discusses the possibility that a peak shift might occur when a former adaptive peak disappears due to environmental change. However another idea that has not received attention is that relatively rare temporary environmental situations may lead to temporary changes in the phenotypic mean or variation in a population. An example of this occurred when controlled burnings were reintroduced in the Ozark Mountains in the 1990s. Prior to the reintroduction of the controlled burns collard lizards were restricted to open glades in the forest, with no measurably migration among glades (Hutchison and Templeton 1999). After the burns were reintroduced this cleared away the undergrowth, and allowed migration among glades (Templeton et al. 2001). The result was larger population sizes, increased migration rates, and the occupation of glades that were too small to support viable independent populations of collard lizards (Templeton et al. 2001; Brisson et al. 2003). It is quite possible that the evolutionary dynamics for collard lizards could be dramatically affected by rare intermittent burnings. In this case the environmental disturbance could affect the shifting balance process both by introducing intermittent migration followed by periods of isolation, and by temporarily changing the environment and with that change altering the selective pressures and the phenotypic and genetic variance components.

In conclusion then, at present we do not have a good understanding of the probability of a peak shift occurring. The probability of a peak shift is a complex function of the genetics and environment of a species. Changes in demography will affect the partitioning of total genetic variance into within and between population components, and within populations it will change the partitioning into additive and non-additive genetic variance components. Population bottlenecks can change the phenotypic variance, which can change the probabilities of peak shifts, and patterns of environmental variation may cause temporary shifts in genetic and phenotypic variance components and selective pressures that may facilitate peak shifts.

The preceding discussion might imply that peak shifts should be common, and yet this is obviously not the case. Most species retain moderately constant phenotypes through time, and only rarely do we see anything that can be construed as a peak shift. The main reason that peak shifts don't occur more often is probably that most species most of the time are reasonably well adapted to their current niche. Thus, in most circumstances selection is mostly stabilizing around the current peak for a population, and populations and migration are probably large enough to lock them on to a particular peak. With respect to Gavrilets's model, it is likely that we do see drift around a neutral genetic plane at the genetic level. An example of this would be the morphologically homologous structures that often have very different genetic and even developmental origins (Wagner 1989). However, considering only phenotypic evolution, in many cases there will be low enough dimensionality that there truly are peaks and valleys, and Wright's adaptive topography with peaks and valleys will be more appropriate than Gavrilets's holey landscape metaphor. If this is the case then populations may have to cross an adaptive valley to reach a different phenotypic adaptive peak. Although, due to gene interaction and changes in phenotypic variance and the environment, the issues raised for the SBT may not be as serious as originally thought, they nevertheless still exist, and argue that peak shifts should be rare. Finally, a peak shift is a phenomenon that occurs at the metapopulation level, and as Wright (1968) observed, the likelihood of any given population undergoing a peak shift is very low. In all likelihood peak shifts will be rare, and unless we are monitoring an entire metapopulation the chance of observing such a peak shift would be even rarer.

References

Arnold, S. J. and Wade, M. J. (1984). On the measurement of natural and sexual selection: Theory. Evolution, 38, 709–718.

Aspi, J., Jåkålåniemi, A., Tuomi J, Siikamäki P. (2003). Multilevel phenotypic selection on morphological characters in a metapopulation of *Silene tatarica*. Evolution, 57, 509–517.

Barton, N. H. (1992). On the spread of new gene combinations in the third phase of Wright's shifting-balance. Evolution, 46, 551–557.

Barton, N. H. and Charlesworth, B. (1984). Genetic revolutions, founder events and speciation. Annual Review of Ecology and Systematics, 15, 133–164.

Bar-Yam, Y. (1997). Dynamics of Complex Systems. Westview Press, Boulder, CO.

Boyd, L. H. and Iversen, G. R. (1979). Contextual analysis: Concepts and statistical techniques. Wadsworth, Belmont, CA.

Brisson, J. A., Strasburb, J. L., and Templeton, A. R. (2003). Impact of fire management on the ecology of collard lizard (*Crotaphytus collaris*) populations living on the Ozark Plateau. Animal Conservation, 6, 247–254.

Carlborg, Ö. and Haley, C. (2004). Epistasis: too often neglected in complex trait studies? Nature Reviews Genetics, 5, 618–625.

Cheverud, J. M. and Routman, E. J. (1995). Epistasis and its contribution to genetic variance components. Genetics, 139, 1455–1461.

Cheverud, J. M. and Routman, E. J. (1996). Epistasis as a source of increased additive genetic variance at population bottlenecks. Evolution, 50, 1042–1051.

Coyne, J. A., Barton, N. H., and Turelli, M. (1997). Perspective: A critique of Sewall Wright's shifting balance theory of evolution. Evolution, 51, 643–671.

Coyne, J. A., Barton, N. H., and Turelli, M. (2000). Is Wright's shifting balance process important in evolution? Evolution, 54, 306–317.

Crow, J. F., Engels W. R., and Denniston, C. (1990). Phase three of Wright's shifting balance theory. Evolution, 44, 233–247.

Drury, D. W. and Wade, M. J. (2011). Genetic variation and co-variation for fitness between intra-population and inter-population backgrounds in the red flour beetle, *Tribolium castaneum*. Journal of Evolutionary Biology, 24, 168–176.

Eldakar, O. T., Wilson, D. S., Dlugos, M. J., and Pepper, J. W. (2010). The role of multilevel selection in the evolution of sexual conflict in the water strider *Aquarius remigis*. Evolution, 64, 3183–3189.

Falconer, D. S. and Mackay, T. (1996). Introduction to Quantitative Genetics. Longman, Essex.

Fenster, C. B., Galloway, L. F., and Chao, L. (1997). Epistasis and its consequences for the evolution of natural populations. Trends in Ecology & Evolution, 12, 282–286.

Fisher, R. A. (1918). The correlations between relatives on the supposition of Mendelian inheritance. Transactions of the Royal Society of London B, 143, 103–113.

Fisher, R. A. (1930). The Genetical Theory of Natural Selection. Oxford University Press, Oxford.

Gavrilets, S. (1997). Evolution and speciation on holey adaptive landscapes. Trends in Ecology & Evolution, 12, 307–312.

Gavrilets, S. (1999). A dynamical theory of speciation on holey adaptive landscapes. American Naturalist, 154, 1–22.

Goodnight, C. J. (1987). On the effect of founder events on the epistatic genetic variance. Evolution, 41, 80–91.

Goodnight, C. J. (1988). Epistasis and the effect of founder events on the additive genetic variance. Evolution, 42, 441–454.

Goodnight, C. J. (1995). Epistasis and the increase in additive genetic variance: implications for phase 1 of Wright's shifting balance process. Evolution, 49, 502–511.

Goodnight, C. J. (2000a). Quantitative trait loci and gene interaction: the quantitative genetics of metapopulations. Heredity, 84, 587–598.

Goodnight, C. J. (2000b). Modeling gene interaction in structured populations. In J. B. Wolf, E. D. Brodie III, and M. J. Wade (eds.) Epistasis and the evolutionary process. Oxford University Press, New York, pp. 213–231.

Goodnight, C. J. (2004). Gene interaction and selection. Plant Breeding Reviews (J. Janick ed.) 24, 269–290.

Goodnight, C. J., Schwartz, J. M., and Stevens, L. (1992). Contextual analysis of models of group selection, soft selection, hard selection and the evolution of altruism. American Naturalist, 140, 743–761.

Goodnight, C. J. and Wade, M. J. (2000). the ongoing synthesis: a reply to Coyne, Barton and Turelli. Evolution, 54, 317–324.

Hanski, I. and Gaggiotti, O. (eds.) (2004). Ecology, Genetics and Evolution of Metapopulations. Elsevier Academic Press, Burlington MA.

Hedrick, P. W. (2005). Genetics of Populations, 3rd edn. Jones and Bartlett, Publishers, London.

Heisler, L. and Damuth, J. D. (1987). A method for analyzing selection in hierarchically structured populations. American Naturalist, 130, 582–602.

Hutchison, D. W. and Templeton, A. R. (1999). Correlation of pairwise genetic and geographic distance measures: inferring the relative influences of gene flow and drift on the distribution of genetic variability. Evolution, 53, 1898–1914.

Lande, R. (1988). Genetics and demography in biological conservation. Science, 241, 1455–1460.

Levins, R. (1969). Some demographic and genetic consequences of environmental heterogeneity for biological control. Bulletin of the Entomological Society of America, 15, 237–240.

Lopez-Fanjul, C. and Villaverde, A. (1989). Inbreeding increases genetic variance for viability in *Drosphila melanogaster*. Evolution, 43, 1800–1804.

Okasha, S. (2006). Evolution and the levels of selection. Oxford University Press, New York.

Phillips, P. C. (1993). Peak shifts and polymorphism during phase three of Wright's shifting-balance process. Evolution, 47, 1733–1743.

Pray, L. A. and Goodnight, C. J. (1997). The effect of inbreeding on phenotypic variance in the red flour beetle *Tribolium castaneum*. Evolution, 51, 308–313.

Provine, W. B. (2001). The origins of theoretical population genetics. University of Chicago Press, Chicago, IL.

Sgró, C. M. and Hoffman, A. A. (1998). Effects of stress combinations on the expression of additive genetic variation for fecundity in *Drosophila melanogaster*. Genetical Research Cambridge, 72, 13–18.

Sgró, C. M. and Hoffman, A. A. (2004). Genetic correlations, tradeoffs and environmental variation. Heredity, 93, 241–248.

Stevens, L., Goodnight, C. J., and Kalisz, S. (1995). Multilevel selection in natural populations of *Impatiens capensis*. American Naturalist, 145, 513–526.

Templeton, A. R., Robertson, R. J., Brisson, J. A., and Strasburb, J. L. (2001). Disrupting evolutionary processes: The effect of habitat fragmentation on collard lizards in the Missouri Ozarks. Proceedings of the National Academy of Sciences, USA, 98, 5426–5432.

Wade, M. J. (1978). A critical review of the models of group selection. Quarterly Review of Biology, 53, 101–114.

Wade, M. J. and Goodnight, C. J. (1991). Wright's Shifting Balance Theory: An experimental study. Science, 253, 1015–1018.

Wade, M. J. and Goodnight, C. J. (1998). The theories of Fisher and Wright: when nature does many small experiments. Evolution, 54, 1537–1553.

Wade, M. J., and Goodnight, C. J. (2006). Cytonuclear epistasis: two-locus random genetic drift in hermaphroditic and dioecious species. Evolution, 60, 643–659.

Wagner, G. P. (1989). The biological homology concept. Annual Review of Ecology and Systematics, 20, 51–69.

Whitlock, M. C. (1995). Variance-induced peak shifts. Evolution, 49, 252–259.

Whitlock, M. C. and Fowler, K. (1996). The distribution among populations in photypic variance with inbreeding. Evolution, 50, 1919–1926.

Weinig, C., Johnston, J., Willis, C. G., and Maloof, J. N. (2007). Antagonistic multilevel selection on size and architecture in variable density settings. Evolution, 61, 58–67.

Williams, G. C. (1966). Adaptation and Natural Selection. Princeton University Press, Princeton, NJ.

Willis, J. H. and Orr, A. H. (1993). Increased heritable variation following population bottlenecks: The role of dominance. Evolution, 47, 949–956.

Wright, S. (1931). Evolution in Mendelian populations. Genetics, 16, 93–159.

Wright, S. (1968). Evolution and Genetics of Populations. Vol. I. Genetic and Biometric Foundations. University of Chicago Press, Chicago, IL.

PART III

Applications: Microevolutionary Dynamics, Quantitative Genetics, and Population Biology

CHAPTER 7

Fluctuating Selection and Dynamic Adaptive Landscapes

Ryan Calsbeek, Thomas P. Gosden, Shawn R. Kuchta, and Erik I. Svensson

7.1 Introduction

Sewall Wright, in his original version of the Adaptive Landscape concept, focused on a scenario in which he assumed that genotypic fitness was constant through time (Wright 1932).[1] If the past several decades of empirical research have taught us anything about the role of natural selection in the wild, however, it is that selection frequently varies in space and time (Kingsolver et al. 2001; Siepielski et al. 2009), although the evolutionary importance of short-term ecological fluctuations is less clear (Kingsolver and Diamond 2011). Consequently, the original and simplified view of a constant Adaptive Landscape needs to be updated somewhat to account for environmental fluctuations. In this chapter, we discuss the importance of spatial and temporal variability in selection and its evolutionary implications, especially in relation to frequency-dependent natural selection, sexual selection, density dependence, predation, and other forms of biotic interaction.

Perhaps the most dramatic shift in our conceptualization of the Adaptive Landscape since Wright has been a move away from genotype-oriented fitness surfaces, which represent genotype × genotype (fitness epistasis) interactions, towards fitness surfaces of phenotypes and phenotype × phenotype interactions (i.e. interactions between quantitative traits (Arnold et al. 2001)). This conceptual shift from genotypes to phenotypes traces back to the insightful and important contribution of G. G. Simpson (1944), one of the architects of the Evolutionary Synthesis (Mayr and Provine 1980). Simpson was the first to formally introduce phenotype × phenotype interactions as a conceptual link between micro- and macroevolution (Arnold et al. 2001). He envisioned a topographically simple Adaptive Landscape, limited to a few peaks, upon which populations moved about through the processes of natural selection and genetic drift. Though Simpson's phenotypic landscape is sometimes considered highly divergent from Wright's original vision, it is actually closely related to the epistatic landscape at the genotypic level, since there are peaks with valleys in between them. The peaks and valleys in Simpson's phenotypic space represent correlational selection (Brodie 1992), which is a higher-level, trait-based form of epistasis (Wolf and Brodie 1998; Sinervo and Svensson 2002). In other words, epistasis at the genotypic level has an ecological counterpart in correlational selection at the phenotypic level (Sinervo and Svensson 2002). Although the relationship between genotype and phenotype is complex and highly non-linear (see Chapter 3), knowledge about the genetic architecture of traits is not required for the Simpsonian landscape to be useful in clarifying the relationship between phenotype and fitness

---

[1] Note that Wright did actually also consider dynamic landscapes (frame C in his original fig. 4, Wright 1932) and went on to develop models of frequency-dependent selection. Thus he certainly did not deny that there could be situations in which genotypic fitnesses varied between generations though these ecologically more realistic scenarios have received less attention (see Chapter 2).

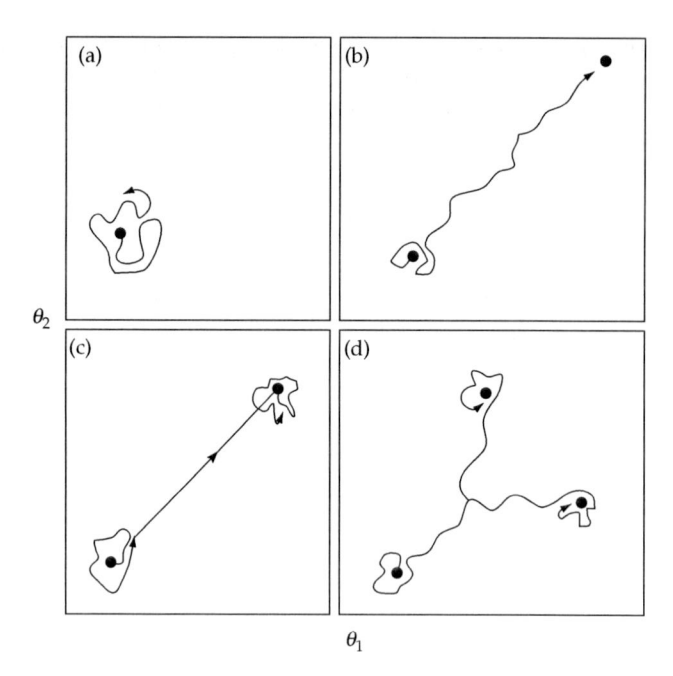

Figure 7.1 Peak movements and evolution on dynamic Adaptive Landscapes. Each panel illustrates the process of peak movement under a range of scenarios, from (a) random drift, (b) a period of stasis followed by directional evolution, (c) a second period of stasis following a peak shift (as in the previous panel) and (d) ecological speciation resulting from divergent selection leading to two optima. Modified and redrawn from Arnold et al. (2001).

(Chapter 13). Naturalists and morphologists have long been aware that traits are not selected in isolation. Rather, they function jointly as integrated phenotypes. Indeed, most of what we know about evolution within the framework of the Adaptive Landscape has come from Simpson's phenotypic version, owing to the severe empirical constraints of studying selection in genotypic space, especially in natural populations. While recent advances in computing and genomics are starting to alleviate that constraint (Pitt and Ferre-D'Amare 2010; McCandlish 2011), the problems are far from resolved. A more detailed discussion of empiricism using true genotype × genotype landscapes is beyond the scope of this chapter (but see McCandlish 2011; Chapter 17).

While the Wrightian and Simpsonian Adaptive Landscape share a related conceptual origin, they operate at different levels of biological organization. In addition, the conceptual shift away from a stationary landscape (Wrightian) to a more dynamic landscape involving peak movement (Simpsonian) has profound implications for the study of character evolution (Fig. 7.1). Indeed, Arnold et al. (2001) viewed this difference as "so profound that it is fair to call it a paradigm shift" (p. 10). Dynamic landscapes represent a twist on Wright's Adaptive Landscape concept, a modification more towards the ecological perspective espoused by Fisher's fundamental theorem of natural selection (Fisher 1958; Frank and Slatkin 1992; see also Chapter 4). Here we critically review current empirical evidence for spatial and temporal variability in selection, and then consider the implications for the study of evolution via Adaptive Landscapes.

7.2 Empirical support for shifting landscapes

Given the fundamental biological and conceptual differences between the Wrightian and Simpsonian landscapes, an important question to ask is: which model is more in line with existing empirical evidence? Though the question is far from fully resolved, increasing empirical evidence suggests that Adaptive Landscapes may be more labile in many systems than was originally thought (Kingsolver et al. 2001; Bell 2010). For instance, a recent review of the literature revealed just how variable selection can be through both space and time (Siepielski et al. 2009). Based on more than 5000 estimates of selection, Siepielski et al. (2009) conclude that reversals in the direction of selection are not only common, but that the greater the selection pressure, the higher the inter-annual standard deviation in selection. These findings suggest that strong selection is rarely chronic. Similar patterns were evident, though somewhat less well-supported, for changes in the modes of selection (e.g. stabilizing selection changing to disruptive selection or vice versa; Siepielski et al. (2009), see also Hoekstra et al. (2001), Kingsolver et al. (2001), and Hereford et al. (2004)). More recently, work by Uyeda and colleagues (2011) have shown that these short-term evolutionary fluctuations are best represented as bounded oscillations due to movements of adaptive peaks within a relatively stable adaptive "zone". Permanent shifts in the Adaptive Landscape are only predicted to occur every million years or so, resulting in large-scale spurts of phenotypic change (for more detail see Chapter 13). Consequently, one of the most useful frameworks for modeling the evolutionary process—the Adaptive Landscape—is being recast as a moving target for selection. However, these results notwithstanding, some precaution should be taken when interpreting these findings. Due to the large sample sizes (hundreds to thousands) that are needed to estimate selection parameters with much confidence (Kingsolver et al. 2001) it may be the case that much of the variation reported by Siepielski et al. (2009) reflects stochastic variation in parameter estimates, which was not controlled for in their analysis. The extent of this problem has yet to be fully addressed (Chapter 9; Kingsolver and Diamond 2011), but in this review we accept the general conclusions from Siepielski et al. (2009): selection across space and time is far less predictable than the founders of the Adaptive Landscape concept anticipated. As Thompson (2005) described it, adaptive peaks and valleys are akin to sailboats bobbing up and down on waves in the ocean.

Perhaps the best-known study of temporal variation in natural selection comes from Peter and Rosemary Grant's work on Galapagos finches. On one of the Galapagos islands (Daphne Major), the Medium ground finch (*Geospiza fortis*) is subject to temporally shifting selection pressures on beak dimensions (e.g. bill depth) and body size. Selection is imposed by variation in the features of seeds that form the basis of their diet during the dry season (Grant and Grant 2002). Cracking hard seeds requires deeper and stronger beaks, whereas smaller beaks process smaller seeds more easily. In flush years, when there is an abundance of small seeds, smaller beak sizes are favored. In the drought of 1977, however, individuals with larger beaks were favored because they were most successful at cracking the large, hard fruits of *Tribulus cistoides*. By contrast, in the drought of 2003–2004 individuals with smaller beaks exhibited better survival, since the colonization of the large ground finch (*G. magnirostris*) on Daphne Major had decreased access to the seeds of *T. cistoides* (Grant and Grant 2009).This difference in the impacts of drought on selection highlights the ecological dependence of fitness functions. The Grant's studies of natural selection in the Medium ground finch represent a monumental effort: in one paper they reported on annual variability in selection over a 30-year period (now closer to 40 years!) (Grant and Grant 2002). Their conclusion was that the direction of selection was highly variable, with evolutionary change unpredictable over the long term. Neither the shape of the fitness landscape, nor the finch's position on it, could have been predicted 1 year in advance, much less three decades hence (Fig. 7.2).

Do these unpredictable, fluctuating selection pressures suggest that the Adaptive Landscape concept is of limited value, as fitness surfaces are not sufficiently stable for there to exist a single Adaptive Landscape? It is important here to realize that the body of research demonstrating fluctuating selection comes from estimates of selection gradients in natural populations, the analyses of which are founded on the Adaptive Landscape concept. Indeed, the procedure of estimating selection gradients invokes an underlying Adaptive Landscape from which parameters can be estimated (Lande and Arnold 1983; Arnold et al. 2001). Therefore, the claims made by some philosophers that biologists should abandon the Adaptive Landscape in favour of "formal modeling" (Kaplan 2008) is flawed, since the output from formal models (e.g. selection parameters) must describe *something*, in this case the slopes and curvatures of an underlying Adaptive Landscape. The study of phenotypic selection in natural populations owes its entire existence to the Adaptive Landscape concept, whose foundations were made by Simpson and Wright (Arnold

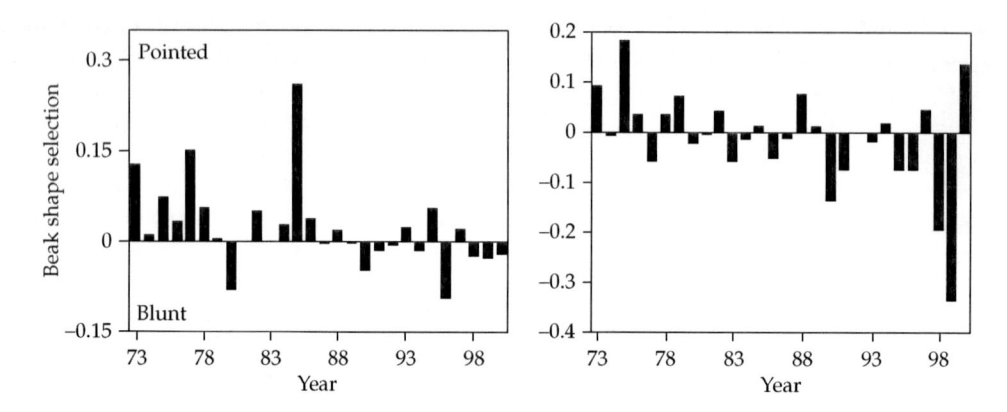

Figure 7.2 Variable selection acting on beak shape in two species of Galapagos finches over 30 years (Grant and Grant 2002). Inter-annual fluctuations were common, and long-term evolutionary trends were unpredictable based on annual patterns. Left panel *G. fortis*, Right panel *G. scandens*.

et al. 2001). Rejecting the concept of the Adaptive Landscape, which some authors have done (Pigliucci 2008), disregards the enormous contributions that studies of natural selection have made to evolutionary biology, thereby throwing the baby out with the bath water.

7.2.1 Establishing criteria for demonstrating variation in selection

The preceding review of evidence for dynamic Adaptive Landscapes may give the impression that all selection is highly variable. Although we do suggest that selection is indeed a more dynamic process than traditionally assumed, we do not claim that this applies to all selection regimes. A particular Adaptive Landscape may well be static under some scenarios. For example, female choice might lead to chronic directional selection for elaborate male traits in certain directions that do not vary much between generations (Andersson 1994; Chapter 8). Species recognition cues might also be under strong stabilizing selection, which should promote landscape stability. Whenever populations are well-adapted to stable environmental conditions, selection may be negligible as populations are already close to an adaptive peak, with a rather flat surface over a large trait region (Hendry and Gonzalez 2008). Thus, analyses should be carried out to assess whether measured fluctuations in selection are real, or if they simply reflect sampling error in the data.

Most studies of fitness surfaces trace back to the early 1980s and publication of the classic paper by Russell Lande and Steve Arnold (1983), one of the most cited papers in evolutionary biology, and John Endler's book *Natural selection in the Wild* (1986). Following publication of these works, the statistical techniques of measuring selection differentials and gradients were widely incorporated into the evolutionary ecologist's toolbox. Recent years have witnessed substantial analytical progress towards better statistical and mathematical descriptions of Adaptive Landscapes, the quantification of slopes and curvatures of fitness surfaces, and graphic visualizations of the surfaces (Lande and Arnold 1983; Phillips and Arnold 1989; Schluter 1988; Schluter and Nychka 1994; Brodie et al. 1995; Blows and Brooks 2003; Blows 2007; see also Chapter 9). Many of these techniques have become popular among evolutionary ecologists, although in most cases the most sophisticated analytical methods remain restricted to laboratory studies of *Drosophila* (e.g. Chenoweth and Blows 2005, 2006; McGuigan et al. 2005; McGuigan et al. 2008; but see Bentsen et al. 2006; Gerhardt and Brooks 2009). This is due to the enormous logistical problems involved in measuring fitness and estimating quantitative genetic parameters in the wild. More recently, theoretical models have been developed to deal with questions about adaptive peak shifts, peak stability, and evolutionary stasis (Gavrilets 1999; Arnold et al. 2001; Estes and Arnold 2007; see also Chapter 13).

Many selection estimates are based on small numbers of individuals, which results in high variance around point estimates of selection. Brodie et al. (1995) recommend that hundreds of individuals need to be measured before accurate estimates of non-linear selection are possible. This problem is exacerbated in studies interested in detecting differences in space and time, as "random effects" variables require large sample sizes to achieve reasonable levels of statistical power (Chapter 9). Few studies approach sufficiently robust sample sizes, and the first step in assessing whether selection is truly variable may be to ask whether spatial or temporal "variation" in selection actually reflects sampling variance rather than biologically meaningful changes in selection. For example, to infer among-group differences requires within group replication such that any confounding effects of sampling error can be statistically separated from the group effects of interest. In addition, phenotypic distributions must be similar among study populations to ensure that the peaks and valleys of the Adaptive Landscape occur within the trait space of each population (Chapter 9, section 9.3.3). This can be a particularly challenging task for field biologists. Moreover, it is important to remember that any aspect of the Adaptive Landscape that falls outside the phenotypic range of the study populations cannot be measured (Fisher et al. 2009; Chapters 8 and 12), though this can be addressed using phenotypic engineering experiments (Sinervo and Huey 1990; Sinervo and Basolo 1996; Svensson et al. 2007).

Finally, and most importantly, to understand whether observed shifts in selection are biologically meaningful, it is important to elucidate the ecological drivers of fluctuations in selection. While the development of novel analytical and statistical techniques, as well as innovative theory of peak movement and stability, have led to fundamental advances in the study of Adaptive Landscapes, empirical investigations of the causal mechanisms driving such dynamism lag behind these technical and analytical developments. In addition, although experimental evolutionists working on powerful model systems like *E. coli* and *D. melanogaster* have made substantial contributions to what we know about how natural selection is translated into an evolutionary response (Lenski and Travisano 1994; Vasi et al. 1994; Kassen et al. 2004; Schoustra et al. 2009; Chapter 11), study of the ecological context of selection remains a problem that must be addressed by empirical investigations of natural populations. The vast majority of selection studies fail to identify or investigate the causal agents of selection, and more detailed and mechanistic studies of natural populations in which fluctuations in selection can be convincingly coupled to ecological causes are needed. Ultimately, understanding the dynamics of Adaptive Landscapes, which are deeply rooted in ecological theory, will require collaboration between ecologists and evolutionary biologists. Next, we focus on the influence of biotic interactions on the Adaptive Landscape.

Ecological causes of variability in selection
The view we advocate in this chapter is that biotic interactions are important causes of fluctuations in selection. While climate and other abiotic factors can play a direct role in the selective process (Grant and Grant 2002; Siepielski et al. 2009), here we emphasize that differences between generations in population density, the intensity of competition, predation risk, and other biotic factors are most often the proximate causal agents of selection, with abiotic factors influencing selection only indirectly (section 7.7). For instance, a changing climate might lead to reduced or increased population sizes, which in turn will lead to changes in the intensity of density-dependent selection. Generally speaking, short-term fluctuations in selection between generations are unlikely to solely reflect climatic factors, which change over decades, rather than years, as in typical selection studies. In addition, biotic interactions are also likely to be more localized in their effects than are abiotic impacts.

7.3 Frequency-dependent selection

When an individual's fitness depends on its interactions with other individuals in a population, it is possible that selection will be frequency-dependent (Ayala and Campbell 1974). Frequency-dependent selection occurs when the average fitness of an individual of a particular morph varies depending on its relative frequency within a population possessing multiple alternative morphs (i.e. the populations are polymorphic; reviewed in Sinervo and Calsbeek 2006). Here "alternative morphs" may refer to different behavioral strategies for reproduction or territoriality among conspecifics, different reproductive morphologies, differences in defenses to combat predation and disease, and so on. These variable forms are most often studied when compartmentalized into discrete morphs because this makes empirical work much more tractable, but continuous phenotypic variation is equally as likely to be subjected to frequency-dependent selection (Kopp and Hermisson 2006). Whatever the cause of the ecological interaction, frequency-dependence violates the assumption of unchanging genotypic fitnesses found in early models of the Adaptive Landscape (Arnold et al. 2001).

A special form of frequency dependent selection is negative frequency-dependence, whereby an individual's fitness is highest when its "morph" is rare in a population (Ayala and Campbell 1974). Negative frequency-dependent selection usually results from either interactions between species, such as predation, parasitism, or competition, or between genotypes within species. The first written discussion of this form of selection came in the late 1800s when Edward Bagnall Poulton suggested that the observed pupal dimorphism of the Geometrid moths *Ephyris pendularia* (now *Cyclophora albipunctata*) and *E. omicronaria* (now *Cyclophora annulatu*) was maintained by frequency-dependent predation on the more common color morph (Poulton 1884). Unfortunately, Poulton's ground-breaking study went largely unnoticed (Allen and Clarke 1984).

Much of the early work on the maintenance of polymorphism through frequency-dependent selection was based on the assumption that different morphs would reach a stable equilibrium and achieve equal fitness (Ayala and Campbell 1974), perhaps by employing alternative evolutionary stable strategies (Maynard Smith 1982). However, several studies have since noted that in many systems, morphs do not exhibit equal fitness (Ryan et al. 1992; Lank et al. 1995; Sinervo and Lively 1996; Alonzo and Warner 2000; Yoshida et al. 2003; Jones et al. 2009). This is because when negative frequency-dependence operates on a population, alternative types need not have equal fitness at any given point in time. It has been suggested that negative frequency-dependence is common, and often leads to cyclical selection dynamics between generations (Sinervo and Calsbeek 2006).

Among the best-known examples of negative-frequency dependent cycles in selection pressure comes from the study of side-blotched lizards in California's Great Central Valley (Sinervo and Lively 1996; Sinervo et al. 2000), where populations of small (ca. 55 mm), mostly annual lizards inhabit sandstone outcroppings on grassy hillsides. Each outcropping forms a semi-isolated island of suitable habitat surrounded by uninhabitable grassland. Within each subdivided population of lizards, males and females display one of six throat-color patterns (orange, blue, yellow, blue/orange, blue/yellow, and yellow/orange) and throat color is correlated with male mating behaviors. It has been shown that the frequency of alternative throat colors cycles among years due to negative frequency dependent selection in both sexes. The throat color cycles in males and females occur at different rates because these cycles are generated by differences in fitness arising from alternative territorial and mating behaviors in males (Sinervo and Lively 1996), but alternative life history strategies in females (Sinervo et al. 2000).

Similar non-transitive fitness differences have been observed in a diversity of systems including marine isopods (Shuster and Wade 1992), damselflies (Svensson et al. 2005), and tristylous plants (Eckert et al. 1996). Frequency-dependent selection also maintains diversity in *E. coli* bacteria (Kerr et al. 2002), whose cyclical interactions are dictated by susceptibility and resistance to colicin toxins. Interestingly, additional insights gleaned from studies of bacteria include the importance of abiotic factors—such as the physical structure available to populations—in maintaining negative frequency dependence. Similarly, the importance of physical structure in the environment for adaptive differentiation has been repeatedly demonstrated in *Pseudomonas fluorescens* (Korona et al. 1994; Rainey and Travisano 1998) Thus, variation in fitness may generally depend on more than just the biotic interactions among individuals, and studies of evolution on Adaptive Landscapes should also consider the interacting role of abiotic factors in generating variation in fitness through changes in ecological opportunity (e.g. spatial structure leads to resource competition).

7.4 Density-dependence

During the lengthy debate preceding the Modern Synthesis, Fisher pointed out that the mean fitness in a population must necessarily fluctuate around zero or else a population's numbers would either be forever growing or shrinking, with catastrophic consequences (Fisher 1958). Darwin had ealier used this same point to explain the inescapability of natural selection:

> There is no exception to the rule that every organic being naturally increases at so high a rate, that if not destroyed, the earth would soon be covered by the progeny of a single pair... The elephant is reckoned to be the slowest breeder of all known animals... [yet] at the end of the fifth century there would be alive fifteen million elephants, descended from the first pair (Darwin 1859, p. 64).

Ever since Darwin, an entire field of evolutionary ecology has burgeoned around the very interesting question of what happens during these "fluctuations around zero" (Roughgarden 1971; Anderson and May 1978; Felsenstein 1979; Mueller and Ayala 1981; Mueller 1988a; Sinervo et al. 2000; Calsbeek and Smith 2007; Einum et al. 2008). The study of density-dependent selection aims to understand how changes in population density alter the makeup of the Adaptive Landscape.

As noted earlier, the Adaptive Landscape, while perhaps the most useful heuristic in evolutionary biology (Provine 1986), has inspired countless offshoot analogies. Consider the following metaphor that incorporates density dependence into the Adaptive Landscape (see Chapter 14). According to Rosenzweig (1978), density-dependent selection results in a moving, dynamic fitness surface much like a "sphagnum bog" with floating mats of vegetation. As weight (density-dependent competition) is pressed onto the sphagnum, the mats are pushed below the waters surface (the fitness peak is depressed). When the weight is alleviated, the mats rise back up (the fitness peak is restored). For instance, when a population of individuals sharing a common resource increases in numbers, the fitness peak is gradually depressed. Such a process might lead to increased competition, which in turn generates disruptive selection through intense competition between those similar phenotypes that are close to the mean value of the population (Dieckmann and Doebeli 1999).

Although several studies have shown peak depression by competition in the lab (Mueller 1988a, 1988b; Borash et al. 1998), empirical evidence of the process in the field remains limited. One recent example was provided by Bolnick (2004), who showed that three-spine stickleback, *Gasterosteus aculeatus*, may be subject to density-dependent disruptive selection driven by competition for food resources. Stickleback exhibit disparate trophic morphologies associated with alternative benthic and littoral life histories. Fish that live in open waters (benthics) tend to have long gill-rakers and smaller mouths than fish from shallower (littoral) communities. These morphological differences reflect trade-offs in feeding strategies in the two niches (Schluter 1995). Resource competition is strongest among individuals of similar morphology, and when population densities were experimentally increased, the strength of disruptive selection within populations was also increased (Bolnick 2004).

Similarly, Calsbeek and Smith (2007) used wild populations of the brown anole (*Anolis sagrei*) in the Bahamas to show that the strength and form of selection acting on body size changed in a predictable manner with changes in population density. In contrast to the results presented by Bolnick (2004), who found that selection favored smaller stickleback, anoles experienced increasingly intense directional selection for larger body sizes as population density increased. Competitive ability in lizards is often mediated by larger body size (Hews 1988; Stamps and Krishnan 1995), and the monotonic advantage of large size may explain the lack of congruence with the stickleback study.

A related study, also using lizards, provided further experimental evidence for the flexibility of the Adaptive Landscape in the context of density-dependence. Svensson and Sinervo (2000) used

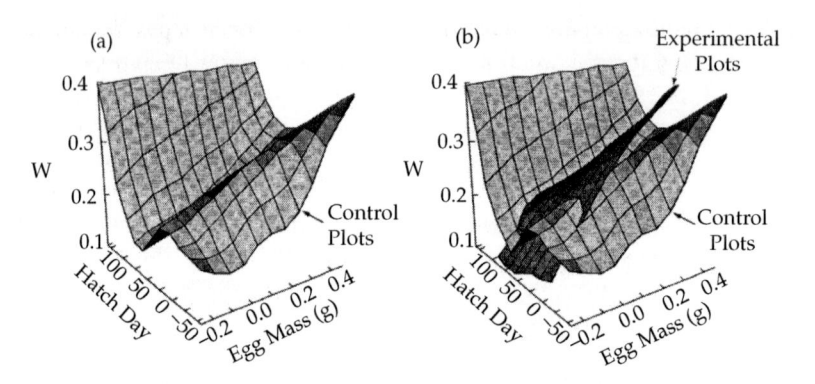

Figure 7.3 Experimental exploration of an Adaptive Landscape by egg size manipulation in side blotched lizards. Panel (a) shows the Adaptive Landscape for lizard progeny on control plots. This fitness landscape is characterized by disruptive selection for early and late hatching dates, and a general trend favoring progeny from larger eggs. Data from experimental plots (b) reveal that variation in survival is driven by temporal variation in density dependent natural selection (redrawn from Svensson and Sinervo 2000). The additional fitness surface in panel (b) is for late-season progeny that were released on plots that were free of competitors, through experimental removal of hatchlings. Though the two surfaces intersect as small egg sizes, the fitness surface for late-season progeny is higher in elevation and has a steeper slope than the surface for control plots.

experimental manipulations of egg-size and local selective environments in side-blotched lizards (*Uta stansburiana*) to test the hypothesis that density-dependent natural selection favored larger hatchling body sizes owing to greater competitive ability. By removing yolk from a subset of eggs produced by each female lizard, the authors generated experimentally reduced hatchling body sizes (see also Sinervo and Huey 1990; Sinervo et al. 1992). In a complementary manipulation, they ablated a portion of the developing follicles in reproductive females, which caused them to lay gigantized eggs. The results of this experiment revealed that experimentally manipulated hatchling body sizes "filled in" a fitness valley in the Adaptive Landscape (Fig. 7.3). Alleviating this selective valley was only made possible by the simultaneous experimental manipulation of both the targets of selection (egg size) as well as the selective agent (hatchling densities). Visualization of the fitness surface showed changes both in elevation and in slope following experimental manipulation (Fig. 7.3). The change in elevation illustrates the overall negative effect of high density on mean population fitness. The change in slope revealed that, contrary to expectations, selection on egg size increased after older hatchlings from early clutches were removed (Fig. 7.3). This likely reflects a shift away from asymmetric competition between cohorts of younger and older hatchlings. The latter have a substantial size and age advantage, and their removal created a situation in which small differences in egg size and competitive ability replaced these age asymmetries in affecting selection on egg size and hatching time (Svensson and Sinervo 2000). This "double-manipulation" study thus illustrates the degree to which Adaptive Landscapes are dynamic rather than static, and how a biotic factor (competition) can dramatically alter the shape of an adaptive surface.

7.5 Competition, predation, or both?

The evidence for density-dependent selection and its effects on fitness surfaces indicates that competitive differences can mold the shape of an Adaptive Landscape. When Alfred Lord Tennyson (1850) wrote that nature is "red in toothe and claw," he envisioned an important role for predation in nature (though he didn't specifically see the world through the lens of selection theory). It seems a matter of intuition that predation should be a seminal force in molding the topography of Adaptive Landscapes.

A large number of now classical studies in the British ecological genetics tradition, ranging from the function of banding patterns in *Cepea* snails to predation by birds on alternatively pigmented

Peppered moths (*Biston betularia*) in post-industrial Europe, have illustrated the importance of predators in driving natural selection (Haldane 1955; Ford 1964). Compared to studies of intraspecific competition, which can be relatively easy to measure, it is rare for predator–prey interactions to be clearly identified or experimentally manipulated in field studies of selection. This is because predation is difficult to observe in nature, and in most cases once a prey item is consumed, it is beyond measurement. Despite these difficulties, there are some excellent examples of predator-mediated selection, including experimental studies of selection on stickleback armor (Reimchen and Nosil 2002; Vamosi and Schluter 2002; Rundle et al. 2003b), selection on horn length in the Flat-tailed horned lizard, *Phyrnosoma mcalli* (Young et al. 2004), selective predation on wing morphology and wing coloration in calopterygid damselflies (Svensson and Friberg 2007), and reciprocal co-evolution between predatory garter snakes (*Thamnophis sirtalis*) and their newt prey items (genus *Taricha*; Brodie 1989; Feldman et al. 2009; Feldman et al. 2010; Weese et al. 2010). Studies of host–parasite coevolution, or plant–herbivore evolution, could also be considered a form of predator–prey relationship with the potential for profound effects on the stability of the Adaptive Landscape (Benkman 2003).

Perhaps more interesting than asking simply whether predation does or does not influence the Adaptive Landscape would be to quantify the relative importance of predation and competition when they operate simultaneously (DiBattista et al. 2011). Predation can interact with competition-mediated divergence by slowing divergence through amelioration of competitive interactions (Meyer and Kassen 2007), or by reinforcing the divergence stemming from competition (Rundle et al. 2003a; Nosil and Crespi 2006). For example, Meyer and Kassen (2007) beautifully demonstrated the joint influence of competition and predation in the adaptive diversification of the bacterium, *Pseudomonas fluorescens*. In the absence of predators, competition for niche space led to repeatable adaptive differentiation of *P. fluorescens* into alternative morphologies. These alternative forms were adapted to different niches in liquid media, and their differentiation took place within 7–10 days from a single genetic stock population. Competition-mediated divergence, however, was slowed significantly when bacterial densities where reduced due to the addition of the ciliated predator *Tetrahymena thermophila*.

Additional evidence that competition can be more influential than predation in shaping Adaptive Landscapes comes from another study of lizards (*Anolis sagrei*) on islands in the Bahamas. Calsbeek and Cox (2010) created three different predator treatments by enshrouding two entire islands in the Bahamas with bird-proof netting, adding snakes to two other islands and monitoring a third pair of unmanipulated islands as controls. At the same time, lizard densities on these islands were manipulated, allowing them to compare the strength of density-dependent natural selection in the context of different predation regimes. Although predator treatments had no discernable effect on natural selection, increasing population densities led to increased directional selection favoring larger body size, longer limb lengths, and higher running endurance (Fig. 7.4). Calsbeek and Cox (2010) noted that these results do not necessarily exclude predators as a possible force in sculpting the Adaptive Landscape for anoles. Indeed, another study of the same lizard species provided evidence that predators can alter selection on both behavior and morphology (Losos et al. 2004). However, when competition and predation were compared in the same experimental framework, competition turned out to be the stronger selective agent (Calsbeek and Cox 2010). Thus, the effect of predation on selection depends on the interactions and strengths of other ecological factors, and can either oppose competition-driven selection or strengthen it (Meyer and Kassen 2007; Calsbeek and Cox 2010).

7.6 Fluctuations in sexual selection: intergenerational changes in male–male competition and mate choice

A simplifying assumption shared by nearly all of the early models of sexual selection is that female mate preferences are constant (Lande 1980; Andersson 1994) and have led many studies of sexual selection to infer a more or less static Adaptive

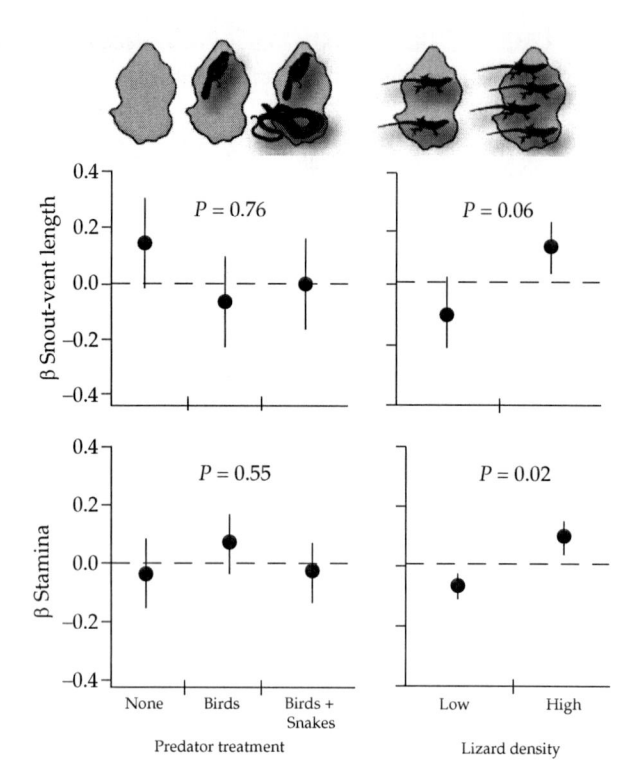

Figure 7.4 Competition is relatively more important than predation as a causal agent of selection acting on *Anolis* lizards in the Bahamas. Panels show selection differentials acting on size (snout-vent-length; top panels) and running endurance (stamina; bottom panels) as a function of predator treatments (left) and population density (right). Whereas predator treatments did not influence the selection, increased population density tended to increase the intensity of directional selection on both traits (modified from Calsbeek and Cox, 2010).

Landscape in which the same trait is consistently favored across generations. However, recent empirical work has demonstrated that both female preferences and male mating success can be highly labile across generations, populations and geographic areas (Qvarnstrom et al. 2000; Qvarnstrom et al. 2006; Gosden and Svensson 2008; Chaine and Lyon 2009; Rundle et al. 2009; Chapter 8). A similar assumption of persistent selection regimes is also implicitly made in many studies of male–male competition—intrasexual selection (Le Boeuf 1974; Rosenqvist 1990; McGlothlin et al. 2005). Similar to the studies of natural selection discussed earlier, the strength and form of sexual selection can fluctuate dramatically (Gosden and Svensson 2008; Chaine and Lyon 2009). Such variation can lead to patterns of selection on male traits (for example) that change over short spatial and temporal scales, reversals in the direction and magnitude of selection, and unpredictability of selection between generations (Fig. 7.5). The underlying causes of these changes in female preferences and male-male competition across generations are not yet fully understood in any system. Sexual selection is known to be a strong and widespread force in natural populations (Kingsolver et al. 2001), yet there is a paucity of examples of contemporary evolution in sexual traits in the field (Svensson and Gosden 2007). The solution to this apparent paradox is not fully understood, but it is quite likely that there are ecological limits to the evolution of sexual traits, due to the opposing antagonistic effects of natural selection (Svensson and Friberg 2007; Hine et al. 2011). Any fluctuations in natural selection that impose antagonistic effects on the strength of sexual selection might also indirectly cause fluctuations in the net effects of sexual selection. Interactions might become even more complicated when females show plasticity in their mating preferences based on intersexual interactions with con- and heterospecific males (Svensson et al. 2010).

Although the majority of empirical work on sexual selection has concentrated on female preferences, there is strong theoretical backing for the

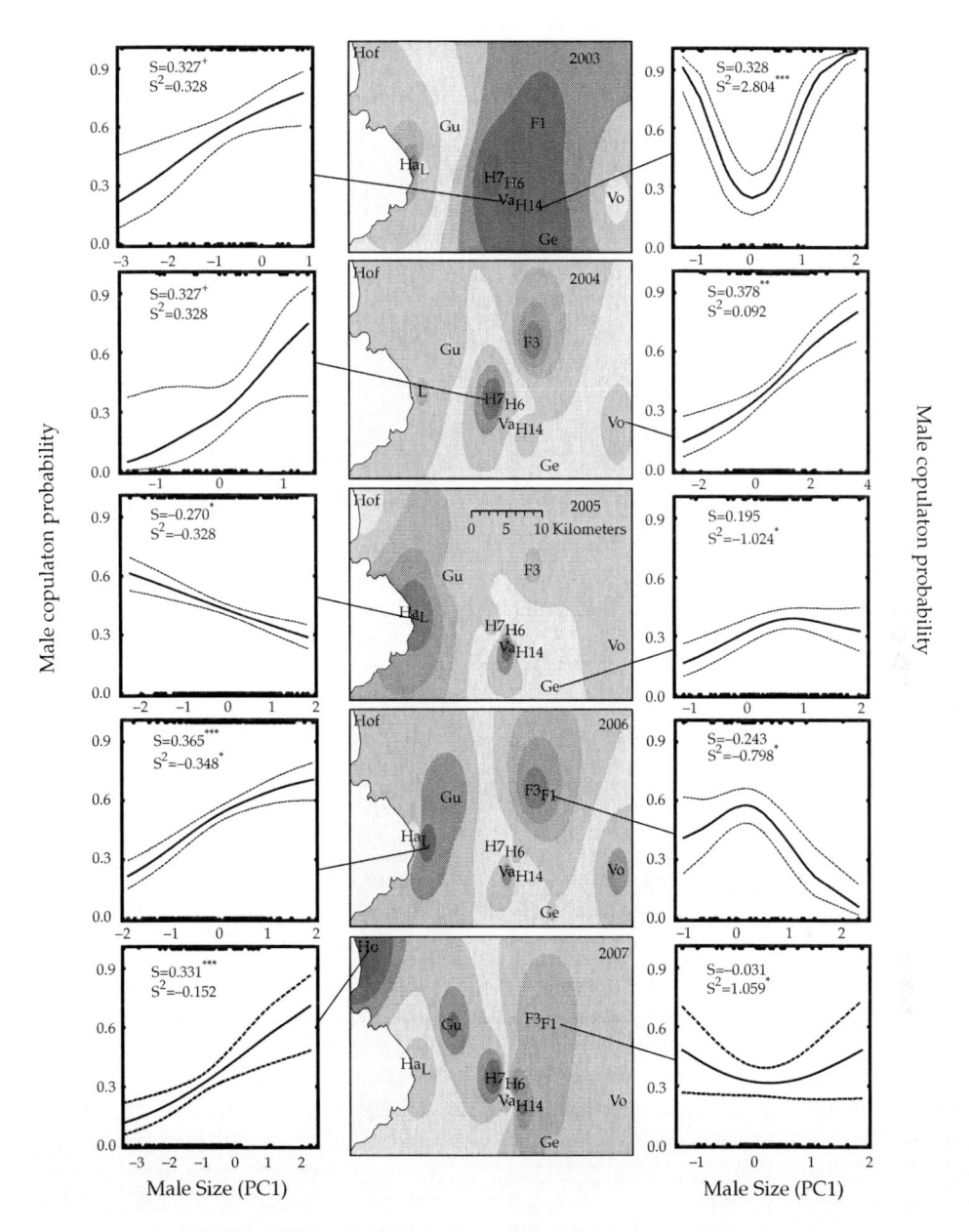

FLUCTUATING SELECTION AND DYNAMIC ADAPTIVE LANDSCAPES   99

Figure 7.5 Patterns of spatial and temporal variation in selection acting in male damselflies (*Ischnura elegans*) in coastal Sweden. Left and right columns depict representative cubic spline regressions of male copulation probability vs. body size. Central figures illustrate the total pattern of variation using heat maps, in which colouration and intensity represent positive and negative selection differentials respectively (from Gosden and Svensson, 2008). See Plate 1 for colour figure; red and dark green represent positive and negative differentials respectively.

importance of male mate choice (Johnstone et al. 1996; Kokko and Johnstone 2002; Hardling et al. 2008), with active male mate choice having been demonstrated in several studies (Amundsen 2000; Bonduriansky 2001). Males, contrary to popular belief, are often choosy. The lack of empirical work on male mating preferences means we know little about how male preferences and female sexual display traits may fluctuate. However, recent empirical work on the two spotted goby (*Gobiusculus flavescens*), a fish known to exhibit mutual mate choice (Amundsen and Forsgren 2003), demonstrates dynamic sex roles where the strength of sexual selection in one sex is highly contingent upon the operational sex ratio (OSR) (Forsgren et al. 2004). When the number of sexually competing males is higher than females, the mating system will largely be driven by female choice, however, if the OSR becomes female biased, then the mating system shifts to one driven by male mate choice (Forsgren et al. 2004).

Another area that is largely unexplored in the context of fluctuating Adaptive Landscapes is sexual conflict (Parker 1979; Gavrilets et al. 2001; Chapman et al. 2003), which has become a prominent area of research within evolutionary ecology in recent years (Arnqvist and Rowe 2002). This work suggests that there should be an evolutionary tug-of-war between the sexes (Rice and Holland 1997; Rice and Chippindale 2001), as genetic correlations between homologous characters in males and females creates a situation in which selection acting on one sex, leads to correlated responses on the same traits in the opposite sex (Lande 1980). In Adaptive Landscape terminology, selection on one sex pulls the other sex off its adaptive peak (Rice and Chippindale 2001). Indeed, growing empirical evidence suggests that the genes that produce high fitness males are often different from the genes that produce high fitness females (Chippindale et al. 2001; Bonduriansky and Chenoweth 2009). Such male benefit/female detriment alleles (and vice versa) impose a gender load on mating systems (Arnqvist and Tuda 2010). The sexually antagonistic selection pressures that result from this phenomenon can be resolved through the evolution of either sexual dimorphism or biased fertilizations (Bonduriansky and Rowe 2005; Fedorka and Zuk 2005; Calsbeek and Bonneaud 2008; Cox and Calsbeek 2009, 2010). Empirical studies estimating the strength of genetic correlations between homologous traits in males and females have confirmed the presence of constraints on the evolution of sexual dimorphism (Bonduriansky and Rowe 2005; Chenoweth and Blows 2003). Despite this flux of empirical work (Arnqvist et al. 2000; Pitnick et al. 2001; Chapman et al. 2003; Day and Bonduriansky 2004; Bonduriansky and Rowe 2005; Fedorka and Zuk 2005; Pischedda and Chippindale 2006; Calsbeek and Bonneaud 2008; Bonduriansky and Chenoweth 2009; reviewed in Cox and Calsbeek 2009), the long-term consequence of such antagonistic interactions are as yet unknown. If the resolution on intersexual conflict is slow, as many suspect, dynamic fluctuations in the fitness surface of the sexes is expected over both temporal and spatial scales. Understanding how the many facets of sexual selection interact to influence trait expression, preference and the stability of the Adaptive Landscape is an extremely useful avenue for future research (Cornwallis and Uller 2010).

7.7 Abiotic environmental factors, fluctuating selection, and the limits of ecological speciation

The indirect effects of climate and other abiotic factors are likely to influence selection on phenotypic traits and the shape of fitness surfaces. For instance, in ectotherms temperature can have a strong influence on survival (e.g. Huey et al. 1989; Angilletta et al. 2000), which alters the strength of density-dependent natural and sexual selection by indirectly changing demographic parameters. One consequence of the influence of abiotic conditions on biotic interactions is that spatially variable weather conditions and microclimatic variation lead to unpredictable social interactions, creating geographic mosaics in natural or sexual selection (Thompson 2005; Gosden and Svensson 2008). In combination with gene flow between populations, this can lead to rapid shifts in selection between generations as well as between different spatial localities (Thompson 2005).

Stochastic fluctuations in the strength and mode of selection have implications for diversification and species formation, particularly for so-called ecological speciation, which holds that divergent selection plays a major role in species formation (Candolin et al. 2007). In the last decade, ecological speciation has come to be regarded by many as perhaps the most prominent mechanism of speciation (e.g. Schluter 2000; Nosil et al. 2002), however, the prevalence of ecological speciation has recently been challenged. One set of criticisms focuses on the role of niche conservatism (Seehausen et al. 2008), or the observation that closely related species regularly share a common niche (Wiens and Graham 2005). Ecological niche similarity between populations combined with biogeographic events can lead to the formation of evolutionary lineages recognizable as species (Buckley et al. 2010; Kozak and Wiens 2010; Wiens et al. 2010), which is a radically different mechanism of speciation than is ecological speciation, where divergent selection between different niches promotes divergence. Another set of criticisms focuses on the fact that most speciation processes—including "non-ecological" processes such as polyploidy, uniform selection, or even genetic drift—always have ecological underpinnings. As ecologically-mediated processes are shared by virtually all mechanisms of species formation, the term ecological speciation may be misleading (Sobel et al. 2010).

Here we would like to put forward a third problem for ecological speciation: fluctuations in the shape of fitness surfaces. As discussed earlier, meta-analyses of selection studies indicate that fluctuations in the sign and magnitude of natural selection are common (Siepielski et al. 2009; Kingsolver and Diamond, 2011). In contrast, ecological speciation, even when fast, requires dozens, hundreds or even thousands of generations of consistent divergent selection to be completed. If selection is not consistently different between divergent populations, the fitness valleys separating adaptive peaks (i.e. divergent selection) could disappear and inhibit the process of divergence (i.e. "speciation reversal"; Taylor et al. 2006; Seehausen et al. 2008). Much of our own recent work has revealed that selection fluctuates dramatically between generations, and that these fluctuations are caused by changing population densities, densities of predators, or the intensity of sexual or natural selection (Svensson et al. 2001; Svensson and Sinervo 2004; Calsbeek and Smith 2007; Gosden and Svensson 2008; Calsbeek et al. 2009). Exactly how dynamic the Adaptive Landscape is remains an unresolved quantitative question. However, it seems to be the case that in most study systems stable fitness surfaces are the exception, not the rule, at least when selection pressures have their origin in biotic interactions. How likely is it that Adaptive Landscapes mediated by biotic interactions are sufficiently constant over the time frames required for ecological speciation to take place? One possible answer is that fluctuations in selection, though common, only slow the rate of evolution and, in the long term, have limited impact on cumulative directional selection (Kingsolver and Diamond, 2011). Nevertheless, insights from selection studies in natural populations currently present a challenge for proponents of the ubiquity of ecological speciation by consistent divergent selection.

7.8 Population variants: genetic morphs and phenotypic plasticity

Adaptive Landscape theory posits that interactions among traits impact organismal performance, such that some combinations of traits lead to higher fitness than others (Simpson 1944). Such landscapes can be topographically rugged, and when a population occupies a fitness peak it is prevented from ascending another higher peak if that peak is isolated by the presence of a fitness valley. As pointed out by West-Eberhard (2003, p. 394), the Adaptive Landscape perspective on trait evolution contrasts somewhat with that of Darwin, who thought that complex structures could be created by numerous small steps. Transposed into the Adaptive Landscape context, Darwin's argument suggests that fitness landscapes are relatively flat. If true, consistent directional selection over vast spans of time can create structures of enormous complexity. As discussed earlier, however, empirical investigations of selection commonly find that selection is idiosyncratic in space and time. Although it remains conceivable that there may be consistent trends over longer time scales, recent work suggest that bounded short-term evolutionary fluctuations, followed by rapid bursts of large-scale phenotypic change are common across several different taxa (Uyeda et al, 2011).

The classical view of the Adaptive Landscape considers situations in which traits are genetically determined and expressed in all individuals. Such trait invariance is an overgeneralization, however, as individuals within a population might often express multiple and phenotypically discrete trait combinations, or morphs. For instance, as described earlier, multiple genetically-determined morphs coexist within populations of the side-blotched lizard, *Uta stansburiana* (Sinervo and Lively 1996; Corl et al. 2010). These morphs represent alternative territorial and life history strategies, and as such are subject to frequency and density dependent selection. Here, one can envision a rugged landscape with the morphs struggling to ascend divergent peaks that are shifting in their elevations based on frequency and density dependent effects (Svensson and Sinervo 2000). Correlational selection maintains successful combinations of morphology, coloration, behavior, and life history traits (Sinervo and Svensson 2002). However, since recombination is constantly eroding the genetic correlations that are built up by selection each generation, morphs are forced to track relatively distant peaks. In reality, morphs may never reach these distant peaks and hence fail to become "optimally" adapted. By contrast, populations that become fixed for a single morph are released from the genetic load arising from frequency-dependent selection, and this relaxes the genetic constraints on their phenotypic evolution. A consequence is that morphs may rapidly evolve in response to novel selective pressures and ascend an alternative adaptive peak, perhaps even resulting in rapid "morphic speciation" (West-Eberhard 1989; Corl et al. 2010).

Multiple peaks within a population can also have their origin in phenotypic plasticity, whereby a single genotype has the capacity to code for the environmentally sensitive production of alternative phenotypes (DeWitt and Scheiner 2004). When plasticity is adaptive, a genotype × environment interaction has evolved to maximize fitness across a variable array of ecological contexts (Pigliucci 2001; Wolf et al. 2004). As a consequence, alternative phenotypes occupy different positions on the Adap-

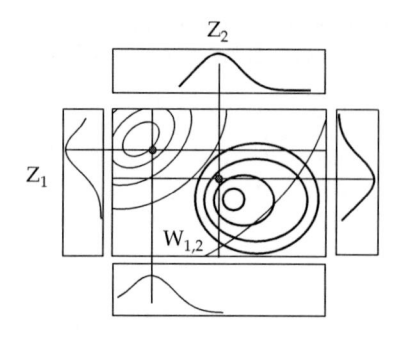

Figure 7.6 (See also Plate 1.) Adaptive Landscape illustrating the effects of phenotypic plasticity for two characters under correlational selection in two environmental contexts. Z denotes the values for traits 1 and 2. The outer boxes provide the average fitness of trait values, with light lines characterizing Environmental Context 1, and heavy lines Environmental Context 2. The central panel represents the adaptive topography in both environmental contexts, and illustrates the divergent optima. Grey circles locate population means on the Adaptive Landscape and illustrate a scenario in which populations are tracking different peaks, yet have failed to evolve optimum trait values.

tive Landscape. A plastic response, in effect, results in an instantaneous peak shift. Adaptive phenotypic plasticity is likely to evolve when the shape of the Adaptive Landscape is heterogeneous, yet predictable given environmental cues (Fig. 7.6). For instance, in many species of frogs, tadpoles raised in the presence of chemical cues from dragonfly larvae develop expanded tail fins and smaller heads (Smith and Van Buskirk 1995), and these induced phenotypes increase the probability of tadpole survival in the presence of free-ranging dragonfly larvae (Van Buskirk and Relyea 1998; Van Buskirk and McCollum 2000a, 2000b).

Adaptive plasticity in response to predators has been identified in a number of other systems (Pigliucci 2001; Laforsch and Tollrian 2004; Crispo 2008). Many models of phenotypic plasticity assume that the ecological context cannot itself evolve, as is the case for abiotic factors. For example, variable light levels have driven the evolution of a shade avoidance response in many plants (Schmitt 1993). Under many circumstances, however, the ecological context (the environmental component of a G × E interaction) is itself genetically determined, and can thus evolve (Wolf et al. 2004). The induction of predator-induced phenotypes in tadpoles, for instance, should impact selection on traits related to predation efficacy in dragonfly larvae. The evolution of dragonfly larvae that occurs as a response to adaptive plasticity in tadpoles could shift the adaptive peak of the tadpoles, leading to further evolution in tadpole larvae. Reciprocal coevolutionary interactions can thus lead to dynamic feedbacks between the plastic species and it's ecological context, resulting in an unstable Adaptive Landscape with constantly moving phenotypic optima (Brodie and Ridenhour 2003; Brodie et al. 2005; Thompson 2005).

In addition to reciprocal selection, populations that express plastic responses to environmental cues may also experience selection stemming from trade-offs associated with producing multiple phenotypes from a single genotype. That is, alternative phenotypes may be displaced from their respective adaptive peaks by developmental trade-offs (West-Eberhard 2003). As in the fixation of a single morph in a polymorphic population, when a population fixes on a phenotype (i.e. developmental canalization), developmental constraints are alleviated. This release may lead to rapid character evolution or even species formation because populations are free to more accurately track their respective adaptive peaks (West-Eberhard 2005; Corl et al. 2010).

7.9 Conclusions and future directions

The theme of this chapter is selection in natural populations is highly variable across space and time—at least on the time scale of the standard research grant (Siepielski et al. 2009). We argue that traits associated with biotic interactions are especially likely to result in shifting Adaptive Landscapes. We have discussed the potential ecological factors that generate variability in selection pressures through space and time, and how these in turn affect the Adaptive Landscape. These ecological factors include frequency- and density-dependence, variability in patterns of mate choice driven by sexual selection and sexual conflict, dynamic evolutionary changes resulting from interactions with predators and competitors, and the environmental causes behind the plastic production of phenotypes. Taken together, empirical studies have ushered in a view of the Adaptive Landscape as a highly dynamic entity, with peaks and valleys sloshing about like so many waves in a bathtub.

This ecological view of the Adaptive Landscape is somewhat different from the original, more static view developed by Sewall Wright in 1932. Many of the ideas we have outlined in this chapter are in the early stage of development and much work remains to be done, both by theoreticians and empiricists, to develop a more complete understanding of the impact on a dynamic landscape on the diversification process. Additional empirical work specifically designed to identify and measure the agents of natural and sexual selection in the wild is badly needed. Studies should aim to establish the variability or consistency in selection by keeping in mind a number of potential pitfalls that can easily derail such efforts (see also Chapter 9). First, studies should clearly identify the biotic drivers across populations and years and demonstrate their importance using experimental manipulations (e.g. Bradshaw and Schemske 2003; Marchinko 2009). Second, every effort should be made to work with large samples as estimates of selection are subject to a high degree of sampling error. Owing in part to the fact that fitness estimates often have non-normal sampling distributions (e.g. live vs. dead), selection coefficients are likely to be highly unstable when sample sizes are small. Finally, while it is clear that Adaptive Landscapes commonly exhibit dynamic features, more work needs to be done to quantify variation in selection in space and time, including improved efforts to separate random fluctuations around zero (sampling error) from biologically relevant, ecologically mediated changes in the Adaptive Landscape. We suggest that such studies incorporate population and/or generation as random effects as is done in random coefficient models (Chapter 9), to ensure that inferences can be made beyond any particular study.

References

Allen, J. A. and Clarke, B. C. (1984). Frequency-dependent selection – Homage to poulton, E.B. Biological Journal of the Linnean Society, 23, 15–18.

Alonzo, S. H. and Warner, R. R. (2000). Female choice, conflict between the sexes and the evolution of male alternative reproductive behaviours. Evolutionary Ecology Research, 2, 149–170.

Amundsen, T. (2000). Why are female birds ornamented? Trends in Ecology & Evolution, 15, 149–155.

Amundsen, T. and Forsgren, E. (2003). Male preference for colourful females affected by male size in a marine fish. Behavioral Ecology and Sociobiology, 54, 55–64.

Anderson, R. M. and May, R. M. (1978). Regulation and stability of host-parasite population interactions. 1. Regulatory processes. Journal of Animal Ecology, 47, 219–247.

Andersson, M. (1994). Sexual Selection. Princeton University Press, Princeton, NJ.

Angilletta Jr., M. J., Winters, R. S., and Dunham, A. E. (2000. Thermal effects on the energetics of lizard embryos: implications for hatchling phenotypes. Ecology, 81, 2957–2968.

Arnold, S. J., Pfrender, M. E., and Jones, A. G. (2001). The Adaptive Landscape as a conceptual bridge between micro- and macroevolution. Genetica, 112, 9–32.

Arnqvist, G., Edvardsson, M., Friberg, U., and Nilsson, T. (2000). Sexual conflict promotes speciation in insects. Proceedings of the National Academy of Sciences of the United States of America, 97, 10460–10464.

Arnqvist, G. and Rowe, L. (2002). Antagonistic coevolution between the sexes in a group of insects. Nature, 415, 787–789.

Arnqvist, G. and Tuda, M. (2010). Sexual conflict and the gender load: correlated evolution between population fitness and sexual dimorphism in seed beetles. Proceedings of the Royal Society B: Biological Sciences, 277, 1345–1352.

Ayala, F. J. and Campbell, C. A. (1974). Frequency-dependent selection. Annual Review of Ecology Evolution and Systematics, 5, 115–138.

Bell, G. (2010). Fluctuating selection: the perpetual renewal of adaptation in variable environments. Philosophical Transactions of the Royal Society B Biological Sciences, 365, 87–97.

Benkman, C. W. (2003). Divergent selection drives the adaptive radiation of crossbills. Evolution, 57, 1176–1181.

Bentsen, C. L., Hunt, J., Jennions, M. D., and Brooks, R. (2006). Complex multivariate sexual selection on male acoustic signaling in a wild population of Teleogryllus commodus. American Naturalist, 167, E102–E116.

Blows, M. W. (2007). A tale of two matrices: multivariate approaches in evolutionary biology. Journal of Evolutionary Biology, 20, 1–8.

Blows, M. W. and Brooks, R. (2003). Measuring nonlinear selection. American Naturalist, 162, 815–820.

Bolnick, D. I. (2004). Can intraspecific competition drive disruptive selection? An experimental test in natural populations of sticklebacks. Evolution, 58, 608–618.

Bonduriansky, R. (2001). The evolution of male mate choice in insects: a synthesis of ideas and evidence. Biological Reviews, 76, 305–339.

Bonduriansky, R. and Chenoweth, S. F. (2009). Intralocus sexual conflict. Trends in Ecology & Evolution, 24, 280–288.

Bonduriansky, R. and Rowe, L. (2005). Intralocus sexual conflict and the genetic architecture of sexually dimorphic traits in Prochyliza xanthostoma (Diptera: Piophilidae). Evolution, 59, 1965–1975.

Borash, D. J., Gibbs, A. G., Joshi, A., and Mueller, L. D. (1998). A genetic polymorphism maintained by natural selection in a temporally varying environment. American Naturalist, 151, 148–156.

Bradshaw, H. D. and Schemske, D. W. (2003). Allele substitution at a flower colour locus produces a pollinator shift in monkeyflowers. Nature, 426, 176–178.

Brodie, E. D. (1989). Genetic correlations between morphology and antipredator behavior in natural-populations of the garter snake *Thamnophis ordinoides*. Nature, 342, 542–543.

Brodie, E. D. (1992). Correlational selection for color pattern and antipredator behavior in the garter snake *Thamnophis ordinoides*. Evolution, 46, 1284–1298.

Brodie, E.D., III, Feldman, C.R., Hanifin, C.T, Motychak, J.E., Mulcahy, D.G., Williams, B.L., et al. (2005). Parallel arms races between garter snakes and newts involving tetrodotoxin as the phenotypic interface of coevolution. Journal of Chemical Ecology, 31, 343–356.

Brodie, E. D. and Ridenhour, B. J. (2003). Reciprocal selection at the phenotypic interface of coevolution. Integrative and Comparative Biology, 43, 408–418.

Brodie, E. D. III., Moore, A. J., and Janzen, F. J. (1995). Visualizing and quantifying natural selection. Trends in Ecology & Evolution, 10, 313–318.

Buckley, L. B., Davies, T. J., Ackerly, D. D., Kraft, N. J. B., Harrison, S. P., Anacker, B. L., et al. (2010). Phylogeny, niche conservatism and the latitudinal diversity gradient in mammals. Proceedings of the Royal Society B: Biological Sciences, 277, 2131–2138.

Calsbeek, R. and Bonneaud, C. (2008). Postcopulatory fertilization bias as a form of cryptic sexual selection. Evolution, 62, 1137–1148.

Calsbeek, R., Buermann, W., and Smith, T. B. (2009). Parallel shifts in ecology and natural selection in an island lizard. BMC Evolutionary Biology, 9, 3.

Calsbeek, R. and Cox, R. M. (2010). Experimentally assessing the relative importance of predation and competition as agents of selection. Nature, 465, 613–616.

Calsbeek, R. and Smith, T. B. (2007). Probing the Adaptive Landscape on experimental islands: density dependent selection on lizard body-size. Evolution, 61, 1052–1061.

Candolin, U., Salesto, T., and Evers, M. (2007). Changed environmental conditions weaken sexual selection in sticklebacks. Journal of Evolutionary Biology, 20, 233–239.

Chaine, A. S. and Lyon, B. E. (2009). Adaptive plasticity in female mate choice dampens sexual selection on male ornaments in the lark bunting (January, p. 459, 2008). Science, 324, 1143–1143.

Chapman, T., Arnqvist, G., Bangham, J., and Rowe, L. (2003). Sexual conflict. Trends in Ecology & Evolution, 18, 41–47.

Chenoweth, S. F., and Blows, M. W. (2003). Signal trait sexual dimorphism and mutual sexual selection in *Drosophila serrata*. Evolution, 57, 2326–2334.

Chenoweth, S. F. and Blows, M. W. (2005). Contrasting mutual sexual selection on homologous signal traits in *Drosophila serrata*. American Naturalist, 165, 281–289.

Chenoweth, S. F. and Blows, M. W. (2006). Dissecting the complex genetic basis of mate choice. Nature Reviews Genetics, 7, 681–692.

Chippindale, A. K., Gibson, J. R., and Rice, W. R. (2001). Negative genetic correlation for adult fitness between sexes reveals ontogenetic conflict in *Drosophila*. Proceedings of the National Academy of Sciences of the United States of America, 98, 1671–1675.

Corl, A., Davis, A. R., Kuchta, S. R., and Sinervo, B. (2010). Selective loss of polymorphic mating types is associated with rapid phenotypic evolution during morphic speciation. Proceedings of the National Academy of Sciences of the United States of America, 107, 4254–4259.

Cornwallis, C. K. and Uller, T. (2010). Towards an evolutionary ecology of sexual traits. Trends in Ecology & Evolution, 25, 145–152.

Cox, R. M. and Calsbeek, R. (2009). Sexually antagonistic selection, sexual dimorphism, and the resolution of intralocus sexual conflict. American Naturalist, 173, 176–187.

Cox, R. M. and Calsbeek, R. (2010). Cryptic sex-ratio bias provides indirect genetic benefits despite sexual conflict. Science, 328, 92–94.

Crispo, E. (2008). Modifying effects of phenotypic plasticity on interactions among natural selection, adaptation and gene flow. Journal of Evolutionary Biology, 21, 1460–1469.

Darwin, C. (1859). The Origin of species by means of natural selection. Harvard University Press, Cambridge, MA.

Day, T. and Bondurianksy, R. (2004). Intralocus sexual conflict can drive the evolution of genomic imprinting. Genetics, 167, 1537–1546.

DeWitt, T. and Scheiner, S. (2004). Plasticity. Functional and Conceptual Approaches. Oxford University Press, Oxford.

DiBattista, J. D., Feldheim, K. A., Garant, D., Gruber, S. H., and Hendry, A. P. (2011). Anthropogenic disturbance and evolutionary parameters: a lemon shark population experiencing habitat loss. Evolutionary Applications, 4, 1–17.

Dieckmann, U. and Doebeli, M. (1999). On the origin of species by sympatric speciation. Nature, 400, 354–357.

Eckert, C. G., Manicacci, D., and Barrett, S. C. H. (1996). Frequency-dependent selection on morph ratios in tristylous *Lythrum salicaria* (Lythraceae). Heredity, 77, 581–588.

Einum, S., Robertsen, G., and Fleming, I. A. (2008). Adaptive landscapes and density-dependent selection in declining salmonid populations: going beyond numerical responses to human disturbance. Evolutionary Applications, 1, 239–251.

Endler, J. A. (1986). Natural selection in the wild. Princeton University Press, Princeton, NJ.

Estes, S. and Arnold, S. J. (2007). Resolving the paradox of stasis: Models with stabilizing selection explain evolutionary divergence on all timescales. American Naturalist, 169, 227–244.

Fedorka, K. M. and Zuk, M. (2005). Sexual conflict and female immune suppression in the cricket, Allonemobious socius. Journal of Evolutionary Biology, 18, 1515–1522.

Feldman, M. W., Brodie, II, E. D., Brodie, III, E. D., and Pfrender, M. E. (2009). The evolutionary origins of beneficial alleles during the repeated adaptation of garter snakes to deadly prey. Proceedings of the National Academy of Sciences of the United States of America 106, 13415–13420.

Feldman, M. W., Brodie, II, E. D., Brodie, III, E. D., and Pfrender, M. E. (2010). Genetic architecture of a feeding adaptation: garter snake (*Thamnophis*) resistance to tetrodotoxin bearing prey Proceedings of the Royal Society B: Biological Sciences, 277, 3317–3326.

Felsenstein, J. (1979). R-selection and K-selection in a completely chaotic population model. American Naturalist, 113, 499–510.

Fisher, R. (1958). The genetical theory of natural selection. Dover, New York.

Fisher, H. S., Mascuch, S. J, and Rosenthal, G. G. (2009). Multivariate male traits misalign with multivariate female preferences in the swordtail fish, *Xiphophorus birchmanni*. Animal Behaviour, 78, 265–269.

Ford, E. B. (1964). Ecological genetics. Chapman and Hall, London.

Forsgren, E., Amundsen, T., Borg, A. A., and Bjelvenmark, J. (2004. Unusually dynamic sex roles in a fish. Nature, 429, 551–554.

Frank, S. A. and Slatkin, M. (1992). Fisher's fundamental theorum theorem of Natural Selection. Trends in Ecology & Evolution, 7, 92–95.

Gavrilets, S. (1999). A dynamical theory of speciation on holey adaptive landscapes. American Naturalist, 154, 1–22.

Gavrilets, S., Arnqvist, G., and Friberg, U. (2001). The evolution of female mate choice by sexual conflict. Proceedings of the Royal Society B: Biological Sciences, 268, 531–539.

Gerhardt, H. C. and Brooks, R. (2009). Experimental analysis of multivariate female choice in Gray treefrogs (*Hyla versicolor*): Evidence for directional and stabilizing selection. Evolution, 63, 2504–2512.

Gosden, T. P. and Svensson, E. I. (2008). Spatial and temporal dynamics in a sexual selection mosaic. Evolution, 62, 845–856.

Grant, P. R. and Grant, B. R. (2002). Unpredictable evolution in a 30-year study of Darwin's Finches. Science, 296, 707–711.

Grant, P. R. and Grant, B. R. (2009). The secondary contact phase of allopatric speciation in Darwin's finches. Proceedings of the National Academy of Sciences of the United States of America, 106, 20141–20148.

Haldane, J. B. S. (1955). On the biochemistry of heterosis, and the stabilization of polymorphism. Proceedings of the Royal Society B: Biological Sciences, 144, 217–220.

Hardling, R., Gosden, T, and Aguilee, R. (2008). Male mating constraints affect mutual mate choice: Prudent male courting and sperm-limited females. American Naturalist, 172, 259–271.

Hendry, A. P. and Gonzalez, A. (2008). Whither adaptation? Biology & Philosophy, 23, 673–699.

Hereford, J., Hansen, T. F., and Houle, D. (2004). Comparing strengths of directional selection: how strong is strong? Evolution, 58, 2133–2143.

Hews, D. K. (1988). Resource defense and sexual selection on male head size in the lizard Uta-Palmeri. American Zoologist, 28, A52–A52.

Hine, E., McGuigan, K., and Blow, M. W. (2011). Natural selection stops the evolution of male attractiveness. Proceedings of the National Academy of Sciences of the United States of America, 108, 3659–64.

Hoekstra, H. E., Hoekstra, J. M., Berrigan, D., Vignieri, S. N., Hoang, A., Hill, C. E., et al. (2001). Strength and tempo of directional selection in the wild. Proceedings of the National Academy of Sciences of the United States of America, 98, 9157–9160.

Huey, R. B., Peterson, C. R., Arnold, S. J., and Porter, W. P. (1989). Hot rocks and not-so-hot rocks: retreat site selection by garter snakes and its thermal consequences. Ecology, 70, 931–944.

Johnstone, R. A., Reynolds, J. D., and Deutsch, J. C. (1996). Mutual mate choice and sex differences in choosiness. Evolution, 50, 1382–1391.

Jones, L. E., Becks, L., Ellner, S. P., Hairston, N. G., Yoshida, T., and Fussmann, G. F. (2009). Rapid contemporary evolution and clonal food web dynamics. Philosophical Transactions of the Royal Society B: Biological Sciences, 364, 1579–1591.

Kaplan, J. (2008). The end of the Adaptive Landscape metaphor? Biology & Philosophy, 23, 625–638.

Kassen, R., Llewellyn, M., and Rainey, P. B. (2004). Ecological constraints on diversification in a model adaptive radiation. Nature, 431, 984–988.

Kerr, B., Riley, M. A., Feldman, M. W., and Bohannan, B. J. M. (2002). Local dispersal promotes biodiversity in a real-life game of rock-paper-scissors. Nature, 418, 171–174.

Kingsolver, J. G. and Diamond, S. E. (2011). Phenotypic selection in natural populations: What limits directional selection? American Naturalist, 177, 346–357.

Kingsolver, J. G. H., Hoekstra, H. E., Hoekstra, J. M., Berrigan, D., Vignieri, S. N., Hill, C. E., et al. (2001). The strength of phenotypic selection in natural populations. American Naturalist, 157, 245–261.

Kokko, H. and Johnstone, R. A. (2002). Why is mutual mate choice not the norm? Operational sex ratios, sex roles and the evolution of sexually dimorphic and monomorphic signalling. Philosophical Transactions of the Royal Society of London B: Biological Sciences, 357, 319–330.

Kopp, M. and Hermisson, J. (2006). The evolution of genetic architecture under frequency-dependent disruptive selection. Evolution, 60, 1537–1550.

Korona, R., Nakatsu, C. H., Forney, L. J., and Lenski, R. E. (1994). Evidence for multiple adaptive peaks from populations of bacteria evolving in a structured habitat. Proceedings of the National Academy of Sciences of the United States of America, 91, 9037–9041.

Kozak, K. H. and Wiens, J. J. (2010). Niche conservatism drives elevational diversity patterns in Appalachian salamanders. American Naturalist, 176, 40–54.

Laforsch, C. and Tollrian, R. (2004). Inducible defenses in multipredator environments: Cyclomorphosis in *Daphnia cucullata*. Ecology 85:2302–2311.

Lande, R. (1980). Sexual dimorphism, sexual selection, and adaptation in polygenic characters. Evolution, 34, 292–305.

Lande, R. and Arnold, S. J. (1983). The measurement of selection on correlated characters. Evolution, 37, 1210–1226.

Lank, D. B., Smith, C. M., Hanotte, O., Burke, T., and Cooke, F. (1995). Genetic polymorphism for alternative mating behaviour in lekking male ruff *Philomachus pugnax*. Nature, 378, 59–62.

Le Boeuf, B. J. (1974). Male-male competition and mating success in elephant seals. American Zoologist, 14, 163–176.

Lenski, R. E. and Travisano, M. (1994). Dynamics of adaptation and diversification – A 10,000 generation experiment with bacterial-populations. Proceedings of the National Academy of Sciences of the United States of America, 91, 6808–6814.

Losos, J. B., Schoener, T. W., and Spiller, D. A. (2004). Predator-induced behaviour shifts and natural selection in field-experimental lizard populations. Nature, 432, 505–508.

Marchinko, K. B. (2009). Predation's role in repeated phenotypic and genetic divergence of armor in threespine stickleback. Evolution, 63, 127–138.

Maynard Smith, J. (1982). Evolution and the Theory of Games. Cambridge University Press, Cambridge.

Mayr, E., and Provine, W. (1980). The Evolutionary Synthesis: perspectives on the unification of biology. Harvard University Press, Cambridge MA.

McCandlish, D. M. (2011). Visualizing fitness landscapes. Evolution, 65, 1544–1558.

McGlothlin, J. W., Parker, P. G., Nolan, V., and Ketterson, E. D. (2005). Correlational selection leads to genetic integration of body size and an attractive plumage trait in dark-eyed juncos. Evolution, 59, 658–671.

McGuigan, K., Chenoweth, S. F., and Blows, M. W. (2005). Phenotypic divergence along lines of genetic variance. American Naturalist, 165, 32–43.

McGuigan, K., Van Homrigh, A., and Blows, M. W. (2008). Genetic analysis of female preference functions as function-valued traits. American Naturalist, 172, 194–202.

Meyer, J. R. and Kassen, R. (2007). The effects of competition and predation on diversification in a model adaptive radiation. Nature, 446, 432–435.

Mueller, L. D. (1988a). Density-dependent population-growth and natural-selection in food-limited environments – The Drosophila model. American Naturalist, 132, 786–809.

Mueller, L. D. (1988b). Evolution of competitive ability in *Drosophila* by density-dependent natural selection. Proceedings of the National Academy of Sciences of the United States of America, 85, 4383–4386.

Mueller, L. D. and Ayala, F. J. (1981). Trade-off between r-selection and K-selection in *Drosophila* populations. Proceedings of the National Academy of Sciences of the United States of America, 78, 1303–1305.

Nosil, P. and Crespi, B. J. (2006). Experimental evidence that predation promotes divergence in adaptive radiation. Proceedings of the National Academy of Sciences of the United States of America, 103, 9090–9095.

Nosil, P., Crespi, B. J., and Sandoval, C. P. (2002). Host-plant adaptation drives the parallel evolution of reproductive isolation. Nature, 417, 440–443.

Parker, G. A. (1979). Sexual selection and sexual conflict. In M. S. Blum and N. A. Blum (eds.) Sexual selection and reproductive competition in insects. Academic Press, New York, pp. 123–166.

Phillips, P. C. and Arnold, S. J. (1989). Visualizing multivariate selection. Evolution, 43, 1209–1222.

Pigliucci, M. (2001). Phenotypic Plasticity: Beyond Nature and Nurture (Syntheses in Ecology and Evolution). Johns Hopkins University Press, Baltimore, MD.

Pigliucci, M. (2008). Sewall Wright's adaptive landscapes: 1932 vs. 1988. Biology & Philosophy, 23, 591–603.

Pischedda, A. and Chippindale, A. K. (2006). Intralocus sexual conflict diminishes the benefits of sexual selection. PLoS Biology, 4, e356.

Pitnick, S., Miller, G. T., Reagan, J., and Holland, B. (2001). Males' evolutionary responses to experimental removal of sexual selection. Proceedings of the Royal Society B: Biological Sciences, 268, 1071–1080.

Pitt, J. N. and Ferre-D'Amare, A. R. (2010). Rapid construction of empirical RNA fitness landscapes. Science, 330, 376–379.

Poulton, E. B. (1884). Notes upon, or suggested by, the colors, markings, and protective attitudes of certain lepidopterous larvae and pupae, and of a phytophagous hymenopterous larva. Transactions of the Entomology Society of London, 1884, 27–60.

Provine, W. (1986). Sewall Wright and Evolutionary Biology. University of Chicago Press, Chicago, IL.

Qvarnstrom, A., Brommer, J. E., and Gustafsson, L. (2006). Evolutionary genetics—Evolution of mate choice in the wild—Reply. Nature, 444, E16–E17.

Qvarnstrom, A., Part, T., and Sheldon, B. C. (2000). Adaptive plasticity in mate preference linked to differences in reproductive effort. Nature, 405, 344–347.

Rainey, P. B. and Travisano, M. (1998). Adaptive radiation in a heterogeneous environment. Nature, 394, 69–72.

Reimchen, T. E. and Nosil, P. (2002). Temporal variation in divergent selection on spine number in threespine stickleback. Evolution, 56, 2472–2483.

Rice, W. R. and Chippindale, A. K. (2001). Intersexual ontogenetic conflict. Journal of Evolutionary Biology, 14, 685–693.

Rice, W. R. and Holland, B. (1997). The enemies within: intergenomic conflict, interlocus contest evolution (ICE), and the intraspecific Red Queen. Behavioral Ecology and Sociobiology, 41, 1–10.

Rosenqvist, G. (1990). Male mate choice and female-female competition for mates in the pipefish *Nerophis ophiodon*. Animal Behavior, 39, 1110–1115.

Rosenzweig, M. L. (1978). Competition speciation. Biological Journal of the Linnean Society, 10, 275–289.

Roughgarden, J. (1971). Density-dependent natural selection. Ecology 52:453–468.

Rundle, H. D., Chenoweth, S. F., and Blows, M. W. (2009). The diversification of mate preferences by natural and sexual selection. Journal of Evolutionary Biology, 22, 1608–1615.

Rundle, H. D., Vamosi, S. M., and Schluter, D. (2003a). Experimental test of predation's effect on divergent selection during character displacement in sticklebacks. Proceedings of the National Academy of Sciences of the United States of America, 100, 14943–14948.

Rundle, H. D., Vamosi, S. M., and Schluter, D. (2003b). Experimental test of predation's effect on divergent selection during character displacement in sticlebacks. Proceedings of the National Academy of Sciences of the United States of America, 100, 14943–14948.

Ryan, M. J., Pease, C. M., and Morris, M. R. (1992). A genetic polymorphism in the swordtail *Xiphophorus nigrensis*: Testing the prediction of equal fitnesses. American Naturalist, 139, 21–31.

Schluter, D. (1988). Estimating the form of natural selection on a quantitative trait. Evolution, 42, 849–861.

Schluter, D. (1995). Adaptive radiation in sticklebacks: Trade-offs in feeding performance and growth. Ecology, 76, 82–90.

Schluter, D. (2000). The ecology of adaptive radiation. Oxford University Press, New York.

Schluter, D. and Nychka, D. (1994). Exploring fitness surfaces. American Naturalist, 143, 597–616.

Schmitt, J. (1993). Reaction norms of morphological and life-history traits to light availibility in *Impatiens capensis*. Evolution, 47, 1654–1668.

Schoustra, S. E., Bataillon, T., Gifford, D. R., and Kassen, R. (2009). The properties of adaptive walks in evolving populations of fungus. PLoS Biology, 7, e1000250.

Seehausen, O., Takimoto, G., Roy, D., and Jokela, J. (2008). Speciation reversal and biodiversity dynamics with hybridization in changing environments. Molecular Ecology, 17, 30–44.

Shuster, M. and Wade, M. J. (1992). Equal mating success among male reproductive strategies in a marine isopod. Nature, 350, 606–661.

Siepielski, A. M., DiBattista, J. D., and Carlson, S. M. (2009). It's about time: the temporal dynamics of phenotypic selection in the wild. Ecology Letters, 12, 1261–1276.

Simpson, G. G. (1944). The tempo and mode of evolution. National Academic Press, New York.

Sinervo, B. and Basolo, A. L. (1996). Testing adaptation using phenotypic manipulations. In M. R. Rose and G. V. Lauder (eds.) Adaption. Academic Press, New York, pp. 149–185.

Sinervo, B. and Calsbeek, R. (2006). The developmental, physiological, neural, and genetical causes and consequences of frequency-dependent selection in the wild. Annual Review of Ecology Evolution and Systematics, 37, 581–610.

Sinervo, B. and Huey, R. B. (1990). Allometric engineering: an experimental test of the causes of interpopulational differences in locomotor performance. Science, 248, 1106–1109.

Sinervo, B. and Lively, C. M. (1996). The rock-paper-scissors game and the evolution of alternative male reproductive strategies. Nature, 380, 240–243.

Sinervo, B. and Svensson, E. (2002). Correlational selection and the evolution of genomic architecture. Heredity, 89, 329–338.

Sinervo, B., Doughty, P., Huey, R. B., and Zamudio, K. (1992). Allometric engineering: a causal analysis of natural selection on offspring size. Science, 258, 1927–1930.

Sinervo, B., Svensson, E., and Comendant, T. (2000). Density cycles and an offspring quantity and quality game driven by natural selection. Nature, 406, 985–988.

Smith, D. C. and Van Buskirk, J. (1995). Phenotypic design, plasticity, and ecological performance in two tadpole species. American Naturalist, 145, 211–233.

Sobel, J. M., Chen, G. F., Watt, L. R., and Schemske, D. W. (2010). The biology of speciation. Evolution, 64, 295–315.

Stamps, J. A. and Krishnan, V. V. (1995). Territory acquisition in lizards: III. Competing for space. Animal Behaviour, 49, 679–693.

Svensson, E. I. and Friberg, M. (2007). Selective predation on wing morphology in sympatric damselflies. American Naturalist, 170, 101–112.

Svensson, E. I. and Gosden, T. P. (2007). Contemporary evolution of secondary sexual traits in the wild. Functional Ecology, 21, 422–433.

Svensson, E. I., Karlsson, K., Friberg, M., and Eroukhmanoff, F. (2007). Gender differences in species recognition and the evolution of asymmetric sexual isolation. Current Biology, 17, 1943–1947.

Svensson, E. and Sinervo, B. (2000). Experimental excursions on adaptive landscapes: density dependent selection on egg size. Evolution, 54, 1396–1403.

Svensson, E. I. and Sinervo, B. (2004). Spatial scale and temporal component of selection in side-blotched lizards. American Naturalist, 163, 726–734.

Svensson, E., Abbot, J., and Hardling, R. (2005). Female polymorphism, frequency dependence, and rapid evolutionary dynamics in natural populations. American Naturalist, 165, 567–576.

Svensson, E., Sinervo, B., and Comendant, T. (2001). Density-dependent competition and selection on immune function in genetic lizard morphs. Proceedings of the National Academy of Sciences of the United States of America, 98, 12561–12565.

Svensson, E. I., Eroukhmanoff, F., Karlsson, K., Runemark, A., and Brodin, A. (2010). A role for learning in population divergence of mate preferences. Evolution, 64, 3101–3113.

Taylor, E. B., Boughman, J. W., Groenenboom, M. Sniatynski, M. Schluter, D., and Gow, J. L. (2006). Speciation in reverse: morphological and genetic evidence of the collapse of a three-spined stickleback (*Gasterosteus aculeatus*) species pair. Molecular Ecology, 15, 343–355.

Thompson, J. N. (2005). The geographic mosaic of coevolution. University of Chicago Press, Chicago, IL.

Uyeda, J. C., Hansen, T. F., Arnold, S. J., and Pienaar, J. (2011). The million-year wait for macroevolutionary bursts. Proceedings of the National Academy of Sciences, of the United States of America, 108, 15908–15913.

Vamosi, S. M. and Schluter, D. (2002). Impacts of trout predation on fitness of sympatric sticklebacks and their hybrids. Proceedings of the Royal Society B: Biological Sciences, 269, 923–930.

Van Buskirk, J. and McCollum, S. A. (2000a). Functional mechanisms of an inducible defence in tadpoles: morphology and behaviour influence mortality risk from predation. Journal of Evolutionary Biology, 13, 336–347.

Van Buskirk, J. and McCollum, S. A. (2000b). Influence of tail shape on tadpole swimming performance. Journal of Experimental Biology, 203, 2149–2158.

Van Buskirk, J. and Relyea, R. A. (1998). Selection for phenotypic plasticity in Rana sylvatica tadpoles. Biological Journal of the Linnean Society, 65, 301–328.

Vasi, F., Travisano, M., and Lenski, R. E. (1994). Long-term experimental evolution in *Escherichia coli*. 2. Changes in life-history traits during adaptation to a seasonal environment. American Naturalist, 144, 432–456.

Weese, D. J., Gordon, S. P., Hendry, A. P., and Kinnison, M. T. (2010). Spatiotemporal variation in linear natural selection on body color in wild guppies (*Poecilia reticulata*). Evolution, 64, 1802–1815.

West-Eberhard, M. J. (1989). Phenotypic plasticity and the origins of diversity. Annual Review of Ecology, Evolution, and Systematics 20, 249–279.

West-Eberhard, M. J. (2003). Developmental plasticity and evolution. Oxford University Press, New York.

West-Eberhard, M. J. (2005). Phenotypic accommodation: Adaptive innovation due to developmental plasticity. Journal of Experimental Zoology Part B – Molecular and Developmental Evolution, 304B, 610–618.

Wiens, J. J. and Graham, C. H. (2005). Niche conservatism: Integrating evolution, ecology, and conservation biology. Annual Review of Ecology Evolution and Systematics, 36, 5, 19–539.

Wiens, J. J., Ackerly, D. D., Allen, A. P., Anacker, B. L., Buckley, L. B., Cornell, H. V., et al. (2010). Niche conservatism as an emerging principle in ecology and conservation biology. Ecology Letters, 13, 1310–1324.

Wolf, J. B. and Brodie, E. D. (1998). The coadaptation of parental and offspring characters. Evolution, 52, 299–308.

Wolf, J. B., Wade, M. J., and Brodie III, E. D. (2004). The genotype-environment interaction and evolution when the environment contains genes. In T. DeWitt and S. Scheiner (eds.) Phenotypic Plasticity. Functional and Conceptual Approaches. Oxford University Press, Oxford, pp. 173–190.

Wright, S. (1932). The roles of mutation, inbreeding, crossbreeding, and selection in evolution. Proceedings of the Sixth Annual Congress of Genetics, 1, 356–366.

Yoshida, T., Jones, L. E., Ellner, S. P., Fussmann, G. F., and Hairston, N. G. (2003). Rapid evolution drives ecological dynamics in a predator-prey system. Nature, 424, 303–306.

Young, K., Brodie, E. D., and Brodie, III, E. D. (2004). How the horned lizard got its horns. Science, 304, 65.

CHAPTER 8

The Adaptive Landscape in Sexual Selection Research

Adam G. Jones, Nicholas L. Ratterman, and Kimberly A. Paczolt

8.1 Introduction

Ever since the concepts of descent with modification and selection were embraced by natural scientists in the nineteenth and early twentieth centuries (Darwin 1859, 1871; Fisher 1930), naturalists and evolutionary biologists have sought to explain the evolution of conspicuous ornaments and armaments in a bewildering variety of organisms. The coarse explanation for the evolution of most sexually dimorphic traits involved in courtship and mating is that they evolved as a consequence of sexual selection (Darwin 1871), that is, selection arising from competition for access to mates or fertilization opportunities (Darwin 1871; Andersson, 1994; Jones and Ratterman 2009). The realization that sexual selection is involved in producing elaborate traits is not a sufficiently complete description, however, as this coarse explanation doesn't fully address how sexual selection produces trait elaboration, why different populations or species have evolved different degrees of sexual dimorphism, or what factors prevent traits involved in mating competition from becoming infinitely large and complex. To answer these questions, a useful first step is to realize that sexual selection is, after all, a form of selection, and we must try to understand sexual selection in the context of formal selection theory. In this chapter, we consider the process of sexual selection in light of the concepts of Adaptive Landscapes and individual selection surfaces, which describe the relationship between relative fitness and phenotypic traits. First, we provide an intuitive explanation of the logic underlying the application of Adaptive Landscapes to sexual selection research. Second, we summarize a particularly illuminating subset of studies, which serve to illustrate the utility of the Adaptive Landscape concept in sexual selection research. Finally, we suggest some areas in the realm of sexual selection research to which the Adaptive Landscape concept could be profitably employed in future studies.

8.2 Sexual selection *is* selection

Part of the appeal of the study of sexual selection is that it occupies a fascinating position at the nexus between evolutionary biology and behavior. However, this unusual position is a double-edged sword. On one hand, the interdisciplinary nature of the study of sexual selection allows scientists to bring a wide array of tools to bear in a multifaceted, intellectually stimulating research arena. On the other hand, it can be easy to lose sight of the fact that sexual selection research should be conducted in the context of the theoretical and analytical machinery that has emerged from decades of research on natural and artificial selection. Consequently, many behavioral studies conducted under the auspices of sexual selection research are very difficult to relate directly to the action of sexual selection in natural populations or to the theoretical expectations of explicit models describing the dynamics of traits under selection.

For example, many species are amenable to study in the laboratory, and laboratory-based studies of mating preferences or male–male competition can elucidate aspects of behavior that are certainly relevant to the process of sexual selection (Bischoff et al. 1985; Klump and Gerhardt 1987; Rosenthal and

Evans 1998). Indeed, much of our understanding of sexually selected traits is based on such studies (Andersson 1994). However, if our goal is to understand the evolution of traits involved in male–male combat or mate choice, including the evolution of the preferences themselves, then we must take care to study these phenomena in a way that can be related to events occurring in natural populations. Many challenges arise in this endeavor, because myriad environmental and social factors have the potential to affect the nature of the interactions between rival males or to prevent individuals from successfully acting on their preferences. In addition, laboratory studies often isolate and study male–male interactions separately from female preferences, for very good reasons, but these different sources of selection on traits and preferences must somehow be combined to produce a measure of total selection on the phenotypes of interest (Hunt et al. 2009). Thus, to understand sexual selection as an evolutionary phenomenon, we must find a way to relate mating behavior and mating preferences to selection on particular traits in natural populations (Andersson 1982; Endler 1986; Grether 1996; Kingsolver et al. 2001; Kruuk et al. 2002; Serbezov et al. 2010). Such an endeavor will require carefully designed laboratory studies of intra- and intersexual selection, possibly involving both pre- and postcopulatory events, interpreted in light of studies of mating dynamics in natural populations whenever possible. The Adaptive Landscape concept provides an excellent framework in which to pursue these goals, because it provides explicit connections between the phenotypic traits (including behavioral traits) that play a central role in sexual selection and the mathematical machinery of selection theory, which tells us how evolving populations are expected to respond to selection (Hunt et al. 2009).

8.3 The Adaptive Landscape and the individual selection surface

As discussed by other authors in this volume (Chapter 2), the concept of the Adaptive Landscape traces back to Sewall Wright (1931, 1932). In a three-dimensional representation of Wright's Adaptive Landscape, elevation on the landscape represents "adaptiveness" and the other axes represent allele frequencies. This Adaptive Landscape serves as the basis of Wright's "shifting balance" model (Chapter 6) and is not especially useful in the study of the evolution of quantitative phenotypic traits (Arnold et al. 2001). Because most sexually selected traits are quantitative, with multiple loci and environmental effects contributing to variation among individuals within populations, we need a formulation of the Adaptive Landscape that can be used with such traits.

Building on George G. Simpson's "adaptive zones" (Simpson 1953), Russell Lande (1976) provided an explicit definition of the Adaptive Landscape for quantitative traits. In Lande's version, which remains the modern manifestation of the concept (Arnold et al. 2001), elevation reflects mean population fitness, and the other axes represent population mean phenotypic values for various traits (Fig. 8.1). The peaks on these landscapes correspond to local optima in phenotypic space, and natural selection will usually drive populations to climb these peaks until the multivariate phenotypic mean coincides with the nearest optimum (Lande 1976, 1979). This representation of selection on quantitative traits is appealing, especially in the context of the evolution of multiple traits, because it provides an intuitively understandable conceptualization of how natural selection operates in natural populations. Simpson's Adaptive Landscape thus plays a central role in selection theory and in the development of empirical approaches for the study of selection (Lande and Arnold 1983).

Despite its intuitive appeal and theoretical utility, however, a strict application of the Adaptive Landscape concept has limitations. For example, the true Adaptive Landscape is difficult, if not impossible, to measure, because such an exercise would require data on mean population fitness as a function of the population mean phenotype across a wide range of possible phenotypic means. In practice, most studies of selection measure aspects of the "individual selection surface" (Fig. 8.1), which describes average individual fitness as a function of phenotypic values of individuals (Kirkpatrick 1982; Phillips and Arnold 1989; Whitlock 1995; Schluter 2000). These individual selection surfaces provide

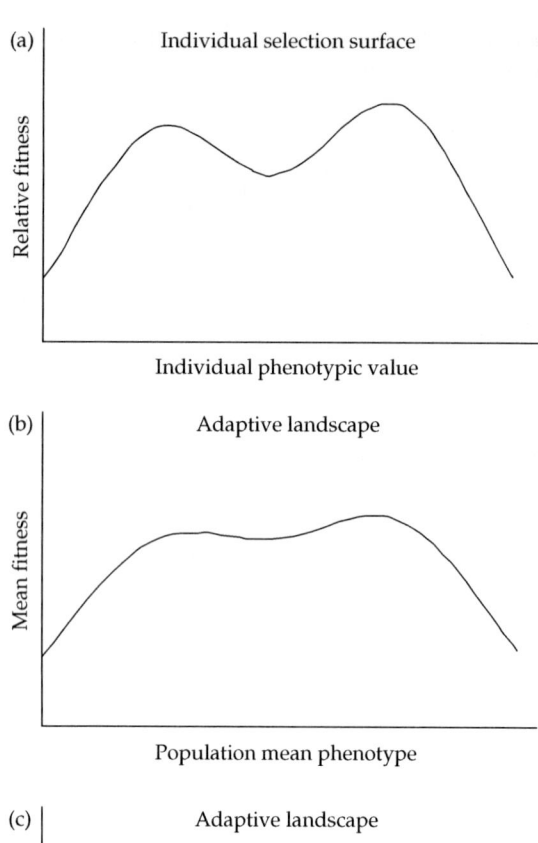

Figure 8.1 A single-trait individual selection surface compared to the Adaptive Landscape. The individual selection surface (a), also called the individual fitness surface, relates expected relative fitness to individual trait values. In contrast, the Adaptive Landscape (b, c) shows the relationship between population mean fitness and population mean phenotype. In this figure, the two Adaptive Landscapes have an underlying individual selection surface identical to the one depicted in the top panel (a). However, the shape of the Adaptive Landscape depends on the phenotypic distribution in the population, and Adaptive Landscapes tend to be less jagged than individual selection surfaces. In the middle panel (b), the population has a relatively small amount of phenotypic variance compared to the bottom panel (c). An increase in phenotypic variation has a smoothing effect on the Adaptive Landscape, such that in the bottom panel (c) the two peaks in the underlying individual selection surface are not apparent from inspection of the Adaptive Landscape. In studies of sexual selection, usually only the individual selection surface is measurable, even though the Adaptive Landscape is the central principle in selection theory. This apparent disconnect between theory and empiricism actually is not a major problem, because features of the individual selection surface indicate the shape of the Adaptive Landscape near the population mean. Reproduced from Whitlock (1995) with permission from John Wiley and Sons.

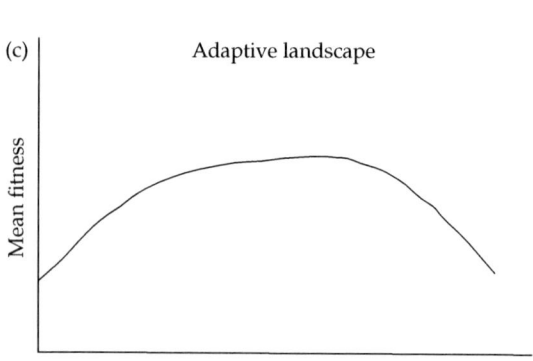

insights into the shape of Adaptive Landscape in the vicinity of the population mean (Arnold et al. 2001), but they are far from a complete characterization of an Adaptive Landscape across a wide swath of phenotypic space, as Adaptive Landscapes often are depicted for heuristic purposes (Wright 1932; Simpson 1953; Kirkpatrick 1982; Mallet 2007). Compounding this problem is the fact that individual selection surfaces are virtually impossible to measure in regions of phenotypic space that are outside of the range of phenotypic variation already present in the population. However, the fact that the Platonic ideal of the Adaptive Landscape is empirically inaccessible does not render the concept useless for the study of evolution.

Indeed, individual selection surfaces provide the necessary information for the diagnosis of the nature and causes of selection acting in populations. For example, measurement of the selection surface near the population mean provides an estimate of the magnitude of directional and non-linear (i.e. stabilizing or disruptive) selection (Lande and Arnold 1983). If the population is on a steep slope, then it is subject to directional selection, whereas a population at a peak or in a valley would be experiencing stabilizing or disruptive selection, respectively (Phillips and Arnold 1989; Brodie et al. 1995; Hunt et al. 2009). The individual selection surface is also useful for the characterization of patterns of correlational selection, where the fitness of an individual depends on particular combinations of trait values (Lande and Arnold 1983; Brodie 1992; Benkman 2003). Thus, empirical and theoretical studies of individual selection surfaces, with a particular emphasis on the factors determining the shape of the surface, the position of the population mean on the surface, and causes of different surfaces in different environments, have the potential to answer fundamental questions in the study of sexual selection.

8.4 Peculiarities of the Adaptive Landscape in sexual selection research

Sexual selection complicates the application of the Adaptive Landscape concept for two reasons: first, both intra- and intersexual selection contain an element of frequency dependence (Lande 1981) and, second, in the case of intersexual selection, selection on the ornament can result in a correlated response that changes mate preferences and alters the Adaptive Landscape (Lande 1981; Mead and Arnold 2004). In the former case, frequency dependence simply means that an individual's absolute fitness depends on the distribution of the trait in the population (Lande 1981). This situation holds for both intra- and intersexual selection, because in either case the fitness of an individual of the sexually selected sex depends upon the phenotypes of the other members of the population (also see Chapter 6 for a discussion of how social interactions affect Adaptive Landscapes). If we take intersexual selection as an example, then an open-ended female preference for males with more extreme ornaments will result in directional selection on ornamentation, so the individual selection surface will give the appearance of a steep slope. Indeed, if some of the variation in ornamentation is due to additive genetic effects, then in the absence of opposing forces the trait will respond to selection (Lande 1976), resulting in an increase in mean ornamentation. However, mean population fitness, which is determined by female fecundity, will remain unchanged, so the population will not climb the Adaptive Landscape. Rather, the slope will remain steep while sliding in the direction of the change in the phenotypic mean. Under some circumstances, particularly when sexual conflict is involved, the mean population fitness may actually decrease as a consequence of sexual selection, even though the population appears to be scaling an adaptive peak on a per-generation basis (Lande 1976). Metaphorically, this situation can be envisioned as a mountaineer climbing a peak on an island that's sinking into the ocean. From the mountaineer's perspective, a peak climbing process is occurring. However, if we observe the situation from a distance, we can see that the gains in elevation are illusory (see Chapter 4 for a more complete discussion of frequency dependence in the context of Adaptive Landscapes).

The second complexity in the application of the Adaptive Landscape to sexual selection research occurs when mate choice contributes to the selection acting on traits. If a population is characterized by additive genetic variation in mating preferences and additive genetic variation in the traits upon which the preferences act, then the process of mate choice will result in an intersexual genetic correlation between ornaments and preferences (Fisher 1930; Mead and Arnold 2004). Indeed, this genetic correlation is a key component of the Fisherian runaway sexual selection process: selection on the trait (due to mate choice) causes the distribution of mating preferences to change as a correlated response to selection, sometimes resulting in a positive feedback loop that leads to greater and greater trait elaboration (Fisher 1930; Lande 1981). However, this situation also means that selection on the trait changes the Adaptive Landscape for the trait by changing the distribution of mating preferences in the population. Once more, the metaphor of a static Adaptive Landscape, upon which a population struggles mightily to scale a nearby fitness peak, seems to be an incomplete description of the process of sexual selection.

Despite these apparent problems, the concept of Adaptive Landscapes and, more specifically, individual selection surfaces can be applied fruitfully to sexual selection research. The first step is to abandon the notion of static Adaptive Landscapes. This idea is not new, as Simpson (1953) envisioned adaptive zones as dynamic products of the environment. Thus, the dynamic landscape concept applies to traits subject to natural selection (Arnold et al. 2001) as well as those subject to sexual selection. One implication of this idea is that we need not concern ourselves with features of the Adaptive Landscape that lie far away from the population mean, as those features may change substantially before the population has time to evolve toward them. Hence, we should focus on features of the Adaptive Landscape in the immediate vicinity of the population mean, and these features can be understood by quantifying the individual selection surface.

Once we embrace the dynamic Adaptive Landscape, then the study of sexual selection becomes remarkably similar to the study of natural selection. For example, we can use the individual selection surface to investigate the strength of directional selection acting on traits by estimating linear selection gradients, β, through standard multiple regression (Lande and Arnold 1983) or logistic regression models (Janzen and Stern 1998). Higher order selection gradients (γ) can be estimated to quantify the strength of non-linear and correlational selection (Lande and Arnold 1983). In some cases, canonical analysis (i.e. rotation of axes) can help, particularly in the estimation of non-linear selection (Phillips and Arnold 1989; Blows and Brooks 2003). Selection surfaces can also be visualized by other means, through the use of cubic splines (Schluter 1988) or projection pursuit regression (Schluter and Nychka 1994; Blows et al. 2003), for example, to provide graphical depictions of the relationship between trait values or trait combinations and fitness. A more in-depth treatment of the statistical approaches and challenges associated with the study of Adaptive Landscapes can be found elsewhere in this volume (Chapter 9). Importantly, many of these methods of quantifying selection can be related to selection theory in a way that allows inferences to be drawn regarding the response to selection and the maintenance of genetic variation over short evolutionary timescales. Thus, scientists studying sexual selection can access a very powerful inferential toolkit if they are willing to take advantage of the Adaptive Landscape concept.

8.5 Achievements of the Adaptive Landscape in sexual selection research

The concepts of the Adaptive Landscape and the individual selection surface have already been used to address many questions in sexual selection research and these applications are paying dividends. In this section, we summarize some of the progress that has been made so far on this front, but the subtext is that wider adoption of this way of thinking would result in even more profound insights. Thus far, application of the Adaptive Landscape school of thought has led to advances in understanding the causes of sexual selection, the nature of selection acting on traits and mating preferences, the realization that sexual selection is amazingly variable over time and space, and the importance of treating sexual selection as a multivariate problem. However, the use of individual selection surfaces in the study of sexual selection is a relatively young enterprise, so our review reflects the descriptive state of a field that has yet to establish generalities on many important fronts.

8.5.1 The Bateman gradient as the cause of sexual selection

One type of individual selection surface of special relevance to sexual selection is the Bateman relationship (Bateman 1948; Arnold and Duvall 1994; Jones and Ratterman 2009), which describes expected reproductive success, defined as the number of offspring produced by an individual, as a function of individual mating success, measured as the number of times an individual mates successfully during the breeding season of interest. In this sense, the Bateman relationship is an individual selection surface with mating success as the trait and reproductive success as the measure of fitness. The Bateman gradient (Arnold and Duvall 1994; Andersson and Iwasa 1996), which is the linear regression of reproductive success on mating success, measures the strength of directional selection on mating success. Thus, if the Bateman gradient is steep, an increase in mating success confers an increase in fitness and any trait correlated with mating success will be under directional precopulatory sexual selection (Arnold and Duvall 1994; Jones 2009). If the Bateman gradient is shallow, however, then precopulatory sexual selection on all traits will be weak (Arnold and Duvall 1994; Jones 2009). It's worth mentioning that the Bateman gradient is not precisely the same as a selection gradient on a typical phenotypic trait, because the trait "mating success" is a product of an individual's competitive ability, the ecological setting for mating competition (Emlen and Oring 1977), and the social environment (Chapter 6). Consequently, the response of the trait "mating success" to selection may not be easily predicted from the breeder's equation. Nevertheless, the consideration of the Bateman gradient in an individual selection surface context has substantial heuristic value.

The significance of the Bateman relationship is that it provides a quantifiable measure of the potential for mating patterns in a particular population to produce sexual selection on males and females (Arnold and Duvall 1994; Jones 2009). Thus, by focusing on the Bateman gradient, researchers can begin to diagnose the factors responsible for differences in Bateman gradients between the sexes, among populations, and among species. For example, in species such as fruit flies and rough-skinned newts, sexual selection acts more strongly on males than on females, and males have steeper Bateman gradients compared to females (Bateman 1948; Arnold and Duvall 1994; Jones et al. 2002). In both of these species, male reproductive success appears to be limited primarily by access to mates, whereas female reproductive success is limited by the rate at which females can produce eggs (Fig. 8.2). Not all species show this pattern, however. For example, in some species of pipefish, which are characterized by male pregnancy, sexual selection acts more strongly on females than on males, and females have a steeper Bateman gradient compared to that of males (Jones et al. 2000; Fig. 8.2). Ultimately, male pregnancy is the cause of this reversal of sexual selection and the steeper Bateman gradients of female pipefish, as the substantial male investment in parental care associated with this unique reproductive mode makes males a limiting resource for reproduction in many pipefish populations (Berglund and Rosenqvist 1993). The bottom line is that the Bateman gradient provides a useful framework for relating mating patterns in populations to the expected intensity of sexual selection. Consequently, studies focusing on the factors shaping the Bateman relationship have the potential to reveal the ecological and social settings responsible for different intensities of sexual selection between the sexes, among populations, and among species (Jones and Ratterman 2009).

8.5.2 The quantification of directional sexual selection on phenotypic traits

Perhaps the most useful and widespread application of the individual selection surface in sexual selection research concerns the measurement of the actual intensity of directional selection on phenotypic traits arising from competition for access to mates or fertilization opportunities. Many such studies have been conducted, involving dozens of taxa in a wide variety of taxonomic groups (see reviews in Endler 1986; Kingsolver et al. 2001; and Hunt et al. 2009), and they have revealed directional selection on a wide variety of phenotypic traits as a consequence of both intra- and intersexual selection. Here, we highlight a handful of representative examples from the plethora of excellent studies that have been published.

Intrasexual selection refers to sexual selection arising from competitive interactions between individuals of the same sex, a phenomenon that Darwin described as "The Law of Battle" (Darwin 1871). This process has resulted in the evolution of some exceptionally dimorphic creatures, such as elephant seals, red deer, stag beetles and so forth, and in many cases research involving the quantification of selection has explicitly shown that contemporary sexual selection is acting on secondary sexual traits (Kingsolver et al. 2001). For example, coho salmon develop a hooked snout, a dorsal hump, and striking red coloration during the breeding season, and each of these traits is more pronounced in males than in females (Fleming and Gross 1994). By

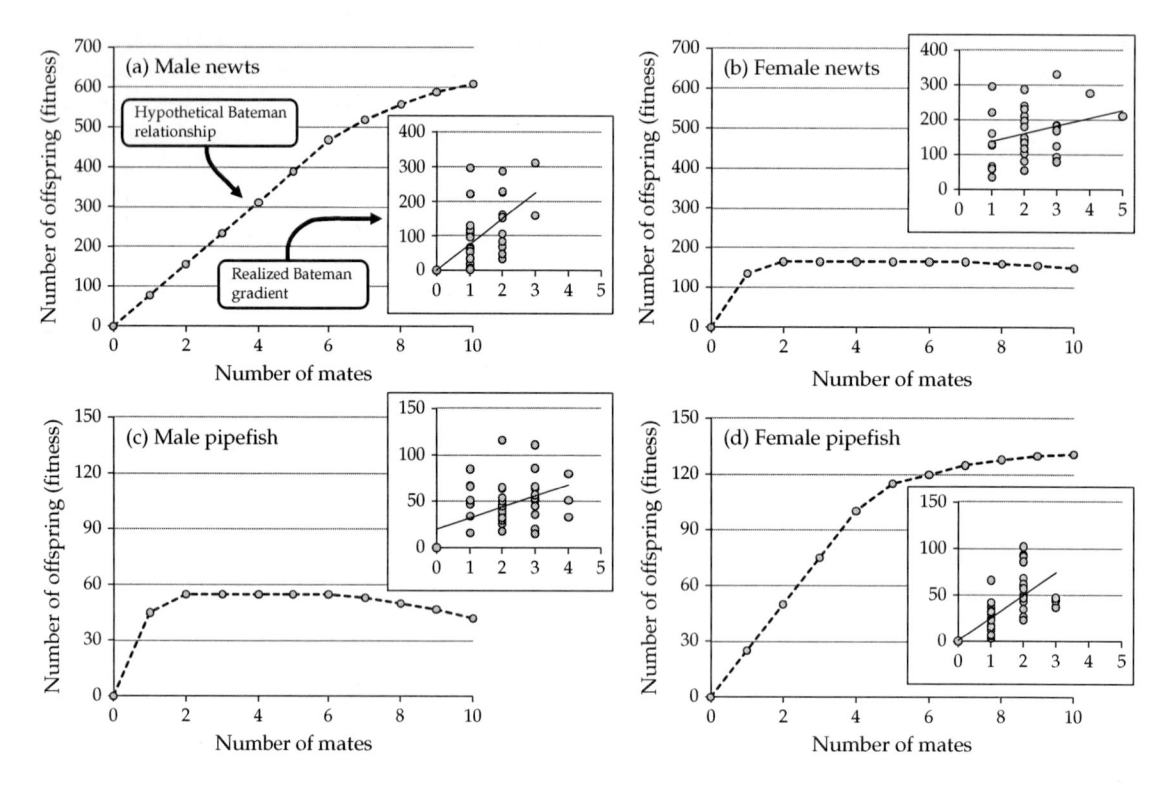

Figure 8.2 Bateman relationships and Bateman gradients for rough-skinned newts and broad-nosed pipefish. The top panels (a, b) show hypothesized Bateman relationships, which are essentially individual selection surfaces for mating success (i.e. number of mates), of male and female rough-skinned newts. Male newts provide no parental care and are capable of mating with many females per breeding season, so the number of offspring they sire presumably increases roughly linearly with their number of mates. Eventually, however, the relationship will level off when the males start running out of time or sperm. Female newts, on the other hand, seem to acquire enough sperm to fertilize all of their eggs in just a few matings. If sexual conflict occurs, we might even expect female fitness to drop somewhat when a female has many more mates than the optimal number. The insets show Bateman gradients, which can be heuristically described as the slope of the Bateman relationship at the population mean. This definition is not precisely correct, since the Bateman relationship is a discontinuous function, but it illustrates that the Bateman gradient is the linear selection gradient on mating success. In newts, males have steeper Bateman gradients than females, and sexual selection acts more strongly on males than on females in this species. In contrast, broad-nosed pipefish are sex-role reversed, with selection acting more strongly on females than on males. As predicted, females (d) exhibit steeper Bateman gradients than do males (c). In terms of Bateman relationships, males seem to be able to fill their brood pouches after one or a few matings (c), whereas females continue to benefit from mating until they start to run out of ripe ova (d).

observing spawning activity in experimental populations and using estimates of fertilization success as a measure of fitness, Fleming and Gross (1994) estimated selection gradients on male traits. They concluded that both male body size and male snout length were subject to strong positive directional sexual selection (Fig. 8.3), because larger males with longer snouts were better fighters. Similarly, a transplantation experiment involving territorial males of the bluehead wrasse showed that directional intrasexual selection favored larger body size and longer pectoral fins in males (Warner and Schultz 1992). From these examples and many other such studies, we can conclude that directional selection on secondary sexual traits arising from intrasexual selection is common, at least in terms of the components of fitness related to mating and reproductive success (Hunt et al. 2009).

Intersexual selection has arguably been studied even more intensively than intrasexual selection and numerous empirical studies have confirmed its occurrence (Kingsolver et al. 2001; Chenoweth and

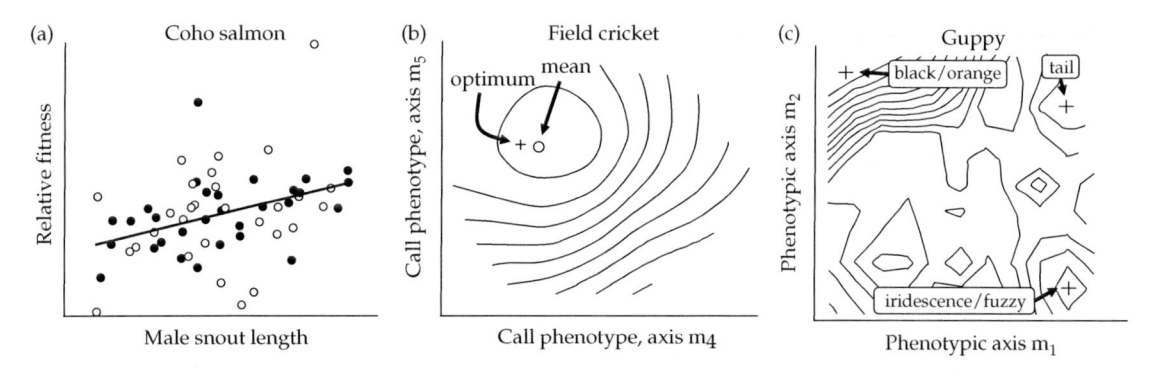

Figure 8.3 Sample individual selection surfaces that have been estimated from various types of organisms. In coho salmon (a), males use their hooked snouts to fight rival males during the breeding season, and males with longer snouts enjoy greater mating success than males with shorter snouts. Closed circles are hatchery males and open circles are wild salmon. In this case, the pattern of linear selection implies directional selection for longer snouts in both artificial and natural settings. In contrast, a study of sexual selection on song characteristics in field crickets (b) found that selection on the male's call included a substantial stabilizing component. Contours in this figure represent changes in attractiveness, which increases as trait values move closer to the single optimum (+). The traits in this case are composite traits similar to principal components, except that the loadings were determined from the shape of the selection surface rather than from the shape of the phenotypic distribution. Interestingly, the mean call in wild crickets is very close to the single peak on the individual selection surface (b). A similar analysis of selection in guppies resulted in a very different individual selection surface for male sexually selected traits (c). In guppies, there are at least three distinct combinations of traits that produce attractive males: males with large amounts of black and orange coloration (upper left +, black/orange), males with large tails (upper right +, tail), and males with large amounts of iridescence and fuzzy black areas (lower right +, iridescence/fuzzy). Interesting, male phenotypes are distributed across this entire individual selection surface (not shown), with many individuals located in the relatively low fitness center. Thus, guppies seem to be experiencing both disruptive and correlational sexual selection on male traits. Graphs are reproduced from (a) Fleming and Gross (1994), (b) Brooks et al. (2005), and (c) Blows et al. (2003), with permission from John Wiley and Sons.

Blows 2006). Intersexual selection refers to sexual selection arising from mate choice, and research on this topic has generated countless examples of directional selection on elaborate traits. As noted earlier, mating preferences often are amenable to laboratory study, but the most informative studies relate these preferences to actual patterns of parentage in nature. For example, Sheldon and Ellegren (1999) used a microsatellite-based parentage analysis in the socially monogamous collared flycatcher to show that extra-pair fertilizations were distributed in such a way as to generate a selection gradient on two male secondary sexual traits, forehead and wing patch size. Many other studies have investigated mating preferences to such a degree that there can be little doubt that intersexual selection is a potent source of directional sexual selection (see reviews in Kingsolver et al. 2001 and Hunt et al. 2009).

One remaining challenge for studies of directional intra- and intersexual selection is to integrate data on male-male combat and female choice with the goal of describing total sexual selection acting on sexually selected traits (Hunt et al. 2009). A non-zero selection gradient indicates a relationship between trait values and fitness, or some component of fitness, but the selection gradient by itself does not distinguish intrasexual selection from intersexual selection. Rather, the diagnosis of female choice, male-male combat, and the integration of these two processes requires careful experimental design and analysis (Hunt et al. 2009). An additional step, beyond the integration of intra- and intersexual selection, will involve measurement of the combined effects of precopulatory sexual selection, postcopulatory sexual selection, sexual conflict, and natural selection on trait evolution. Such an integration of selection pressures may be difficult to achieve in any particular system, but it remains a worthy goal of empirical sexual selection research.

8.5.3 Non-linear individual selection surfaces

If sexual selection always resulted in directional selection on individual, uncorrelated traits, then there would hardly be a need for the Adaptive Landscape concept, as a selection coefficient coupled with an estimate of heritability would provide sufficient information from which to predict the response to selection. However, patterns of selection can be much more complex. In some cases, sexually selected traits may be experiencing non-linear selection, that is, the individual selection surface may be curved, either concavely or convexly (Brodie et al. 1995). This curvature can result in situations where both linear and non-linear selection act simultaneously on the population, depending on the position of the population mean relative to the peaks and valleys on the selection surface (Arnold et al. 2001). For the sake of simplicity, we will refer to the case in which a population mean is near a peak on a curved selection surface as stabilizing selection and the case in which the population is near a valley on a curved selection surface as disruptive selection (Brodie et al. 1995; Hunt et al. 2009). Thus far, only a small fraction of the studies of sexually selected traits have focused on non-linear selection (Kingsolver et al. 2001), but these studies suggest that stabilizing and disruptive selection may be more common than expected.

From a strictly theoretical perspective, stabilizing selection usually seems more likely than disruptive selection, because selection is expected to keep the population mean in the vicinity of adaptive peaks. Even though genetic drift may cause the population mean to depart from the optimum, directional selection will correct these departures by pushing the population back toward the peak. Despite this expectation, however, both stabilizing and disruptive selection have been documented in the sexual selection literature, and too few studies exist for the assertion of generalities. One compelling example of stabilizing sexual selection comes from a study of mating preferences for call characteristics in a field cricket, *Teleogryllus commodus*. Brooks et al. (2005) constructed artificial calls that varied with respect to five call parameters and used playback experiments to quantify female preferences. Their estimate of the shape of the individual selection surface arising from mate choice revealed a single peak for the call characteristics subject to non-linear selection. Their results also showed that the average call in the population was within half a standard deviation of the optimum (Fig. 8.3). Bentsen et al. (2006) followed up these laboratory experiments with a field study, confirming that their findings in the lab were realized in a natural setting. A number of other studies, usually focusing on the component of fitness arising from mating preferences, have similarly detected stabilizing selection on secondary sexual traits (Butlin et al. 1985; Fairbairn and Preziosi 1996; Chenoweth et al. 2007). From these results we can conclude that sexually selected traits are likely to experience directional or stabilizing selection or both simultaneously, but the stabilizing component (i.e. quadratic selection) has been largely neglected by empirical studies until recently, probably due to the large sample sizes needed to detect statistically significant evidence of non-linear selection.

Disruptive selection is another archetypal form of selection that arises in discussions of the Adaptive Landscape (Phillips and Arnold 1989). In this case, the population mean is near a valley on the selection surface, and several studies have revealed such a scenario for sexually selected traits. A study on female mating preferences in guppies (*Poecilia reticulata*), for example, revealed that the fitness surface for a suite of male traits was complex (Fig. 8.3), with at least three distinct peaks of male attractiveness (Blows et al. 2003). However, the distribution of male phenotypes did not coincide with any of the peaks, suggesting that the disruptive selection imposed by female choice could be responsible for the high levels of phenotypic polymorphism in secondary sexual traits in guppies. Similarly, a study involving lazuli buntings, *Passerina amoena*, found that males with either the brightest or the dullest plumage had higher fitness compared to intermediate males (Greene et al. 2000), suggesting that at least some of the selection on plumage characteristics is disruptive. In some extreme cases, disruptive sexual selection probably accounts for the evolution of alternative male mating strategies (Fleming and Gross 1994). For example, in some species, certain males may defend territories or nests while others sneak copulations or mimic females (Taborsky 1994). The different strategies can be correlated with male morphology, and more than one strategy may simultaneously achieve high reproductive success in the same population. However, males with intermediate phenotypes perform poorly in all strategies, resulting in an overall pattern of disruptive selection.

The topic of disruptive sexual selection also frequently arises in the speciation literature, and in this context disruptive selection can be a manifestation of species recognition mechanisms or selection against hybridization among phenotypically distinct populations (Seehausen and van Alphen 1999; Naisbit et al. 2001). These phenomena highlight an important relationship between stabilizing selection and disruptive selection, which becomes obvious from the perspective of the individual selection surface. When two distinct populations or species are each characterized by stabilizing selection on a sexually selected trait, admixture can produce a population in which the individual selection surface has two peaks, and most hybrids will fall into the low-fitness valley between the peaks. Thus, we find that stabilizing and disruptive selection are important parts of the sexual selection process, and the study of these modes of selection requires the application of the Adaptive Landscape concept. Unfortunately, these types of non-linear selection have languished in relative obscurity due to a strong historical bias toward the study of directional sexual selection, but the situation is beginning to change.

8.5.4 Correlational selection

If disruptive and stabilizing selection didn't make the situation complex enough, there's now a growing realization that correlational selection is also important in sexual selection. Correlational selection refers to the situation in which an interaction between phenotypic traits makes a contribution to an individual's fitness (Lande and Arnold 1983; Brodie 1992). In a sexual selection context, this situation probably most commonly occurs when females prefer particular combinations of traits in males. In crickets, for example, females prefer certain combinations of call structure and calling effort (Bentsen et al. 2006), whereas female guppies prefer particular combinations of orange and black coloration, tail size and iridescence in their mates (Blows et al. 2003; Fig. 8.3). In one enlightening example, female swordtails have been shown to prefer males with large bodies, vertical bars, and small dorsal fins (Fisher et al. 2009). The catch is that such males don't exist in nature, because fin size is positively correlated with body size. Thus, sexual selection arising from mating preferences appears to be acting in a direction in trait space with very little phenotypic variation, a situation that also seems to be occurring in *Drosophila bunnanda* (Van Homrigh et al. 2007). These conclusions were made possible because the authors simultaneously studied multiple traits, an approach that explicitly or implicitly relies on the concept of the Adaptive Landscape. Thus, growing evidence suggests that future progress in the study of sexual selection will require careful application of the concept of individual selection surfaces to multivariate suites of traits and preferences. Such an undertaking will likely entail large, quantitative studies of mating patterns in experimental or observational settings that can be related to events occurring in natural populations.

8.5.5 Selection on mating preferences

To this point we've focused on the phenotypic traits that evolve as a consequence of intra- or intersexual selection, and an abundance of evidence suggests that multiple traits often evolve in concert. However, there is yet another dimension to sexual selection, because mating preferences themselves are expected to evolve as a correlated response to selection on ornaments (Lande 1981; Mead and Arnold 2004). Thus, to understand the evolution of mating preferences, which is a critical goal since mating preferences drive the process of intersexual selection and appear be a large component of total sexual selection in many systems (Hunt et al. 2009), we must consider direct as well as indirect selection. The individual selection surface provides the best conceptual arena in which to accomplish this goal. Unfortunately, the quantification of a selection surface for mating preferences is an extremely difficult problem, because the relationship between a particular individual's mating preferences and relative fitness can be very difficult to measure. These difficulties can be appreciated by first realizing that the characterization of average female preferences within a population represents a substantial empirical challenge (Wagner 1998; Ritchie

2000; Shaw and Herlihy 2000). An assessment of the selection surface for female preferences would require data on individual preference functions for a large number of females plus an assessment of relative fitness for each of those females. Such a task is extremely daunting, and only a handful of studies have managed to obtain relevant data. For example, Rundle et al. (2007) measured the relative fitness of females mating with males that were determined to be attractive in three rounds of choice tests by other females. Females mating with more attractive males had increased fitness due to sexual selection favoring their sons, as well as natural selection favoring increased longevity in their sons but not daughters, in a competitive environment. Other studies have shown that females of various species can increase relative fitness by exercising mate choice (e.g. Hine et al. 2002; Gowaty et al. 2003; Head et al. 2005), but most studies produce far from a complete characterization of a selection surface for mating preferences, so this topic remains a substantial challenge for sexual selection research.

8.5.6 Temporal and spatial variation in sexual selection

Consideration of the many modes of sexual selection that seem to be operating and the multivariate nature of the process indicate that sexual selection may be more complex than we assumed two or three decades ago. Similarly, recent studies focusing on selection surfaces over time and space suggest that the complexity within populations may be the tip of the iceberg. In natural populations of an ambush bug, *Phymata americana*, for instance, patterns of multivariate selection on a sexually dimorphic color pattern varied considerably over time (Punzalan et al. 2010). In concordance with mating system theory, the variation in selection appears to be related to changes in population density and the sex ratio. Indeed, temporal variation in the strength and direction of selection seems to be a common feature of both natural and sexual selection (Siepielski et al. 2009). If selection varies temporally, then it also seems reasonable to predict that selection will vary spatially. The sexual selection literature confirms this suspicion. In a gargantuan study of sexual selection in twelve populations of a damselfly (*Ischnura elegans*) across 5 years, Gosden and Svensson (2008) found that the nature of selection acting on male traits varied considerably over space and time (Fig. 8.4). Studies of other taxa have obtained similar results (Arnqvist 1992; Blanckenhorn et al. 1999; Chaine and Lyon 2009; Chapter 7). At some level, these results might elicit feelings of despair among students of sexual selection, as temporal and spatial variation add more layers of complexity to an already complex process. However, such variation should be seen as an opportunity, because variation in sexual selection intensity, social conditions, and the ecological backdrop of mating competition provide excellent opportunities for the types of comparative studies that will be required to sort out many of the details of the sexual selection process.

8.5.7 Integrating precopulatory and postcopulatory sexual selection

Thus far in the sexual selection literature, the concept of individual selection surfaces has been applied most frequently to the study of precopulatory sexual selection, which explains why we have dealt almost entirely with such selection in this chapter. However, a focus on selection surfaces also offers great promise for the study of postcopulatory sexual selection, which occurs when, for example, the gametes of multiple individuals compete within the same mate for access to fertilization opportunities (Parker 1970; Thornhill 1983; Birkhead and Pizzari 2002; Eberhard 2009). The application of individual selection surfaces to postcopulatory sexual selection offers additional challenges compared to the study of precopulatory sexual selection alone, but such studies will be critical if the goal is to obtain a comprehensive understanding of the sexual selection process.

At face value, the application of the selection surface concept to postcopulatory processes is no different than its application to precopulatory sexual selection. Ultimately, both exercises hinge upon describing the relationship between trait values and relative fitness. However, several major challenges arise in the study of postcopulatory sexual selection. First, the identification and measurement of traits that are the targets of postcopulatory sexual

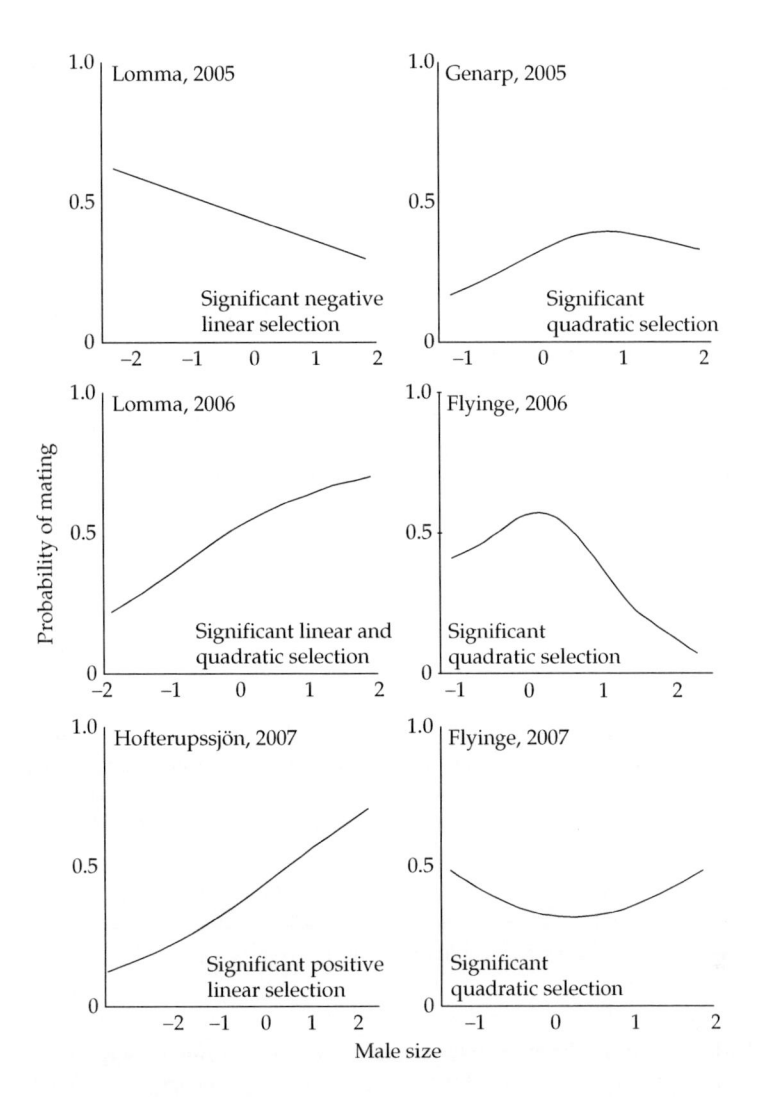

Figure 8.4 Evidence for spatial and temporal variation in sexual selection in a damselfly. Gosden and Svensson (2008) studied 12 populations of a damselfly over 5 years, and a subset of their dataset is shown here. Individual selection surfaces depicting the mating success component of fitness were amazingly variable. All of the populations were within a few dozen kilometers of one another, yet they often showed very different patterns of selection within a year (compare rows in this figure). In addition, the same population often displayed opposing patterns of selection in different years. For example, the population in Lomma, Sweden, showed a pattern of negative directional selection in 2005 (upper left panel) that reversed to positive directional selection in 2006 (middle left panel). Similarly, the Flyinge population seemed to be experiencing stabilizing selection in 2006 (middle right panel) but disruptive selection in 2007 (bottom right panel). Redrawn from Gosden and Svensson (2008), with permission from John Wiley and Sons.

selection can be extremely difficult in some systems (Birkhead 1998; Snook 2005). For instance, postcopulatory sexual selection often targets characteristics of the gametes, substances transferred along with the gametes during mating, morphology of reproductive structures, subtle aspects of behavior, and other traits that tend to be difficult to measure (Birkhead and Pizzari 2002). Second, the contributions of precopulatory and postcopulatory sexual selection can be difficult to cleanly separate from one another. For example, a common observation in the study of sexual selection might be that a male trait is positively correlated with the number of offspring sired, which would indicate the trait is under sexual selection. However, to quantify the relative roles of pre- and postcopulatory processes would require quantification of male mating success, preferably including information about the number of sperm transferred per successful mating, as well as data regarding the success of those sperm in competition with other sperm after mating (Birkhead and Pizzari 2002). These data, while not impossible to collect, are not easily obtained for most systems.

A third challenge in the study of postcopulatory sexual selection concerns the measurement of fitness. If postcopulatory sexual selection results in different numbers of offspring for successful compared to unsuccessful males, then the measurement of fitness is relatively straightforward. However, postcopulatory sexual selection also can result in differences in offspring quality, which can be extremely difficult to quantify and integrate into a measure of relative fitness (Neff and Pitcher 2005). This problem also applies to some degree to the measurement of fitness in precopulatory sexual selection, but it seems especially problematic for the study of postcopulatory sexual selection. A final challenge in the study of postcopulatory sexual selection is to separate intrasexual selection, usually in the form of sperm competition, from intersexual selection, or cryptic female choice (Pitnick and Brown 2000; Eberhard 2000; Birkhead 2000). Because postcopulatory sexual selection typically occurs within the female's reproductive tract, it is notoriously difficult to separate male from female effects on fertilization success. These challenges make the study of postcopulatory processes difficult, but successful studies have the potential to contribute substantially to our overall understanding of sexual selection as an evolutionary process.

A few studies have taken the plunge and applied the concept of selection surfaces to the study of postcopulatory sexual selection. For example, Hall et al. (2008) used selection surfaces to quantify postcopulatory sexual selection in the black field cricket, *Teleogryllus commodus*, under conditions that promoted or removed the potential for sexual conflict. In this species, sexual conflict occurs as a consequence of males harassing their recent mates to prevent them from removing recently deposited spermatophores. The results indicated that sexual conflict dramatically reduced the overall strength of sexual selection on males by inhibiting the females' ability to exercise postcopulatory choice and changed the shape of the individual selection surface for male body size and call characteristics. Thus, the selection surface approach was useful in this case both to verify the existence of postcopulatory sexual selection and to establish a link between sexual selection and sexual conflict. Even though we have largely ignored sexual conflict in this chapter due to space constraints, the study of postcopulatory sexual selection is one arena in which the connection between sexual selection and conflict is especially obvious and important (Arnqvist and Rowe 2005), and the study by Hall et al. (2008) indicates that sexual conflict, too, is an area of research that can be enriched by the use of a selection surface approach.

8.6 Conclusions and future directions

The application of Adaptive Landscapes and individual selection surfaces in the study of sexual selection has transformed the field over the last several decades. The strength of the approach is that it provides a rigorous method for the quantification and comparison of the nature of sexual selection in a common currency that is comparable across systems. Currently, we remain primarily in a descriptive phase, in which studies are revealing a baffling array of types of selection on secondary sexual traits. The studies that have been conducted so far suggest that directional selection is common, as expected from theoretical treatments of sexual selection. However, larger studies making better use of selection surfaces indicate that stabilizing, disruptive and correlational selection may be more common than we originally anticipated. Even though we've treated different modes of selection separately, all of these types of selection (i.e. linear, quadratic and correlational) may occur simultaneously in a given population, so the characterization of such complex patterns necessitates a focus on individual selection surfaces. This realization underscores the need to confront sexual selection as a multivariate problem, in which the simultaneous evolution of multiple armaments, ornaments, and preferences acts together to produce or constrain changes in the phenotype over evolutionary time. Furthermore, sexual selection varies temporally and spatially, a situation that at once provides novel challenges and opportunities.

The future of the Adaptive Landscape in sexual selection research looks very promising. The field has already started to move beyond descriptive studies to ask questions about why Adaptive Landscapes behave the way they do, but many challenges remain. For example, with respect to mating system evolution, we need to investigate the factors that shape Bateman relationships and gradients. With respect to sexual selection on particular traits, we can begin to ask what factors are responsible for changes in the shape of the selection surface over time and space. By addressing these issues, we will be able to answer some of the big questions that remain in the field of sexual selection. For instance, why are some species dramatically altered by sexual selection while their close relatives are not? Is the sexual selection process more commonly stopped by opposing natural selection or by an erosion of genetic variance in key directions in phenotypic space? What explains the evolution of mating preferences? Precisely how does sexual selection contribute to other major evolutionary processes, such as speciation and extinction? And, how do precopulatory and postcopulatory sexual selection interact to produce total sexual selection on particular traits? Modern studies are beginning to answer these questions, but we are far from generalities. The powerful inference framework provided by the Adaptive Landscape and individual selection surface provides a clear path forward in the quest for answers to these and other difficult questions in the study of sexual selection.

References

Andersson, M. (1982). Female choice selects for extreme tail length in a widowbird. Nature, 299, 818–820.

Andersson, M. (1994). Sexual Selection. Princeton University Press, Princeton, NJ.

Andersson, M. and Iwasa, Y. (1996). Sexual selection. Trends in Ecology & Evolution, 11, 53–58.

Arnold, S. J. and Duvall, D. (1994). Animal mating systems—A synthesis based on selection theory. American Naturalist, 143, 317–348.

Arnold, S. J., Pfrender, M. E. and Jones, A. G. (2001). The Adaptive Landscape as a conceptual bridge between micro- and macroevolution. Genetica, 112, 9–32.

Arnqvist, G. (1992). Spatial variation in selective regimes—sexual selection in the water strider *Gerris odontogaster*. Evolution, 46, 914–929.

Arnqvist, G. and Rowe, L. (2005). Sexual Conflict. Princeton University Press, Princeton, NJ.

Bateman, A. J. (1948). Intra-sexual selection in *Drosophila*. Heredity, 2, 349–368.

Benkman, C. W. (2003). Divergent selection drives the adaptive radiation of crossbills. Evolution, 57, 1176–1181.

Bentsen, C. L., Hunt, J., Jennions, M. D. and Brooks, R. (2006). Complex multivariate sexual selection on male acoustic signaling in a wild population of *Teleogryllus commodus*. American Naturalist, 167, E102–E116.

Berglund, A. and Rosenqvist, G. (1993). Selective males and ardent females in pipefishes. Behavioral Ecology and Sociobiology, 32, 331–336.

Birkhead, T. R. (1998). Cryptic female choice: criteria for establishing female sperm choice. Evolution, 52, 1212–1218.

Birkhead, T. R. (2000). Defining and demonstrating postcopulatory female choice—again. Evolution, 54, 1057–1060.

Birkhead, T. R. and Pizzari, T. (2002). Postcopulatory sexual selection. Nature Reviews Genetics, 3, 262–273.

Bischoff, R. J., Gould, J. L. and Rubenstein, D. I. (1985). Tail size and female choice in the guppy (*Poecilia reticulata*). Behavioral Ecology and Sociobiology, 17, 253–255.

Blanckenhorn, W. U., Morf, C., Mühlhäuser, C. and Reusch, T. (1999). Spatiotemporal variation in selection on body size in the dung fly *Sepsis cynipsea*. Journal of Evolutionary Biology, 12, 563–576.

Blows, M. W. and Brooks, R. (2003). Measuring nonlinear selection. American Naturalist, 162, 815–820.

Blows, M. W., Brooks, R. and Kraft, P. G. (2003). Exploring complex fitness surfaces: Multiple ornamentation and polymorphism in male guppies. Evolution, 57, 1622–1630.

Brodie, E. D. III. (1992). Correlational selection for color pattern and antipredator behavior in the garter snake *Thamnophis ordinoides*. Evolution, 46, 1284–1298.

Brodie, E. D. III, Moore, A. J. and Janzen, F. J. (1995). Visualizing and quantifying natural selection. Trends in Ecology & Evolution, 10, 313–318.

Brooks, R., Hunt, J., Blows, M. W., Smith, M. J., Bussière, L. F. and Jennions, M. D. (2005). Experimental evidence for multivariate stabilizing sexual selection. Evolution, 59, 871–880.

Butlin, R. K., Hewitt, G. M. and Webb, S. F. (1985). Sexual selection for intermediate optimum in *Chorthippus brunneus* (Orthoptera: Acrididae). Animal Behavior, 33, 1281–1292.

Chaine, A. S. and Lyon, B. E. (2009). Adaptive plasticity in female mate choice dampens sexual selection on male ornaments in the lark bunting. Science, 319, 459–462.

Chenoweth, S. F. and Blows, M. W. (2006). Dissecting the complex genetic basis of mate choice. Nature Reviews Genetics, 7, 681–692.

Chenoweth, S. F., Petfield, D., Doughty, P. and Blows, M. W. (2007). Male choice generates stabilizing selection on a female fecundity correlate. Journal of Evolutionary Biology, 20, 1745–1750.

Darwin, C. (1859). On the Origin of Species by Means of Natural Selection. John Murray, London.

Darwin, C. (1871). The Descent of Man and Selection in Relation to Sex. John Murray, London.

Eberhard, W. G. (2000). Criteria for demonstrating postcopulatory female choice. Evolution, 54, 1047–1050.

Eberhard, W. G. (2009). Postcopulatory sexual selection: Darwin's omission and its consequences. Proceedings of the National Academy of Sciences of the United States of America, 106, 10025–10032.

Emlen, S. T. and Oring, L. W. (1977). Ecology, sexual selection, and the evolution of mating systems. Science, 197, 215–223.

Endler, J. A. (1986). Natural Selection in the Wild. Princeton University Press, Princeton, NJ.

Fairbairn, D. J. and Preziosi R. F. (1996). Sexual selection and the evolution of sexual size dimorphism in the water strider, *Aquarius remigis*. Evolution, 50, 1549–1559.

Fisher, R. A. (1930). The Genetical Theory of Natural Selection. Clarendon, Oxford.

Fisher, H. S., Mascuch, S. J. and Rosenthal, G. G. (2009). Multivariate male traits misalign with multivariate female preferences in the swordtail fish, *Xiphophorus birchmanni*. Animal Behavior, 78, 265–269.

Fleming, I. A. and Gross, M. R. (1994). Breeding competition in a Pacific salmon (coho: *Oncorhynchus kisutch*): Measures of natural and sexual selection. Evolution, 48, 637–657.

Gosden, T. P. and Svensson, E. I. (2008). Spatial and temporal dynamics in a sexual selection mosaic. Evolution, 62, 845–856.

Gowaty, P. A., Drickamer, L. C. and Schmid-Holmes, S. (2003). Male house mice produce fewer offspring with lower viability and poorer performance when mated with females they do not prefer. Animal Behavior, 65, 95–103.

Greene, E., Lyon, B. E., Muehter, V. R., Ratcliffe, L., Oliver, S. J. and Boag, P. T. (2000). Disruptive sexual selection for plumage coloration in a passerine bird. Nature, 407, 1000–1003.

Grether, G. F. (1996). Sexual selection and survival selection on wing coloration and body size in the rubyspot damselfly *Hetaerina americana*. Evolution, 50, 1939–1948.

Hall, M. D., Bussière L. F., Hunt, J. and Brooks, R. (2008). Experimental evidence that sexual conflict influences the opportunity, form and intensity of sexual selection. Evolution, 62, 2305–2315.

Head, M. L., Hunt, J., Jennions, M. D. and Brooks, R. (2005). The indirect benefits of mating with attractive males outweigh the direct costs. PLOS Biology 3, e33.

Hine, E., Lachish, S., Higgie, M. and Blows, M. W. (2002). Positive genetic correlation between female preference and offspring fitness. Proceedings of the Royal Society B: Biological Sciences, 269, 2215–2219.

Hunt, J., Breuker, C. J., Sadowski, J. A. and Moore, A. J. (2009). Male-male competition, female mate choice and their interaction: determining total sexual selection. Journal of Evolutionary Biology, 22, 13–26.

Janzen, F. and Stern, H. S. (1998). Logistic regression for empirical studies of multivariate selection. Evolution, 52, 1564–1571.

Jones, A. G. (2009). On the opportunity for sexual selection, the Bateman gradient and the maximum intensity of sexual selection. Evolution, 63, 1673–1684.

Jones, A. G., Arguello, J. R. and Arnold S. J. (2002). Validation of Bateman's principles: a genetic study of sexual selection and mating patterns in the rough-skinned newt. Proceedings of the Royal Society B: Biological Sciences, 269, 2533–2539.

Jones, A. G. and Ratterman, N. L. (2009). Mate choice and sexual selection: What have we learned since Darwin? Proceedings of the National Academy of Sciences of the United States of America, 106, 10001–10008.

Jones, A. G., Rosenqvist, G., Berglund, A., Arnold, S. J. and Avise, J. C. (2000). The Bateman gradient and the cause of sexual selection in a sex-role-reversed pipefish. Proceedings of the Royal Society B: Biological Sciences, 267, 677–680.

Kingsolver, J. G., Hoekstra, H. E., Hoekstra, J. M., Berrigan, D., Vignieri, S. N., Hill, C. E., et al. (2001). The strength of phenotypic selection in natural populations. American Naturalist, 157, 245–261.

Kirkpatrick, M. (1982). Quantum evolution and punctuated equilibria in continuous genetic characters. American Naturalist, 119, 833–848.

Klump, G. M. and Gerhardt, H. C. (1987). Use of nonarbitrary acoustic criteria in mate choice by female gray tree frogs. Nature, 326, 286–288.

Kruuk, L. E. B., Slate, J., Pemberton, J. M., Brotherstone, S., Guinness, F. and Clutton-Brock, T. (2002). Antler size in red deer: Heritability and selection but no evolution. Evolution, 56, 1683–1695.

Lande, R. (1976). Natural selection and random genetic drift in phenotypic evolution. Evolution, 30, 314–334.

Lande, R. (1979). Quantitative genetic analysis of multivariate evolution, applied to brain:body size allometry. Evolution, 33, 402–416.

Lande, R. (1981). Models of speciation by sexual selection on polygenic traits. Proceedings of the National Academy of Sciences of the United States of America, 78, 3721–3725.

Lande, R. and Arnold, S. J. (1983). The measurement of selection on correlated characters. Evolution, 37, 1210–1226.

Mallet, J. (2007). Hybrid speciation. Nature, 446, 279–283.

Mead, L. S. and Arnold, S. J. (2004). Quantitative genetic models of sexual selection. Trends in Ecology & Evolution, 19, 264–271.

Naisbit, R. E., Jiggins, C. D. and Mallet, J. (2001). Disruptive sexual selection against hybrids contributes to speciation between *Heliconius cydno* and *Heliconius melpomene*. Proceedings of the Royal Society B: Biological Sciences, 268, 1849–1854.

Neff, B. D. and Pitcher, T. E. (2005). Genetic quality and sexual selection: an integrated framework for good genes and compatible genes. Molecular Ecology, 14, 19–38.

Parker, G. A. (1970). Sperm competition and its evolutionary consequences in the insects. Biological Reviews, 45, 525–568.

Phillips, P. C. and Arnold, S. J. (1989). Visualizing multivariate selection. Evolution, 43, 1209–1222.

Pitnick, S. and Brown, W. D. (2000). Criteria for demonstrating female sperm choice. Evolution, 54, 1052–1056.

Punzalan, D., Helen Rodd, F. and Rowe, L. (2010). Temporally variable multivariate sexual selection on sexually dimorphic traits in a wild insect population. American Naturalist, 175, 401–414.

Ritchie, M. G. (2000). The inheritance of female preference functions in a mate recognition system. Proceedings of the Royal Society B: Biological Sciences, 267, 327–332.

Rosenthal, G. G. and Evans, C. S. (1998). Female preference for swords in *Xiphophorus helleri* reflects a bias for large apparent size. Proceedings of the National Academy of Sciences of the United States of America, 95, 4431–4436.

Rundle, H. D., Ödeen A., and Mooers, A. Ø. (2007). An Experimental test for indirect benefits in *Drosophila melanogaster*. BMC Evolutionary Biology, 7, 36.

Schluter, D. (1988). Estimating the form of natural selection on a quantitative trait. Evolution, 42, 849–861.

Schluter, D. (2000). The Ecology of Adaptive Radiation. Oxford University Press, Oxford.

Schluter, D. and Nychka, D. (1994). Exploring fitness surfaces. American Naturalist, 143, 597–616.

Seehausen, O. and van Alphen, J. J. M. (1999). Can sympatric speciation by disruptive sexual selection explain rapid evolution of cichlid diversity in Lake Victoria? Ecolgy Letters, 2, 262–271.

Serbezov, D., Bernatchez, L., Olsen, E. M. and Vøllestad, L.A. (2010). Mating patterns and determinants of individual reproductive success in brown trout (*Salmo trutta*) revealed by parentage analysis of an entire stream living population. Molecular Ecology, 19, 3193–3205.

Shaw, K. L. and Herlihy, D. P. (2000). Acoustic preference functions and song variability in the Hawaiian cricket *Laupala cerasina*. Proceedings of the Royal Society B, 267, 577–584.

Sheldon, B. C. and Ellegren, H. (1999). Sexual selection resulting from extrapair paternity in collared flycatchers. Animal Behavior, 57, 285–298.

Siepielski, A. M., DiBattista, J. D. and Carlson, S. M. (2009). It's about time: the temporal dynamics of phenotypic selection in the wild. Ecology Letters, 12, 1261–1276.

Simpson, G. G. (1953). The Major Features of Evolution. Columbia University Press, New York.

Snook, R. R. (2005). Sperm in competition: not playing by the numbers. Trends in Ecology & Evolution, 20, 46–53.

Taborsky, M. (1994). Sneakers, satellites, and helpers—parasitic and cooperative behavior in fish reproduction. Advances in the Study of Behavior, 23, 1–100.

Thornhill, R. (1983). Cryptic female choice and its implications in the scorpionfly *Harpobittacus nigriceps*. American Naturalist, 122, 765–788.

Van Homrigh, A, Higgie, M., McGuigan, K. and Blows, M.W. (2007). The depletion of genetic variance by sexual selection. Current Biology, 17, 528–532.

Warner, R. R. and Schultz, E. T. (1992). Sexual selection and male characteristics in the bluehead wrasse, *Thalassoma bifasciatum*: Mating site acquisition, mating site defense, and female choice. Evolution, 46, 1421–1442.

Wagner, W. E. (1998). Measuring female mating preferences. Animal Behavior, 55, 1029–1042.

Whitlock, M. C. (1995). Variance-induced peak shifts. Evolution, 49, 252–259.

Wright, S. (1931). Evolution in Mendelian populations. Genetics, 16, 97–159.

Wright, S. (1932). The roles of mutation, inbreeding, crossbreeding and selection in evolution. Proceedings of the Sixth Annual Congress of Genetics, 1, 356–366.

CHAPTER 9

Analyzing and Comparing the Geometry of Individual Fitness Surfaces

Stephen F. Chenoweth, John Hunt, and Howard D. Rundle

9.1 Introduction

Evolutionary quantitative genetic theory integrates two fundamental relationships: the individual fitness surface, summarizing the pattern of covariance between multiple traits and relative fitness, and a variance-covariance matrix, **G**, that summarizes the joint influences of additive genetic effects and frequencies of segregating variants on phenotypic variation (Lande 1979). The empirical study of individual fitness surfaces, launched in part by Wright's (1931, 1932) concept of the fitness landscape, was greatly stimulated by the classic contribution from Lande and Arnold (1983). Their framework for estimating linear and non-linear selection gradients using multiple regression provided an approximation to the individual fitness surface that was expressed in a common currency. The Lande and Arnold (1983) approach has facilitated the estimation and comparison of selection across species, traits, and fitness components to the extent that over a quarter-century on, we have an appreciation that phenotypic selection is both relatively common in nature (Hoekstra et al. 2001; Kingsolver et al. 2001) and strong to the extent that it represents a significant proportion of the variance in fitness (Hereford et al. 2004). Phenotypic selection also tends to vary spatially (Schluter 2000; Gosden and Svensson 2008), temporally (Siepielski et al. 2009, Chapter 7), and often between the sexes (Cox and Calsbeek 2009, Chapter 8).

Empirical questions that demand characterization of individual fitness surfaces range from simply identifying the components of phenotypic variation that mediate fitness differences and what causes selection on these characters, to predicting microevolutionary change. That variation in selection matters is also clear: changes in phenotypic selection within and among groups have obvious implications for the tempo and direction of trait evolution, the origins of phenotypic diversity within (e.g. sexual dimorphism, distinct morphotypes) and among populations/species (Schluter 2000; Coyne and Orr 2004), and the occurrence of prolonged evolutionary stasis in many characters (Estes and Arnold 2007; Chapters 14, 16). Variation in phenotypic selection may also affect other fundamental and related issues in evolutionary genetics including the maintenance of genetic variance (Hedrick 1986; Gillespie and Turelli 1989; Turelli and Barton 2004; Hedrick 2006), the evolution of genetic architecture (Bulmer 1980; Turelli 1988; Kawecki 2000; Jones et al. 2004), the genetic basis of adaptation (Bell 2010), and the homogeneity of Adaptive Landscapes (Arnold et al. 2001; see also Chapter 12). Fitness surfaces and their variation are also central to quantifying the extent and nature of maladaptation (i.e. adaptive inaccuracy; Chapter 10).

Although the approaches for explicitly characterizing selection are not new, their uptake by empiricists in multivariate form has been relatively poor. Wright (1935) appreciated the reality that selection acts simultaneously on multiple rather than single traits (Walsh and Lynch 2011) and more recently, the value in considering microevolution as a multivariate problem has been re-emphasized (Blows 2007; Walsh and Blows 2009; Houle 2010). We suspect that the roadblock to the uptake of multivariate analyses of selection by empiricists has in part been due to their somewhat complex nature. Our goal in this chapter is to provide an overview of statistical and geometric approaches available for the multivariate analysis of phenotypic selection. Many of the approaches we outline have been covered in greater depth in other reviews (Phillips and Arnold 1989; Brodie et al. 1995), as have some of the finer technical aspects that have arisen with their implementation (Mitchell-Olds and Shaw 1987; Blows and Brooks 2003; Hereford et al. 2004; Stinchcombe et al. 2008). With interest growing in their application, we attempt to provide practical guidance for issues arising along the way, and to highlight areas where additional work is needed. First, we develop the analysis of selection as an individual fitness surface within a single population at a single point in time. Second, we address the comparison of fitness surfaces between different types of groups such as populations, sexes, or times. Finally, we show two ways in which the multivariate geometry of fitness surfaces can be integrated with quantitative genetic analyses to provide additional insight, first with respect to understanding the nature of genetic variance under selection and second to provide a predictive framework for characterizing phenotypic divergence in complex phenotypes.

9.2 Analysis of the individual fitness surface within a single population

9.2.1 Approximating the individual fitness surface using multiple regression

While a variety of techniques exist to describe the strength and form of selection within a population (e.g. Crespi and Bookstein 1989; Kingsolver and Schemske 1991; Brodie and Janzen 1996; Shaw and Geyer 2010), the most widely used and influential is the ordinary least-squares regression approach of Lande and Arnold (1983) that provides the best linear or quadratic approximation to the individual fitness surface. Following this approach, the variance-standardized ($\sim N(0,1)$) phenotypic values of n continuous traits, z_n, are regressed against relative fitness, w, for a sample of individuals:

$$\omega = \alpha + \sum_{i=1}^{n} \beta_i z_i + \varepsilon \qquad (9.1)$$

where α is the intercept and β_i represents the partial regression coefficient for trait z_i and ε is the random error component. The partial regression coefficients represent the standardized directional selection gradients and they estimate the linear contribution of a given trait to fitness when holding the effects of all other traits being examined constant. The vector β, containing all β_i values, therefore represents the direction of the steepest uphill slope from the population mean on the individual fitness surface (Lande and Arnold 1983).

For non-linear selection, a separate second-order regression model is fitted that includes quadratic (z_{ii}^2) and cross-product ($z_i z_j$) terms that together constitute non-linear selection gradients:

$$\omega = \alpha + \sum_{i=1}^{n} \beta_i z_i + \sum_{i=1}^{n} \frac{1}{2} \gamma_{ii} z_i^2 + \sum_{i=1}^{n} \sum_{j=i+1}^{n} \gamma_{ij} z_i z_j + \varepsilon \quad (9.2)$$

Although not interpreted, the linear terms are included in this model to remove their influence when examining how non-linear selection affects the variances and covariances of traits (Lande and Arnold 1983). The γ_{ii} coefficients quantify the direct effect of non-linear selection on the traits and therefore describe the curvature of the fitness surface along the individual trait axes (Lande and Arnold 1983). A positive γ_{ii} indicates concave selection (i.e. curved upwards) and a negative γ_{ii} indicates convex selection (i.e. curved downwards), and these can be interpreted as disruptive and stabilizing selection respectively if a fitness minimum/maximum occurs within the range of observed trait values (Mitchell-Olds and Shaw 1987; Phillips and Arnold 1989). Importantly, the implementation of equation 9.2 in most statistical software packages underestimates the strength of quadratic selection by a half and therefore it is necessary to double the γ_{ii} coefficients to gain a true representation of the curvature of the fitness surface (Stinchcombe et al. 2008). The γ_{ij} coefficient represents the direct effect of correlational selection on the covariance between z_i and z_j that will, in theory, favor traits becoming either positively ($+\gamma$) or negatively ($-\gamma$) correlated (Lande and Arnold

1983; Cheverud 1984). Although much attention has been given to the distinction between the γ_{ii} and γ_{ij} coefficients, referred to respectively as quadratic and correlational selection gradients, as we discuss below this distinction may be somewhat artificial, depending on how the original traits were defined as opposed to some fundamental difference in the geometry of the fitness surface (Walsh and Lynch 2011).

9.2.2 Analyzing non-linear selection through canonical analysis

While the interpretation of linear selection operating on phenotypic traits is relatively straightforward, the interpretation of non-linear selection can be more challenging. This is in part because, given n traits, there are $n(n + 1)/2$ individual γ coefficients. The interpretation of non-linear selection can be simplified, however, through a canonical analysis of the γ matrix (Phillips and Arnold 1989). A canonical analysis rotates γ to define a set of new multivariate traits, similar to a principal components analysis, that are aligned with the major axes of non-linear selection on the fitted surface (Box and Draper 1987). As γ is symmetric, the necessary rotation can be located through matrix diagonalization as (Phillips and Arnold 1989):

$$\Lambda = \mathbf{M}\gamma^T\mathbf{M}, \qquad (9.3)$$

where $\mathbf{M}$ is an orthogonal matrix whose columns are the eigenvectors of γ, normalized to unit length, and Λ is a matrix with the eigenvalues (λ) of γ on its diagonal (and zeros on all the off-diagonal elements) that describes the curvature of the fitness surface. This rotation condenses non-linear selection onto the quadratic terms, with correlational selection on the original traits now represented as the relationships between these traits and each eigenvector of $\mathbf{M}$ (Blows 2007). For example, large loadings for two traits on a given eigenvector experiencing strong non-linear selection indicates correlational selection on the two original traits.

Writing $\theta = \mathbf{M}^T\beta$ and $y = \mathbf{M}^T z$, equation 9.2 can now be rewritten after canonical analysis as:

$$\omega = \alpha + \sum_{i=1}^{n} \theta_i y_i + \sum_{i=1}^{n} \frac{1}{2}\lambda_i y_{ii}^2 + \varepsilon, \qquad (9.4)$$

This is referred to as the "A canonical form" of the system (Box and Draper 1987). The θ_i terms measure the slope of the surface from the origin ($\bar{z} = 0$) along the rotated axes described by the transformed variables, y_i. The signs of λ_i determine the type of fitted quadratic surface and their magnitudes describe its curvature (Phillips and Arnold 1989). If all the λ_i are negative then convex (stabilizing) selection is operating on trait combinations, suggesting a multivariate peak on the fitness surface. Conversely, if all the λ_i are positive then concave (disruptive) selection is operating on trait combinations and the fitness surface is best described as a multivariate bowl. If there is a mixture of positive and negative λ_i, the fitness surface will be a multivariate saddle. In each instance, the larger the magnitude of $|\lambda_i|$, the more curved the fitness surface will be along that axis in trait space.

Although not often done, the system can also be placed in "B canonical form" (Box and Draper 1987) by locating the stationary point on the surface (z_0), shifting the origin of the surface to it and then rotating the axes using the transformation: $y = \mathbf{M}^T(z - z_0)$. This B canonical form transformation should be used to characterize the system if a stationary point is located within the phenotypic range of the traits being examined (Phillips and Arnold 1989), whereas the A canonical form should be used in the absence of a stationary point (i.e. if the signs of λ_i are different), if the stationary point is outside the range of traits being examined, or when linear selection is strong relative to quadratic selection (Phillips and Arnold 1989). Other than the removal of the linear terms, the interpretation of λ_i in form A and B are identical (Box and Draper 1987; Phillips and Arnold 1989). However, using the eigenvectors in $\mathbf{M}$, the original trait values can be placed into canonical space and a y score can be determined for each individual in the data set. Once in canonical space, equation 9.4 can be used to estimate θ and λ for each of the major axes of the fitness surface, as well as their standard errors, using the double regression method of Bisgaard and Ankenman (1996).

Despite the considerable advantages associated with the canonical analysis of γ in aiding the interpretation of non-linear selection, this technique has been considerably under-utilized in the literature (Blows and Brooks 2003). This is problematic as interpreting individual elements of γ, rather than the system as a whole, can seriously underestimate the strength and importance of non-linear selection (Blows and Brooks 2003). For example, Kingsolver et al.'s (2001) review of published estimates of individual elements of γ tended to show that non-linear selection, particularly stabilizing and disruptive selection, were generally weak in natural populations. However, canonical analysis on a subset of these studies shows that the largest eigenvector of γ (median $\lambda = 0.55$) was larger than the largest quadratic coefficient for any individual trait (median $\gamma = 0.37$) (Blows and Brooks 2003) indicating, as expected, that selection generally targets combinations of the traits that investigators choose to measure (Blows 2007).

9.2.3 Testing the significance of selection gradients and vectors of selection

Statistical analysis often has two discrete stages, parameter estimation and significance testing, and these may be based on different assumptions and require different statistical procedures (Mitchell-Olds and Shaw 1987). This is particularly true for multiple regression-based selection analyses where the response variable, relative fitness, may often have a non-normal distribution (e.g. binary outcomes like live/die), and the predictors themselves (i.e. the phenotypic traits) may also deviate substantially from normality. While Lande and Arnold (1983) demonstrated that their least-squares approach can be used to estimate selection gradients irrespective of the underlying distribution of fitness or phenotypic traits, it is not always appropriate to test the statistical significance of these gradients using this approach.

A variety of different parametric and non-parametric statistical approaches can be used to circumvent this issue and have been discussed in detail elsewhere. If relative fitness is a binary measure (e.g. survived vs. died, mated vs. unmated) or a count (e.g. number of offspring), in some cases the normal distribution may provide a sufficient approximation (e.g. if sample sizes are large and the probability of either outcome is roughly equal (Zar 1984)). However, a preferable alternative is to employ a generalized linear model that uses a link function with an appropriate distribution (e.g. binomial, Poisson); for example a logistic regression (Mitchell-Olds and Waller 1985; Janzen and Stern 1998) or probit analysis (Price 1984). Such models are generally fit using maximum likelihood (ML) with significance determined using a likelihood ratio test.

Alternative approaches under deviations from normality include non-parametric resampling techniques such as the jacknife or bootstrap methods (Wu 1986; Mitchell-Olds and Shaw 1987; Manly 1997). A particularly useful non-parametric approach for testing the significance of selection gradients is to employ a permutation test (Manly 1997), where real point estimates of the gradients are estimated for the data (β_{test}) and compared to those when fitness is shuffled across individuals in the data set (β_{random}) using Monte Carlo simulation (see Brooks et al. (2005) for an application of this approach). For each phenotypic trait, the number of times that β_{random} exceeds β_{test} is determined and expressed as a proportion (p) of the total number of iterations, which can be used to determine the significance of each selection gradient (Manly 1997).

The permutation test can be used to test the significance of linear and non-linear selection along the eigenvectors of **M** using the "double regression" method (Bisgaard and Ankenman 1996). Even though canonical analysis minimizes the effects of correlational selection, a full model (including correlational terms) must be fitted to test the significance of λ values using this approach (Bisgaard and Ankenman 1996). Reynolds et al. (2010) recently showed that the double regression method for testing the significance of λ may actually lead to high false positive rates when testing against the null hypothesis of no selection (i.e. $\lambda = 0$). This means that the null expectation of the individual eigenvalues being equal to zero will only occur if the variance of the non-linear selection gradients is equal to zero, which will only occur when the sample sizes approach infinity. Thus canonical analysis can suggest curvature on the fitness surface when the eigenstructure of γ actually only reflects random error. The extent of non-linear selection is therefore likely to be over-estimated by the double-regression approach (Reynolds et al. 2010).

To correct this problem, Reynolds et al. (2010) developed a permutation test (and accompanying R code) that maintains the correct type I error rate. In this procedure, fitness is randomly shuffled against trait values and a new canonical analysis is performed prior to fitting the second regression model. This approach creates a null distribution of selection gradients taken from a population with no association between trait and fitness. Crucially, this test concerns the existence of particular ordered eigenvalues in **M**, defined not by the vectors of selection themselves but their relative rank in size compared with other eigenvalues. Consequently, the same eigenvectors, and therefore specific combinations of phenotypic traits, will not necessarily be tested at each iteration. If one is interested in a specific vector of selection, this test is inappropriate because it violates the assumption that all permutations within a series are exchangeable under the null hypothesis. It is, however, a valid test of the overall strength of non-linear selection. This contrasts with the double regression approach (Bisgaard and Ankenman 1996) in which the significance of an eigenvalue corresponding to a specific eigenvector of **M** is being tested. In practice, careful consideration should be paid as to which hypothesis is of interest and that it is tested appropriately. Questions like "Is there significant non-linear selection occurring overall?" can be answered using the Reynolds approach. However, questions such as "Does a specific linear combination of original traits experience more non-linear selection than expected by chance?" require a different approach. In this case it is necessary to adapt the permutation procedure of Reynolds et al. (2010) to hold a specific eigenvector constant and generate a null distribution of selection coefficients relating *specifically to that vector* under the null hypothesis of no association between trait and fitness. Note that similar issues surrounding violation of the exchangeability assumption in permutation tests have recently been addressed within the context of tests of the dimensionality of genetic and phenotypic variance-covariance matrices (Hine and Blows 2006; Pavlicev et al. 2009).

9.2.4 Issues and caveats

In addition to the issues just outlined with respect to statistical tests of selection gradients, approximating the fitness surface using ordinary least-squares carries with it a number of assumptions and potential caveats. Here we provide an overview of those that we consider particularly pertinent. Taken collectively, these issues highlight the fact that because conventional selection analyses are correlative, they should always be interpreted with a degree of caution and, whenever possible, be supported by manipulative experiments.

Selection is most often estimated using standing phenotypic variation in a population, which can result in two statistical issues: multicollinearity among the traits and the infrequent occurrence of rare phenotypes. In the context of a selection analysis, multicollinearity occurs when the phenotypic traits being examined are highly correlated. Multicollinearity can be detected using the variance inflation factor (VIF) and, while its presence does not alter the overall significance or explanatory ability of model (i.e. R^2), it makes the estimates of the individual selection gradients unreliable (Mitchell-Olds and Shaw 1987). Multicollinearity is best dealt with by one of two approaches that reduce or eliminate it either via a rotation of the original phenotypic trait axes or by the removal of one or more traits from the analysis. Although principal components analysis has often been used in the former approach, a canonical rotation may be preferable in the case of non-linear selection because the canonical axes are extracted specifically to identify the multivariate phenotypes targeted by selection. Principal components, in comparison, are not and may therefore diffuse the true targets of selection across multiple new traits (Shaw and Geyer 2010; Walsh and Lynch 2011). With respect to the latter approach, the targeted removal of traits based on VIFs may alleviate multicollinearity, although more formal techniques, such as model selection via information-theoretic criteria (e.g. Akaike's information criterion), may also be useful in this respect and provide the additional benefit of a greater reduction in the number parameters being estimated (Shaw and Geyer 2010).

Selection is expected to remove rare, low fitness phenotypes from the population. Selection analyses that rely on phenotypic variance therefore often suffer from a suboptimal coverage of phenotypic space, with few or no individuals of certain phenotypes and numerous individuals of others. This can both reduce the statistical power to infer selection and bias the results via rare outliers. This can be addressed via two types of manipulations, both of which increase trait variance (See Fig. 9.1). The first involves "natural" manipulations such as altered diet (Anholt 1991), changes in maternal provisioning (Svensson and Sinervo 2000), hybridization (Schluter 1994), or others that serve to increase the frequency of otherwise rare phenotypes. Such manipulations, however, do not typically break trait covariances and also have the potential to directly impact organismal performance, necessitating careful controls. The second form of manipulation involves the construction of synthetic phenotypes. While this is may not be possible in all systems, it is moderately straightforward for some, such as acoustic traits (see Box 9.1 Brooks et al. 2005; Bentsen et al. 2006; Gerhardt and Brooks 2009). In addition to improving phenotypic coverage, such manipulations can also reduce any multicollinearity among traits, providing experimental (instead of simply correlative) evidence for the targets of selection. The latter is important because selection gradients from observational studies alone can be strongly influenced by selection on unmeasured traits (Mitchell-Olds and Shaw 1987) or by environmentally induced covariances between traits and fitness (Mitchell-Olds and Shaw 1987; Rausher 1992; Stinchcombe et al. 2002). The latter issue can also be dealt with by estimating genotypic selection (i.e. selection on breeding values, Stinchcombe et al. 2002), although this is best done by estimating the genetic correlation between a trait and fitness within a single quantitative genetic model (Hadfield et al. 2010).

Finally, it is widely acknowledged that, while multivariate regression provides the best linear or quadratic approximation of the true fitness surface (Lande and Arnold 1983; Phillips and Arnold 1989), it has the potential to be misleading if the true surface takes a more complex form involving, for instance, multiple peaks or valleys (Schluter and Nychka 1994). The common approach to deal with this is to visualize the fitness surface via a non-parametric approach that does not constrain it to conform to any particular mathematical shape. Such approaches include the cubic or thin-plate spline for one or two traits respectively (Schluter 1988; Green and Silverman 1994; Blows et al. 2003), and for more than two traits projection pursuit regression can be used to identify cross-sections of the surface in the directions of the strongest selection, allowing its visualization in reduced form via splines (Schluter and Nychka 1994). While such techniques provide a flexible approach to visualizing the geometry of fitness surfaces, they suffer two disadvantages. First, the ruggedness of a spline is determined by the value of the smoothing parameter λ that minimizes the generalized cross-validation (GCV) score (Craven and Wahba 1979). Unfortunately, in practice substantial subjectivity in the choice of λ often remains after this procedure, which can result in very different surfaces being produced. Second, selection gradients cannot be obtained from such surfaces so they cannot be used to predict evolutionary responses (Lande and Arnold 1983).

9.3 Comparing fitness surfaces among groups

Despite substantial attention being given to characterizing the strength and form of phenotypic selection in nature, we know much less about how and why it varies, in particular with respect to multivariate fitness surfaces (Arnold et al. 2001; Punzalan et al. 2010). Variation in selection may be spatial (i.e. among geographic populations; e.g. Arnqvist 1992; Gosden and Svensson 2008; Rundle et al. 2008), temporal (i.e. across time within a population, Box 9.2, Siepielski et al. 2009, Chapter 7), or it may involve different classes of individuals within a population (e.g. males and females, juveniles and adults, Schluter and Smith 1986; Chenoweth and Blows 2005). Selection may also vary among treatment levels in an experiment (e.g. Rundle et al. 2009). Although a diverse range of possible group types exist among which selection may vary, in each case we require a set of statistical techniques with which to formally compare the geometry of selection surfaces among them and to characterize any differences that exist. In contrast to the large body

132   PART III APPLICATIONS: MICROEVOLUTION, QUANTITATIVE GENETICS, AND POPULATION BIOLOGY

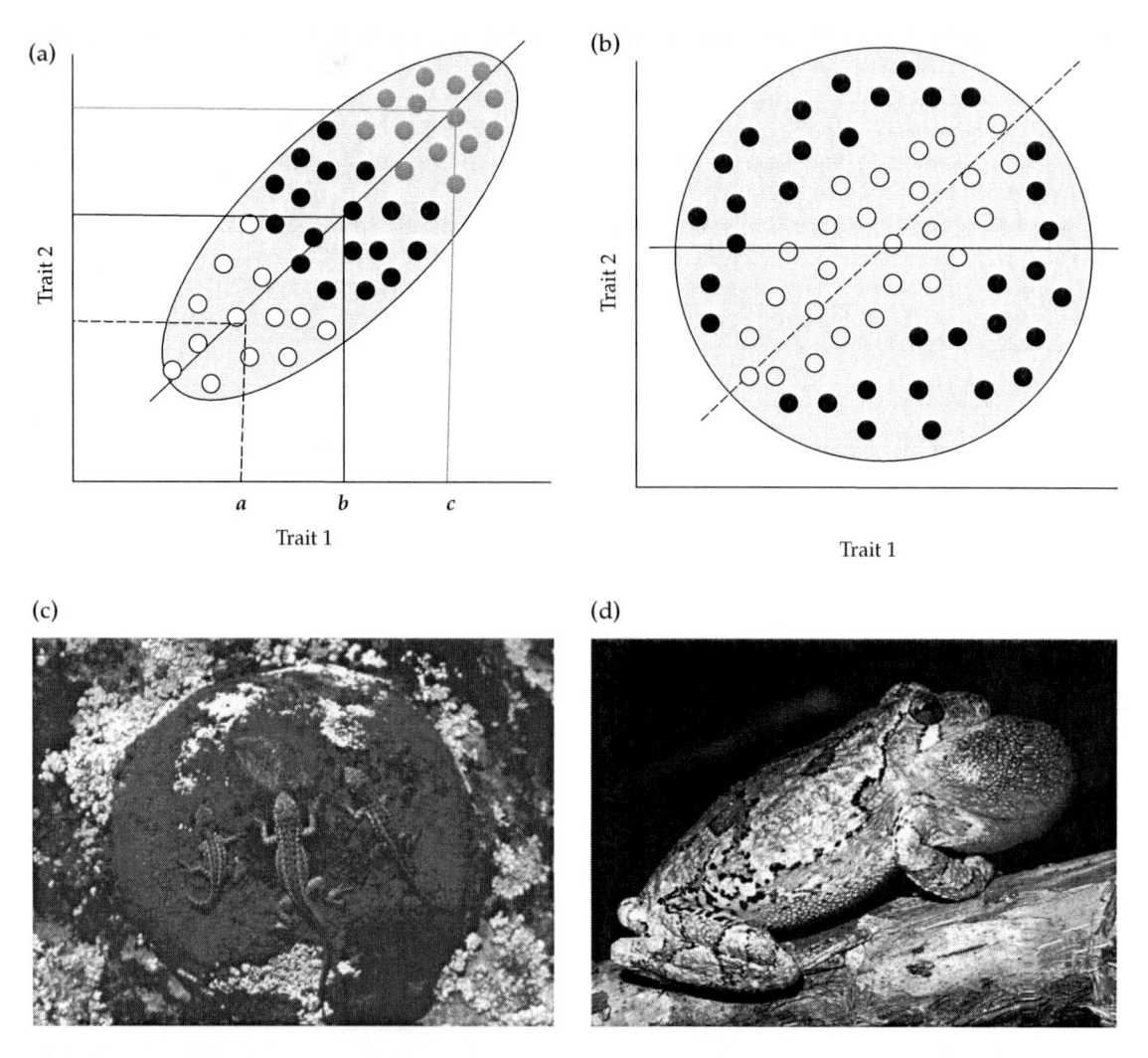

Figure 9.1 Using phenotypic manipulation to study selection. A particularly challenging aspect to the measurement of selection is that because selection has already shaped the distribution of traits, extreme phenotypes are rare and may therefore contribute little information. A further complication is that traits are often highly correlated leading to multicollinearity (see text). Two types of phenotypic manipulations can be used to address these issues. (a) *Natural manipulations* can be used to either decrease (*a*, open symbols) or increase (*c*, gray symbols) the size of the traits relative to the un-manipulated mean (*b*, closed symbols). This approach increases the frequency of rare phenotypes in the population, and may or may not alter the covariance structure between different traits. (b) *Synthetic phenotypes* can also be constructed by sampling phenotypes from a known distribution and re-assembling them at random. This approach not only increases the frequency of rare phenotypes but also breaks the natural pattern covariance between traits that generates multicollinearity. (c) Hatchling size variation in the side-blotched lizard (*Uta stansburiana*) that results from a natural manipulation of egg volume (photo courtesy of Erik Svensson). Removing contents of the egg with a syringe reduces offspring size; Svensson and Sinervo (2000) used this approach to show that natural selection on egg size differs with the intensity of competition. (d) A male gray treefrog (*Hyla versicolor*) producing an advertisement call to attract females to mate (photo courtesy of Joshua Schwartz). Gerhardt and Brooks (2009) produced synthetic calls that varied in five call parameters (pulse rate, pulse number, call period and low and high peak frequency) using specialist acoustic software. These five call parameters were sampled from a natural distribution and re-assembled at random to produce a synthetic call. Each synthetic call was then played back to females in a mate choice arena, in competition against a standard call with mean call parameters for the population. This approach enabled Gerhardt and Brooks (2009) to show that female mate choice exerted both directional and stabilizing sexual selection on male call structure.

Box 9.1 Multivariate stabilizing selection on call structure in the field cricket, *Teleogryllus commodus*

Like most orthopterans, males of the Australian black field cricket, *Teleogryllus commodus*, produce an advertisement call that they broadcast to attract a mate. This characteristic of the mating system has seen field crickets become an important model for the evolution of female choice, as well as the structure of the male acoustic signals that females target (Greenfield 2002). Despite this, very few studies have actually quantified the strength and form of multivariate sexual selection imposed on male calls by female mate choice. In fact, the majority of studies have taken a univariate approach to measuring selection where one component of the call is varied, while all other components are held constant. This approach ignores the natural pattern of covariance that exists between different call components; likely a result of the fact that the same fixed structures on the male forewing produce the call during stridulation. Furthermore, most studies taking this univariate approach only use dichotomous treatments (i.e. high versus low frequency) meaning they can, at best, only detect the operation of linear selection (Hunt et al. 2009) even though non-linear selection on male call structure may be expected due to constraints imposed by the female neurosensory system (Ryan and Wilczynski 1988; Ryan et al. 1992; Greenfield 2002).

To examine the strength and form of sexual selection operating on male call structure in *T. commodus*, Brooks et al. (2005) used a manipulative study. They started by measuring the call structure of 15 males and used this to estimate the mean and standard deviation for five individual call components (Fig. 9.2). Each of the individual call components were then sampled at random from their natural distribution and reformed (at random) to produce 300 unique calls where the call components were uncorrelated (i.e. no multicollinearity) and the full range of phenotypic space was covered. Four separate virgin females were then tested to each call in a series of binary choice trials where the focal call was tested against a control call that consisted of the mean call components in the population.

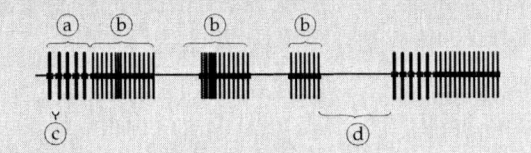

Figure 9.2 Schematic representation of the advertisement call of *Teleogryllus commodus*. Each call consists of a chirp (a) followed by one or more trills (b) in this species. Brooks et al. (2005) experimentally manipulated the number of pulses (CPN), the duration of the interval between pulses in the chirp (CIPD, c), the number of trills (TN), the intercall duration (ICD, d) and the dominant frequency of the call (DF, not shown). All other aspects of the call were held constant. Redrawn from Brooks et al. (2005), with permission from John Wiley and Sons.

Multivariate selection analysis (Lande and Arnold 1983) was then conducted, based on a bootstrap re-sampling method to account for the fact that each call was tested using multiple females, to estimate the vector of linear selection gradients (β) and matrix of non-linear selection gradients (γ) estimated (Table 9.1). Visual inspection of the γ matrix reveals that it is difficult to determine the overall

Table 9.1 The vector of standardized linear gradients (β) and the matrix of standardized quadratic and correlational selection gradients (γ) for male call structure components in *T. commodus*. Note that the quadratic gradients were not doubled in the original study but have been done so here. The significance of selection gradients were tested using permutation test with 9999 iterations (see text): *$P<0.05$, **$P<0.01$, ***$P<0.001$

	β	CPN	CIPD	TN	ICD	DF
CPN	0.007	0.012				
CIPD	−0.003	0.017	−0.012			
TN	0.015	0.019	0.039	−0.080		
ICD	−0.214***	−0.022	−0.036	0.086*	−0.160**	
DF	0.059	0.024	−0.031	0.041	−0.013	−0.094*

Continued

Box 9.1 (Continued)

Table 9.2 The **M** matrix of eigenvectors derived from the canonical analysis of γ. The linear (θ_i) and quadratic (λ_i) gradients of selection acting along these eigenvectors are provided in the last two columns. We tested the significance of selection along these eigenvectors using a permutation test based on the double regression method of (Bisgaard and Ankenman 1996) but note that the results were qualitatively the same when the permutation test of Reynolds et al. (2010) was used. Randomization test: *$P < 0.05$, **$P < 0.01$, ***$P < 0.001$

	M					Selection	
	CPN	CIPD	TN	ICD	DF	θ_i	λ_i
m_1	0.800	0.497	0.305	−0.057	0.130	0.028	0.035
m_2	−0.446	0.806	−0.001	−0.091	−0.377	−0.017	−0.003
m_3	−0.302	0.003	0.776	0.500	0.240	−0.082	−0.019
m_4	−0.257	0.208	−0.102	−0.405	0.846	0.144***	−0.108**
m_5	0.068	0.244	−0.543	0.758	0.258	−0.160***	−0.240***

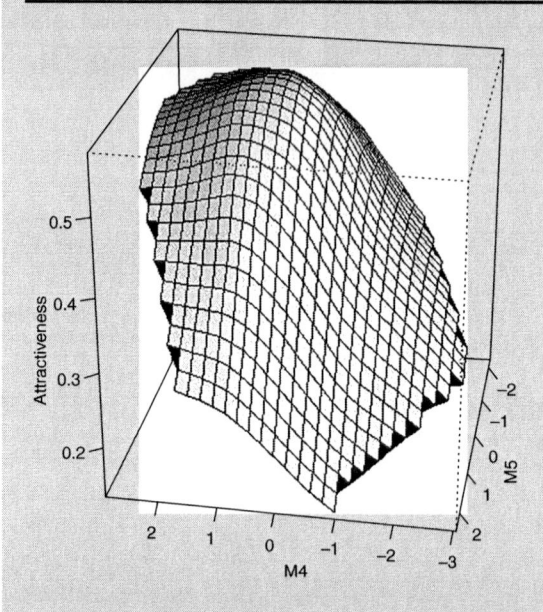

Figure 9.3 Thin-plate spline perspective view visualization of the fitness surface on the two major axes of non-linear selection, m_4 and m_5. The thin-plate spline was estimated using the *Tps* function in R (version 9.2.1), using the value of the smoothing parameter, λ, that minimized the GCV score. Redrawn from Brooks et al. (2005), with permission from John Wiley and Sons.

pattern of sexual selection operating on male call structure, as a mixture of positive and negative quadratic and correlational gradients exist. However, canonical analysis of γ provides a clear pattern of multivariate stabilizing sexual selection operating on male call structure, with 4 of the 5 eigenvalues (including the two with statistically significant selection) being negative (Table 9.2). This can be visualized in Fig. 9.3 that presents the thin-plate spline for the two major axes of non-linear sexual selection on call structure.

Brooks et al. (2005) then tested the prediction that, in the absence of frequency-dependent selection, the population mean phenotype should evolve uphill towards the peak of the fitness surface (Simpson 1953; Lande 1979; Lande and Arnold 1983). In order to test this prediction, it was necessary to calculate the stationary point on the individual fitness surface (Fig. 9.3), as well as the confidence region for this point, and then determine whether the population mean call structure resides within this confidence region. The stationary points on the major axes of non-linear sexual selection were estimated at 1.44 and −0.67 for m_4 and m_5, respectively, by differentiating the equation for the rotated fitness surface (Phillips and Arnold 1989) (Fig. 9.4a). The 95% confidence region for this surface was also estimated using the algebraic formulation of Box and Hunter (1954), implemented in the MAPLE software package provided by Del Castillo and Cahya (2001) (Fig. 9.4b).

Visual inspection of Fig. 9.4 reveals two important points. First, there is clear evidence that, for all four treatments, the mean call structures of males from this population reside in close proximity to the stationary point of the fitness surface (Fig. 9.4a) and within the 95% confidence region of this stationary point (Fig. 9.4b). In fact, the call structure of males from four independent samples from this population (three different dietary regimes and wild caught males) were all within half a standard deviation from the stationary point and MANOVA shows that they do not differ significantly along these two dominant eigenvectors comprising the fitness surface (m_4 and m_5). This therefore provides strong evidence for the prediction that male call structure has converged on the peak of the fitness surface in this population. Second, despite mean call structures being aligned with the peak of the fitness surface, there was still considerable individual

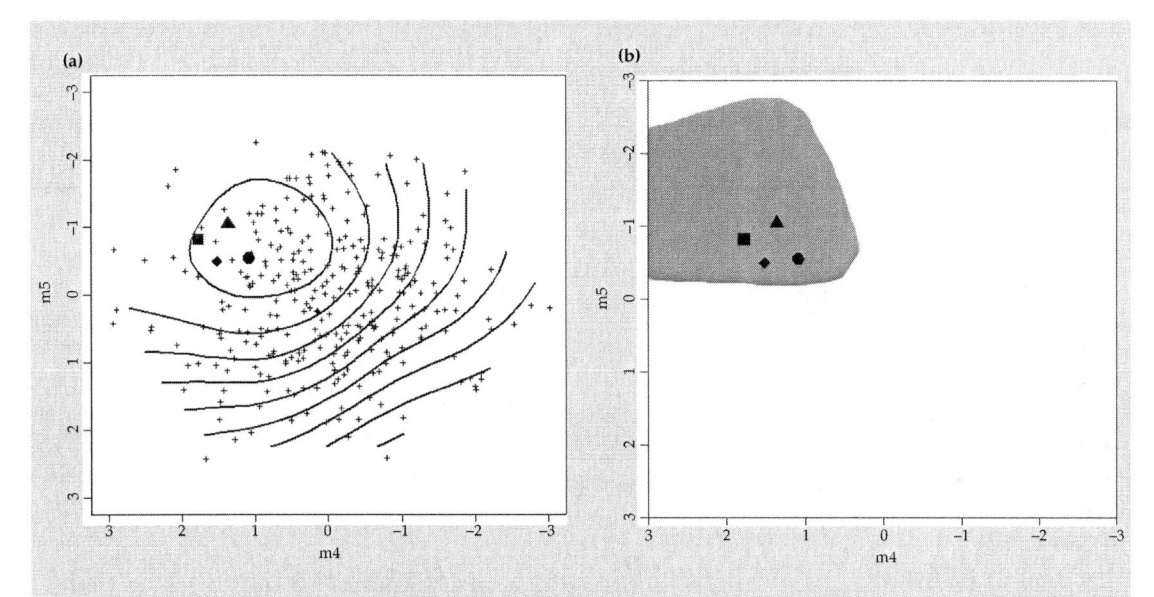

Figure 9.4 (a). Thin-plate spline contour map visualization of the fitness surface on the two major axes of non-linear selection, m_4 and m_5, showing the distribution of artificial call structure values on which the surface is based (+) and the mean call values of the four independent samples of males (wild caught males λ, high protein diet ■, medium protein diet υ and low protein diet ●). (b) The Box-Hunter (1954)) conditional 95% confidence region (shaded area) for the position of the stationary point on the fitness surface and the mean call values of the four independent samples of males. Redrawn from Brooks et al. (2005), with permission from John Wiley and Sons.

variation in male call structure with numerous males from each of the four samples falling outside the 95% confidence region of the stationary point (Fig. 9.4b). This suggests that males are not necessarily constrained to this area of the fitness surface. It is also in agreement with Mitchell-Olds and Shaw's (1987) definition of "true" stabilizing selection where the stationary point of the fitness surface exists within the range of phenotypic data.

Box 9.2 Temporal variation in selection in harlequin bugs

The native Australian harlequin bug, *Dindymus versicolor*, is a pest of fruit crops but also an ideal system to study selection in the wild because pairs remain in copula for up to 3 days. To examine temporal variation in sexual selection on male morphology, one of us (J. Hunt) collected random samples of 138, 141, and 130 males from three different times during the breeding season. Males were scored as to whether they were in copula at the time of collection (1 vs. 0 for mated or not) and subsequently measured for four morphological traits: antenna length, head length, elytra width, and body mass (Fig. 9.5). Prior to analyses, body mass was *ln*-transformed and five individuals were removed as multivariate outliers. Within each time, traits were standardized ($\sim N(0,1)$) and mating success was transformed to a relative measure (by dividing each score by the mean for that time period). Standardized linear selection gradients were then estimated via least-squares separately within each sampling time (Table 9.3) and likelihood ratio tests comparing a full generalized linear model with a reduced one lacking the four traits (performed using the GENMOD procedure with a logistic link function in SAS v. 9.2) revealed that linear sexual selection was highly significant overall in all three of the sampling times ($d.f. = 4, P < 0.0001$ in each case) and explained a remarkable 56.0–73.5% of the variance in male mating success ($r^2_{adjusted}$). Because the three sampling

Continued

Box 9.2 (*Continued*)

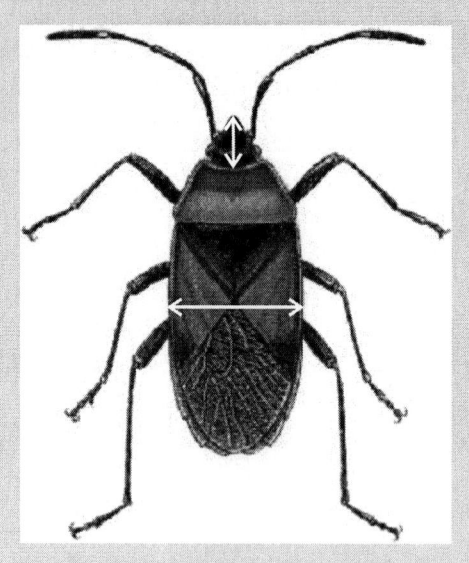

Figure 9.5 Male *Dindymus versicolor*. Depicted are two of the traits measured: head length and elytra width. Not shown are antenna length (total length after being removed from the animal and straightened using dissecting pins) and body mass.

times constitute a random draw from the population of possible times during the breeding season, among-group variation in selection was tested using a first-order multi-trait random regression (with the traits standardized globally across times and using the absolute mating scores). Significance of the dominant axis (i.e. first eigenfunction) of the covariance matrix of random regression coefficients was tested using factor analytic modeling as implement in the MIXED procedure of SAS (Hine and Blows 2006; McGuigan et al. 2008). The intercept was included as a random effect in both models when conducting this likelihood ratio test, thereby accounting for any variance in the copulation rate among times. This dimension approached statistical significance although did not achieve it at the standard $\alpha = 0.5$ level (likelihood ratio test; $\chi^2 = 7.60$, $d.f. = 4$, $P = 0.107$), providing weak evidence at best for among-sample variation in linear sexual selection on these traits. While it is possible that selection varies little temporally, low power from minimal replication (i.e. only three sampling times) likely also contributes to this non-significant result, demonstrating the empirical challenge of making statistically robust inferences with respect to random group effects.

In contrast to a random effect, if these samples represented three particular times of interest that would be sampled again should the study be repeated, and inference was to be restricted to these three times, they would constitute a fixed effect. As an example of such an analysis, a categorical group effect term representing sampling time was added to the standard first-order polynomial regression model estimating linear sexual selection, along with the four trait × sampling time interactions. The fit of this model was compared to a reduced model lacking these four interaction terms via a likelihood ratio test (as implemented using the GENMOD procedure with a logistic link function in SAS). Linear sexual selection on these four traits varied significantly overall among these three samples ($\chi^2 = 17.13$, $d.f. = 8$, $P = 0.029$), although in the absence of replicate selection estimates within each time, sampling error is confounded with any time effect and these cannot be separated. Note that none of the individual trait × sampling time interactions were significant in isolation, although antenna length × sampling time approached so (Table 9.3). This is not unusual and simply indicates that the trait vector along which selection differs the most among samples is a composite (i.e. linear combination) of the measured traits. To characterize the differences in selection in trait space, a 4 × 4 symmetrical inter-group (co)variance matrix (**B**) was calculated from the standardized selection gradients within each population. The first eigenvector of **B** ($\mathbf{b}_{max}$) explained most (81.3%) of the among-group variance in sexual selection and represented a trait

Table 9.3 Vectors of linear sexual selection on four traits in male harlequin bugs as estimated during three separate times during the breeding season, with significance as determined from a generalized linear model. Also shown is the significance of the among-group variation in selection for each trait from a fixed effect analysis and the first eigenvector of the among-group variation in trait space ($\mathbf{b}_{max}$).

Trait	$\beta_{sampling\ period}$ (P)			Fixed effect	
	β_1	β_2	β_3	P (trait × group)	$\mathbf{b}_{max}$
antenna length	0.422 (<0.001)	0.300 (0.004)	0.245 (<0.001)	0.075	−0.509
head length	0.207 (<0.001)	0.189 (0.008)	0.427 (<0.001)	0.111	0.771
elytra width	−0.108 (0.083)	−0.022 (0.455)	0.004 (0.635)	0.483	0.347
ln(body mass)	0.314 (<0.001)	0.226 (<0.001)	0.333 (<0.001)	0.978	0.163

combination that contrasted antenna length with head length and to a lesser extent, the other two traits (Table 9.3). Why selection may vary most temporally on this trait is a topic for further study. To characterize differences in selection in group space, a similarity matrix of vector correlations was assembled from the standardized selection gradients. Vector correlations were high, ranging from 0.865 to 0.987. The first eigenvector of this matrix accounted for the majority (94.9%) of the among-group variance in sexual selection and had almost identical loadings for all three groups (first eigenvector: [0.579 (time 1), 0.589 (time 2), 0.564 (time 3)]), indicating that the three sampling times contribute roughly equally to a single axis along which selection differs among these groups

of statistical literature surrounding the estimation and analysis of fitness surfaces within populations, surprisingly little attention has been given to this topic.

9.3.1 Testing among-group variation

A key issue in any statistical comparison of fitness surfaces among groups is whether the groups represent fixed or random effects. This is not always straightforward and confusion persists around this topic in general (Littell et al. 1996). A statistical effect is fixed if the chosen levels represent all levels of interest of the factor. Inference for fixed effects concerns the mean responses at the various factor levels and cannot be extrapolated to levels not included in the study. Also, if the study were to be repeated, the same levels of the fixed factor would normally be used. An effect is random if the chosen levels represent a random sample drawn from a larger population of potential levels. Random effects are also exchangeable, meaning that any set of levels from the larger population of inference could have been used in the study. With a random effect, the inference is about the variance of the underlying population (e.g. the among-group variance).

Fixed groups
When groups are a fixed effect, variation in selection among them can be tested using a sequential model-building approach for response surface designs (Draper and John 1988; Chenoweth and Blows 2005). The approach is straightforward and entails adding a categorical term representing the group effect to the first-order multiple regression used to estimate selection, with the group × trait interactions testing whether selection on the trait in question varies among groups. For linear selection, the model essentially becomes an ANCOVA:

$$w = \beta_0 + G + \sum_{i=1}^{n} \beta_i z_i + \sum_{i=1}^{n} \alpha_i z_i G + \varepsilon, \quad (9.5)$$

where w is an individual's relative fitness, β_0 is the intercept, G is the group effect (not to be confused with the genetic variance-covariance matrix **G**), z_i refers to value of the ith trait, n represents the number of traits, and ε is unexplained error. The partial regression coefficients, β_i, represent the linear selection gradients on the n traits and the $\alpha_i z_i G$ interaction term allows selection to vary among groups. For multiple traits, the fit of this full model can be compared with that of a reduced model lacking the $\alpha_i z_i G$ interaction terms, providing a test of whether the model that includes the group structure is a better fit to the data than the reduced model. This is equivalent to a single overall test for variation in the fitness surface and can be done by fitting the models via least squares and then employing a partial F-test (Bowerman and O'Connell 1990.) to compare the unexplained sums of squares (SSE) from the two models:

$$F_{a,b} = \frac{SSE_{reduced} - SSE_{complete}/a}{SSE_{complete}/b}, \quad (9.6)$$

where a is the difference in the degrees of freedom between the complete and reduced models (note that this is not equal to the number of terms that differ between the two models when there are more than two group levels) and b is the degrees of freedom for $SSE_{complete}$. An alternative approach for unbalanced data is to fit the models via maximum likelihood (ML) and then use a likelihood ratio test to compare them. In this case, the difference in −2(log likelihood) values of the two models is χ^2 distributed with degrees-of-freedom equal to the difference in the number of model parameters between the full and reduced models. Note that when models differ in their fixed effect structure such as these, ML and not restricted maximum likelihood (REML) must be used when comparing them (Pinheiro and Bates 2000). This is because REML essentially performs a ML estimation of the transformed data and altering the fixed effect structure changes this transformation (Littell et al. 1996). The use of information-based metrics like Akaike's or Schwarz's criteria does not circumvent this problem because these are based on the −2(log likelihood) values.

For non-linear selection, the approach is analogous and involves adding a term representing the fixed effect of group to the second-order multiple regression used to estimate the fitness surface:

$$w = \beta_0 + G + \sum_{i=1}^{n} \beta_i z_i + \sum_{i=1}^{n} \alpha_i z_i G$$
$$+ \sum_{i=1}^{n}\sum_{j=1}^{n} \gamma_{ij} z_i z_j + \sum_{i=1}^{n}\sum_{j=1}^{n} \alpha_{ij} z_i z_j G + \varepsilon. \quad (9.7)$$

Here, the $\alpha_{ij} z_i z_j G$ terms allow non-linear selection, including both quadratic and correlational selection, to vary among groups. Again, a single overall test for variation in the fitness surfaces among groups involves comparing the fit of this full model to one lacking the $\alpha_{ij} z_i z_j G$ terms via either a partial F or likelihood ratio test. Linear selection is allowed to vary among groups in both the full and reduced model (i.e. the $\alpha_i z_i G$ terms remain), meaning that it is the additional contribution of variation in non-linear selection alone that is being tested. Any combination of nested models may be compared via partial F or likelihood ratio tests, for example a full model allowing linear and non-linear selection to differ among groups could be compared to a reduced one in which both form of selection are invariant, thereby providing a single test for among-group variation in total selection. It is also possible to separately test for variation in quadratic and correlational selection among groups (e.g. Chenoweth and Blows 2005), although as discussed earlier, the distinction between these may be somewhat arbitrary and depend more on how the traits are defined as opposed to representing some fundamental difference in the geometry of the fitness surfaces. Although not directly indicated by our notation in equation 9.7 because it does not affect the significance of tests for group-based differences, if one wishes to produce quadratic selection gradients as per equation 9.2, a coefficient of ½ will apply to the quadratic terms. More complex linear models are possible in theory, although with multiple traits and numerous interactions, a loss of degrees-of-freedom may become an issue. Also, if relative fitness is non-normally distributed, generalized linear models can be fit using ML (see section 9.2.3) and likelihood ratio tests can be performed as described above.

Random groups

When groups are a random effect, random coefficient mixed models (a.k.a. "random regression") can be employed to model variation in selection among groups. This can be viewed as an extension of the ANCOVA-based approach above in which the regression coefficients for the continuous covariates (i.e. the selection gradients on the traits) are not assumed to be fixed effects but rather to represent a random sample from a population of possible coefficients (Littell et al. 1996). The multivariate random coefficient model for testing for variation in selection among groups can be written as:

$$\mathbf{w} = \beta_0 + \mathbf{Xb} + \delta^{(G)} + \varepsilon, \quad (9.8)$$

where $\mathbf{w}$, a column vector of the individual fitness scores, is modeled as a function of the fixed effect intercept (β_0) and group-wide regression coefficients (**b**) for the set of continuous traits (**X**). $\delta^{(G)}$ represents the random effect (co)variances in these regression coefficients among groups and is assumed to be normally distributed with a mean of zero and variance σ^2. The model can be fit via REML and a likelihood ratio test can be used to compare it with a restricted model. In the case of a single trait, the restricted model would be one lacking the $\delta^{(G)}$ term, thereby testing the significance of the among-group variance in the regression coefficients. For a multi-trait model, the random regression coefficients form a covariance matrix and hypothesis testing must focus on its dominant axes (termed eigenfunctions, Meyer and Kirkpatrick 2005; McGuigan et al. 2008). There are various ways to do this, a straightforward one of which employs the factor analytic modeling approach implemented in the MIXED procedure of SAS (SAS Institute, Cary, NC). Via factor analytic modeling, reduced rank covariance matrices can be specified at the group level and compared via a series of nested likelihood ratio tests (McGuigan and Blows 2007). Of fundamental interest is the significance of the first eigenfunction representing the single trait combination having the greatest among-group variance in directional selection, thereby providing a test for among-group variance in selection. For non-linear selection, a quadratic version of this model would include the group-wide quadratic regression coefficients as fixed effects along with the random effect (co)variances in these regression coefficients among groups. However, a multi-trait quadratic random coefficient model has, to our knowledge, not been applied before in such a context and this is therefore an area in need of development.

When relative fitness does not have a normal distribution, a generalized linear mixed model may be employed. However, as likelihood calculations can be slow or sometimes impossible in such models, they are normally fit using an approximation of the true likelihood (e.g. pseudo- or penalized quasi-likelihood, Laplace approximations, or Markov chain Monte Carlo algorithms). This complicates model comparison because likelihood ratio tests are generally not possible. Although model estimation and comparison can be a challenging endeavor, generalized linear mixed models provide a flexible approach for analyzing non-normal data when random effects are present. The application of such models in ecology and evolution is therefore a topic of active development (Bolker et al. 2009) and recent software packages aim to make the flexibility of such analyses more widely accessible (e.g. Hadfield 2010).

9.3.2 Characterizing among-group differences

Given statistically significant differences in selection among groups, it may be of interest to then explore how the fitness surfaces differ. Two distinct approaches are possible that characterize the differences in either trait (i.e. phenotypic) or group space. The former is well grounded in quantitative genetic theory (e.g. see section 9.4) whereas the latter is not, although the goal here is to interpret statistically demonstrated differences and not to make predictions. The general approach is the same for each: a symmetrical matrix is created that characterizes variation in selection gradients either among traits (trait space) or among groups (group space) and then the structure of this matrix explored to provide insight into how selection varies.

Trait space

Generalizing from Zeng (1988) and see Chenoweth et al. (2010), variation among groups in the slopes of the individual fitness surfaces, represented by the vector of directional selection gradients (β) on the n traits within each group, can be characterized by the $n \times n$ symmetrical inter-group (co)variance matrix **B**. The diagonal elements of **B** represent the variances among groups in the directional selection gradients on each of the n traits and the off-diagonal elements represent the covariances of directional selection gradients among groups for each bivariate trait combination. Diagonalization of **B** yields a set of eigenvalues and eigenvectors that describe its major axes (i.e. principal components), providing insight into the structure of among-group variation in multitrait space. The first eigenvector of **B**, for example, describes the single composite trait (i.e. linear combination of original traits) for which directional selection differs the most among groups, or in other words, on which divergent selection is strongest. Subsequent eigenvectors explain progressively less of the among-group variance in directional selection. The dimensionality or rank of this matrix refers to the number of positive eigenvalues and provides insight into the number of composite traits (i.e. phenotypic dimensions) among which directional selection differs among groups.

For non-linear selection, the γ matrix of quadratic and correlational selection gradients provides an estimate of the curvature of the fitness surface that is independent of (i.e. uncorrelated with) its slope (i.e. with β; Lande and Arnold 1983). It may therefore be possible to compare non-linear selection among groups via an analogous approach to that for linear selection, involving the construction of an inter-group (co)variance matrix from the individual elements of γ within each population. Such an approach has recently been used to characterize the among-population (co)variance structure of genetic variance-covariance (i.e. **G**) matrices (Hine et al. 2009). However, as far as we are aware this has not been attempted with respect to variation in non-linear selection among groups and is therefore an important topic for future work.

Group space
Comparisons in group space follow the same basic approach as those in trait space in that a symmetrical matrix is assembled and then its eigenstructure is determined. In this case, however, the rows and columns of this $g \times g$ matrix correspond to the g groups, with the individual elements representing the overall similarity of linear or non-linear selection between a particular pair of groups. For linear selection, the vector correlation, calculated as the dot-product of the standardized β vectors from two groups, provides a straightforward, bounded measure of the overall similarity between them in the direction of linear selection on the entire suite of traits, ranging from −1 (selection vectors oriented 180° from one another) to + 1 (coincident vectors), with zero indicating orthogonal vectors. For non-linear selection, the approach is potentially more complex because a number of possible metrics exist for comparing two matrices. For example, Arnold et al. (2001) suggest the Flury (1988) hierarchy, a commonly used approach for the comparison of **G** matrices. However, this technique was originally intended for product-moment matrices and may be ill-suited for comparing γ's. A simpler geometric approach, first developed by Krzanowski (1979), may be appropriate however, in part because it makes no restrictive assumptions about the matrices being compared (Blows et al. 2004; Blows 2007). This latter approach provides a single, bounded measure of the similarity of two matrices by comparing the orientation of subspaces defined by a subset (up to half) of the principal components of each matrix. Details of this method are provided in Blows et al. (2004) and its application to γ matrices in particular is described in Rundle et al. (2008).

As with the approach within trait space, structure of these among-group similarity matrices for linear and non-linear selection can be summarized via diagonalization. The first eigenvector in this case represents the dominant axis of among-group variation in linear or non-linear selection respectively. For example, two populations with similar loadings for this vector contribute to among group variance in selection to a similar degree. The decline in eigenvalues indicates the extent to which among-group variation in selection is dominated by a single type of difference or multiple ones. The structure of these similarity matrices can also be explored via their association (e.g. matrix correlation) with specific model matrices specifying particular patterns of variation (e.g. habitat types), with significance assessed via a randomization procedure (e.g. Mantel's test, see Rundle et al. 2008). Such an approach could likewise apply to a trait space matrix (i.e. **B**), although we are not aware of any published studies that do so.

Finally, in a recent application of a group space approach, Hohenlohe and Arnold (2010) developed a maximum likelihood framework to determine the dimensionality of a group space matrix constructed from sexual isolation measures among populations (as estimated from mating trials), with the goal of inferring the number of independent traits underlying sexual isolation. Here, traits are unknown but insight can be gained via the dimensionality of the among-group similarity matrix that characterizes the pattern of mating compatibilities. The extension of this ML framework more generally to the problem of among-group differences in selection could be a promising exercise.

9.3.3 Issues and caveats

The statistical comparison of fitness surfaces among groups is an extension of the same model used to estimate these surfaces within individual populations and therefore shares all assumptions and caveats with it (see section 9.2.4). Statistical power may be an additional concern given the complexity of the models being fit to the data and the practical challenge of collecting sufficient data with which to estimate individual fitness surfaces in multiple groups. This may be particularly acute in the presence of correlational selection because a canonical analysis cannot be used within each group (this would generate different traits among groups, preventing their comparison). Among-group comparisons must therefore be performed within the original trait space (or some consistent transformation of it). Power may tend to be lower when the number of groups is small and groups are treated as random rather than fixed because the variance components will be poorly estimated.

Estimates of selection are dependent on the phenotypic trait distributions (Lande and Arnold 1983; Phillips and Arnold 1989), meaning that comparisons among groups are only biologically meaningful when these distributions do not differ substantially, otherwise different parts of what could be the same fitness surface are being estimated within each group. Potential solutions for dealing with such differences could include restricting comparisons to overlapping trait values or modifying the trait distributions prior to estimating selection. More generally, however, as with other multivariate analyses, the independent variables (i.e. traits) in the above models are assumed to follow a multivariate normal distribution. This assumption is necessary because in most applications of selection gradient analysis the predictors (traits) are random rather than fixed variables. Making the assumption of multivariate normality among all predictors and response variables allows one to use the same, simpler, regression equations that are suitable for fixed predictors (Rencher 2002). Thus homogeneity of phenotypic (co)variances (i.e. the **P** matrix) among groups would therefore appear to be a necessary assumption to proceed with among-group statistical comparisons. Although ensuring homogeneity of **P** among groups may be challenging in some cases, it may be relatively straightforward in others. For example, in a comparison of female mate preferences for male pheromones among 12 populations of *Drosophila serrata*, Rundle et al. (2009) ensured the constancy of **P** among groups by presenting females from every population with a choice among a standard set of males.

Another issue concerning phenotypic traits is whether and how to standardize them prior to undertaking among-group comparisons. The choice of scale has been shown to have a large influence on multivariate selection and quantitative genetic analyses in general (Hansen and Houle 2008; Simonsen and Stinchcombe 2010) and this may be particularly acute in among-group comparisons (e.g. because standardization may remove among-group differences in trait means and variances). Although variance-standardized selection gradients are commonly employed when comparing selection, these may have some undesirable properties in comparison with mean-standardized gradients (Hereford et al. 2004). It is also possible that the most appropriate approach with respect to statistical vs. biological inference may not be the same. The choice of standardization and the scale at which it is applied (e.g. across vs. within groups) is a complex topic that has not been sufficiently addressed with respect to among-group comparison.

Finally, it is important to note that the inference of among-group differences, whether fixed or random, requires replication within groups. In the absence of such replication (i.e. with only one fitness surface estimated per group), sampling error cannot be separated from group-effects because there is no estimable within-group variation in the selection gradients. In such a situation, the application of the above methods can demonstrate significant differences among samples but cannot attribute such differences to a group effect per se. For example, Chenoweth and Blows (2005) demonstrated significant differences between two fitness surfaces within a single population of *D. serrata*. While these surfaces represented sexual selection on contract pheromones in males and females respectively, sex-differences were confounded with sampling variation because only a single estimate of the surface was made in each sex. The inference of between-sex differences was only possible in a later study (Rundle and Chenoweth 2011), in which male and female fitness surfaces were estimated in multiple replicate populations. Unfortunately, when appropriate replication is performed, dealing with it statistically may be challenging given the complexity of the resulting models (e.g. replicates represent random effects nested with the fixed or random group effect).

9.4 Integrating multivariate selection analyses with evolutionary genetics

While describing natural selection and how it varies is an important endeavor in-and-of itself, a key strength of the quadratic approximation approaches we have outlined and built upon thus far is that the estimated parameters are integrated with evolutionary quantitative genetic theory, thereby permitting the data to be used in further analyses. In this section, we outline two areas where the integration of fitness surface data with genetic analyses has been instrumental. In the first example, we show how a deeper appreciation of how selection affects genetic variance for quantitative traits has been gained though the consideration of multivariate linear and non-linear selection. In the second example, we show how estimates of multivariate selection taken from sets of conspecific populations have been combined with genetic (co)variance estimates to forge a framework for understanding how variation in selection drives phenotypic divergence in the face of genetic constraints. Our central focus in these examples is to demonstrate how a consideration of the geometry of fitness surfaces, quantified as linear and non-linear selection gradients via the Lande and Arnold (1983) approach, has led to new insights.

9.4.1 Model misspecification and the nature of genetic variance under selection

The maintenance of genetic variation in quantitative traits in the face of strong selection has been an enduring problem for quantitative genetics that remains unsolved to this day (Barton and Turelli 1989; Barton and Keightley 2002). Mutation-selection-balance models for the maintenance of genetic variation envision standing variation as reflecting mutation introducing new variants and natural selection either fixing or removing them (Bulmer 1989). A fundamental problem with such models has been that the amount of standing genetic variance observed vastly exceeds expectations given estimates of mutation and stabilizing selection (Johnson and Barton 2005). A requisite assumption of such models is that selection, both stabilizing and directional, reduces genetic variance. In the two cases we outline next, a comprehensive analysis of multivariate selection has been instrumental in showing how selection shapes standing genetic variation in general, and in suggesting a potential resolution to the perceived problem of excess variation.

First, if stabilizing selection depletes genetic variance, a negative correlation should develop among traits between the strength of stabilizing selection and the amount of genetic variance. Building upon the finding of multivariate stabilizing selection on male cricket calls by Brooks et al. (2005) (see Box 9.1), Hunt et al. (2007) performed a quantitative genetic analysis of the same call traits using a half-sib breeding design in this population. By scoring the progeny for the traits defined by the canonical axes of the fitness surface, they demonstrated that the canonical axes experiencing the strongest stabilizing selection had the lower additive genetic variance (Fig. 9.6a and b), whereas those subject to weaker selection tended to have more genetic variance (Fig. 9.6c and d). Although apparent in a univariate consideration of the traits, the observed relationship between the amount of genetic variance and the strength of stabilizing selection was far stronger for the canonical axes of the fitness surface (Hunt et al. 2007), suggesting that the traits defined by the canonical analysis may have provided a more suitable model for capturing the true targets of stabilizing selection than the quadratic selection gradients estimated for the original traits.

Second, in sexual selection research there has been debate surrounding the maintenance of genetic variance in male display traits. Because these traits are often subject to persistent and relatively strong directional selection through female choice, they are expected to have low genetic variance. However, univariate genetic variances are not unusually low for male display traits (Pomiankowski and Moller 1995). This finding spawned several theories for how genetic variance can be maintained in such traits (Pomiankowski and Moller 1995; Rowe and Houle 1996). An alternative view to this problem is simply that the targets of selection have not been accurately captured via single trait analyses. For example, via multivariate analyses of sexual selection on male contact pheromone displays in two species in the montium subgroup of *Drosophila*, Blows and colleagues have shown that little genetic variance is present along the specific vector of directional selection (i.e. β),

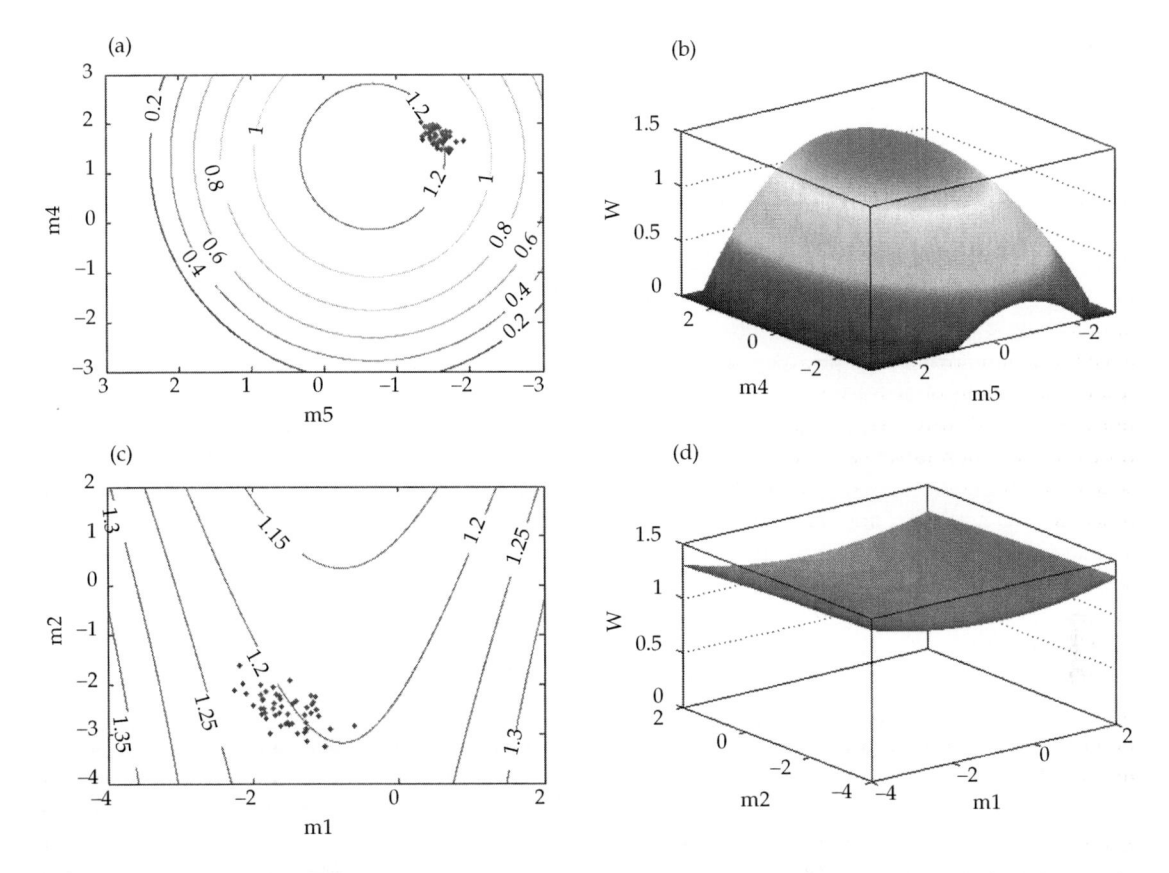

Figure 9.6 (See also Plate 2). Association between the strength of multivariate stabilizing selection and additive genetic variance for five call traits in the black field cricket, *T. commodus* after Hunt et al. (2007). (a) A tighter clustering of breeding values was observed for traits canonical axes under stronger stabilizing selection (b). (c) Contour plot showing predicted breeding values (BLUPs) for two canonical axes under weak non-linear selection (d). Reproduced with permission from the Genetics Society of America.

and that the vast majority of it lies almost orthogonal to this (Blows et al. 2004; Hine et al. 2004; Van Homrigh et al. 2007; Chenoweth et al. 2010). This depletion of genetic variance in the direction of selection is not evident, however, when analyses are performed on single traits.

The described examples show how marginally more sophisticated statistical analyses of selection—multivariate vs. univariate or canonical analysis of a multivariate model—has provided novel insight into how selection may shape standing genetic variance within populations. More broadly, they illustrate how univariate analyses of these problems entail a misspecification of the model of selection, and how the multivariate analyses may bring us closer to a more realistic view of the true targets of selection.

9.4.2 The contribution of selection and genetic constraints to adaptive divergence

Divergent selection is a central element to many issues in evolutionary biology including the origin of phenotypic diversity and new species (Schluter 2000). However, because divergence in multiple traits need not always proceed in the direction favored by selection (Lande 1979), of particular interest has been understanding the extent to which patterns of phenotypic divergence between populations and closely related species reflect past selection, a compromise between divergent selection and genetic constraints, or genetic drift (Schluter 1996; McGuigan et al. 2005). In particular, comparisons of the direction of phenotypic divergence among populations with axes of genetic variance available within populations, have found that divergence often closely aligns with the major axes of additive genetic variance (Mitchell-Olds 1996; Schluter 1996). This suggests that either drift dominates selection or that genetic constraints commonly bias the response to natural selection along lines of greatest genetic variance. However, in some cases, divergence appears poorly aligned with these major axes of genetic variation, suggesting that natural selection has been able to overcome genetic constraints (McGuigan et al. 2005).

A major limitation of such comparative approaches is the potential confounding of genetic drift and selection as causes of any association between axes of genetic variance and divergence (Phillips et al. 2001). Under genetic drift, the pattern of phenotypic divergence among populations is expected to be proportional to the level of genetic variance (Lande 1979), and any association between the major axes of genetic variance and divergence may therefore be the product of neutral divergence. Further, as selection itself is unmeasured, variation in selection is inferred from the direction of divergence itself, making it difficult to isolate the impact of genetic constraints. However, Lande (1979) showed through the multivariate version of the breeders' equation, $\Delta z = \mathbf{G}\beta$, how the response to selection of a multivariate phenotype (Δz) is biased away from the direction of selection (β) by the pattern of genetic variances and covariances among traits ($\mathbf{G}$). Thus, if both of these could be measured for a set of populations, the direct contribution of constraints and selection to divergence could be isolated.

Building upon a theoretical framework originally developed by Zeng (1988) and independently derived by Hansen and Martins (1996) and Felsenstein (1988), Chenoweth et al. (2010) showed how estimates of both $\mathbf{G}$ and β from multiple populations experiencing divergent selection can be integrated to estimate the extent to which population divergence has been driven by variation in selection, genetic constraints or both. The approach involves expressing the pattern of mean trait divergence among populations in the form of a covariance matrix (termed $\mathbf{D}$, Lande 1979) and then the relative contributions of among-population variation in selection (as described by the $\mathbf{B}$ matrix) and genetic constraints (as described by $\mathbf{G}$) can be evaluated via their association with $\mathbf{D}$. In addition, by calculating two among population variance-covariance matrices from the predicted response to selection using the multivariate breeders' equation (one that assumes uniform genetic constraints across populations and another that allows the genetic constraints themselves to differ), the contribution of divergence in genetic constraints can be evaluated. Applying this approach to nine populations of *D. serrata*, Chenoweth et al. (2010) estimated both directional sexual selection, β, and $\mathbf{G}$ for eight contact pheromones (cuticular hydrocarbons). These pheromones had diverged among the populations in mean in a manner that could not be accounted for by genetic drift alone (Chenoweth and Blows 2008), implicating divergent selection. The authors found that spatial variation in sexual selection alone was a poor predictor of divergence, accounting for only 10% of the similarity in $\mathbf{D}$. However, when genetic constraints were explicitly incorporated into the prediction, the association improved significantly to 51%. Allowing genetic constraints to vary among populations provided little additional explanatory power, suggesting that trait means have diverged more rapidly than their genetic (co)variance structure (or that divergence in the latter mirrors the pattern of trait means, Hine et al. 2009). The key point is that a multivariate treatment of selection and genetic constraints among populations can be integrated to begin to bridge microevolutionary theory for the response to selection with patterns of phenotypic divergence.

Although it was possible to explain up to 51% of the phenotypic divergence among populations as a consequence of both sexual selection and the pattern of genetic covariance between the traits, variation in natural selection may account for additional divergence. In particular, insect cuticular hydrocarbons serve not only as contact pheromones but are also involved in ecologically important functions such as maintaining water balance. Clinal variation

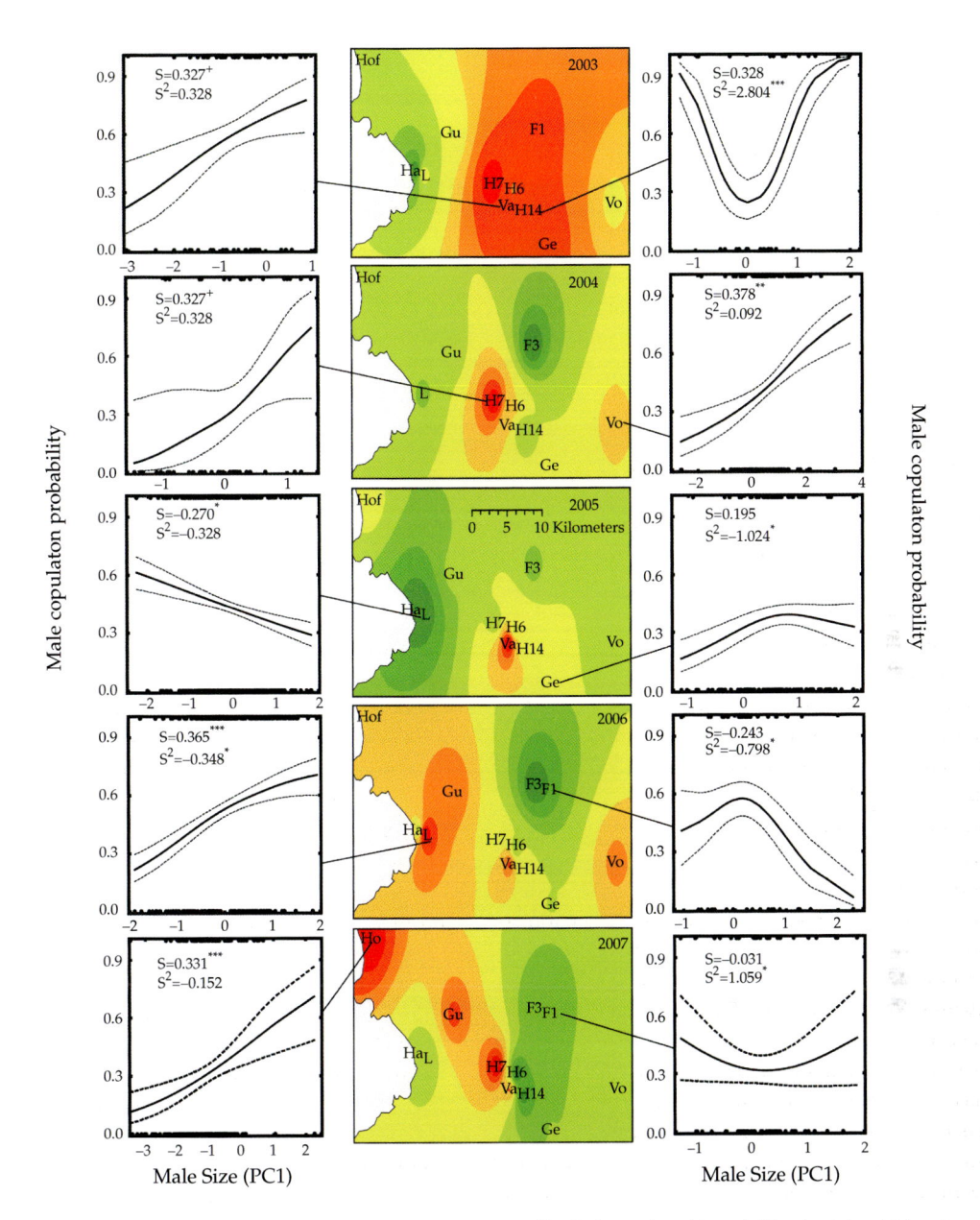

Plate 1 Patterns of spatial and temporal variation in selection acting in male damselflies (*Ischnura elegans*) in coastal Sweden. Left and right columns depict representative cubic spline regressions of male copulation probability vs. body size. Central figures illustrate the total pattern of variation using heat maps, in which red and dark green represent positive and negative selection differentials respectively (from Gosden and Svensson, 2008). See also Figure 7.5, page 99.

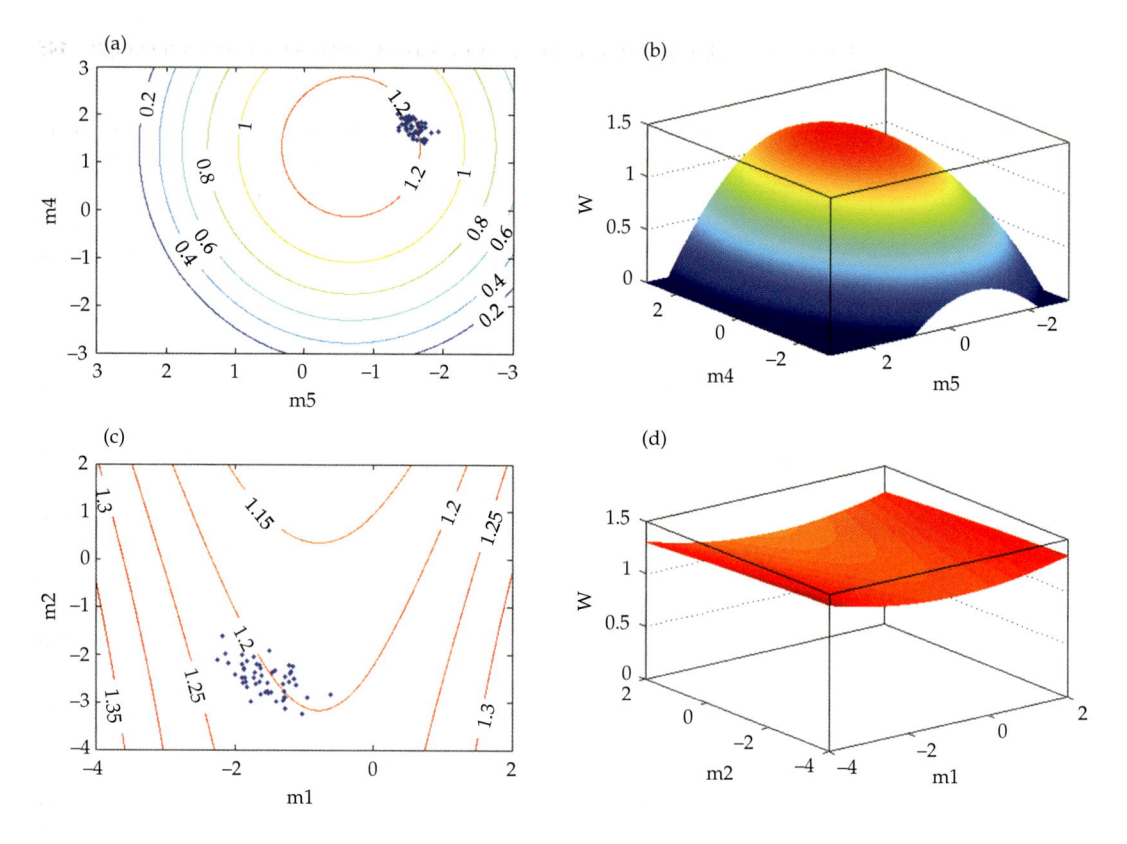

Plate 2 Association between the strength of multivariate stabilizing selection and additive genetic variance for five call traits in the black field cricket, *T. commodus* after Hunt et al. (2007). (a) contour plot showing predicted breeding values (BLUPs) for two canonical axes under weak non linear selection (b). (c) A tighter clustering of breeding values was observed for traits canonical axes under stronger stabilizing selection (d). Reproduced with permission from the Genetics Society of America. See also Figure 9.6, page 143.

Plate 3 Burnet moth mimicry. The top row depicts mimetic forms: red peucedanoid (left), red ephialtoid (middle), and yellow ephialtoid (right) of the variable burnet moth *Zygaena ephialtes*. In the bottom row are two models; the six-spot burnet *Zygaena filipendulae* (left) and the nine-spotted moth *Syntomis phegea* (right) are presumed models for the red peucedanoid and the yellow ephialtoid mimetic forms, respectively. The red ephialtoid form might have been an intermediate in an evolutionary transition from red peucedanoid to yellow ephialtoid; the form might have an advantage if both *Z. Filipendulae* and *S. Phegea* are present. Images derive from photos by Clas-Ove Strandberg of hand painted illustrations in Boisduval (1834) and Hübner (1805), obtained with permission from the Library of the Royal Swedish Academy of Sciences, deposited in Stockholm University Library. Original illustrations are (top left) nr. 8, (top middle) nr. 5, (top right) nr. 6, (bottom left) nr. 10 on Plate 55 of Boisduval (1834), and (bottom right) nr. 100 on Plate 20 of Hübner (1805). Some original species names and identities vary from those in our illustration. See also Figure 16.4, page 266.

in abiotic factors such as temperature and humidity likely contribute to among population divergence in CHCs (Frentiu and Chenoweth 2010) and accounting for this could strengthen the association between selection and divergence. In addition, as with all retrospective selection analyses, genetic covariances and selection were estimated from contemporary populations; if these have varied temporally during evolutionary divergence, this will weaken the association between selection and divergence.

9.5 Future directions

Lande and Arnold's (1983) approximation of the individual fitness surface has facilitated significant insight into the nature of phenotypic selection (Hoekstra et al. 2001; Kingsolver et al. 2001). However, our coverage here has revealed some areas where further development is needed. A consistent criticism leveled at the quadratic approximation has been that it forces either a linear or quadratic fit to the data such that the gradients may misrepresent the true fitness surface if it is more complex. This may occur when fitness deviates significantly from a Gaussian distribution (Mitchell-olds and Shaw 1987), when there are multiple fitness peaks (Schluter 1988), and when the population mean resides a long way from the optimum (Shaw and Geyer 2010). While fitness surface visualizations have been seen as one way to address this (see section 9.2.4; Schluter and Nychka 1994; Schluter 1998), Shaw and colleagues have developed a new class of "aster" models that incorporate both heterogeneous distributions of fitness components and the operation of selection throughout the life history (Geyer et al. 2007; Shaw et al. 2008; Shaw and Geyer 2010). Unlike many previous alternatives to the quadratic approximation, it is possible to obtain estimates of selection gradients from aster models that are suitable for microevolutionary prediction. Insufficient attention has also been given to alternative approaches that may be able to accommodate complex surfaces within the context of a quadratic approximation. For example, mixture models, a generic class of statistical model that can deal with observations from a mixture of distinct but unknown groups (McLachlan and Peel 2000), could potentially be applied to model an unknown mixture of quadratic relationships within a data set.

Comparison of selection between groups is an area where much work remains to be done—some of the more complex analyses we have shown here still require further development. In particular, we should pay careful attention to estimating the sampling variances on the parameters of interest. For example, for higher dimensional problems, individual elements of matrices are likely of little informative value; hypothesis tests should focus on the summaries of these such as eigenvalues and vectors. Nevertheless, an inherent advantage of the Lande and Arnold (1983) least-squares approach is that such comparison can be done within a rigorous statistical framework. The likelihood approaches developed by Hohenlohe and Arnold (2010) represent a step forward in this regard. In the end, perhaps the greatest reason for pursuing this approach is that it allows integration with quantitative genetic theory, providing the opportunity to extend analyses from a simple description of selection and how it differs to addressing evolutionary genetic hypotheses such as the maintenance of genetic variance and the processes underlying phenotypic divergence. Such analyses are data intensive, however, and we urge the continued development of systems in which the dual study of quantitative genetic variation and multivariate selection is feasible. Finally, although multivariate approaches to the analysis of selection are an improvement, they are by no means a panacea, and they remain sensitive to the critical problem in any correlational selection analysis that arises from unmeasured traits. When selection is directly occurring on an unmeasured but phenotypically (or genetically) correlated trait that is not included in a multivariate model, the selection observed is termed "apparent." The observation of widespread pleiotropy underlying quantitative trait variation implies that apparent selection may be more the norm rather than the exception in these types of studies (Johnson and Barton 2005). Indeed Hansen (Chapter 13) suggests that this could be a major component of observed univariate selection. Understanding the extent to which observed phenotypic selection reflects real versus apparent selection is perhaps the major challenge for future study of selection (Takana 2010). Recently, empirical approaches to do so have emerged (McGuigan et al. 2011); paradoxically some of these techniques share a great deal with the Lande and Arnold (1983) approach.

References

Anholt, B. R. (1991). Measuring selection on a population of damselflies with a manipulated phenotype. Evolution, 45, 1091–1106.

Arnold, S. J., Pfrender, M. E., and Jones, A. G. (2001). The Adaptive Landscape as a conceptual bridge between micro- and macroevolution. Genetica, 112, 9–32.

Arnqvist, G. (1992). Spatial variation in selective regimes—sexual selection in the water strider *Gerris odontogaster*. Evolution, 46, 914–929.

Barton, N. H., and Keightley, P. D. (2002). Understanding quantitative genetic variation. Nature Reviews Genetics, 3, 11–21.

Barton, N. H., and Turelli, M. (1989). Evolutionary quantitative genetics—how little do we know? Annual Review of Genetics, 23, 337–370.

Bell, G. (2010). Fluctuating selection: the perpetual renewal of adaptation in variable environments. Philosophical Transactions of the Royal Society B: Biological Sciences, 365, 87–97.

Bentsen, C. L., Hunt, J., Jennions, M. D., and Brooks, R. (2006). Complex multivariate sexual selection on male acoustic signaling in a wild population of *Teleogryllus commodus*. American Naturalist, 167, E102–E116.

Bisgaard, S., and Ankenman, B. (1996). Standard errors for the eigenvalues in second-order response surface models. Technometrics, 38, 238–246.

Blows, M. W. (2007). A tale of two matrices: multivariate approaches in evolutionary biology. Journal of Evolutionary Biology, 20, 1–8.

Blows, M. W., and Brooks, R. (2003). Measuring nonlinear selection. American Naturalist, 162, 815–820.

Blows, M. W., Brooks, R., and Kraft, P. G. (2003). Exploring complex fitness surfaces: Multiple ornamentation and polymorphism in male guppies. Evolution, 57, 1622–1630.

Blows, M. W., Chenoweth, S. F., and Hine, E. (2004). Orientation of the genetic variance-covariance matrix and the fitness surface for multiple male sexually selected traits. American Naturalist, 163, E329–E340.

Bolker, B. M., Brooks, M. E., Clark, C. J., Geange, S. W., Poulsen, J. R., Stevens, M. H. H., et al. (2009). Generalized linear mixed models: a practical guide for ecology and evolution. Trends in Ecology & Evolution, 24, 127–135.

Bowerman, B. L., and O'Connell, R. T. (1990). Linear statistical models: an applied approach. Duxbury Press, Belmont, CA.

Box, G. E. P., and Draper, N. R. (1987). Empirical Model-Building and Response Surfaces. Wiley, New York.

Box, G. E. P., and Hunter, J. S. (1954). A confidence region for the solution of a set of simultaneous equations with an application to experimental design. Biometrika, 41, 190–199.

Brodie, E. D., and Janzen, F. J. (1996). On the assignment of fitness values in statistical analyses of selection. Evolution, 50, 437–442.

Brodie, E. D., Moore, A. J., and Janzen, F. J. (1995). Visualizing and quantifying natural-selection. Trends in Ecology & Evolution, 10, 313–318.

Brooks, R., J. Hunt, Blows, M. W., Smith, M. J., Bussiere, L. F., and Jennions, M. D. (2005). Experimental evidence for multivariate stabilizing sexual selection. Evolution, 59, 871–880.

Bulmer, M. G. (1980). Effects of selection on genetic-variability. Heredity, 45, 141–141.

Bulmer, M. G. (1989). Maintenance of genetic-variability by mutation selection balance—a childs guide through the jungle. Genome, 31, 761–767.

Chenoweth, S. F., and Blows, M. W. (2005). Contrasting mutual sexual selection on homologous signal traits in *Drosophila serrata*. American Naturalist, 165, 281–289.

Chenoweth, S. F., and Blows, M. W. (2008). Qst meets the G matrix: the dimensionality of adaptive divergence in multiple correlated quantitative traits. Evolution, 62, 1437–1449.

Chenoweth, S. F., Rundle, H. D., and Blows, M. W. (2010). The contribution of selection and genetic constraints to phenotypic divergence. American Naturalist, 175, 186–196.

Cheverud, J. M. (1984). Quantitative genetics and developmental constraints on evolution by selection. Journal of Theoretical Biology, 110, 155–171.

Cox, R. M., and Calsbeek, R. (2009). Sexually antagonistic selection, sexual dimorphism, and the resolution of intralocus sexual conflict. American Naturalist, 173, 176–187.

Coyne, J. A., and Orr, H. A. (2004). Speciation. Sinauer Associates, Inc. Publishers, Sunderland, MA.

Craven, P., and Wahba, G. (1979). Smoothing noisy data with spline functions –etimating the correct degree of smoothing by the method of generalized cross-validation. Numerische Mathematik, 31, 377–403.

Crespi, B. J., and Bookstein, F. L. (1989). A path-analytic model for the measurement of of selection on morphology. Evolution, 43, 18–28.

Del Castillo, E., and Cahya, S. (2001). A tool for computing confidence regions on the stationary point of a response surface. American Statistician, 55, 358–365.

Draper, N. R., and John, J. A. (1988). Response-surface designs for quantitative and qualitative variables. Technometrics, 30, 423–428.

Estes, S., and Arnold, S. J. (2007). Resolving the paradox of stasis: Models with stabilizing selection explain evolutionary divergence on all timescales. American Naturalist, 169, 227–244.

Felsenstein, J. (1988). Phylogenies and quantitative characters. Annual Review of Ecology and Systematics, 19, 445–471.

Flury, B. (1988). Common Principal Components and Related Multivariate Models. Wiley, New York.

Frentiu, F. D., and Chenoweth, S. F. (2010). Clines in cuticular hydrocarbons in two *Drosophila* species with independent population histories. Evolution, 64, 1784–1794.

Gerhardt, H. C., and Brooks, R. (2009). Experimental analysis of multivariate female choice in gray treefrogs (*Hyla versicolor*): Evidence for directional and stabilizing selection. Evolution, 63, 2504–2512.

Geyer, C. J., Wagenius, S., and Shaw, R. G. (2007). Aster models for life history analysis. Biometrika, 94, 415–426.

Gillespie, J. H., and Turelli, M. (1989). Genotype-Environment Interactions and the Maintenance of Polygenic Variation. Genetics, 121, 129–138.

Gosden, T. P., and Svensson, E. I. (2008). Spatial and temporal dynamics in a sexual selection mosaic. Evolution, 62, 845–856.

Green, P. J., and Silverman, B. W. (1994). Nonparametric Regression and Generalised Linear Models. Chapman and Hall, London.

Greenfield, M. D. (2002). Signallers and receivers: mechanisms and evolution of arthropod communication. Oxford University Press, Oxford.

Hadfield, J. (2010). MCMC Methods for multiresponse generalized linear mixed models: The MCMCglmm R Package. Journal of Statistical Software, 33:1–22.

Hadfield, J. D., Wilson, A. J., Garant, D., Sheldon, B. C., and Kruuk, L. E. B. (2010). The Misuse of BLUP in Ecology and Evolution. American Naturalist, 175, 116–125.

Hansen, T. F., and Houle, D. (2008). Measuring and comparing evolvability and constraint in multivariate characters. Journal of Evolutionary Biology, 21, 1201–1219.

Hansen, T. F., and Martins, E. P. (1996). Translating between microevolutionary process and macroevolutionary patterns: The correlation structure of interspecific data. Evolution, 50, 1404–1417.

Hedrick, P. W. (1986). Genetic-polymorphisms in heterogeneous environments—a decade later. Annual Review of Ecology and Systematics, 17, 535–566.

Hedrick, P. W. (2006). Genetic polymorphism in heterogeneous environments: The age of genomics. Annual Review of Ecology Evolution and Systematics, 37, 67–93.

Hereford, J., Hansen, T. F., and Houle, D. (2004). Comparing strengths of directional selection: How strong is strong? Evolution, 58, 2133–2143.

Hine, E., and Blows, M. W. (2006). Determining the effective dimensionality of the genetic variance-covariance matrix. Genetics, 173, 1135–1144.

Hine, E., Chenoweth, S. F., and Blows, M. W. (2004). Multivariate quantitative genetics and the lek paradox: Genetic variance in male sexually selected traits of *Drosophila serrata* under field conditions. Evolution, 58, 2754–2762.

Hine, E., Chenoweth, S. F., Rundle, H. D., and Blows, M. W. (2009). Characterising the evolution of genetic variance using genetic covariance tensors. Philosophical Transactions of the Royal Society B: Biological Sciences, 364, 1567–1578.

Hoekstra, H. E., Hoekstra, J. M., Berrigan, D., Vignieri, S. N., Hoang, A., Hill, C. E., et al. (2001). Strength and tempo of directional selection in the wild. Proceedings of the National Academy of Sciences of the United States of America, 98, 9157–9160.

Hohenlohe, P. A., and Arnold, S. J. (2010). Dimensionality of mate choice, sexual isolation, and speciation. Proceedings of the National Academy of Sciences of the United States of America, 107, 16583–16588.

Houle, D. (2010). Numbering the hairs on our heads: The shared challenge and promise of phenomics. Proceedings of the National Academy of Sciences of the United States of America, 107, 1793–1799.

Hunt, J., Blows, M. W., Zajitschek, F., Jennions, M. D., and Brooks, R. (2007). Reconciling strong stabilizing selection with the maintenance of genetic variation in a natural population of black field crickets (*Teleogryllus commodus*). Genetics, 177, 875–880.

Hunt, J., Breuker, C. J., Sadowski, J. A., and Moore, A. J. (2009). Male-male competition, female mate choice and their interaction: determining total sexual selection. Journal of Evolutionary Biology, 22, 13–26.

Janzen, F. J., and Stern, H. S. (1998). Logistic regression for empirical studies of multivariate selection. Evolution, 52, 1564–1571.

Johnson, T., and Barton, N. (2005). Theoretical models of selection and mutation on quantitative traits.

Philosophical Transactions of the Royal Society B: Biological Sciences, 360, 1411–1425.

Jones, A. G., Arnold, S. J., and Burger, R. (2004). Evolution and stability of the G-matrix on a landscape with a moving optimum. Evolution, 58, 1639–1654.

Kawecki, T. J. (2000). The evolution of genetic canalization under fluctuating selection. Evolution, 54, 1–12.

Kingsolver, J. G., Hoekstra, H. E., Hoekstra, J. M., Berrigan, D., Vignieri, S. N., Hill, C. E., et al. (2001). The strength of phenotypic selection in natural populations. American Naturalist, 157, 245–261.

Kingsolver, J. G., and Schemske, D. W. (1991). Path analyses of selection. Trends in Ecology & Evolution, 6, 276–280.

Krzanowski, W. J. (1979). Between-groups comparison of principal components. Journal of the American Statistical Association, 74, 703–707.

Lande, R. (1979). Quantitative genetic analysis of multivariate evolution, applied to brain-body size allometry. Evolution, 33, 402–416.

Lande, R., and Arnold, S. J. (1983). The measurement of selection on correlated characters. Evolution, 37, 1210–1226.

Littell, R. C., Milliken, G. A., Stroup, W. W., and Wolfinger, R. D. (1996). SAS System for Mixed Models. SAS Institute Inc., Cary, NC.

Manly, B. (1997). Randomization, Bootstrap and Monte Carlo Methods in Biology. Chapman and Hall, London.

McGuigan, K., and Blows, M. W. (2007). The phenotypic and genetic covariance structure of drosphilid wings. Evolution, 61, 902–911.

McGuigan, K., Chenoweth, S. F., and Blows, M. W. (2005). Phenotypic divergence along lines of genetic variance. American Naturalist, 165, 32–43.

McGuigan, K., Rowe, L., and Blows, M. W. (2011). Pleiotropy, apparent selection and uncovering fitness optima. Trends in Ecology & Evolution, 26, 22–29.

McGuigan, K., Van Homrigh, A., and Blows, M. W. (2008). Genetic analysis of female preference functions as function-valued traits. American Naturalist, 172, 194–202.

McLachlan, G. J., and Peel, D. (2000). Finite Mixture Models: Wiley series in probability and statistics. Wiley, New York.

Meyer, K., and Kirkpatrick, M. (2005). Restricted maximum likelihood estimation of genetic principal components and smoothed covariance matrices. Genetics Selection Evolution, 37, 1–30.

Mitchell-Olds, T. (1996). Pleiotropy causes long-term genetic constraints on life-history evolution in Brassica rapa. Evolution, 50, 1849–1858.

Mitchell-Olds, T., and Shaw, R. G. (1987). Regression-analysis of natural-selection—statistical-inference and biological interpretation. Evolution, 41, 1149–1161.

Mitchell-Olds, T., and Waller, D. M. (1985). Relative performance of selfed and outcrossed progeny in *Impatiens capensis*. Evolution, 39, 533–544.

Pavlicev, M., Wagner, G. P., and Cheverud, J. M. (2009). Measuring evolutionary constraints through the dimensionality of the phenotype: Adjusted bootstrap method to estimate rank of phenotypic covariance matrices. Evolutionary Biology, 36, 339–353.

Phillips, P. C., and Arnold, S. J. (1989). Visualizing multivariate selection. Evolution, 43, 1209–1222.

Phillips, P. C., Whitlock, M. C., and Fowler, K. (2001). Inbreeding changes the shape of the genetic covariance matrix in *Drosophila melanogaster*. Genetics, 158, 1137–1145.

Pinheiro, J., and Bates, D. (2000). Mixed-effects models in S and S-PLUS. Springer Verlag, New York.

Pomiankowski, A., and Moller, A. P. (1995). A Resolution of the Lek Paradox. Proceedings of the Royal Society of London Series B: Biological Sciences, 260, 21–29.

Price, T. D. (1984). Sexual selection on body size, territory and plumage variables in a population of Darwin finches. Evolution, 38, 327–341.

Punzalan, D., Rodd, F. H., and Rowe, L. (2010). Temporally variable multivariate sexual selection on sexually dimorphic traits in a wild insect population. American Naturalist, 175, 401–414.

Rausher, M. D. (1992). The measurement of selection on quantitative traits—biases due to environmental covariances between traits and fitness. Evolution, 46, 616–626.

Rencher, A. C. (2002). Methods of Multivariate Analysis, 2nd edn. Wiley, New York.

Reynolds, R. J., Childers, D. K., and Pajewski, N. M. (2010). The distribution and hypothesis testing of eigenvalues from the canonical analysis of the gamma matrix of quadratic and correlational selection gradients. Evolution, 64, 1076–1085.

Rowe, L., and Houle, D. (1996). The lek paradox and the capture of genetic variance by condition dependent traits. Proceedings of the Royal Society of London Series B: Biological Sciences, 263, 1415–1421.

Rundle, H. D., and Chenoweth, S. F. (2011). Stronger convex (stabilizing) selection on homologous sexual display traits in females than males: a multipopulation comparison in *Drosophila serrata*. Evolution, 65, 893–899.

Rundle, H. D., Chenoweth, S. F., and Blows, M. W. (2008). Comparing complex fitness surfaces: among-population variation in mutual sexual selection in *Drosophila serrata*. American Naturalist, 171, 443–454.

Rundle, H. D., Chenoweth, S. F., and Blows, M. W. (2009). The diversification of mate preferences by natural and sexual selection. Journal of Evolutionary Biology, 22, 1608–1615.

Ryan, M. J., Perrill, S. A., and Wilczynski, W. (1992). Auditory tuning and call frequency predict population-based mating preferences in the cricket frog, *Acris crepitans*. American Naturalist, 139, 1370–1383.

Ryan, M. J., and Wilczynski, W. (1988). Coevolution of sender and receiver—effect on local mate preference in cricket frogs. Science, 240, 1786–1788.

Schluter, D. (1988). Estimating the form of natural-selection on a quantitative trait. Evolution, 42, 849–861.

Schluter, D. (1994). Experimental-evidence that competition promotes divergence in adaptive radiation. Science 266, 798–801.

Schluter, D. (1996). Adaptive radiation along genetic lines of least resistance. Evolution, 50, 1766–1774.

Schluter, D. (2000). The Ecology of Adaptive Radiation. Oxford University Press, Oxford.

Schluter, D., and Nychka, D. (1994). Exploring fitness surfaces. American Naturalist, 143, 597–616.

Schluter, D., and Smith, J. N. M. (1986). Natural-selection on beak and body size in the song sparrow. Evolution, 40, 221–231.

Shaw, R. G., and Geyer, C. J. (2010). Inferring fitness landscapes. Evolution, 64, 2510–2520.

Shaw, R. G., Geyer, C. J., Wagenius, S., Hangelbroek, H. H., and Etterson, J. R. (2008). Unifying life-history analyses for inference of fitness and population growth. American Naturalist, 172, E35–E47.

Siepielski, A. M., DiBattista, J. D., and Carlson, S. M. (2009). It's about time: the temporal dynamics of phenotypic selection in the wild. Ecology Letters, 12, 1261–1276.

Simonsen, A. K., and Stinchcombe, J. R. (2010). Quantifying evolutionary genetic constraints in the ivyleaf morning glory, *Ipomoea hederacea*. International Journal of Plant Sciences, 171, 972–986.

Simpson, G. G. (1953). The major features of evolution. Columbia University Press, New York.

Stinchcombe, J. R., Agrawal, A. F., Hohenlohe, P. A., Arnold, S. J., and Blows, M. W. (2008). Estimating nonlinear selection gradients using quadratic regression coefficients: Double or nothing? Evolution, 62, 2435–2440.

Stinchcombe, J. R., Rutter, M. T., Burdick, D. S., Tiffin, P., Rausher, M. D., and Mauricio, R. (2002). Testing for environmentally induced bias in phenotypic estimates of natural selection: Theory and practice. American Naturalist, 160, 511–523.

Svensson, E., and Sinervo, B. (2000). Experimental excursions on adaptive landscapes: Density-dependent selection on egg size. Evolution, 54, 1396–1403.

Tanaka, Y. (2010). Apparent directional selection by biased pleiotropic mutation. Genetica, 138, 171–723.

Turelli, M. (1988). Phenotypic evolution, constant covariances, and the maintenance of additive variance. Evolution, 42, 1342–1347.

Turelli, M., and Barton, N. H. (2004). Polygenic variation maintained by balancing selection: Pleiotropy, sex-dependent allelic effects and GxE interactions. Genetics, 166, 1053–1079.

Van Homrigh, A., Higgie, M., McGuigan, K., and Blows, M. W. (2007). The depletion of genetic variance by sexual selection. Current Biology, 17, 528–532.

Walsh, B., and Blows, M. W. (2009). Abundant genetic variation + strong selection = multivariate genetic constraints: a geometric view of adaptation. Annual Review of Ecology Evolution and Systematics, 40, 41–59.

Walsh, B., and Lynch, M. (2011). Evolution and Selection of Quantitative Traits. Sinauer, Sunderland, MA.

Wright, S. (1931). Evolution in Mendelian populations. Genetics, 16, 97–159.

Wright, S. (1932). The roles of mutation, inbreeding, crossbreeding and selection in evolution. Proceedings of the Sixth International Congress of Genetics, 1, 356–366.

Wright, S. (1935). Evolution in population in approximate equilibrium. Journal of Genetics, 30, 257–266.

Wu, C. F. J. (1986). Jacknife, bootstrap and other resampling methods in regression analysis—discussion. Annals of Statistics, 14, 1261–1295.

Zar, J. H. (1984). Biostatistical analysis. Prentice Hall, Englewood Cliffs, NJ.

Zeng, Z. B. (1988). Long-term correlated response, interpopulation covariation, and interspecific allometry. Evolution, 42, 363–374.

CHAPTER 10

Adaptive Accuracy and Adaptive Landscapes

Christophe Pélabon, W. Scott Armbruster, Thomas F. Hansen, Geir H. Bolstad, and Rocío Pérez-Barrales

10.1 Introduction

The Adaptive Landscape, first introduced by Wright (1932) and extended to phenotypes by Simpson (1944, 1953), maps the population mean fitness values onto the corresponding mean phenotypic values for multiple traits, identifying fitness peaks, ridges, and valleys. In parallel, the multiple regression of the individual fitness onto the phenotypic values was introduced by Pearson (1903) who defined the individual fitness surface as the surface linking individual survival to the phenotypic value for multiple characters. Later, Lande and Arnold (1983) showed that the two concepts of Adaptive Landscape and individual fitness surface were closely related. Indeed, under certain assumptions, the directional and quadratic selection gradients defined by the Adaptive Landscape equal the population average slope and curvature of the individual fitness surface (Lande and Arnold 1983; Phillips and Arnold 1989; Chapter 9). When these gradients are linked to quantitative genetic information, summarized by the genetic variance matrix **G**, They can be used to predict the evolutionary trajectory of populations (Lande 1979; Lande and Arnold 1983; Schluter 1996, 2000; Arnold et al. 2001; Blows 2007).

By identifying fitness peak(s), the Adaptive Landscape has also become a tool for testing adaptation as optimality (Orzack and Sober 2001). When a fitness optimum is identified, one can test for adaptation by assessing how much populations differ from the optimum phenotype. In this "optimality" approach, the degree of maladaptation of a population is usually measured as the distance between the population mean phenotype and the nearest fitness peak (Roff 1981; Krebs and McCleery 1984; Herre 1987; Mitchell and Valone 1990; Parker and Maynard-Smith 1990; Crespi 2000; Orzack and Sober 2001; Hendry and Gonzalez 2008). However, by considering only the mean and not the population variance when quantifying maladaptation, optimality tests may have underestimated the degree of maladaptation, because the correspondence between the population mean and the fitness peak does not guarantee that individuals in the population are well adapted (Orzack and Sober 1994a, 1994b). Maladaptation is therefore affected by: (1) the distance between the population mean and the fitness peak, and (2) by the scatter of the individuals around the population mean. Despite these being well-known components of the genetic load in population genetics (Wright 1935; Haldane 1937; Lynch and Lande 1993; Lande and Shannon 1996), the individual-variation component has rarely been considered in studies of trait adaptation (Orzack and Sober 1994a, 1994b).

Armbruster et al. (2004) suggested, in analogy with the concepts of bias, precision, and accuracy in statistical estimation theory, that trait adaptation should be seen as an accuracy rather than as a simple correspondence between the population mean and the optimum. In statistics, inaccuracy refers to the expected squared deviation of an estimate from the true value of the parameter. It is a sum of the squared bias and the imprecision, where the bias is the expected deviation of the estimate from the parameter, and the imprecision is the variance of the estimate. Analogously, adaptive inaccuracy could be defined in terms of the expected deviation of individual trait values from the optimum. This gives us a concept of trait maladaptation that would be similar to the concept of genetic load. If we assume a quadratic fitness function around the optimum, the adaptive inaccuracy can then be decomposed into a mean trait deviation from the optimum (a bias), and an adaptive imprecision due to trait variation across individuals. Specifically,

Inaccuracy = Bias^2 + Imprecision,

where imprecision is the variance in the trait. Distinguishing between these two sources of maladaptation is crucial for understanding trait adaptation because they are influenced by different mechanisms and may respond differently to selection. Fitness could be increased by reducing either bias or imprecision. However, if the response to selection on bias can be easily predicted (Lande 1979), the response to selection on precision, that is, on variance, is much harder to predict (Barton and Turelli 1987). Furthermore, the evolution of precision is not just a question of short-term changes in levels of genetic variance, but involves the longer-term evolution of the capacity to generate and maintain genetic, environmental and developmental variation (Wagner and Altenberg 1996; Hansen 2006; Pélabon et al. 2010). Indeed, it is unclear how efficiently selection can act on variational properties of the genome related to precision, such as robustness and canalization (Proulx and Phillips 2005; Hansen 2011). These considerations underscore the importance of identifying and quantifying bias and imprecision as sources of maladaptation.

Previous applications of adaptive accuracy have been limited to simple univariate fitness functions with an assumed negative quadratic shape around a theoretically defined optimum. For example, in order to estimate the part of adaptive imprecision that could be due to developmental imprecision (i.e. phenotypic variation resulting from developmental noise), Hansen et al. (2006) assumed the average left minus right trait value to be optimal in bilateral traits, and estimated developmental imprecision from fluctuating asymmetry. They found that developmental imprecision is very variable across traits but can often generate a substantial part of the total adaptive imprecision. Pélabon and Hansen (2008) further tested the hypothesis that directional asymmetry in insect wing size could be an adaptation by using fluctuating asymmetry to estimate the adaptive imprecision in directional asymmetry. This adaptive imprecision was so large that an adaptive function for directional asymmetry could be ruled out since most individuals would be far away from any putatively optimal value.

Applying the concept of adaptive accuracy to pollination ecology, Armbruster et al. (2004, 2009a, 2009b) used theoretical expectations to define different, but interrelated, adaptive optima for the positions of anthers and stigmas in terms of male and female floral functions. The optimum for male function (deposition of pollen on pollinators) was estimated as the stamen position that resulted in pollen placement on the pollinator at the expected point of contact with the stigmas. In turn, the optimum for female function (receipt of pollen from pollinators) was estimated as the stigma position that resulted in contact with pollinators at the expected location of pollen placement. Importantly, phenotypic variation in the position of the anthers and the stigmas generated variation in the optimum that, in turn, contributed to the adaptive imprecision of each trait. Comparing pollination accuracy among four genera with different levels of phenotypic integration revealed that the degree of inaccuracy varied considerably across species and genera. In two genera, ontogenetic modification of the position of the anthers and stigmas generated further variation in the optimum that increased the bias. On the other hand, inaccuracy tended to decrease with increasing integration among floral whorls (Armbruster et al. 2009a). In some species, inaccuracy was increased by conflicting selection pressures such as avoidance of self pollination through herkogamy (spatial separation of stigma and anther). Indeed, self-compatible species that lack separation of the sexual functions in time (no dichogamy) tended to have higher bias, and therefore inaccuracy, as a possible mean to decrease auto pollination (Armbruster et al. 2009b). Variation in phenotypic optima was also considered by Dvorak and Gvozdik (2010) who quantified the adaptive accuracy of female newts (*Ichthyosaura alpestris*) when choosing the temperature of oviposition sites. They showed that the adaptive accuracy of the female choice for oviposition site was negatively affected by the temporal variation in water temperature.

All these studies have been based on theoretically assumed optima rather than on optima explicitly estimated from phenotypic selection studies. Furthermore, as pointed out by Gilchrist and Kingsolver (2001), there are problems with testing optimality in the context of single traits in presence of strong genetic or phenotypic correlations. First, the fitness peak may be misidentified if one or several correlated traits influencing fitness are not included in the analysis (Lande and Arnold 1983; Walsh and Blows 2009; Chapter 9). Second, trait variation may appear maladaptive when it is in fact this variation that maintains populations in the region of high fitness, as when correlated traits covary along a fitness ridge. These considerations underscore the necessity to measure multiple correlated traits when studying adaptation and to extend the concept of adaptive accuracy to multivariate analyses.

In this chapter, we first derive a multivariate version of adaptive accuracy and then show how empirically estimated fitness functions describing Adaptive Landscapes can be used to estimate the adaptive inaccuracy of multiple traits and its different components. We illustrate this approach with an example taken from the morphology and pollination of the blossoms of *Dalechampia* vines (Euphorbiaceae).

10.2 Theory

10.2.1 Adaptive accuracy of a population

Consider an Adaptive Landscape for a trait vector $\mathbf{z} = \{z_1, \ldots, z_n\}$ in the vicinity of an optimal state vector $\boldsymbol{\theta} = \{\theta_1, \ldots, \theta_n\}$. This landscape can then be described as a fitness function

$$W(\mathbf{z}; \boldsymbol{\theta}) = W(z_1, \ldots, z_n; \theta_1, \ldots, \theta_n), \quad (10.1)$$

with maximum at $\mathbf{z} = \boldsymbol{\theta}$. Analogously to the genetic load in population genetics, the degree of maladaptation of an individual with phenotype $\mathbf{z}$ can be defined by:

$$L(\mathbf{z}; \boldsymbol{\theta}) = (W(\boldsymbol{\theta}; \boldsymbol{\theta}) - W(\mathbf{z}; \boldsymbol{\theta}))/W(\boldsymbol{\theta}; \boldsymbol{\theta}), \quad (10.2)$$

which measures the proportional reduction in fitness from the maximum possible fitness in a given environment. The average population load is not, however, generally equal to the load of the average individual; that is $E[L(\mathbf{z}; \boldsymbol{\theta})] \neq L(E[\mathbf{z}]; \boldsymbol{\theta})$, where $E[]$ denotes expectation. Instead,

$$E[L(\mathbf{z}; \boldsymbol{\theta})] = L(E[\mathbf{z}]; \boldsymbol{\theta}) + (E[L(\mathbf{z}; \boldsymbol{\theta})] - L(E[\mathbf{z}]; \boldsymbol{\theta})). \quad (10.3)$$

We can take this average load as a general definition of the adaptive inaccuracy of a population, and we see that it can be partitioned into two components, one due to the population mean diverging from the optimum, the bias, and another due to variation among individuals, the imprecision:

Load due to "bias" : $L(E[\mathbf{z}]; \boldsymbol{\theta})$,

Load due to "imprecision" : $E[L(\mathbf{z}; \boldsymbol{\theta})] - L(E[\mathbf{z}]; \boldsymbol{\theta})$.

If an estimated fitness landscape is available, each of these components can be obtained from equation 10.3 by computing the fitness of the mean phenotype, $W(E[\mathbf{z}], \boldsymbol{\theta})$, the fitness of the optimal phenotype, $W(\boldsymbol{\theta}, \boldsymbol{\theta})$, and the mean fitness of the population, $E[W(\mathbf{z}, \boldsymbol{\theta})]$.

In the vicinity of the optimum, the fitness function will be concave (negative second derivatives), and the expected fitness is less than the fitness of the average individual (the Jensen inequality). In this situation, we can describe the dependence of the load on population variation and patterns of stabilizing selection with a Taylor approximation around the optimum. Consider first the case of a single trait, z. Then

$$E[L(z; \theta)] \approx -(\gamma/2)(E[z] - \theta)^2 - (\gamma/2)\text{Var}[z], \quad (10.4)$$

where $\gamma = \partial^2 \ln W(\theta; \theta)/\partial z^2$ is a quadratic selection gradient measured at the optimum (where it has to be negative). To derive this we have used the fact that $\partial W(\theta; \theta)/\partial z = 0$, which follows from the assumption that θ is an optimum. The load due to bias is proportional to the squared deviation of the mean from the optimum (squared "bias"), and the load due to imprecision is proportional to the variance of the trait. We can extend this to multiple traits as follows

$$E[L(\mathbf{z}; \boldsymbol{\theta})] \approx -(1/2)\Sigma_i \Sigma_j \gamma_{ij}\big((E[z_i] - \theta_i)(E[z_j] - \theta_j)$$
$$+ \text{Cov}[z_i, z_j]\big), \quad (10.5)$$

where the summations are over all traits, and $\gamma_{ij} = \partial^2 \ln W(\theta, \theta)/\partial z_i \partial z_j$ is the quadratic selection gradient measuring correlated stabilizing selection on the two traits *i* and *j* (Lande and Arnold 1983). In matrix notation this is

$$E[L(\mathbf{z};\theta)] \approx -(1/2)\left((E[\mathbf{z}]-\theta)\Gamma(E[\mathbf{z}]-\theta)^T + \mathrm{Tr}[\Gamma \mathbf{P}]\right), \quad (10.6)$$

where Γ is the quadratic selection matrix with the γ_{ij} as elements, $\mathbf{P}$ is the population variance matrix with $\mathrm{Cov}[z_i, z_j]$ as elements, and Tr is the trace function. This simply shows that the load is more affected by deviations in directions of morphospace where stabilizing selection is stronger.

10.2.2 Adaptive accuracy of the genotype

Just as population adaptation is better quantified as accuracy than as a fit of the population mean to the optimum, the adaptedness of a genotype is also better quantified as accuracy than as a fit of its average realized phenotype to the optimum (Fig. 10.1). This idea was developed in Hansen et al. (2006), where they showed that the accuracy of adaptation of a single genotype could be decomposed into the squared deviation of its target phenotype from the optimum (a bias) and the variance in the developmental realization of this target (an imprecision). Expressed as a load, this is

$$E[L(\mathbf{z};\theta)|g] = L(E[\mathbf{z}|g];\theta)$$
$$+ \left(E[L(\mathbf{z};\theta)|g] - L(E[\mathbf{z}|g];\theta)\right), \quad (10.7)$$

where "|g" means that the expectation is conditional on the genotype g. A quadratic approximation around the optimum then gives

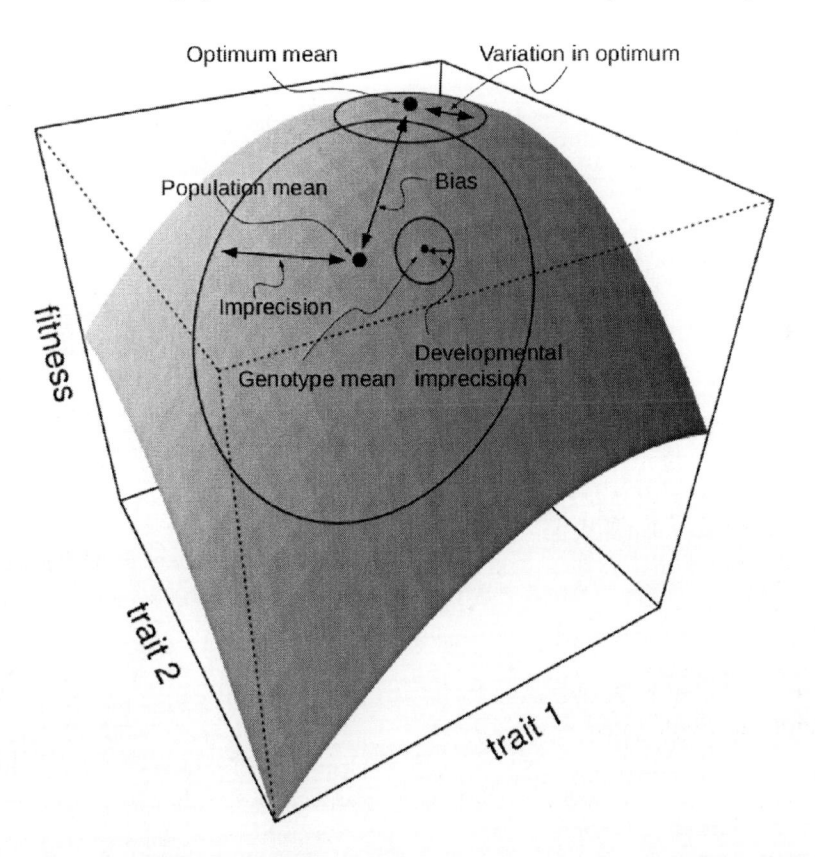

Figure 10.1 Adaptive inaccuracy and its different components in a multivariate morphospace. In this Adaptive Landscape, the position of the optimum may vary due to microenvironmental variation (*Variation in optimum*). The target phenotype is the average phenotype reached by a particular genotype in a particular distribution of environments (*Genotype mean*). Developmental imprecision refers to the imprecision resulting from the within-individual variation.

$$E[L(\mathbf{z};\theta)|g] \approx -(1/2)((E[\mathbf{z}|g] - \theta)\Gamma(E[\mathbf{z}|g] - \theta)^T$$
$$+ \text{Tr}[\Gamma \mathbf{P}_{z|g}]), \quad (10.8)$$

where $\mathbf{P}_{z|g}$ is the variance in phenotype over developmental realizations of the genotype. By taking the expectation of this over the genotypes in the population, we recover the population load in equation 10.6, and we can see that the expectation of $\text{Tr}[\Gamma \mathbf{P}_{z|g}]$ over genotypes is simply one component of the total imprecision in the population. The developmental imprecision will be nonzero whenever there is variation in the realization of the phenotype due to developmental noise or microenvironmental variation. Hansen et al. (2006) found that developmental imprecision could constitute a substantial fraction (mean 14.7%; median 6.15%) of the total imprecision load for many traits and populations.

10.2.3 Maladaptation due to variation in the fitness landscape

In many cases individuals in a population may not experience the exact same fitness function. This variation in the fitness function may induce maladaptation in a manner similar to trait variation. For example, the adaptive accuracy of pollen deposition on a pollinator depends both on the variation in the placement of pollen due to anther position and on the variation in where the stigma can pick it up (Armbruster et al. 2009a). Hence, variation in stigma position generates variation in the individually optimal anther position and conversely, variation in anther position generates variation in the individually optimal stigma position. Mating partners are not usually perfectly matched even if their averages are. We can represent the effects of variation in the optimum on the load as follows:

$$E_\theta E_z[L(\mathbf{z};\theta)] = L(E_z[\mathbf{z}]; E_\theta[\theta])$$
$$+ E_z[L(\mathbf{z}; E_\theta[\theta])] - L(E_z[\mathbf{z}]; E_\theta[\theta])$$
$$+ E_\theta[L(E_z[\mathbf{z}];\theta)] - L(E_z[\mathbf{z}]; E_\theta[\theta])$$
$$+ E_\theta E_z[L(\mathbf{z};\theta)] - (E_z[L(\mathbf{z}; E_\theta[\theta])] - L(E_z[\mathbf{z}]; E_\theta[\theta]))$$
$$- (E_\theta[L(E_z[\mathbf{z}];\theta)] - L(E_z[\mathbf{z}]; E_\theta[\theta]))$$
$$- L(E_z[\mathbf{z}]; E_\theta[\theta]), \quad (10.9)$$

where E_θ signifies expectation with respect to a variable optimum as experienced by individuals in the population. The first term on the right-hand side is the load due to a mismatch between the average phenotype and the average optimum (bias). The term on the second line is the load due to variation (imprecision) in the phenotype. The term on the third line is the load due to variation in the optimum, and the rest is due to interactions between the variation in the phenotype and the variation in the optimum. With a quadratic approximation for both the phenotype and the optimum around the average optimum, this becomes

$$E_\theta E_z[L(\mathbf{z};\theta)] = -(1/2)((E[\mathbf{z}] - E[\theta])\Gamma(E[\mathbf{z}] - E[\theta])^T$$
$$+ \text{Tr}[\Gamma \mathbf{P}] + \text{Tr}[\Theta \mathbf{V}_\theta] + \text{Tr}[\vartheta \mathbf{C}_{z\theta}]), \quad (10.10)$$

where Θ is a matrix with elements $\partial^2 \ln W(E[\theta]; E[\theta])/\partial \theta_i \partial \theta_j$, ϑ is a matrix with elements $\partial^2 \ln W(E[\theta]; E[\theta])/\partial z_i \partial \theta_j$, $\mathbf{V}_\theta$ is a variance matrix with elements $\text{Cov}[\theta_i, \theta_j]$, and $\mathbf{C}_{z\theta}$ is a covariance matrix with upper and lower triangular elements $\text{Cov}[z_i, \theta_j]$ and $\text{Cov}[\theta_i, z_j]$, respectively, and zero along the main diagonal. The three trace functions represent the effects of, respectively, variation in the phenotype, variation in the optimum, and covariation between phenotypes and the optima, all weighted with the elasticity of the fitness function to changes in phenotype and optimum in the relevant directions (technically the traits need be mean standardized to make this an elasticity; van Tienderen 2000; Hereford et al. 2004). A univariate version of this equation assuming no correlation between phenotypes and optima was given in Armbruster et al. (2009a).

10.2.4 Inaccuracy of focal traits and marginal Adaptive Landscapes

Including many traits improves the accuracy of the fitness surface, because it helps identifying indirect selection on traits that results from selection on correlated characters (Lande and Arnold 1983; Gilchrist and Kingsolver 2001; Chapter 9). This eventually results in a more accurate identification of the position of the optimum, and therefore a better estimation of the population maladaptation. Nevertheless, it remains interesting to estimate and compare the contribution of single traits or combination of traits to the population maladaptation.

Estimating the inaccuracy, bias and imprecision of specific traits, or combinations of traits, while excluding the contribution of other traits can be achieved by calculating the three functions $W(E[z]; \theta)$, $E[W(z; \theta)]$, and $W(\theta; \theta)$ for the trait of interest, while keeping the other traits either at their fitness optimum or at their mean (i.e. giving all individuals the value of the optimum phenotype or the value of the mean phenotype for the fixed traits). By fixing the value of the non-focal trait(s), we remove any bias and imprecision associated with these traits, and we use the marginal Adaptive Landscape at this specific point to calculate the inaccuracy of the remaining trait or combination of traits. As shown in Fig. 10.2, fixing a trait at its optimum value (θ) or at its mean phenotypic value $E[z]$ may have strong consequences on the shape of the marginal Adaptive Landscape from which we subsequently

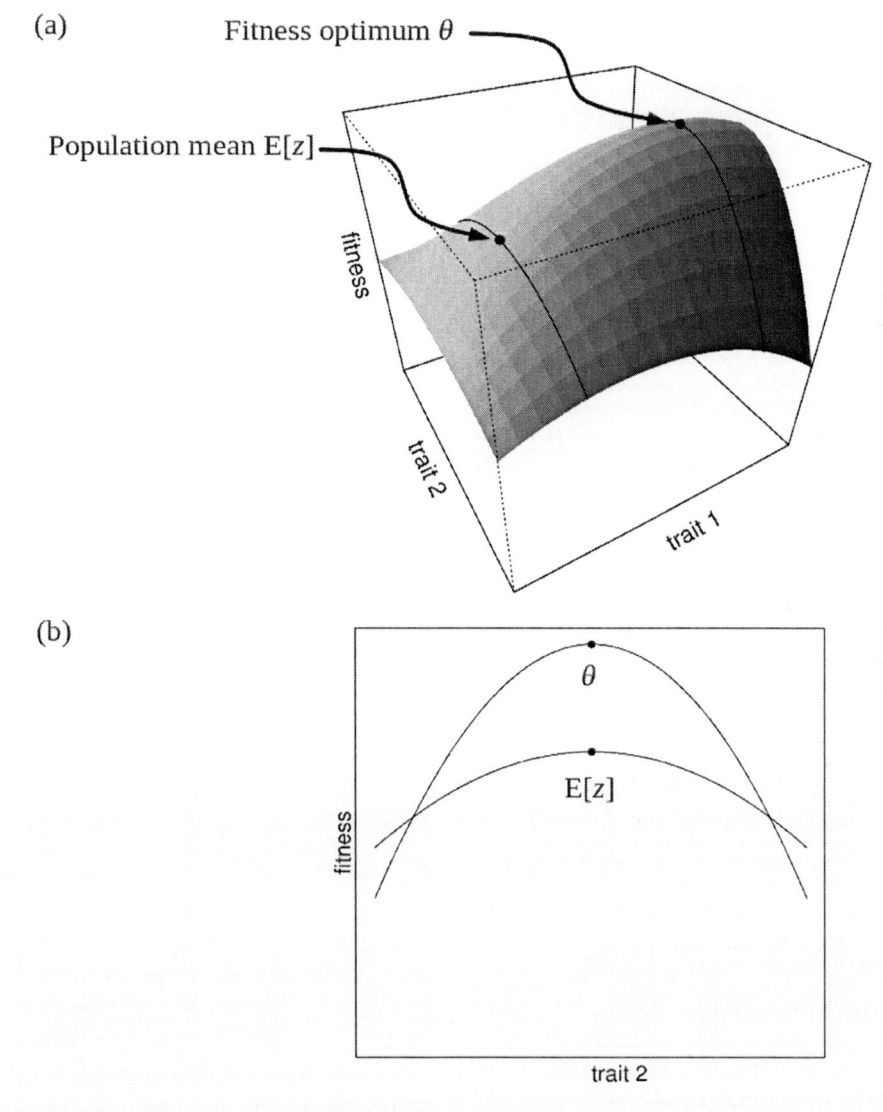

Figure 10.2 Marginal Adaptive Landscapes: a) Adaptive landscape for two traits with correlational selection. Both traits are under stabilizing selection, but the correlational selection generates interaction between the two selective processes implying a change in the strength of stabilizing selection on trait 2 depending of the phenotypic value of trait 1. The fitness optimum and the population mean for the two traits are reported on the adaptive surface. b) Fitness functions for trait 2 when trait 1 is fixed either at the optimum (θ) or at the population mean $E[z]$. These fitness functions are also reported in (solid lines). The interaction between the fitness functions of the two traits affects the shape and optimum value of the marginal Adaptive Landscapes of trait 2.

estimate maladaptation. These two alternatives are also conceptually different.

By fixing non-focal traits at their fitness optimum (θ), we estimate what would be the adaptive inaccuracy of the focal trait if the other traits were perfectly adapted. This approach first assumes that the fitness surfaces of the non-focal traits are properly estimated as well as the possible correlated selection between the focal and non-focal traits. It also assumes that the differences between the fitness optimum and the mean trait value observed for the non-focal traits truly reflect maladaptation (bias). These assumptions may be problematic, however. For example, when directional selection is observed, the optimum value of the trait does not lie within the range of the population. In our empirical example in section 10.3, we used the point with the highest fitness within the range of the population to estimate the optimum phenotype, θ. However, the marginal Adaptive Landscape at this specific point may be poorly defined due to the scarcity of observations at the limits of the phenotypic distribution. Furthermore, although many studies of phenotypic selection in the field fail to find evidence for stabilizing selection (Kingsolver et al. 2001), this may be due to failure or error in incorporating relevant traits or in the modeling of correlational selection (Arnold and Wade 1984; Blows and Brooks 2003; Blows 2007; Estes and Arnold 2007; Stinchcombe et al. 2008; Walsh and Blows 2009).

Alternatively, by fixing non-focal traits at their mean values, we assume that the population is perfectly adapted for these traits, and we use the marginal Adaptive Landscape at this specific point (E[z] in Fig. 10.2) to calculate the inaccuracy of the focal trait or combination of traits. With this approach, one can remove the influence of traits for which the fitness function is uncertain when estimating fitness surfaces and maladaptation.

With both methods, if the contributions of the different traits to the population maladaptation are independent (no correlational selection), the total inaccuracy and its different components should equal the sum of the individual inaccuracies. If, however, the fitness function of one trait or combination of traits depends on the phenotypic value of other traits, as is the case for adaptive ridges (Fig. 10.2; see also Armbruster and Schwaegerle 1996), the inaccuracy estimated using all traits simultaneously will not equal the sum of the inaccuracies estimated for each trait considered separately. These discrepancies in adaptive inaccuracy may be generated by discrepancies in bias or in imprecision. If, for example, the total bias is lower than the sum of the individual biases, it means that the point defined by the joint trait means lies in a region of higher fitness than for the means of each trait considered separately. Discrepancy in imprecision, in turn, indicates variation in the alignment of the vector of trait variation with an adaptive ridge. When the total imprecision is smaller than the sum of the individual imprecision, it indicates that correlation between traits is adaptive and follows an adaptive ridge. Conversely, total imprecision larger than the sum of the individual imprecision indicates maladaptive interaction between traits.

10.3 An example from the pollination ecology of *Dalechampia*

10.3.1 Phenotypic selection study in *Dalechampia scandens*

Dalechampia scandens L. (Euphorbiaceae) is a vine widely distributed in the Neotropics from Mexico to Argentina. It presents bisexual "blossoms" (pseudanthial inflorescences: i.e. clusters of flowers forming flower-like structures), with a cluster of three pistillate flowers situated below a cluster of ten staminate flowers (Fig. 10.3; see also Webster and Armbruster 1991). The male and female flower clusters are subtended by two involucral bracts that have a protective role in some species, closing at night to protect the flower and during the period of fruit maturation. A gland producing terpenoid resin (in most species) is associated with the staminate flowers. The resin produced by the gland varies in color between species and is collected by species of megachilid and apid bees for use in nest construction. The amount of resin offered to bees, estimated by the area of the secreting surface of the gland, determines the type and size of bees that visit the blossoms (Armbruster 1985, 1986, 1988). Furthermore, the distances between the resin gland and the stigma (GSD) and between the resin gland

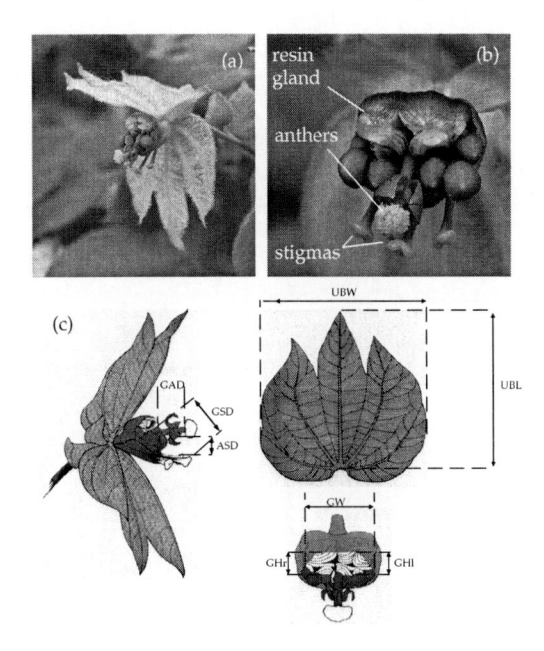

Figure 10.3 Blossom of *D. scandens*: a) General view (photo C. Pélabon), b) close view of the male and female flower clusters with the resin-producing gland (photo P. H. Olsen). The first (terminal) male flower is open while the other male flowers are closed. Note the presence of the transparent resin on the surface of the gland. c) Drawing representing the functional and morphological traits measured in the phenotypic selection study and included in the fitness function. The gland-stigma distance (GSD) is the average of the shortest distances between the gland and the tip of each of the three styles. The gland-anther distance (GAD) is the shortest distance between the gland and the cyme of the terminal male flower. The anther-stigma distance (ASD) is the shortest distance between the terminal male flower and the stigmas. We also measured the area of the gland (GA) as the gland width (GW) multiplied by the average height of the two half left and right gland (GHr, GHl). The upper bract area (UBA) is measured as the product of the bract length (UBL) × bract width (UBW).

and the anther (GAD) determine how efficient bees of different sizes will be in pollinating the blossoms (Armbruster 1988, 1990), while the distance between the anther and the stigma (ASD) during the bisexual phase affects the ability of the blossom to self-pollinate (Armbruster 1988; Fig. 10.3). The open morphology of the *Dalechampia* blossom allows pollinators to directly assess the quantity of reward offered. However, the relative gland and bract sizes vary across species, as does coloration of both bracts and resin. This variation suggests that the use of bracts and glands by pollinators to find blossoms and assess the quantity of reward offered may differ among *Dalechampia* species.

In order to better understand selective pressures shaping *Dalechampia* blossoms, we conducted phenotypic selection studies on different *Dalechampia* species (Armbruster et al. 2005; Bolstad et al. 2010; Pérez-Barrales et al. 2012). The data used here to estimate a fitness function and to calculate the different components of adaptive inaccuracy were collected in a population of *D. scandens* near Veracruz, Mexico in early fall 2006 (see Pérez-Barrales et al. 2012 for a more complete analysis of these data). In this population, the blossoms were pollinated by female *Hypanthidium* cf. *melanopterum* (Megachilidae), which collected resin. These blossoms were also attacked by the larvae of a small seed-eating weevil (Curculionidae), which markedly reduced seed production (32% of the blossoms attacked). We built the fitness function using causal relationships between the traits and components of the blossom fitness as in structural-equation modeling. However, in contrast to the classical approach of structural-equation modeling, we allowed the different paths to be represented by complex non-linear functions with non-normally distributed errors (Fig. 10.4, see Bolstad et al. 2010 for details of the approach). Both attraction and pollination traits were included in the fitness function. Attraction traits included those that measure either the advertisement (the area of the upper bract, UBA) or the reward (the area of the resin-producing gland, GA) for pollinators. Traits related to pollination function included those that reflect the fit between the blossom and the pollinator, that is, traits affecting pollen deposition on, and pick up from the pollinator (GAD and GSD, respectively), and a trait that reflects the probability of self pollination (ASD; see Fig. 10.3c for a depiction of the traits and the measurements). The blossom fitness was estimated as the number of seed produced. In order to distinguish between the effects of pollinators and seed predators on the blossom fitness, we also recorded pollen load on the stigmas at the end of the female phase (before the male flowers opened), the seed set (i.e. before predation), and the number of seeds damaged by seed predators (see Pérez-Barrales et al. 2012 for more details).

This example may appear as primarily heuristic because only a subset of fitness components was estimated over a short period of time compared

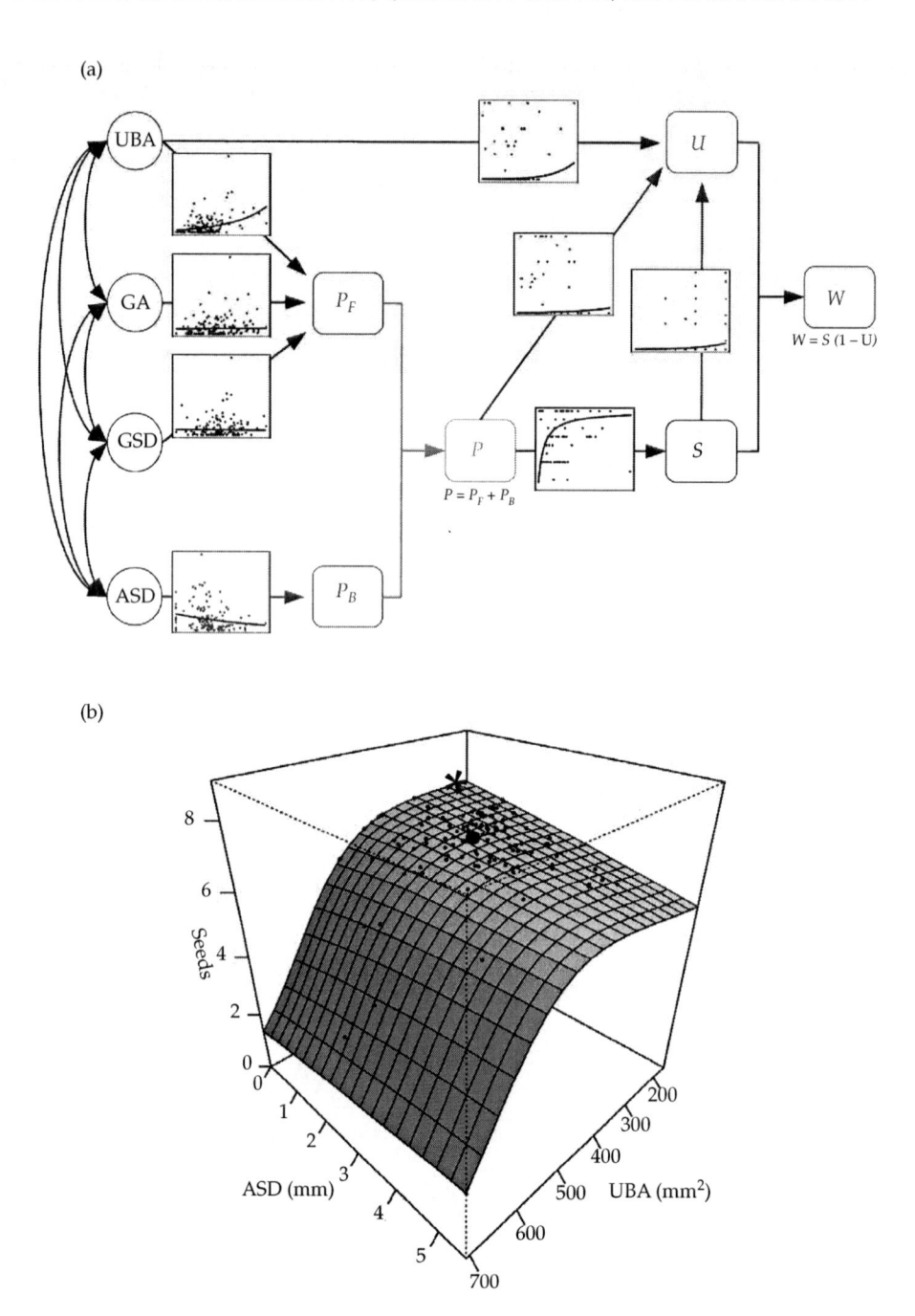

Figure 10.4 Fitness function and adaptive surface of the *D. scandens* blossom. a) Path diagram explaining the functional and causal relationships between the different traits used to build the fitness function (P_F: pollen load at the end of the female phase; P_B: the pollen deposited during the first day of the bisexual phase; P: total pollen load at the end of the first day of the bisexual phase = $P_F + P_B$; S: number of seed set; U: seed predation; W: total number of seed produced). The observed relationships between the different variables with the fitted functions are presented for each path (predictor variable at the origin of the arrow; response variable at the end of the arrow). b) Fitness surface describing the relationship between the variation in blossom fitness and the two traits with the largest effect on this variation. The blossom fitness is expressed in number of seeds produced (large black dot: population mean; black star: point of highest fitness; UBA: upper bract area; ASD: anther-stigma distance).

to the individual lifetime fitness. Indeed, neither male fitness (pollen export) nor lifetime seed production could be assessed in this perennial species. Additionally, temporal or geographical changes in the pollinator fauna may create fluctuating selection gradients over time and space (Herrera et al. 2006) that will strongly affect the location of adaptive optima. Furthermore, fitness of individual plants was not available because *Dalechampia* often grow in clumps in which it is not easy to distinguish single individuals. Identifying selection at the blossom level is not necessarily similar to identifying selection at the individual level if the actors generating selection behave differently at these two levels, for example if the cues used by pollinators for choosing blossoms within plant are different from the cues used for choosing blossoms among plants (cf. Langbein and Lichtman 1978; Heisler and Damuth 1987). Nevertheless, selection at the blossom level probably captures most of the pollinator and seed-predator mediated selection acting at the plant level (see Bolstad et al. 2010 for further discussion). Thus, the adaptive surface estimated here is a reliable representation of the selective pressures on blossom traits generated by the effects of pollinators and seed predators on seed production. The adaptive inaccuracy, bias and imprecision estimated here can therefore be interpreted as sources of maladaptation relative to the particular selection agents identified in this study.

10.3.2 The Adaptive Landscape

Fitness function and adaptive surfaces
Blossom fitness was affected primarily by two traits, the area of the upper bract (UBA) and the distance separating the stigma from the anther (ASD; Fig. 10.4). The effect of bract area on blossom fitness resulted from conflicting selection generated by pollinators and seed predators. On one hand, an increase in bract area augmented pollen load on the stigmas during the female phase, presumably via an increase in the number of visit by pollinators (Fig. 10.4a, path between UBA and P_F). On the other hand, an increase in bract area also resulted in a higher rate of seed predation (Fig. 10.4a, path between UBA and U), suggesting that seed predators also used bract size when "deciding" where to deposit their eggs. These conflicting selection pressures resulted in net stabilizing selection on bract area (Pérez-Barrales et al. 2012). Variation in area of the resin gland (GA) did not detectably affect pollinator attraction or blossom fitness (Fig. 10.4a, path between GA and P_F). Interestingly, this pattern of selection on signaling and reward traits is similar to the pattern observed in *D. ipomoeifolia*, also pollinated by a megachilid bee (Armbruster et al. 2005), but contrasts with that observed in *D. schottii*, for which the pollinator (euglossine bees) preferentially visited blossoms with large glands, and the variation in bract size had only a limited effect on the visitation rate and pollen deposition on the stigmas (Bolstad et al. 2010).

The effect of ASD on blossom fitness was mediated by the contribution of autogamous self pollination (pollen falling from the anther onto the stigmas), to the total pollen load during the bisexual phase, which in turn affected seed production. A decrease in the distance between anther and stigma increased the rate of deposition of self-pollen (Fig. 10.4a, path between ASD and P_B). The effect of ASD on fitness was most probably overestimated, however, because we ignored offspring quality. Armbruster and Rogers (2004) found evidence of inbreeding depression in *D. scandens*, indicating that a smaller ASD may decrease seed quality. Therefore, including seed quality in the estimation of the fitness function should increase the optimal ASD. The Adaptive Landscape described in Fig. 10.4b may be representative only of years when pollinators are scarce. We can expect that temporal fluctuations in pollinator abundance and visitation rate will affect the costs and benefits of self pollination (Eckert et al. 2006), and therefore the shape of the Adaptive Landscape. Although we observed quadratic selection on bract size, the optimum phenotype was outside the observed ranges for other traits. Here, we took the optimum to be the trait values with highest fitness within the observed trait range (Fig. 10.4b).

10.3.3 Adaptive inaccuracy

Inaccuracy, bias, and imprecision of the population
Adaptive inaccuracy, bias and imprecision overall and for each trait or combination of traits are

Table 10.1 Inaccuracy, bias and imprecision calculated for the mean fitness surface for each trait and combination of traits in *Dalechampia scandens*. The maximum fitness in number of seeds set is 7.30 seeds. Inaccuracy, bias, and imprecision are given in number of seeds. The percentage for the inaccuracy is given in percentage reduction of the maximum fitness. For the bias and imprecision, the percentages give the relative contribution of the two sources of maladaptation to the inaccuracy. The 95% confidence intervals were obtained by parametric bootstrapping (Davison and Hinkley 1997). We resampled the different fitness components (pollen load, seed set, and surviving seeds) 1000 times and reestimated the parameters of the fitness function. This produced 1000 different fitness-surface estimates, which we used to recalculate the parameters of interest and their 95% confidence intervals.

Trait	Inaccuracy load # seeds		Bias load # seeds		Imprecision load # seeds	
All	0.66 (0.39; 1.68)	9.0%	0.46 (0.26; 1.43)	69.7%	0.20 (−0.03, 0.41)	30.3%
UBA	0.38 (0.10; 1.70)	5.2%	0.13 (0.00; 1.33)	34.1%	0.25 (−0.01, 0.59)	65.9%
GA	0.01 (0.00; 0.44)	0.2%	0.01 (0.00; 0.44)	100.4%	0.00 (−0.01, 0.01)	−0.4%
GSD	0.01 (0.00; 0.36)	0.1%	0.01 (0.00; 0.37)	100.2%	0.00 (−0.01, 0.01)	−0.2%
ASD	0.50 (0.01; 1.08)	6.8%	0.49 (0.00; 1.07)	98.1%	0.01 (0.00, 0.03)	1.9%
UBA and GA	0.39 (0.15; 1.40)	5.3%	0.14 (0.04; 1.06)	36.8%	0.24 (−0.02, 0.52)	63.2%
UBA and GSD	0.38 (0.14; 1.43)	5.2%	0.14 (0.04; 1.09)	35.8%	0.24 (−0.01, 0.47)	64.2%
UBA and ASD	0.63 (0.22; 1.60)	8.6%	0.42 (0.06; 1.21)	66.0%	0.21 (−0.01, 0.64)	34.0%
GA and GSD	0.02 (0.01; 0.64)	0.2%	0.02 (0.01; 0.65)	100.4%	0.00 (−0.02, 0.01)	−0.4%
GA and ASD	0.52 (0.07; 1.19)	7.2%	0.51 (0.06; 1.17)	98.1%	0.01 (−0.02, 0.04)	1.9%
GSD and ASD	0.51 (0.06; 1.22)	7.0%	0.50 (0.05; 1.20)	98.0%	0.01 (−0.02, 0.04)	2.0%
UBA; GA and GSD	0.39 (0.18; 1.38)	5.3%	0.15 (0.08; 1.09)	38.4%	0.24 (−0.02, 0.40)	61.6%
UBA; GA and ASD	0.65 (0.31; 1.60)	8.9%	0.44 (0.15; 1.28)	68.3%	0.21 (−0.02, 0.52)	31.7%
UBA; GSD and ASD	0.64 (0.32; 1.58)	8.7%	0.43 (0.15; 1.30)	67.5%	0.21 (−0.02, 0.48)	32.5%
GA; GSD and ASD	0.53 (0.14; 1.37)	7.3%	0.52 (0.12; 1.37)	98.0%	0.01 (−0.03, 0.05)	2.0%

reported in Table 10.1. The total adaptive inaccuracy represented a reduction of 0.66 seeds (9%) from the population maximum fitness of 7.30 seeds. About 30% of this was generated by the bias, the remaining 70% being due to imprecision. These estimates illustrate that maladaptation can be seriously underestimated if the effects of population variation are ignored. The two traits generating most of the adaptive inaccuracy were the ASD and the bract size. For ASD the contribution was almost entirely due to the bias (98%), while for bract size it was mainly due to imprecision (66%). The 95% confidence intervals for the inaccuracy and its two components are also reported in Table 10.1. These are large due to the imprecision of our fitness function. This illustrates the difficulties of estimating fitness surfaces with enough precision.

Decomposing the inaccuracy into contributions of single traits further revealed that the total inaccuracy differed from the sum of the individual trait inaccuracies (e.g. Inaccuracy$_{UBA}$ + Inaccuracy$_{ASD}$ > Inaccuracy$_{Total}$: 0.38 seeds + 0.50 seeds > 0.66 seeds). Similarly, inaccuracy of groups of traits sometimes differed from the sum of the inaccuracies for each single trait. These discrepancies can go in either directions (i.e. total inaccuracy can be larger or smaller than the sum of the individual inaccuracies), and may be generated by discrepancies in either the bias or the imprecision.

For the area of the resin-producing gland (GA) and GSD, adaptive imprecision was small and negative. Negative adaptive imprecision is generated by the convexity (positive second derivative) of the adaptive surface, which favors phenotypic variation (Layzer 1980).

Adaptive accuracy of the genotype
Within-individual variation can reduce the fitness of an individual genotype compared to the fitness of its "target" phenotype (Fig. 10.1). To illustrate this point, we used greenhouse measurements from a population of *D. scandens* in which two blossoms per individual had been measured (Hansen et al. 2003; Pélabon et al. 2004a). We applied these measurements to the adaptive surface obtained from the field study and estimated the contribution of the within-individual variation to the adaptive inaccuracy by use of equation 7. These estimates are presented in Table 10.2, along with an assessment of the total inaccuracy of the greenhouse population based on the fitness surface from the field study. Although the inaccuracy of the greenhouse population was similar to the field population, the bias was larger and the precision higher. This makes sense as the greenhouse population may be adapted to a slightly different environment, thereby increasing bias, and as the benign greenhouse environment may reduce individual (and within-individual) variation, thereby increasing precision. The within-individual (developmental) variation accounted for a substantial part of the population imprecision. For example, for upper bract area, developmental imprecision generated 19% of the inaccuracy, comparable to the 24% generated by among-individual variation. We believe that the relative importance of developmental imprecision observed here does not result from the limited environmental variation in the greenhouse, because Pélabon et al. (2011) found that within-individual variation was at least as high as among-individual variation in gland area and bract length in a greenhouse population of *D. scandens* subjected to environmental manipulations.

Within-individual variation measured across modules (here blossoms) in plants comprises several sources of variation, and can be larger and different from the variation estimated via fluctuating asymmetry (Pélabon et al. 2004b). This may explain why the contribution of within-individual variation observed here is larger than in most other studies of developmental variation in animals reviewed in Hansen et al. (2006), where within-individual variation was estimated by fluctuating asymmetry. Measurement error can also inflate the component of adaptive imprecision attributed to within-individual variance. Although measurement error generally represent only a fraction of the total phenotypic variance (from 0.15% for UBW to 4.5% for GSD in the *Dalechampia* blossom; Hansen et al. 2003), its contribution to the individual imprecision may be particularly important if the within-

Table 10.2 Components of the adaptive inaccuracy of *D. scandens* blossoms including within-individual variation as a source of imprecision.

Trait	Inaccuracy		Bias		Imprecision		Individual imprecision	
All	0.81	11.1%	0.77	95.0%	0.02	2.1%	0.02	2.9%
UBA	0.31	4.3%	0.18	56.5%	0.08	24.8%	0.06	18.7%
GA	0.01	0.1%	0.01	100.5%	0.00	−0.4%	0.00	−0.1%
GSD	0.01	0.1%	0.01	100.1%	0.00	−0.2%	0.00	0.1%
ASD	0.96	13.1%	0.95	99.7%	0.00	0.3%	0.00	0.0%

individual variance is limited. In *Dalechampia* blossoms, the contribution of the measurement variance to the individual variance ranged from 2% in the upper bract width to more than 50% in the GSD (Hansen et al. 2003). However, the measurement error of the traits with the largest effect on inaccuracy remained relatively low and should not have strongly influenced our results. Nevertheless, the contribution of measurement error to individual imprecision may be important in some cases and should not be ignored.

The estimates presented here show that adaptive inaccuracy is strongly affected by the absence of a fitness optimum in the range of the population, as when traits are under directional selection. The occurrence of directional selection on floral traits is not surprising, at least empirically, and many phenotypic selection studies have revealed directional selection on such traits (e.g. Galen 1996; Campbell et al. 1996; Johnson and Steiner 1997; Maad and Nilsson 2004; Armbruster et al. 2005; Benitez-Vieyra et al. 2006; Sletvold et al. 2010). Furthermore, the prevalence of directional selection seems a general pattern in phenotypic selection studies (Kingsolver et al. 2001; Hereford et al. 2004). At face value, these estimates suggest that populations are generally poorly adapted, with strong adaptive inaccuracies primarily generated by the bias, a result at odd with the general stasis observed in natural populations, given the evolvabilities observed for most quantitative traits (Hansen and Houle 2004; Estes and Arnold 2007; Hansen et al. 2011; Chapter 13). Although one can argue that identification of selection is often incomplete, preventing the location of the true fitness optima, the relatively low frequency of negative quadratic estimates of selection gradient suggests that populations are often far from their optimum (Estes and Arnold 2007). Alternatively, spatial or temporal variation in the Adaptive Landscape may generate overall stabilizing selection (Bell 2010).

In the next section, we first show how adaptive inaccuracy of specific traits can be estimated while controlling for the effects of other traits. We then show an example of how variation in the Adaptive Landscape can be taken into account when calculating adaptive inaccuracy.

Marginal Adaptive Landscapes

In our study, the distance between anther and stigma strongly influenced the blossom fitness, because shorter ASDs promote self-pollination during the bisexual phase and therefore ensure seed production when pollinators are rare. This directional selection may be overestimated, however, because our fitness function does not account for inbreeding depression. In order to estimate the adaptive inaccuracy of the different traits, independently of the effects of ASD on the adaptive surface, we fixed ASD at its mean (i.e. assuming that ASD was perfectly adapted in this population) and recalculated the inaccuracy, bias, and imprecision for the remaining traits. These estimates are presented in Table 10.3.

Compared with the previous estimates, inaccuracy was substantially reduced (0.24 seeds instead of 0.66, a 3.3% reduction in fitness instead of the initial 9%) and primarily resulted from maladaptation in bract size. Moreover, imprecision in bract size (UBA) became the major source of maladaptation, bias in bract size representing a reduction of only 0.008 seeds from a maximum of 7.30 seeds (ca. 2% of the reduction in fitness), while imprecision

Table 10.3 Adaptive inaccuracy and its different components for *D. scandens* when anther-stigma distance is fixed at its mean (all individuals have been given the population mean ASD for their anther–stigma distance). The maximum fitness achievable for a blossom in this landscape was 7.30 seeds.

Trait	Inaccuracy		Bias		Imprecision	
All	0.24	3.3%	0.05	19.4%	0.19	80.6%
UBA	0.21	2.9%	0.00	1.9%	0.21	98.1%
GA	0.03	0.4%	0.03	100.3%	0.00	−0.3%
GSD	0.01	0.2%	0.01	100.1%	0.00	−0.1%

represented a reduction of 0.21 seeds (ca. 98% of the fitness reduction).

Inaccuracy and variation of the Adaptive Landscape
Spatial and temporal variations in the environment commonly affect the topography of the Adaptive Landscape (Svensson and Gosden 2007; Siepielski et al. 2009; Bell 2010; Chapters 7 and 9). In our study, blossoms were measured on different patches of plants with one to several entangled individuals. Patches differed in their visitation rate due possibly to microenvironmental differences such as exposure to solar radiation. In turn, differences in visitation rate affected the relative influence on the blossom fitness of signaling and rewarding traits on the one hand, and traits affecting the probability of self pollination on the other hand. A study on *D. schottii* showed that blossoms belonging to patches with low visitation rate depended more strongly on self pollination to produce seeds, while variation in traits attracting or rewarding visitors had weaker effects on the blossom fitness (Bolstad et al. 2010).

In Fig. 10.5, we depict the Adaptive Landscapes corresponding to a good, a poor, and the average pollination environments. Variation in the Adaptive Landscape across patches of plants was essentially generated by changes in the effect of the bract length on the blossom fitness, with larger bracts increasing the blossom fitness in good environments (high pollinator visitation and low seed-predation rates) while they were decreasing fitness in poor environments (low pollinator-visitation and high seed-predation rates).

The adaptive inaccuracy and its components, including the ones generated by variation in the Adaptive Landscape, are presented in Table 10.4. Microenvironmental variation in the fitness landscape increased the inaccuracy of the population from 0.66 seeds to 1.87 seeds. This resulted primarily from variation in the position of the optimum, the environment-specific bias, which contributed to the inaccuracy in a similar proportion as the total bias. The contribution of the environment-specific imprecision was generally modest, and sometimes

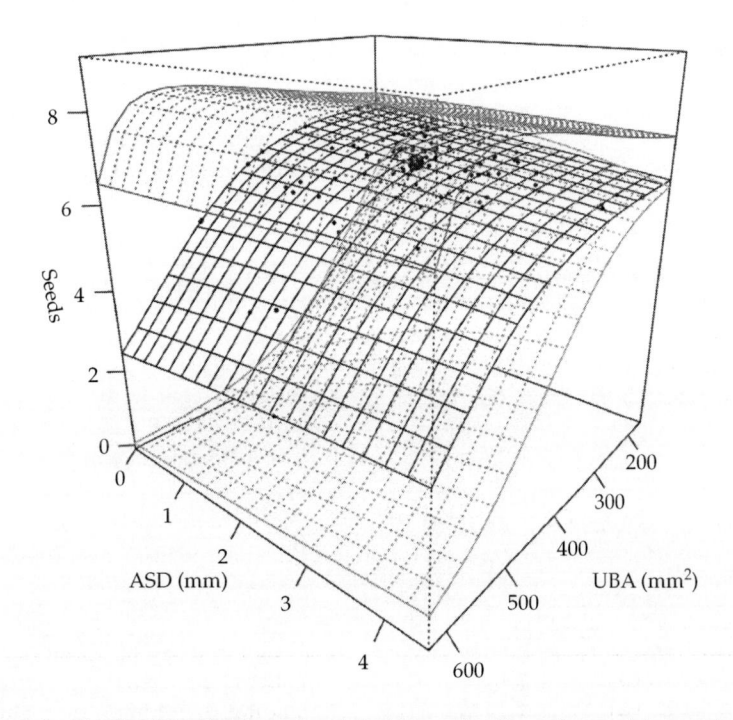

Figure 10.5 Fitness surfaces describing the relationship between the variation in *D. scandens* blossom fitness and the two traits with the largest effect on this variation in a good (dark grey surface), the average (black surface) and a poor (light grey surface) environment. The quality of the environment is defined by the blossom average fitness in the environment (large black dot: population mean; UBA: upper bract area; ASD: anther-stigma distance).

Table 10.4 Components of adaptive inaccuracy including environmental variation, that is, changes in both the position of the fitness peak and topography of the fitness surface in *D. scandens*. The maximum fitness achievable for a blossom in the average landscape is 7.30 seeds. Adaptive surfaces for each patch were estimated using the predicted effects for the different patches obtained from mixed-effects linear models fitted with REML where patch was entered as random effect. We estimated the inaccuracy as the difference between the maximum fitness $W(\theta, \theta)$ and environment-specific mean fitness, which is the average of the fitness of the individuals in their respective environment. Bias and imprecision were calculated as before in the average environment. We estimated the environment-specific bias as the difference between the bias and the average of the biases calculated in each microenvironment (line 3, equation 10.9). Finally, the imprecision due to variation in Adaptive Landscape was estimated by subtracting the bias, the imprecision and the environment-specific bias from the total inaccuracy.

Traits	Inaccuracy		Bias		Imprecision		Environment-specific bias		Environment-specific imprecision	
All	1.87	24.4%	1.12	59.8%	0.18	9.4%	0.39	20.7%	0.19	10.1%
UBA	1.52	19.9%	0.60	39.6%	0.24	15.9%	1.02	67.0%	−0.34	−22.6%
GA	0.58	7.6%	0.42	71.7%	−0.01	−2.3%	−0.19	−32.8%	0.37	63.4%
GSD	0.29	3.7%	0.07	24.5%	0.00	0.0%	0.14	48.9%	0.08	26.6%
ASD	0.09	1.1%	0.17	196.8%	0.00	−1.6%	−0.07	−78.3%	−0.01	−16.8%

negative, suggesting that the imprecision calculated in each environment was lower than the imprecision calculated in the average environment, in other words, that individuals were adapted to their local environment.

Temporal variation in the Adaptive Landscape is also highly prevalent and should be taken into account when assessing adaptive accuracy. Indeed, this is what is done when fitnesses from successive selective events are multiplied together to give a combined fitness landscape (Levins 1962; Haldane and Jayakar 1963). Inaccuracy, bias, and imprecision can be computed on such a combined fitness landscape in the same way as they are computed on a simple fitness landscape.

It should be kept in mind that shift in the direction of selection due to varying Adaptive Landscape (e.g. a moving optimum) may generate disruptive selection and favor phenotypic variation (Siepielski and Benkman 2009). Therefore, not all variation is necessarily maladaptive. Genetic variation determines the short-term evolvability of populations and phenotypic variance may be adaptive in varying environment either as bet-hedging strategy (Simons and Johnston 1997) or because it increases adaptability (Lee and Gelembiuk 2008). These remarks underscore the importance of characterizing spatial and temporal fluctuations in the Adaptive Landscape in order to understand the adaptive or maladaptive contribution of genetic and phenotypic variation, and the possible mechanisms acting on this variation.

10.4 Final remarks

The concept of adaptive accuracy allows more refined measurement of maladaptation by considering not only the relation of a population or individual mean phenotype to a fixed optimum, but also the effects of variation across individuals and environments or in the developmental realization of a genotype. Our previous models of adaptive inaccuracy have been based on a simple quadratic fitness function for a single trait. Here we have extended the concept to an arbitrary multitrait fitness function. This approach can be applied to study adaptation together with sophisticated fitness-surface modeling in the field (Schluter 1988; Crespi and Bookstein 1989; Phillips and Arnold 1989; Armbruster 1990; Wade and Kaliz 1990; Schluter and Nychka 1994; Janzen and Stern 1998; Scheiner et al. 2000; Svensson and Sinervo 2000; Shaw and Geyer 2010; Bolstad et al. 2010). It should be kept in mind, however, that the validity of interpreting fitness surfaces and population distribution in terms of adaptive inaccuracy, bias and imprecision strongly depends on the quality of the estimated fitness function and trait(s) distribution. Building a full fitness surface is challenging, regarding both the necessary knowledge to include the proper fitness components and the amount of data required for building accurate fitness surfaces. Furthermore, the natural distribution of quantitative traits generally limits the range over which precise fitness function can be estimated. Although phenotypic manipulations (Cresswell 2000; Svensson and Sinervo 2000; Campbell 2009) may improve the precision of the parameters describing the fitness function, and may remove spurious correlation between phenotype and fitness (Mitchell-Olds and Shaw 1987; Rausher 1992; Scheiner et al. 2002), this methodology remains limited regarding the number of traits that can be simultaneously manipulated. Of course, constraining assumptions about the distribution and the shape of fitness function (e.g. normal distribution and quadratic fitness function) will decrease the amount of data necessary to estimate the parameters of fitness surfaces. However, there is little reason to think that adaptive surfaces are simple and smooth (Schluter 1988; Blows et al. 2003; Bentsen et al. 2006), or stable in time and space (Svensson and Sinervo 2004; Svensson and Gosden, 2007; Calsbeek et al. 2009; Hereford 2009; Siepielski et al. 2009; Bell 2010). Our understanding of adaptation therefore depends on long-term data on natural selection and proper estimation of Adaptive Landscapes. Nevertheless our conceptual approach of testing adaptation supports the idea that the Adaptive Landscape, as defined by Simpson (1944, 1953) and further developed by Lande and Arnold (1983) and others, is not only a heuristic device to illustrate the effects of natural selection, but also a powerful tool to quantify the levels of adaptation and maladaptation of phenotypic traits.

Acknowledgments

The authors would like to thank L. Antonsen and two anonymous referees for commenting an earlier draft of this chapter. Matt L. Carlson made the drawings in Fig. 10.3. Our research was supported by NSF Grant DEB-0444157 to T.F.H. and DEB-0444745 to W.S.A.

References

Armbruster, W. S. (1985). Patterns of character divergence and the evolution of reproductive ecotypes of *Dalechampia scandens* (Euphorbiaceae). Evolution, 39, 733–752.

Armbruster, W. S. (1986). Reproductive interaction between sympatric *Dalechampia* species: are natural assemblages "random" or organized? Ecology, 67, 522–533.

Armbruster, W. S. (1988). Multilevel comparative analysis of morphology, function, and evolution of *Dalechampia* blossoms. Ecology, 69, 1746–1761.

Armbruster, W. S. (1990). Estimating and testing the shapes of adaptive surfaces: the morphology and pollination of *Dalechampia* blossoms. American Naturalist, 135, 14–31.

Armbruster, W. S., Antonsen, L. and Pélabon, C. (2005). Phenotypic selection on *Dalechampia* blossoms: honest signalling affects pollination success. Ecology, 86, 3323–3333.

Armbruster, W. S., Hansen, T. F., Pélabon, C., Pérez-Barrales, R. and Maad, J. (2009a). The adaptive accuracy of flowers: measurement and microevolutionary patterns. Annals of Botany, 103, 1529–1545.

Armbruster, W. S., Pélabon, C., Hansen, T. F. and Bolstad, G. H. (2009b). Macroevolutionary patterns of pollination accuracy: a comparison of three genera. New Phytologist, 183, 600–617.

Armbruster, W. S., Pélabon, C., Hansen, T. F. and Mulder, C. P. H. (2004). Floral integration, modularity, and precision: distinguishing complex adaptations from genetic constraints. In M. Pigliucci and K. Preston (eds.) Phenotypic integration: studying the ecology and evolution of complex phenotypes. Oxford University Press, Oxford, pp. 23–49.

Armbruster, W. S. and Rogers, D. G. (2004). Does pollen competition reduce the cost of inbreeding? American Journal of Botany, 91, 1939–1943.

Armbruster, W. S. and Schwaegerle, K. E. (1996). Causes of covariation of phenotypic traits among populations. Journal of Evolutionary Biology, 9, 261–276.

Arnold, S. J., Pfrender, M. E. and Jones, A. G. (2001). The Adaptive Landscape as a conceptual bridge between micro- and macroevolution. Genetica, 112–113, 9–32.

Arnold, S. J. and Wade, M. J. (1984). On the measurement of natural and sexual selection: theory. Evolution, 38, 709–719.

Barton, N. H. and Turelli, M. (1987). Adaptive landscapes, genetic distance and the evolution of quantitative characters. Genetical Research Cambridge, 49, 157–173.

Bell, G. (2010). Fluctuating selection: the perpetual renewal of adaptation in variable environments. Philosophical Transactions of the Royal Society B, 365, 87–97.

Benitez-Vieyra, S., Medina, A. M., Glinos, E. E. and Cocucci, A. A. (2006). Pollinator-mediated selection on floral traits and size of floral display in *Cyclopogon elatus*, a sweat bee-pollinated orchid. Functional Ecology, 20, 948–957.

Bentsen, C. L., Hunt, J., Jennions, M. D. and Brooks, R. (2006). Complex multivariate sexual selection on male acoustic signaling in a Wild Population of *Teleogryllus commodus*. American Naturalist, 167, E102–E116.

Blows, M. W. (2007). A tale of two matrices: multivariate approaches in evolutionary biology. Journal of Evolutionary Biology, 20, 1–8.

Blows, M. W. and Brooks, R. (2003). Measuring nonlinear selection. American Naturalist, 162, 815–820.

Blows, M. W., Brooks, R. and Kraft, P. G. (2003). Exploring complex fitness surfaces: multiple ornamentation and polymorphism in male guppies. Evolution, 57, 1622–1630.

Bolstad, G. H., Armbruster, W. S., Pélabon, C., Pérez-Barrales, R. and Hansen, T. F. (2010). Direct selection at the blossom level on floral reward by pollinators in a natural population of *Dalechampia schottii*: full-disclosure honesty? New Phytologist, 188, 370–384.

Calsbeek, R., Buermann, W. and Smith, T. B. (2009). Parallel shifts in ecology and natural selection in an island lizard. BMC Evolutionary Biology, 9, 3.

Campbell, D. R. (2009). Using phenotypic manipulations to study multivariate selection of floral traits associations. Annals of Botany, 103, 1557–1566.

Campbell, D. R., Waser, N. M. and Price, M. V. (1996). Mechanisms of hummingbird-mediated selection for flower width in *Ipomopsis aggregata*. Ecology, 77, 1463–1472.

Crespi, B. J. (2000). The evolution of maladaptation. Heredity, 84, 623–629.

Crespi, B. J. and Bookstein, F. L. (1989). A path-analytic model for the measurement of selection on morphology. Evolution, 43, 18–28.

Cresswell, J. E. (2000). Manipulation of female architecture in flowers reveals a narrow optimum for pollen deposition. Ecology, 81, 3244–3249.

Davison, A. C. and Hinkley, D. V. (1997). Bootstrap methods and their application. Cambridge University Press, New York.

Dvorak, J. and Gvozdik, L. (2010). Adaptive accuracy of temperature oviposition preferences in newts. Evolutionary Ecology, 24, 1115–1127.

Eckert, C. G., Samis, K. E. and Dart, S. (2006). Reproductive assurance and the evolution of uniparental reproduction in flowering plants. In L. D. Harder and S. C. H. Barrett (eds.) Ecology and Evolution of Flowers. Oxford University Press, Oxford, pp. 183–203.

Estes, S. and Arnold, S. J. (2007). Resolving the paradox of stasis: models with stabilizing selection explain evolutionary divergence on all timescales. American Naturalist, 169, 227–244.

Galen, C. (1996). Rates of floral evolution: adaptation to bumblebee pollination in an alpine wildflower *Polemonium viscosum*. Evolution, 50, 120–125.

Gilchrist, G. W. and Kingsolver, J. G. (2001). Is optimality over the hill? In S. H. Orzack and E. Sober (eds.) Adaptationism and optimality. Cambridge University Press, Cambridge, pp. 219–241.

Haldane, J. B. S. (1937). The effect of variation on fitness. American Naturalist, 71, 337–349.

Haldane, J. B. S. and Jayakar, D. (1963). Polymorphism due to selection of varying direction. Journal of Genetics, 58, 237–242.

Hansen, T. F. (2006). The evolution of genetic architecture. Annual Review of Ecology, Evolution, and Systematics, 37, 123–157.

Hansen, T. F. (2011). Epigenetics: Adaptation or contingency? In B. Hallgrimsson, and B. K. Hall (eds.) Epigenetics: Linking genotype and phenotype in development and evolution. University of California Press, San Francisco, CA, pp. 357–376.

Hansen, T. F., Carter, A. J. R. and Pélabon, C. (2006). On adaptive accuracy and precision in natural populations. American Naturalist, 168, 168–181.

Hansen, T. F. and Houle, D. (2004). Evolvability, stabilizing selection and the problem of stasis. In M. Pigliucci and K. Preston (eds.) Phenotypic integration: studying the ecology and evolution of complex phenotypes. Oxford University Press, Oxford, pp. 130–150.

Hansen, T. F., Pélabon, C., Armbruster, W. S. and Carlson, M. L. (2003). Evolvability and genetic constraint in *Dalechampia* blossoms: components of variance and measures of evolvability. Journal of Evolutionary Biology, 16, 754–766.

Hansen, T. F., Pélabon, C. and Houle, D. (2011). Heritability is not evolvability. Evolutionary Biology, 38, 258–277.

Heisler, I. L. and Damuth, J. (1987). A method for analyzing selection in hierarchically structured populations. American Naturalist, 130, 582–602.

Hendry, A. P. and Gonzalez, A. (2008). Whither adaptation? Biology and Philosophy, 23, 673–699.

Hereford, J. (2009). A quantitative survey of local adaptation and fitness trade-offs. American Naturalist, 173, 579–588.

Hereford, J., Hansen, T. F. and Houle, D. (2004). Comparing strengths of directional selection: How strong is strong? Evolution, 10, 2133–2143.

Herre, E. A. (1987). Optimality and selective regime in fig wasp sex-ratios. Nature, 329, 627–629.

Herrera, C. M., Castellanos, M. C. and Medrano, M. (2006). Geographical context of floral evolution: towards an improved research programme in floral diversification. In L. D. Harder and S. C. H. Barrett (eds.) Ecology and

Evolution of Flowers. Oxford University Press, Oxford, pp. 278–294.

Janzen, F. J. and Stern, H. S. (1998). Logistic regression for empirical studies of multivariate selection. Evolution, 52, 1564–1571.

Johnson, S. D. and Steiner, K. E. (1997). Long-tongued fly pollination and evolution of floral spur length in the *Disa draconis* complex (Orchidaceae). Evolution, 51, 45–53.

Kingsolver, J. G., Hoekstra, H. E., Hoekstra, J. M., Berrigan, D., Vignieri, S.N., Hill, C.E., et al. (2001). The strength of phenotypic selection in natural populations. American Naturalist, 157, 245–261.

Krebs, J. R. and McCleery, R. H. (1984). Optimization in behavioural ecology. In J. R. Krebs and N. B. Davies (eds.) Behavioural ecology: an evolutionary approach. Blackwell, London, pp. 91–121.

Lande, R. (1979). Quantitative genetic analysis of multivariate evolution, applied to brain: body size allometry. Evolution, 33, 402–416.

Lande, R. and Arnold, S. J. (1983). The measurement of selection on correlated characters. Evolution, 37, 1210–1226.

Lande, R. and Shannon, S. (1996). The role of genetic variation in adaptation and population persistence in a changing environment. Evolution, 50, 434–437.

Langbein, L. I. and Lichtman, A. J. (1978). Ecological inference. Sage Publications, Beverly Hill, CA.

Layzer, D. (1980). Genetic variation and progressive evolution. American Naturalist, 115, 809–826.

Lee, C. E. and Gelembiuk, G. W. (2008). Evolutionary origins of invasive populations. Evolutionary Applications, 1, 427–448.

Levins, R. (1962). Theory of fitness in a heterogeneous environment. I. The fitness set and adaptive function. American Naturalist, 96, 361–373.

Lynch, M. and Lande, R. (1993). Evolution and extinction in response to environmental change. In G. Kingsolver, P. Kareiva and R. B. Huey (eds.) Evolution and extinction in response to environmental change. Sinauer, Sunderland, MA, pp. 234–250.

Maad, J. and Nilsson, L. A. (2004). On the mechanism of floral shifts in speciation: gained pollination efficiency from tongue- to eye-attachment of pollinia in *Platanthera* (Orchidaceae). Biological Journal of the Linnean Society, 83, 481–495.

Mitchell, W. A. and Valone, T. J. (1990). The optimization research program: studying adaptations by their function. Quarterly Review of Biology, 65, 43–52.

Mitchell-Olds, T. and Shaw, R. G. (1987). Regression analysis of natural selection: statistical inference and biological interpretation. Evolution, 41, 1149–1161.

Orzack, S. H. and Sober, E. (1994a). Optimality models and the test of adaptationism. American Naturalist, 143, 361–380.

Orzack, S. H. and Sober, E. (1994b). How (not) to test an optimality model. Trends in Ecology & Evolution, 9, 265–267.

Orzack, S. H. and Sober, E. (2001). Adaptationism and optimality. Cambridge University Press, Cambridge.

Parker, G. A. and Maynard Smith, J. (1990). Optimality theory in evolutionary biology. Nature, 348, 27–33.

Pearson, K. (1903). Mathematical contributions to the theory of evolution. XI. On the influence of natural selection on the variability and correlation of organs. Philosophical Transactions of the Royal Society London Series A, 200, 1–66.

Pélabon, C., Armbruster, W. S. and Hansen, T. F. (2011). Experimental evidence for the Berg hypothesis: vegetative traits are more sensitive than pollination traits to environmental variation. Functional Ecology, 25, 247–257.

Pélabon, C., Carlson, M. L., Hansen, T. F., Yoccoz, N. G. and Armbruster, W.S. (2004b). Consequences of interpopulation crosses on developmental stability and canalization of floral traits in *Dalechampia scandens* (Euphorbiaceae). Journal of Evolutionary Biology, 17, 19–32.

Pélabon, C. and Hansen, T. F. (2008). On the adaptive accuracy of directional asymmetry in insect wing size. Evolution, 62, 2855–2867.

Pélabon, C., Hansen, T. F., Carlson, M. L. and Armbruster, W. S. (2004a). Variational and genetic properties of developmental stability in *Dalechampia scandens*. Evolution, 58, 504–514.

Pélabon, C., Hansen, T. F., Carter, A. J. R. and Houle, D. (2010). Evolution of variation and variability under fluctuating, stabilizing and disruptive selection. Evolution, 64, 1912–1925.

Pérez-Barrales, R., Bolstad, G. H., Pélabon, C., Hansen, T. F. and Armbruster, W. S. (2012). Pollinators and seed predators generate stabilizing selection on *Dalechampia* blossoms. Manuscript submitted.

Phillips, P. C. and Arnold, S. J. (1989). Visualizing multivariate selection. Evolution, 43, 1209–1222.

Proulx, S. R. and Phillips, P. C. (2005). The opportunity for canalization and the evolution of genetic networks. American Naturalist, 165, 147–162.

Rausher, M. D. (1992). The measurement of selection on quantitative traits: biases due to environmental covariances between trait and fitness. Evolution, 46, 616–626.

Roff, D. (1981). On being the right size. American Naturalist, 118, 405–422.

Scheiner S. M., Donohue, K., Dorn, L. A., Mazer, S. J. and Wolfe L. M. (2002). Reducing environmental bias when measuring natural selection. Evolution, 56, 2156–2167.

Scheiner S. M., Mitchell, R. J. and Callahan, H. S. (2000). Using path analysis to measure natural selection. Journal of Evolutionary Biology, 13, 423–433.

Schluter, D. (1988). Estimating the form of natural selection on a quantitative trait. Evolution, 42, 849–861.

Schluter, D. (1996). Adaptive radiation along genetic lines of least resistance. Evolution, 50, 1766–1774.

Schluter, D. (2000). The ecology of adaptive radiation. Oxford University Press, Oxford.

Schluter, D. and Nychka, D. (1994). Exploring fitness surfaces. American Naturalist, 143, 597–616.

Shaw, R. G. and Geyer, C. J. (2010). Inferring fitness landscapes. Evolution, 64, 1510–1520.

Siepielski, A. M. and Benkma, C. W. (2009). Conflicting selection from an antagonist and a mutualist enhances phenotypic variation in a plant. Evolution, 65, 1120–1128.

Siepielski, A. M., DiBattista, J. and Carlson, S. M. (2009). It's about time: the temporal dynamics of phenotypic selection in the wild. Ecology Letters, 12, 1261–1276.

Simons, A. M. and Johnston, M. O. (1997). Developmental instability as a bet-hedging strategy. Oikos, 80, 401–406.

Simpson, G. G. (1944). Tempo and mode in evolution. Columbia University Press, New York.

Simpson, G. G. (1953). The major features of evolution. Columbia University Press, New York.

Sletvold, N., Grindeland, J. M. and Ågren, J. (2010). Pollinator-mediated selection on floral display, spur length and flowering phenology in the deceptive orchid *Dactylorhiza lapponica*. New Phytologist, 188, 385–392.

Stinchcombe, J. R., Agrawal, A. F., Hohenlohe, P. A., Arnold, S. J. and Blows, M.W. (2008). Estimating nonlinear selection gradients using quadratic regression coefficients: double or nothing? Evolution, 62, 2435–2440.

Svensson, E. and Gosden, T. P. (2007). Contemporary evolution of secondary sexual traits in the wild. Functional Ecology, 21, 422–433.

Svensson, E. and Sinervo, B. (2000). Experimental excursions on adaptive landscapes: density-dependent selection on egg size. Evolution, 54, 1396–1403.

Svensson, E. and Sinervo, B. (2004). Spatial scale and temporal component of selection in sideblotched lizards. American Naturalist, 163, 726–734.

van Tienderen, P. H. (2000). Elasticities and the link between demographic and evolutionary dynamics. Ecology, 81, 666–679.

Wade, M. J. and Kalisz, S. (1990). The causes of natural selection. Evolution, 44, 1947–1955.

Wagner, G. P. and Altenberg, L. (1996). Complex adaptations and evolution of evolvability. Evolution, 50, 967–976.

Walsh, B. and Blows, M. W. (2009). Abundant genetic variation + strong selection = multivariate genetic constraints: a geometric view of adaptation. Annual Review of Ecology, Evolution, and Systematics, 40, 41–59.

Webster, G.L. and Armbruster, W. S. (1991). A synopsis of the neotropical species of *Dalechampia* (Euphorbiaceae). Botanical Journal of the Linnean Society, 105, 137–177.

Wright, S. (1932). The roles of mutation, inbreeding, crossbreeding and selection in evolution. Proceedings of the Sixth International Congress of Genetics, 1, 1365–1366.

Wright, S. (1935). Evolution in populations in approximate equilibrium. Journal of Genetics 30, 257–266.

CHAPTER 11

Empirical Insights into Adaptive Landscapes from Bacterial Experimental Evolution

Tim F. Cooper

11.1 Introduction

Experimental evolution is the laboratory-based study of evolutionary processes. The high degree of control that can be applied to experimental environments allows researchers to examine the influence of variables such as population size, resource complexity, and selective history on patterns of adaptation. Moreover, the ability to keep bacterial populations frozen in a non-evolving state allows evolutionary trajectories to be subsequently examined. As advances in sequencing technology allow genetic analyses to complement phenotypic assays, experimental evolution has the potential to provide a uniquely detailed view of the dynamics of adaptation.

The Adaptive Landscape has had a major role in shaping and motivating many experimental evolution studies. In return, these studies have made important contributions to our understanding of the nature of Adaptive Landscapes. The simplest findings come from studies that have examined the similarity of evolutionary outcomes from initially identical replicate populations. Comparison of resulting evolutionary trajectories has allowed experimenters to examine the repeatability of evolution, a reflection of the relative influence of chance and selection on the outcome of evolution. Other experiments have addressed the evolution of diversity within (usually) initially homogeneous populations—an outcome determined by the availability of distinct adaptive peaks and the existence of ecological conditions that allow the maintenance of subpopulations at different peaks. Experiments can be carefully designed and controlled to assess the effect of environment—e.g. spatial structure, resource complexity—on the likelihood that distinct peaks will be reached. Here I discuss how bacterial experimental evolution studies can and have contributed to our understanding of the form of Adaptive Landscapes.

11.2 The repeatability of evolution

11.2.1 Assessing divergence of replicate populations

Divergence among initially identical populations can be caused by adaptation to different environments or by chance events such as mutation order (Box 11.1) and genetic drift, which may constrain subsequent evolution and promote divergence even among populations evolving in the same environment. Variation between populations has been studied by comparing extant populations and attempting to correlate organismal traits with environmental parameters to infer the action of selection (Harvey and Pagel 1991). Experimental evolution studies complement this approach by allowing evolutionary dynamics to be monitored in real time, allowing controlled tests to examine the effects of environmental conditions on evolutionary processes and outcomes (reviewed in Elena and Lenski 2003; Buckling et al. 2009).

Wright's metaphor of an Adaptive Landscape provides a framework for considering the adaptation and divergence of populations (Wright 1932, 1988). This landscape is typically visualized as a two-dimensional representation of genotype space, onto which a fitness surface is projected (see also Chapters 1 and 2). Natural selection pushes populations toward accessible peaks of high fitness. In the simplest case, the Adaptive Landscape has only one fitness peak and replicate populations will eventually converge on the same solution. However, if interactions exist between mutations, such that the fitness effect of a mutation depends on whether or not another one is already present (i.e. mutations interact epistatically), then the landscape may contain multiple peaks separated by valleys (Whitlock et al. 1995, 2005). In this case, several different fitness peaks might be accessible to a population starting at a given point in genotype space, depending on the order in which beneficial mutations arise and escape loss by random drift. Natural selection inhibits movement of populations between peaks separated by intermediate types having lower fitness; therefore, rugged landscapes increase the probability that replicate populations will diverge toward distinct adaptive peaks.

A number of experimental evolution studies have examined the extent of divergence among replicate populations and thus the availability of distinct adaptive peaks from a particular genetic starting point. The basic design of these experiments is to evolve replicate populations—often founded from a single clone to ensure the populations are initially as similar as possible—in a constant defined environment for enough generations that new mutations have a chance to occur and increase in frequency. In the absence of migration, the magnitude and trend of divergence between replicate populations provides a window into the accessibility of distinct adaptive peaks.

An ongoing long-term evolution experiment, started by Richard Lenski in 1988, exemplifies this kind of experiment. Twelve populations of *Escherichia coli* were started from a common ancestor and independently evolved in a simple defined environment supplemented with a low concentration of glucose as the only carbon source (Lenski et al. 1991). Fitness increased in all populations, relative to their ancestor in the evolution environment (Lenski et al. 1991; Cooper and Lenski 2000). The simplest metric of population divergence, with intuitive and practical appeal, is to compare the fitness response of replicate populations in the selective environment. Doing this, Lenski and colleagues found that the standard deviation in fitness across the replicate populations first rose and then remained steady over at least the first 10,000 generations of evolution, consistent with the populations approaching distinct adaptive peaks of different fitness (Lenski and Travisano 1994) (Fig. 11.1). Nevertheless, it should be emphasized that fitness differences between populations were quite small relative to the average fitness increase relative to the ancestor—a point I return to later in this chapter.

A limitation of using between-population variation in fitness as a measure of population divergence is that it is a conservative metric. Many different phenotypes can result in a similar fitness increase. Thus, even if replicate populations reach and maintain similar fitness, there is no guarantee that they are approaching the

Figure 11.1 Trajectories of relative fitness increase for 12 populations of *E. coli* over their first 10,000 generations of evolution. Lines show a best-fit hyperbolic model of fitness of each population, which was measured relative to the ancestor at 500-generation intervals. The average fitness increase over the first 5000 generations was approximately fivefold greater than the average fitness increase between 5000–10,000 generations, indicating that the rate of adaptation declined. Data was downloaded from http://myxo.css.msu.edu/ecoli/relfit.html and was first reported by Lenski and Travisano (1994) Reproduced with permission from the National Academy of Sciences.

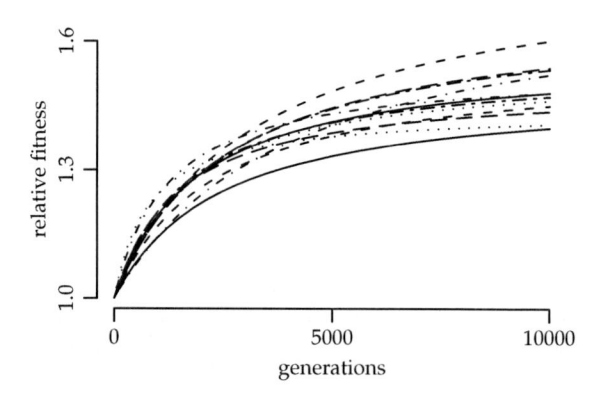

same fitness peak. Fortunately, there is a simple experimental approach that can address this possibility—compare replicate population over more characteristics. Replicate populations may have similar fitness in the selected environment, but be recognized as distinct from one another if more detailed phenotypic profiles are compared. Additional phenotypes that have been examined include correlated responses to alternative non-selected environments (Travisano et al. 1995a; Travisano and Lenski 1996; Maclean and Bell 2003; Ostrowski et al. 2005), metabolic/catabolic profiling (Notley-McRobb and Ferenci 1999; Cooper and Lenski 2000; MacLean et al. 2004) and gene expression (Ferea et al. 1999; Cooper et al. 2003; Fong et al. 2005; Pelosi et al. 2006).

Studies examining between-population variation in fitness between replicate populations evolving in single-resource environments have typically found at least some evidence of divergence (Lenski et al. 1991; Lee et al. 2009; Cooper and Lenski 2010). At the same time though, studies that compare replicate populations over multiple phenotypes typically find that replicate evolving populations exhibit a significant degree of parallelism (Box 11.1), even when fitness differences appear real and consistent (Cooper and Lenski 2000; Cooper et al. 2003; Fong et al. 2005). There are at least two ways to reconcile these apparently contradictory findings. First, the suites of phenotypic traits that correspond to different adaptive peaks may overlap. For this reason, finding parallel phenotypic changes should be interpreted to indicate that replicate populations are moving toward similar, but not necessarily the same, adaptive peaks. Second, practical considerations lead multiple-phenotype studies to emphasize parallel over unique changes because the former represent a simpler and more robust statistical signature.

The development of accessible whole-genome sequencing technology is now allowing a new resolution of comparison between replicate evolved populations. Whereas previous genetic methods, for example candidate gene sequencing and DNA-array approaches, allowed variation to be assessed at a limited number of loci, entire genome sequences of evolved clones can now be obtained. As yet no published study has compared microbial genome sequences from clones evolved in replicate populations, but mutations found by sequencing one evolved clone can be used to identify target loci that were sequenced in evolved clones isolated from independent populations (Velicer et al. 2006; Barrick et al. 2009). Comparing populations evolved in the Lenski long-term experiment, Barrick et al. (2009) found a strong signature of genetic parallelism. Forty-five evolved mutations were found in a clone isolated after 20,000 generations of evolution. Of the gene regions that these mutations occurred in, 14 were sequenced in each of 11 other replicate populations. In only two cases was a mutation found to be unique, and in seven

Box 11.1 Glossary of terms used

Antagonistic pleiotropy (AP)	Mutations that are beneficial in a selected environment but are detrimental in alternative environments.
Parallelism	A signature of significant similarity in evolutionary changes occurring in independently evolving populations.
Mutation accumulation	Accumulation of mutations by random genetic drift. Accumulation of neutral mutations in a selected environment can cause a cost to adaptation if they are detrimental in alternative environments.
Mutation order	The order in which mutation are acquired in an evolved population—determined by analysis of periodically stored samples.
Insertion sequence (*IS*)	A class of mobile genetic elements that is common in bacteria.
Sign epistasis	The sign of the fitness effect of a mutation—i.e. whether it confers a fitness benefit or cost—depends on its genetic background.

cases at least five other populations had a mutation in the same gene region. Assuming that most of the other populations had a similar total number of mutations, this overlap is strikingly significant, though certainly not complete.

Interpreting a finding of between-population divergence among replicate, initially identical, evolved populations raises a difficult question. How can we know whether divergence will be maintained, reflecting some aspect of the landscape topology, or will be temporary, being caused by inevitable differences in the time at which mutations arise and reach a high enough frequency to reliably fix in the population? Even in the absence of distinct adaptive peaks, differences in mutation order will cause transient divergence. Only with the minimal condition of sign epistasis—the reversal of the sign of a mutation's fitness effect depending on its genetic background (Box 11.1)—will the inevitable differences in mutation order cause populations to ascend to different peaks so that divergence will be maintained. The ideal test to distinguish between these possibilities and, thus, determine whether divergence is likely to be transient, is to construct strains with defined combinations of mutations derived from independent populations and estimate the fitness of these strains to test directly for the presence of multiple adaptive peaks (Weinreich et al. 2006; da Silva et al. 2010). Unfortunately, the difficulty of finding, let alone precisely manipulating relevant mutations, makes direct tests of sign epistasis exceedingly difficult. For this reason, indirect approaches that track divergence as relative to overall changes in the phenotype, testing whether divergence declines over time, have been used (Travisano and Lenski 1996).

Rozen et al. (2008) demonstrate a further complication of testing for divergence of replicate populations. If populations are large enough that multiple mutations compete with one another for fixation, population adaptation will tend to be dominated by large-effect mutations because these drive small-effect mutations to extinction. Even if multiple peaks were accessible, populations would tend to follow paths dominated by larger effect mutations. This bias will make evolution more repeatable, and thus the landscape appear smoother, than would be the case if mutations substituted individually.

Rozen et al. (2008) tested this prediction by evolving populations at relatively small (but still large enough that genetic drift was expected to be negligible over the course of the experiment) and large population sizes—if clonal interference constrained the evolutionary paths followed by populations, they expected that between-population variation in fitness would be lower in a large population size treatment. When small and large populations were evolved in a complex environment, they saw this effect. The experiment was relatively short, but the effect was maintained over the course of the experiment.

Finally, an elegant study by Travisano et al. (1995b) provides an example of how the consequences of populations being at different peaks can be addressed. Twelve replicate populations were first evolved in one environment (as part of the Lenski experiment) and then each population was split into four new populations and evolved in a second environment. A comparison of the fitness of the final populations at the beginning and end of the second period of evolution allowed the role of genotype, chance events and selection to be partitioned. Strikingly, despite the intermediate populations differing by more than 60% in their relative fitness in the second environment, these differences were dramatically reduced by the end of the experiment. Apparently, differences in the position of intermediate genotypes on the Adaptive Landscape relevant to the second environment were overwhelmed by subsequent adaptive mutations.

11.2.2 The effect of environment on between-population divergence

A natural extension of experiments that examine population divergence in one environment is to compare divergence of sets of replicate populations evolved in different environments. This approach can be used to assess the effect of a given environmental variable on the ruggedness of the Adaptive Landscape. For example, it might be expected that qualitatively more complex environments—perhaps defined by the number of resources present or the extent of spatial heterogeneity—will promote greater between-population variation, implying that they create more rugged Adaptive Landscapes.

One of the first studies to test the effect of environmental complexity on the ruggedness of the Adaptive Landscape was by Korona et al. (1994). Replicate populations of the bacteria *Alcaligenes eutrophus* (then known as *Comamonas* sp. strain TFD41) were evolved in environments that contained the same limiting resource, but with (agar surface) or without (shaken liquid) spatial heterogeneity. Populations evolved in the spatially heterogeneous environment diverged significantly from one another by 500 generations and this divergence remained present at 1000 generations, suggesting that it would be maintained and therefore that replicate populations had ascended distinct adaptive peaks.

Other studies have manipulated environmental complexity by altering the number of resources present in the environment (Barrett et al. 2005; Barrett and Bell 2006; Cooper and Lenski 2010). In one study where population divergence was tested explicitly, higher diversity was found between populations evolved in environments with alternating resource identity than those evolved only in single component resources (Cooper and Lenski 2010). The simplest explanation for this result is the presence of genetic trade-offs in the ability of an organism to adapt to two resources creating a composite landscape having more fitness peaks than were present in either single resource environment.

11.3 Diversity within populations

An alternative approach to examining the ruggedness of Adaptive Landscapes is to assess the evolution of stable diversity within evolving populations. Finding stable diversity within populations indicates the presence of at least two accessible fitness peaks where each peak corresponds to a phenotype specialized to some suite of environmental niches.

The role of environment on within-population diversity has been the focus of a large number of experimental evolution studies. Many of these studies have examined diversity arising in environments containing a single limiting resource (Turner et al. 1996; Treves et al. 1998; Notley-McRobb and Ferenci 1999; Rozen and Lenski 2000). Surprisingly, even in these simple environments, populations can evolve to multiple peaks, often driven by the production and asymmetric use of some secondary metabolite. Other studies have been designed to test the role of some aspect of environmental heterogeneity on the evolution and maintenance of diversity (Rainey and Travisano 1998; Barrett and Bell 2006; Habets et al. 2006). A prominent example is a series of studies focusing on the role of spatial heterogeneity, physical disturbance and environmental productivity on the evolution of distinct ecotypes in evolving populations of the bacterium *Pseudomonas fluorescens* (Rainey and Travisano 1998; Buckling et al. 2000; Kassen et al. 2005). When a single clone of *P. fluorescens* was inoculated into a spatially structured environment, at least three distinct phenotypes quickly and reproducibly evolved (Rainey and Travisano 1998). Competition experiments demonstrate that all three types are stably maintained through ecological interactions, providing a clear demonstration that a clonal population can evolve to exploit multiple ecological niches. Extensive follow-up studies have examined the mutations underlying one evolved phenotype, giving one of the best available insights into the basis of the evolutionary pathways leading to an adaptive peak (McDonald et al. 2009).

Two *E. coli* systems have been used to study the basis of trade-offs that underlie and maintain within-population diversity (Dykhuizen and Dean 2004; Friesen et al. 2004). Experiments with both systems involve growth of populations in an environment containing two sugars and subsequent examination of the dynamics of a resource use polymorphism. In one experiment diversity evolved *de novo* (Friesen et al. 2004), while in the other ancestral strains were carefully chosen to test polymorphism stability (Dykhuizen and Dean 2004). In both experiments models were developed that could explain the maintenance of the polymorphism, in one case from basic biochemical parameters derived from knowledge of catabolic efficiency of relevant enzymes (Dykhuizen and Dean 2004). Interestingly, as that experiment was continued, the polymorphism was sometimes lost, presumably as a superior generalist arose. Evidently additional adaptive peaks were present, representing genotypes that were able to outcompete the initial specialist genotypes.

11.4 Correlation of fitness across environments

Measuring the fitness of evolved populations in environments other than the one in which they were selected can be used to assess correlations between the corresponding Adaptive Landscapes. These correlations are important because they underlie trade-offs in the ability of a population to simultaneously adapt to multiple environments.

A common approach used to identify fitness trade-offs involves measuring the fitness of individuals in selected and non-selected environments. Reciprocal transplant experiments extend this approach by measuring the fitness of individuals that have evolved in different environments in their own and each others environment. Trade-offs are usually inferred if there is a negative correlation between the fitness of the individuals across environments, that is, if individuals are fitter in the environment in which they were selected (Schluter, 2000). An important caveat to this interpretation is that a negative correlation identifies a cost of adaptation rather than a trade-off per se. In fact, such a cost can be caused by any combination of three underlying population genetic mechanisms: (1) antagonistic pleiotropy (AP; Box 11.1), true trade-offs whereby mutations underlying adaptation in one environment decrease fitness in alternative environments; (2) mutation accumulation (MA; Box 11.1), accumulation of mutations that are neutral in the selected environment but deleterious in alternative environments; (3) by mutations that are generally beneficial, but to different extents in different environments. The distinction between AP and the alternative mechanisms is important because only in the former case will generalist organisms selected in a combination of environments necessarily be less fit in the individual environments than specialist competitors.

In contrast to typical ecological experiments, a significant advantage of experimental evolution studies in addressing costs of adaptation is that the fitness of independently evolved populations can be measured in selected and alternative environments relative to their ancestor. Because of this, it is possible to determine whether fitness in non-selected environments has declined relative to the ancestor, or just not improved as much as in the selected environment, which would reflect a specificity (SP) of selection, but not a trade-off (Fig. 11.2). Several experimental evolution studies have examined the role of trade-offs in adaptation (Bennett et al. 1992; Velicer and Lenski 1999; Cooper and Lenski 2000; Jasmin and Kassen 2007). These studies have often found a cost of adapting in a single environment with respect to fitness in alternative environments, but have not reached consensus on the underlying mechanism. In part this disagreement may be due to differences in methodology—studies that have compared the responses of populations evolving in constant and mixed environments have often found that generalist populations evolved in mixed environments are not subject to the same costs as specialist populations evolved in a constant environment, consistent with MA underlying those costs. By contrast, studies looking at the correlated responses of populations evolved in a constant environment have typically concluded a role for AP.

One possibility for the broadly different conclusions of mixed- and single-resource type studies

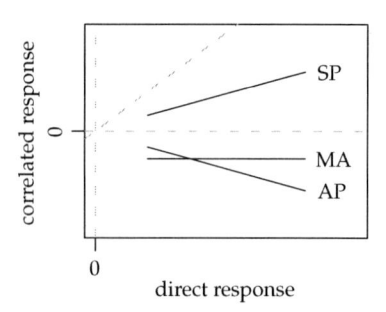

Figure 11.2 Predicted relationship between direct and correlated responses of replicate populations evolved in the same environment under AP, MA, and SP models. AP predicts the slope will be negative because populations with higher direct responses will tend to have fixed more adaptive mutations and therefore have a more severe trade-off in alternative environments. MA predicts a zero slope with negative intercept because mutations that are deleterious in alternative environments accumulate independently of those that contribute to adaptation in the selected environment. SP predicts a positive slope between zero and one because mutations are adaptive in both selection and alternative environments but tend to confer a larger benefit in the selected environment. These models are not exclusive of one another and combinations of mechanisms are likely to underlie fitness responses in real populations. The diagonal dotted line indicates the isocline, where direct and correlated responses are equivalent.

stems from the different selection faced by specialist and generalist populations. For example, imagine two mutations that confer an equivalent adaptation (i.e. the presence of one would make the other redundant) are available to a specialist population evolving in a constant environment, and that only one of the mutations confers a pleiotropic cost in a second environment. In specialist populations it would be a matter of chance which mutation was substituted such that a cost of adaptation would be expected in ~50% of replicate populations. By contrast, in populations selected in a regime alternating between both environments, the non-pleiotropic mutation has a relative advantage and would be substituted in a larger fraction of replicate populations. The effect of this biased selection is that the correlation between fitness landscapes relevant to each single resource environment will not simply predict the constraints faced by populations selected in both environments.

11.5 Form of Adaptive Landscapes around adaptive peaks

A general feature of adaptive trajectories that has emerged from experimental evolution studies is a decelerating rate of adaptation (Lenski et al. 1991; Korona et al. 1994; Barrick et al. 2009; Lee et al. 2009). What does this say about the underlying Adaptive Landscape? On the face of it, two (non-exclusive) mechanisms could cause the rate of adaptation to slow: either the availability or the average benefit of available beneficial mutations could decline as a population becomes better adapted. A third possibility has recently been formalized, beneficial mutations could interact in a way that reduces their marginal benefit when they occur in an already well adapted genotype (Kryazhimskiy et al. 2009). These possibilities can be thought of as affecting the width, slope, and curvature of the route available to an evolving population as it moves toward an adaptive peak, respectively.

Determining the relative contribution of different evolutionary processes to changes in the rate of adaptation requires knowledge of the dynamics of the underlying beneficial mutations. The most complete picture of these dynamics comes from the Lenski long-term *E. coli* experiment. An early analysis of insertion sequence (IS; Box 11.1) mutations found an approximately constant rate of transposition over the first 10,000 generations of the experiment, even though the rate of fitness change declined substantially during this time (Papadopoulos et al. 1999). This finding was surprising, suggesting that genetic and fitness changes did not have a simple correspondence, but was difficult to interpret because it was not clear how many IS mutations were beneficial and how many were neutral or deleterious. A recent analysis using whole-genome sequencing, found a similar pattern—genetic changes accumulated at a constant rate, despite a slowing of the rate of adaptation (Barrick et al. 2009). In the second study three independent lines of evidence were presented that support the interpretation that most of the mutations were probably beneficial. If so, it is unlikely that the decelerating rate of adaptation is caused by a slowing of the beneficial mutation rate, because that would predict a slowdown in mutation substitution. It is more difficult to distinguish between changes in the fitness of available mutations and epistasis as explanations, although a theoretical analysis predicts that epistasis is significant and influential (Kryazhimskiy et al. 2009). In any case, it is clear that combining genotypic and phenotypic data will allow new insight into the form of fitness peaks approached by experimentally evolving populations. Experimental techniques that allow beneficial mutations to be combined so that interactions and direct effects can be quantified directly, promise to help determine underlying mechanisms influencing patterns of adaptation.

Other experiments have focused on one of the most basic aspects of the approach of a population to a fitness peak—the distribution of beneficial mutation effect sizes available to a population (Rozen et al. 2002; Barret et al. 2006; Kassen and Bataillon 2006; Barrick et al. 2010; McDonald et al. 2010). Both of the two general theoretical models of adaptation, Fisher's geometric model and the mutation landscape (ML) model, predict an exponential-like distribution of beneficial mutation fitness effects, independent of the progenitor genotype's fitness (Orr 2005). In practice, however, this prediction is difficult to test in bacterial experiments because of the requirement to identify individuals with beneficial mutations before they are affected by natural selection. One strategy that has been used to overcome this difficulty is to isolate antibiotic resistant clones—these are likely to contain a single mutation—and measure their distribution of fitness effects in an antibiotic free environment (Kassen and Bataillon 2006). Doing this, Kassen and Bataillon (2006) found that a surprisingly high fraction of nalidixic acid resistant *P. fluorescens* mutants (18 of 665 mutants tested) had a fitness benefit relative to their ancestor in an antibiotic free environment and that the distribution of these fitness effects conformed to expectations of the geometric and ML models.

Results of the experiments described here, as well as others carried out in a bacteriophage system (e.g. Rokyta et al. 2005), are consistent with the distribution of mutation effects predicted by the geometric and ML models. It is, however, not clear what the range of alternative theoretical possibilities is. For example, although the geometric and ML models both predict an exponential distribution of mutation effects along an adaptive walk, a recent analysis showed that the same distribution should also be expected during adaptation on landscapes explicitly excluded from the geometric and ML models (Kryazhimskiy et al. 2009). This prediction therefore cannot by itself distinguish between different types of Adaptive Landscapes. Nevertheless, it is clear that microbial systems are contributing much needed data on patterns of short-term adaptation and that this data will continue to inform and refine our understanding of the underlying evolutionary processes.

11.6 Reversibility

Perhaps one of the most practical applications of experimental evolution has been to examine the reversibility of evolution. Will a population shifted from an adaptive peak by selection in a new environment return to the same peak if the original environment is restored? If the original environment contained a single accessible fitness peak, the population would be expected to revert mutations that occurred in the second environment and return to the original genotype. Alternatively, if the original environment contained multiple fitness peaks, new mutations might make alternative peaks available and the population would evolve to a peak different from the one originally occupied.

A simple experimental test of evolutionary reversibility following a transient selection change comes from a study by Schrag and Perrot (1996). They found that exposure of populations of *E. coli* to the antibiotic streptomycin readily selected for resistance mutations. These mutations increased the proof-reading capability of the ribosome, but at the expense of slowing it down. This combination provided a huge fitness advantage in the presence of streptomycin, but a fitness cost in its absence. When clones with resistance mutations were evolved in the absence of streptomycin, fitness quickly improved, but none of 20 replicate populations reverted to streptomycin sensitivity. A failure to find similar fitness improvements in control populations that did not have the resistance mutation suggests that the fitness improvements were due to compensatory mutations that depend on the original resistance mutation to have a beneficial effect—a common conclusion in similar studies (reviewed in Andersson and Hughes 2010). In this case, a simple transition back to the original streptomycin sensitive genotype would involve a fitness valley, with reversion of either the compensatory or resistance mutation alone being deleterious.

11.7 Conclusion

Findings from experimental evolution studies are providing much needed empirical data on the form of the Adaptive Landscapes relevant to evolving populations. Initially identical replicate populations do not remain identical and differences seem likely to be maintained, within-population diversity is often observed, even in apparently simple environments, and populations that are displaced by an environmental change from one position on an Adaptive Landscape are unlikely to return to the same position when the original conditions are reinstated. These observations point to Adaptive Landscapes containing multiple peaks that are accessible to even initially genetically identical replicate populations. Nevertheless, a striking degree of parallelism is often seen when replicate populations are compared across multiple phenotypes or at the level of the genotype. Replicate populations might evolve to different peaks, but the peaks often seem to be quite similar. In the future, a focus of study will be to determine how meaningful the differences are.

How might experimental evolution continue to contribute to our understanding of Adaptive Landscapes? A promising new avenue is the increasing availability of whole-genome sequences from experimentally evolved strains combined with the ability to precisely manipulate mutations to construct specific genotypes. Study of these strains will allow researchers to test specific explanations for general experimental outcomes. For example, two studies have constructed collections of strains that contain all possible combinations of a series of beneficial mutations (Chou et al. 2011; Khan et al. 2011). Fitness measurements of these strains will allow a survey of the distribution of epistasic effects between beneficial mutations and can be used to test for general patterns of epistasis that might systematically influence the trajectory of population adaptation (Kryazhimskiy et al. 2009).

Experimental construction of Adaptive Landscapes will also provide valuable empirical data relevant to general theories examining population evolvability and robustness, which can have major implications for population evolution and long-term success (e.g. Woods et al. 2011). How Adaptive Landscapes can influence evolvability and robustness has been the subject of extensive theoretical investigation (reviewed in Chapter 17), but there is relatively little experimental information available (although see Burch and Chao (2000), and Sanjuan et al. (2007) for work done in viruses). Real Adaptive Landscapes defined by constructed genotypes will represent an important resource—and reality check—to allow analysis of the extent and nature of constraints imposed on natural selection by the specific topology of the Adaptive Landscape (e.g. Weinreich et al. 2006).

Finally, and perhaps most fundamentally, the availability of genotypes that differ by known combinations of mutations combined with the ability to track their subsequent genotypic (through whole genome sequencing) and phenotypic evolution will help move us beyond the Adaptive Landscape as "just" a metaphor and associated mathematical framework (discussed in Chapter 2). With the surface of the Adaptive Landscape "alive" as a mutational network (Chapter 17) with living organisms (or other biological entity) at known mutation combinations, we can design experiments to determine the dependence of key events, for example transitions between distinct fitness peaks, on different parameters. The size of the mutational networks we can construct and examine is vanishingly small compared to the total Adaptive Landscape relevant to even the simplest organism, but it is already clear that they can give important insights into the evolutionary process (Lunzer et al. 2005; Weinreich et al. 2006). Perhaps the metaphor of the Adaptive Landscape cannot capture the high-dimensional dynamics of evolution on these networks, but it certainly inspired the thinking that will.

References

Andersson, D.I. and Hughes, D. (2010). Antibiotic resistance and its cost: is it possible to reverse resistance? Nature Reviews Microbiology, 8, 260–271.

Barrett, R.D. and Bell, G. (2006). The dynamics of evolution in evolving *Pseudomonas* populations. Evolution, 60, 484–490.

Barrett, R.D.H., MacLean, R.G. and Bell, G. (2005). Experimental evolution of *Pseudomonas fluorescens* in simple and complex environments. American Naturalist, 166, 470–480.

Barrett, R.D., MacLean, R.C. and Bell, G. (2006). Mutations of intermediate effect are responsible for adaptation in evolving *Pseudomonas fluorescens* populations. Biology Letters, 22, 236–238.

Barrick, J.E., Yu, D.S., Yoon, S.H., Jeong, H., Oh, T. K., Schneider, D., et al. (2009). Genome evolution and adaptation in a long-term experiment with *Escherichia coli*. Nature, 461, 1243–1247.

Barrick, J.E., Kauth, M.R., Strelioff, C.C. and Lenski, R. E. (2010). *Escherichia coli* rpoB mutants have increased evolvability in proportion to their fitness defects. Molecular Biology and Evolution, 27, 138–1347.

Bennett, A.B., Lenski, R.E. and Mittler, J.E. (1992). Evolutionary adaptation to temperature. I. Fitness responses of *Escherichia coli* to changes in its thermal environment. Evolution, 46, 16–30.

Buckling, A., Kassen, R., Bell, G. and Rainey, P.B. (2000). Disturbance and diversity in experimental microcosms. Nature, 408, 961–964.

Buckling, A., MacLean, R.C., Brockhurst, M.A. and Colegrave, N. (2009). The Beagle in a bottle. Nature, 457, 824–829.

Burch, C.L. and Chao, L. (2000) Evolvability of an RNA virus is determined by its mutational neighbourhood. Nature, 406, 625–628.

Chou, H-H., Chiu, H-C., Delaney, N.F., Segre, D. and Marx, C.J. (2011) Diminishing returns epistasis among beneficial mutations. Science 332, 1190–1192.

Cooper, V.S. and Lenski, R.E. (2000). The population genetics of ecological specialization in evolving *Escherichia coli* populations. Nature, 407, 736–739.

Cooper, T.F. and Lenski, R.E. (2010). Experimental evolution with E. coli in diverse resource environments. I. Fluctuating environments promote divergence of replicate populations. BMC Evolutionary Biology, 10, 11.

Cooper, T.F., Rozen, D.E. and Lenski, R.E. (2003). Parallel changes in gene expression after 20,000 generations of evolution in *Escherichia coli*. Proceedings of the National Academy of Sciences of the United States of America, 100, 1072–1077.

da Silva, J., Coetzer, M., Nedellec, R., Pastore, C. and Mosier, D.E. (2010). Fitness epistasis and constraint on adaptation in a human immunodeficiency virus type 1 protein region. Genetics, 185, 293–303.

Dykhuizen, D.E. and Dean, A.M. (2004). Evolution of specialists in an experimental microcosm. Genetics, 167, 2015–2026.

Elena, S.F. and Lenski, R.E. (2003). Evolution experiments with microorganisms: the dynamics and genetic bases of adaptation. Nature Reviews Genetics, 4, 457–469.

Ferea, T.L., Botstein, D., Brown, P.O. and Rosenzweig, R.F. (1999). Systematic changes in gene expression patterns following adaptive evolution in yeast. Proceedings of the National Academy of Sciences of the United States of America, 96, 9721–9726.

Fong, S.S., Joyce, A.R. and Palsson, B.O. (2005). Parallel adaptive evolution cultures of Escherichia coli lead to convergent growth phenotypes with different gene expression states. Genome Research, 15, 1365–1372.

Friesen, M.L., Saxer, G., Travisano, M. and Doebeli, M. (2004). Experimental evidence for sympatric ecological diversification due to frequency-dependent competition in *Escherichia coli*. Evolution, 58, 245–260.

Habets, M.G.J.L., Rozen, D.E., Hoekstra, R.F. and de Visser, A.G.M. (2006). The effect of population structure on the adaptive radiation of microbial populations evolving in spatially structured environments. Ecology Letters, 9, 1041–1048.

Harvey, P.H. and Pagel, M.D. (1991). The Comparative Method in Evolutionary Biology. Oxford University Press, Oxford.

Jasmin, J-N. and Kassen, R. (2007). On the experimental evolution of specialization and diversity in heterogeneous environments. Ecology Letters, 10, 272–281.

Kassen, R. and Bataillon, T. (2006). Distribution of fitness effects among beneficial mutations before selection in experimental populations of bacteria. Nature Genetics, 38, 484–488.

Kassen, R., Llewellyn, M. and Rainey, P.B. (2005). Ecological constraints on diversification in a model adaptive radiation. Nature, 431, 984–988.

Khan, A.I., Dinh, D.M., Schneider, S., Lenski R.E. and Cooper, T.F. (2011) Negative epistasis between beneficial mutations in an evolving bacterial population. Science 100, 1193–1196.

Korona, R., Nakatsu, C.H., Forney, L.J. and Lenski, R.E. (1994). Evidence for multiple adaptive peaks from populations of bacteria evolving in a structured habitat. Proceedings of the National Academy of Sciences of the United States of America, 91, 9037–9041.

Kryazhimskiy, S., Tkačik, G. and Plotkin, J.B. (2009). The dynamics of adaptation on correlated fitness landscapes. Proceedings of the National Academy of Sciences of the United States of America, 106, 18638–18643.

Lee, M.C., Chou, H.H. and Marx, C.J. (2009). Asymmetric, bimodal trade-offs during adaptation of Methylobacterium to distinct growth environments. Evolution, 63, 2816–2830.

Lenski, R.E. and Travisano, M. (1994). Dynamics of adaptation and diversification: A 10,000-generation experiment with bacteria. Proceedings of the National Academy of Sciences of the United States of America, 91, 6808–6814.

Lenski, R.E., Rose, M.R., Simpson, S.C. and Tadler, S.C. (1991). Long term experimental evolution in *Escherichia coli*. I. Adaptation and divergence during 2,000 generations. American Naturalist, 138, 1315–1341.

Lunzer, M., Miller, S.P., Felsheim, R. and Dean, A.M. (2005) The biochemical architecture of an ancient Adaptive Landscape. Science, 310, 499–501.

MacLean, R.C. and Bell, G. (2003). Divergent evolution during an experimental adaptive radiation. Proceeding of the Royal Society London B, 270, 1645–1650.

MacLean, R.C., Bell, G. and Rainey, P.B. (2004). The evolution of a pleiotropic fitness tradeoff in *Pseudomonas fluorescens*. Proceedings of the National Academy of Sciences of the United States of America, 101, 8072–8077.

McDonald, M.J., Gehrig, S.M., Meintjes, P.L., Zhang, X.X. and Rainey, P.B. (2009). Adaptive divergence in experimental populations of Pseudomonas fluorescens. IV. Genetic constraints guide evolutionary trajectories in a parallel adaptive radiation. Genetics, 183, 1041–1053.

McDonald, M.J., Cooper, T.F., Beaumont, H.J. and Rainey, P.B. (2011). The distribution of fitness effects of new beneficial mutations in *Pseudomonas fluorescens*. Biology Letters, 7, 98–100.

Notley-McRobb, N. and Ferenci, T. (1999). Adaptive mgl-regulatory mutations and genetic diversity in glucose-limiting *Escherichia coli* populations. Environmental Microbiology, 1, 33–43.

Orr, H.A. (2005). Theories of adaptation: what they do and don't say. Genetica, 123, 3–13.

Ostrowski, E.A., Rozen, D.E. and Lenski, R.E. (2005). Pleiotropic effects of beneficial mutations in *Escherichia coli*. Evolution, 59, 2343–2352.

Papadopoulos, D., Schneider, D., Meier-Eiss, J., Arber, W., Lenski, R.E. and Blot, M. (1999). Genomic evolution during a 10,000-generation experiment with bacteria. Proceedings of the National Academy of Sciences of the United States of America, 96, 3807–3812.

Pelosi, L., Kühn, L., Guetta, D., Garin, J., Geiselmann, J., Lenski, R.E., et al. (2006). Parallel changes in global protein profiles during long-term experimental evolution in *Escherichia coli*. Genetics, 173, 1851–1869.

Rainey, P.B. and Travisano, M. (1998). Adaptive radiation in a heterogeneous environment. Nature, 394, 69–72.

Rokyta, D.R., Joyce, P., Caudie, S.B. and Wichman, H.A. (2005). An empirical test of the mutational landscape model of adaptation using a single-stranded DNA virus. Nature Genetics, 37, 441–444.

Rozen, D.E. and Lenski, R.E. (2000). Long-term experimental evolution in Escherichia coli. VIII. Dynamics of a balanced polymorphism. American Naturalist, 155, 24–35.

Rozen, D.E., de Visser, J.A. and Gerrish, P.J. (2002). Fitness effects of fixed beneficial mutations in microbial populations. Current Biology, 12, 1040–1045.

Rozen, D.E., Habets, M.G.J.L., Handel, A. and de Visser, J.A.G.M. (2008). Heterogeneous adaptive trajectories of small populations on complex fitness landscapes. PLoS One, 3, e1715.

Sanjuán, R., Cuevas, J.M., Furió, V., Holmes, E.C. and Moya, A. (2007) Selection for robustness in mutagenized RNA viruses. PLoS Genetics, 3, e93.

Schluter D. (2000) The Ecology of Adaptive Radiation. Oxford University Press, Oxford.

Schrag, S.J. and Perrot, V. (1996). Reducing antibiotic resistance. Nature, 381, 120–121.

Travisano, M. and Lenski, R.E. (1996). Long-term experimental evolution in *Escherichia coli*. IV. Targets of selection and the specificity of adaptation. Genetics, 143, 15–26.

Travisano, M., Vasi, F. and Lenski, R.E. (1995a). Long-term experimental evolution in Escherichia coli III. Variation among replicate populations in correlated responses to novel environments. Evolution, 49, 189–200.

Travisano, M., Mongold, J.A., Bennett, A.F. and Lenski, R.E. (1995b). Experimental tests of the roles of adaptation, chance, and history in evolution. Science, 267, 87–90.

Treves, D.S., Manning, S. and Adams, J. (1998). Repeated evolution of an acetate-crossfeading polymorphism in long-term populations of *Escherichia coli*. Molecular Biology and Evolution, 15, 789–797.

Turner, P.E., Souza, V. and Lenski, R.E. (1996). Tests of ecological mechanisms promoting the stable coexistence of two bacterial genotypes. Ecology, 77, 2119–2129.

Velicer, G.J. and Lenski, R.E. (1999). Evolutionary trade-offs under conditions of resource abundance and scarcity: Experiments with bacteria. Ecology, 80, 1168–1179.

Velicer, G.J., Raddatz, G., Keller, H., Deiss, S., Lanz, C., Dinkelacker, I., et al. (2006). Comprehensive mutation identification in an evolved bacterial cooperator and its cheating ancestor. Proceedings of the National Academy of Sciences of the United States of America, 103, 8107–8112.

Weinreich, D.M., DeLaney, N.F., DePristo, M.A. and Hartl, D.L. (2006). Darwinian evolution can follow only very few mutational paths to fitter proteins. Science, 312, 111–114.

Weinreich, D.M., Watson, R.A. and Chao, L. (2005). Sign epistasis and genetic constraint on evolutionary processes. Evolution, 59, 1165–1174.

Whitlock, M.C., Philips, P.C., Moore, F.B-G. and Tonsor, S.J. (1995). Multiple fitness peaks and epistasis. Annual Review of Ecology and Systematics, 26, 601–629.

Woods, R.J., Barrick, J.E., Cooper, T.F., Shrestha, U., Kauth, M.R. and Lenski, R.E. (2011) Second-order selection of evolvability in a large *Escherichia coli* population. Science, 331, 1433–1436.

Wright, S. (1932). The roles of mutation, inbreeding, crossbreeding and selection in evolution. Proceedings of the Sixth Annual Congress of Genetics, 1, 356–366.

Wright, S. (1988). Surfaces of selective value revisited. American Naturalist, 131, 115–123.

CHAPTER 12

How Humans Influence Evolution on Adaptive Landscapes

Andrew P. Hendry, Virginie Millien, Andrew Gonzalez, and Hans C. E. Larsson

12.1 Introduction

Most organisms must be reasonably well adapted for their local environments—given that they persist in those environments. Environmental change, however, can alter the relationship between organismal phenotypes and their fitness. Many organisms will be less well adapted for the altered environments and are likely to experience reduced fitness that could lead to population declines and perhaps extirpation or extinction. Other organisms might be well adapted for the altered environments, such as weedy plant species in disturbed habitats. For those species where the effects are negative, the ultimate impact might be ameliorated or arrested through the improved adaptive matching of phenotypes to environments through habitat choice, migration, maternal effects, phenotypic plasticity, and evolution (Bürger and Lynch 1995; Gomulkiewicz and Holt 1995; Orr and Unckless 2008; Chevin et al. 2010). The ultimate outcome will thus depend on the rate and severity of the environmental changes relative to the rate and effectiveness of adaptive responses. These eco-evolutionary dynamics (Kinnison and Hairston 2007) should be particularly important in the case of human-caused environmental disturbances, because these disturbances are particularly rapid and dramatic (Vitousek et al. 1997; Palumbi 2001). Indeed, such disturbances are a frequent cause of both extirpation/extinction (Pimm et al. 1995; Hughes et al. 1997; Blackburn et al. 2004) and phenotypic changes (Stockwell et al. 2003; Hendry et al. 2008; Darimont et al. 2009).

The foregoing discussion of evolution within a particular population or species (anagenesis) can be expanded to consider evolutionary diversification among populations or species (cladogenesis). In particular, colonization of different environments by offshoots of a single ancestral population should expose the different populations to different selective pressures that can cause their adaptive divergence (Simpson 1944; Simpson 1953; Endler 1986). This adaptive divergence should then contribute to the evolution of reproductive isolation between the different populations, a process now called ecological speciation (Schluter 2000b; Rundle and Nosil 2005). Together, these effects drive the adaptive radiation of a single ancestral population into multiple populations/species adapted to use different environments or resources (Schluter 2000b; Losos 2009). Thus, human-caused environmental changes that influence evolution within populations should also influence diversification among populations.

In this chapter, we conceptualize, organize, and exemplify the various ways in which humans influence evolutionary change by reference to the concept of Adaptive Landscapes. We start with an outline of the particular Adaptive Landscape concept we use as a framework. We then dissect some components of Adaptive Landscapes so as to organize the various ways in which evolution on these landscapes might be altered by environmental disturbances. Finally, we use the categories revealed by this dissection to exemplify and discuss the evolutionary impacts of several types of human-caused environmental disturbance, including invasive species, climate change, hunting and harvesting, habitat fragmentation and loss, and changes in habitat quality. These particular categories were chosen to parallel those commonly delineated in conservation biology for problems faced by natural populations (Groom et al. 2006). Other common types of human influences, such as domestication selection, are not considered except in regard to how they influence evolution on an Adaptive Landscape in nature.

12.2 Our Adaptive Landscape

Starting with Wright (Wright 1932), different versions of "Adaptive Landscapes" have been described (Schluter 2000b; Gavrilets 2004; Chapters 2, 3, 7, 13 and 19 in this volume). The version we employ has its origins in Simpson's (1944, 1953) phenotypic Adaptive Landscape, where different phenotypes have different fitness expectations. These landscapes are rooted in the relationship between individual fitness and individual trait values, with an example shown in Fig. 12.1. The conversion of these individual fitness landscapes into true "Adaptive Landscapes" involves the conversion of individual trait and fitness values to population mean trait and fitness values. That is, Adaptive Landscapes depict the relationship between population mean fitness and population mean trait values, assuming a given variance in those trait values (Lande 1976, 1979; Schluter 2000b; Arnold et al. 2001; Hendry and Gonzalez 2008; Chapters 8, 9, and 13). Adaptive Landscapes will be smoother than their underlying individual fitness landscapes because individual fitness is averaged across the entire population for each possible mean trait value. Adaptive Landscapes of this sort are difficult to formally construct for real environments because doing so requires the specification of mean fitness for each possible mean phenotype—and the entire range of possible phenotypic combinations never exists in nature (see also Chapter 8). Because no estimate of mean fitness is possible for non-existent phenotypes, empirically-constructed Adaptive Landscapes are inevitably only partial. Regardless of this logistical constraint on their formal quantification, Adaptive Landscapes must exist in the sense that particular phenotype combinations will inevitably correspond to particular fitness values under a specific set of environmental conditions.

Adaptive Landscapes are n-dimensional, where n represents the number of phenotypic traits, but are most easily illustrated in two or three dimensions. No matter the number of dimensions, the elevation (height) of the landscape at each phenotypic position is the mean fitness of a population with that mean phenotype. These landscapes tend to have peaks and ridges of high fitness separated by valleys, pits, and moats of low fitness. Using the two-dimensional individual fitness landscape shown in Fig. 12.1 as an illustration, western hemlock and ponderosa pine cones generate reasonably sharp fitness peaks surround by fitness moats, whereas the other cones generate fitness ridges separated by fitness valleys. Adaptive radiation then involves the evolution of different populations or species that come to "reside" on different fitness peaks or ridges, but not in the fitness valleys, pits, or moats (Simpson 1944; Lack 1947; Simpson 1953; Schluter 2000b; Losos 2009; Chapter 13). Fig. 12.2 illustrates this pattern, where the average beak size of Darwin's ground finch species on a given island generally matches what would be expected from the seed types available on that island. Of course, some peaks might never be colonized (i.e. no population specializes there) and some populations might be constrained short of adaptive peaks owing to (for example) high gene flow (Hendry and Taylor 2004; Bolnick and Nosil 2007), developmental and/or functional constraints (Hansen and Houle 2008; Walsh and Blows 2009; Chapters 9 and 13), and temporal environmental fluctuations (Siepielski et al. 2009). Fig. 12.2 is again illustrative in that the match between observed and expected beak sizes is not perfect, and some potential peaks may be unoccupied.

The Adaptive Landscape framework that we adopt goes a step beyond the one just described. The reason is that real Adaptive Landscapes are both frequency- and density-dependent, which is not normally a part of the above framework (Fear and Price 1998; Dieckmann et al. 2004; Rueffler et al. 2006). For example, intraspecific competition dictates that specialization by a population on a given adaptive peak increases competition for that peak, which thereby depresses mean fitness and causes

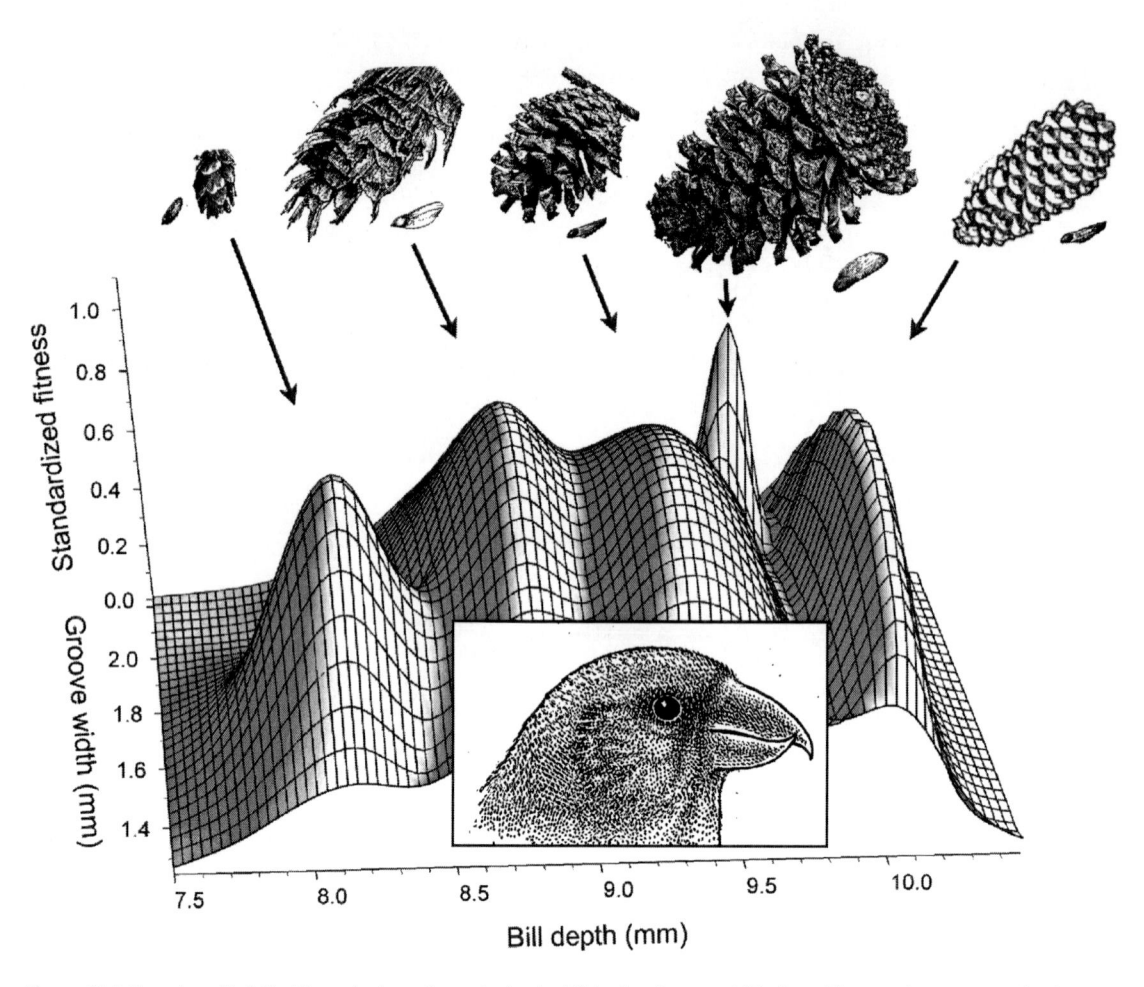

Figure 12.1 The estimated individual fitness landscape for two beak traits (bill depth and groove width) of crossbills in North America, modified from Benkman (2003) and provided by C. Benkman. The first step was to estimate the feeding efficiency of birds with particular beak traits on cones of five different conifer species. Mark-recapture data for wild birds was then used to convert feeding efficiency to survival, which was the measure of individual fitness. The result is a fitness landscape with five peaks, corresponding to the beak characteristics that are best suited for feeding on different conifer cones, which from left to right are western hemlock, Douglas fir, Rocky Mountain lodgepole pine, ponderosa pine, and South Hills lodgepole pine (cones and seeds drawn to scale). Conversion from this individual fitness landscape to a formal Adaptive Landscape would require estimating mean population fitness for hypothetical populations with the corresponding mean trait values (assuming particular trait variances).

the peak to sink. Similarly to density dependence, frequency dependence can alter phenotypic distributions and fitness relationships and thereby also alter Adaptive Landscapes (Fear and Price 1998; Dieckmann et al. 2004; Rueffler et al. 2006; Chapters 7 and 14).

The outcome of adaptive radiation on frequency- and density-dependent (FDD) phenotypic Adaptive Landscapes might be that mean fitness is roughly unity at all phenotypic positions where populations can persist. That is, average individual fitness in stable populations will be near the replacement rate. Mean fitness is then lower at other phenotypic positions to a degree that depends on just how bad those positions are. Thus, fitness peaks are still present on FDD landscapes but they are all approximately the same height with respect to mean population fitness, instead differing in the population sizes they support. Indeed, the Adaptive Landscape in Fig. 12.2 was constructed based

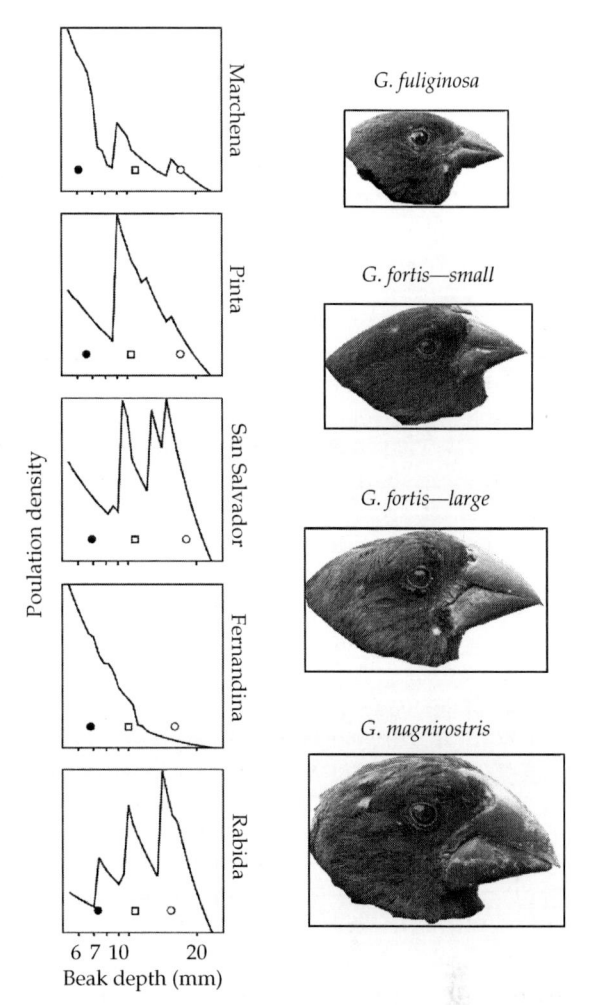

HOW HUMANS INFLUENCE EVOLUTION ON ADAPTIVE LANDSCAPES   183

Figure 12.2 The estimated Adaptive Landscape for beak depth of Darwin's ground finches on five Galapágos islands (provided by D. Schluter). These results and those for more islands are shown in Schluter and Grant (1984). The first step was to calculate the range of seed sizes preferred by ground finches of a given beak depth. The next step was to estimate the biomass of those seeds on each island. The final step was to relate the abundance of seeds of that type to the density of finches of that beak depth that could be supported. The heights of the curves thus represent the density of finches that could be supported for a solitary species with the given beak size. The different symbols below the curves show the average beak sizes of ground finch species found on the same islands: *Geospiza fuliginosa* (closed circle), *Geospiza fortis* (open square), and *Geospiza magnirostris* (open circle). Also shown to scale are representative examples of mature males from each of the first three species, as well as the large and small beak size morphs of *G. fortis* mentioned in the text. These particular images are from birds on Santa Cruz Island. Photos courtesy of Andrew Hendry.

on expected population sizes. Stated another way, a fitness peak that would be higher in the absence of competition will instead hold more individuals in the presence of competition. By contrast, valleys will vary in depth when it comes to mean fitness but will bottom out at zero with respect to the population size they could support—because population persistence is not possible for some phenotypic positions.

Several additional features of Adaptive Landscapes need to be discussed. First, what generates multiple fitness peaks on an Adaptive Landscape? In many cases, different peaks represent alternative habitats or resources (e.g. different host plants). Alternatively, the different peaks could represent alternative phenotypic solutions to problems presented by a single habitat or resource: e.g. alternative jaw structures that have roughly the same performance on a given resource (Alfaro et al. 2005). In this latter case, frequency-dependence comes into play because specialization on a given resource through one phenotypic solution might remove the fitness peaks represented by the other phenotypic solutions for specialization on that resource. The discussion that follows mainly focuses on the first possibility–different peaks are produced by different resources.

Second, a useful distinction might be made between "local" versus "global" Adaptive Landscapes. We suggest a local Adaptive Landscape represents the landscape to which a given lineage in a given area is exposed. A local Adaptive Landscape for a lineage of herbivorous insects might be all plant species within the dispersal range of the members of that lineage, such as all related species on an island or between two mountain ranges. A global Adaptive Landscape might then represent the landscape that is theoretically present across the entire potential geographical range of a lineage—even if that range isn't realized. A global Adaptive Landscape for a lineage of herbivorous insects might be all plant species on Earth. The value of making this local versus global distinction becomes clear through the contemplation of empty adaptive peaks. In the global sense, many peaks that could be occupied by members of a given lineage might remain unoccupied—or might be occupied by a different lineage—simply because the lineage of interest has never encountered that peak. In the present chapter, we usually consider local Adaptive Landscapes but global ones will come into play in the consideration of invasive species.

Third, the movement of populations across Adaptive Landscapes could be the result of genetic change or various forms of non-genetic change: such as phenotypic plasticity, maternal effects, and epigenetics. We will generally discuss the evolutionary component of movement on such landscapes, thereby invoking genetic change, but it is important to recognize the likely importance of non-genetic changes. Indeed, a number of our examples focus specifically on phenotypic change, the genetic contribution to which has not been formally demonstrated. Regardless, most of the concepts we discuss do not depend on genetic versus non-genetic routes to phenotypic change.

12.3 Altering evolution on Adaptive Landscapes

We organize this section around several fundamental ways in which evolution on Adaptive Landscapes can be altered: changes in topography, changes in dimensionality, and phenotypic excursions. (Note that relate to evolution of the Adaptive Landscape as discussed by Hansen in Chapter 13.) By topography, we mean the number, position, gradient, and elevation of surface features such as fitness peaks, ridges, moats, pits, and valleys. By dimensionality, we mean the number of traits on which selection is acting. By phenotypic excursions, we mean changes in the position of a population on a given Adaptive Landscape. These distinctions will be made clear in the following paragraphs, although they are certainly not mutually exclusive. For example, a change in dimensionality can also cause changes in n-dimensional topography and can initiate phenotypic excursions.

12.3.1 Topography

Numbers

One major aspect of topography is the number of surface features. For instance, the number of peaks can change through the emergence of a new peak or the submergence of an old peak. The submergence of a peak might occur if an existing resource disappears, such as through the extinction of a plant species on which a particular insect specializes. The expectation is an initial decline in fitness for the insect owing to emerging maladaptation, which can then cause directional selection toward a new fitness peak (a different plant). If the initial maladaptation is too high or the phenotypic change is too slow, extirpation might result. Alternatively, evolution toward the new peak might permit the population to recover fitness quickly enough to allow persistence, i.e. "evolutionary rescue" (Gomulkiewicz and Holt 1995; Bell and Gonzalez 2009). If the new peak was originally empty, the outcome will be a shift in the population from the lost peak to the new peak. If the new peak was already occupied, the resident population may exclude the "invading" population, the invading population might replace the resident population, or the two populations might merge into a hybrid swarm.

The emergence of a new peak might occur if a new resource is added to the environment, such as a new plant that might be colonized by herbivorous insects. We here distinguish between two cases: the new peak might or might not alter the distinctiveness of existing peaks. In the first case, the new peak might be so close to existing peaks that it eliminates, or at least attenuates, the fitness valley between them. The expectation is a general relaxation of stabilizing selection around the original peaks (and disruptive selection between the peaks) and a resulting spread of phenotypes to occupy the intermediate space (Fig. 12.3a). Possible outcomes could be fusion into a single variable population (if hybridization is possible), competitive exclusion/replacement of one population by the other (if hybridization is not possible), or character displacement to reduce competition (Brown and Wilson 1956; Schluter 2000a). In the second case, a new peak might be positioned far enough away from existing peaks that it does not reduce their distinctiveness. The new peak might then remain unoccupied if it is isolated from occupied peaks by too deep a fitness valley. For instance, biocontrol agents are often so specialized on the intended target species that they rarely evolve the ability to use non-target species (van Klinken and Edwards 2002). Alternatively, some individuals in the original population could have phenotypes close enough to the new peak to experience an increase in fitness. The result can be disruptive selection on the original population that generates phenotypic bimodality, and perhaps eventually, different populations or species specialized on the old and new peaks (Fig. 12.3b). For instance, many native insect species have evolved new host races on introduced plants, while not reducing integrity of the original host races (see section 12.4.1).

Positions

Another major aspect of topography is the position of surface features in phenotypic space. For fitness peaks, positions can change through time when trait combinations that yield the highest local fitness are altered. For example, gradual peak movement can be caused by climate change that shifts optimal reproductive times (discussed later). More abrupt peak movement could be caused by the introduction of a competitor species that favors divergent phenotypes and thereby promotes character displacement. In these cases, the expectation is an initial decline in fitness owing to increasing maladaptation, which generates directional selection and evolution toward the new peak position (Simpson 1944; Simpson 1953; Arnold et al. 2001; Estes and Arnold 2007). Depending on the potential for evolutionary rescue, the population could persist or could decline to the point of extirpation (Bürger and Lynch 1995; Gomulkiewicz and Holt 1995; Orr and Unckless 2008).

Gradient

A third aspect of topography is that the gradient (or slope) of a landscape can range from sharp/steep (the Himalayas) to smooth/shallow (the Pyrenees). A landscape might become smoother if, for example, conditions generally become more benign, such that formerly maladaptive phenotypes no longer suffer so great a disadvantage relative to "optimal" phenotypes. Or smoothing could occur through density and frequency dependence, where specialization on a common resource increases intraspecific competition to the point that selection favors individuals that use less common resources (Dieckmann et al. 2004; Rueffler et al. 2006). Or temporal variation (Siepielski et al. 2009) could mean that the "average" landscape is flatter than that seen in any given year. The expectation if an Adaptive Landscape becomes smoother is that existing populations will experience a decrease in the intensity of directional selection (if they are not at the peak) or stabilizing selection (if they are near the peak). One outcome might be increased drift of the mean phenotype. Another is that populations should become phenotypically more variable—because phenotypes that deviate from the optimum are no longer weeded out to the same extent. Smoothing of the Adaptive Landscape could thus reduce disruptive/divergent selection between populations, potentially initiating fusion/exclusion/replacement as described earlier (Fig. 12.3c). Alternatively, the expectation if an Adaptive Landscape becomes steeper is increased directional (if not at the peak) or stabilizing (if at the peak) selection. This should cause existing populations to increasingly specialize on those peaks, which should promote phenotypic divergence, and reduce the chances of fusion/exclusion/replacement (Fig. 12.3d).

Elevation

A fourth aspect of topography is related to the height and depth of surface features. Following from our earlier description of Adaptive Landscapes, a lower peak (or ridge) should correspond to mean phenotypes with lower growth rates in

Figure 12.3 Changes in the topography of an Adaptive Landscape. All landscapes show mean population fitness (contours) with respect to mean phenotype for two traits (x and y axes). Gray shading shows the distribution of phenotypes. The starting Adaptive Landscape is in the central thick box and has two fitness peaks that are both occupied by populations. Panels (a) and (b) represent an increase in the number of peaks and panels (c) and (d) represent changes in the heights of peaks and the gradients around peaks. In (a), a new peak is added that reduces the distinctiveness of existing peaks and leads to a single variable population. In (b), a new peak is added that does not reduce the distinctiveness of existing peaks and leads to the formation of a new population adapted to the new peak. In (c), the valley between two existing peaks shallows to the point that the two populations are converted into one variable population. In (d), the valley between the two peaks deepens (and the peaks sharpen) so that the two populations are maintained but each becomes less phenotypically variable.

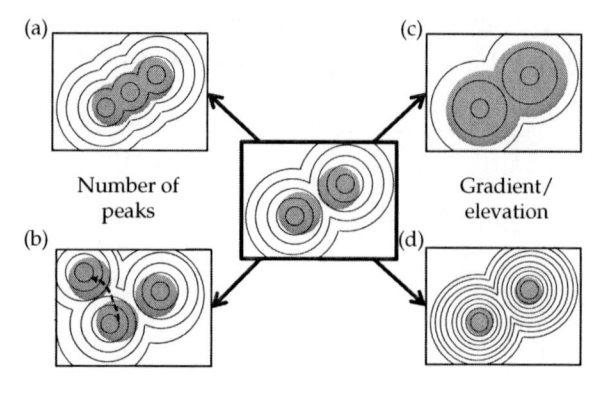

the absence of density dependence or fewer individuals in the presence of density dependence. Lower valleys (or pits or moats) would correspond to mean phenotypes with faster rates of decline. Because lower elevations correspond to smaller population sizes they should also correspond to greater demographic and genetic risks of extirpation, as well as a greater influence of genetic drift. All of these effects could then increase directional selection in any given year and also the temporal variation in selection. Another expectation for a deeper valley is an increased difficulty for populations to shift between peaks. Deeper valleys should also, assuming they remain in the same position relative to peaks of unchanged height, increase the gradient transition to a nearby peak, thus merging into the consequences of gradients discussed in the preceding section.

12.3.2 Dimensionality

The dimensionality of an Adaptive Landscape involves the number of phenotypic traits under selection. These traits need to be at least partially independent (i.e. not too closely linked either genetically or developmentally), otherwise they would represent the same dimension or "module" (Wagner and Altenberg 1996; Hansen 2003; Hansen and Houle 2008; Walsh and Blows 2009; Chapters 9 and 13). Adaptation to different environments (different fitness peaks) will often involve many dimensions, partly because many different selective forces are often present and partly because adaptation to a given selective force may involve many traits. As a (still incomplete) example, different lakes could differ in temperature, predators, parasites, and food types; and adaptation to these factors could involve temperature tolerance, breeding time, antipredator behavior, immune responses, body shape, prey preference, and trophic morphology. Adaptive Landscape dimensionality is expected, and sometimes intended, to change under human influence. In medicine, multidrug cocktails are designed so that the different drugs target very different aspects of a pathogen, and thus require independent mutations for the pathogen to circumvent (Barbaro et al. 2005). In agriculture, the same basic idea is implemented through "toxin stacking" or "pyramiding" of independent insect resistance genes into transgenic crops (Roush 1998; Beckie and Reboud 2009). We will later provide additional examples relating to invasive species and pollution.

Changing landscape dimensionality could have several evolutionary effects. First, it could alter the total distance between existing peaks in n-dimensional phenotype space—e.g. a new dimension of divergent selection between populations already under divergent selection along other dimensions (Fig. 12.4). Second, changing dimensionality could split a single peak into multiple peaks—e.g. a new dimension of divergent selection between populations not already under divergent selection along other dimensions. In both of these cases, one would expect the emergence of novel directional selection and therefore evolution

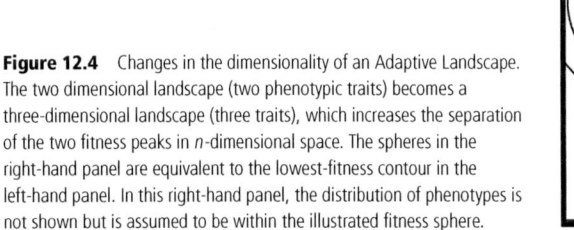

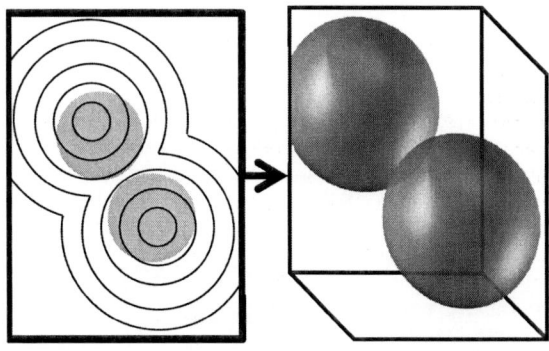

Figure 12.4 Changes in the dimensionality of an Adaptive Landscape. The two dimensional landscape (two phenotypic traits) becomes a three-dimensional landscape (three traits), which increases the separation of the two fitness peaks in n-dimensional space. The spheres in the right-hand panel are equivalent to the lowest-fitness contour in the left-hand panel. In this right-hand panel, the distribution of phenotypes is not shown but is assumed to be within the illustrated fitness sphere.

in new directions. In addition, one would expect an increase in the total strength of divergent selection acting between populations, which should favor increased adaptive divergence and reproductive isolation (Schluter 2000b). It has also been argued that an increase in dimensionality should increase the chances of reproductive isolation even if it doesn't increase the total strength of divergent selection (Nosil et al. 2009). The suggested reason is that more dimensions should increase the number of traits/genes under divergent selection, which should increase the chances that some will contribute to reproductive barriers (Nosil et al. 2009). (Of course, increasing the number of dimensions without increasing total selection should decrease the strength of divergent selection along each dimension, potentially reducing reproductive isolation.)

The discussed points highlight that increasing dimensionality can increase the potential for diversification—but the opposite is also possible. For instance, the new dimension might impose similar selection on populations that were under divergent selection along other axes. This might lead to the evolution of new shared phenotypes that are (for example) favored by females in all environments, potentially reducing reproductive isolation. Also, partial non-independence between dimensions can mean that evolution along a new dimension can cause correlated changes along the other dimensions (Price et al. 2003) that can reduce phenotypic differences. The result could be peak shifts that lead to fusion/exclusion/replacement. Of course, new dimensions could also cause correlated evolution along older dimensions, which thus increases the difference between peaks.

12.3.3 Phenotypic excursions

Here we are referring to situations where phenotypes change for reasons other than alterations to the underlying Adaptive Landscape. As one example, hybrids often have different phenotypes from their parental species (Rieseberg et al. 1999), which places them in new positions on an Adaptive Landscape. A similar effect can attend any change in gene flow between existing populations/species—because gene flow alters average phenotypes (Garant et al. 2007). Phenotypic excursions can also occur through genetic manipulations (domestication, polyploidy, transgenes) or environmentally-induced developmental changes (plasticity, maternal effects, epigenetics). Once phenotypes are altered, directional selection is likely on the new population. This selection may be toward an already occupied peak, perhaps by one of the parental species, which could result in fusion/exclusion/replacement as discussed earlier. Or it might be toward an open adaptive peak (Fig. 12.5). Indeed, hybridization (Rieseberg et al. 2003, Mallet 2007) and plasticity (Fear and Price 1998; Price et al. 2003) are both suggested to be important factors allowing the colonization of new adaptive peaks.

This description of phenotypic excursions invokes a shift in phenotypes on a single, unchanged local Adaptive Landscape. An alternative possibility is that organisms are moved between different local Adaptive Landscapes, such as through

Figure 12.5 Excursions on an Adaptive Landscape. In this case, the Adaptive Landscape does not change but phenotypes are altered (e.g., through hybridization) so that a new population pops up in a new position on the landscape (center panel). Assuming it persists, this new population then evolves toward the nearest fitness peak (right-hand panel).

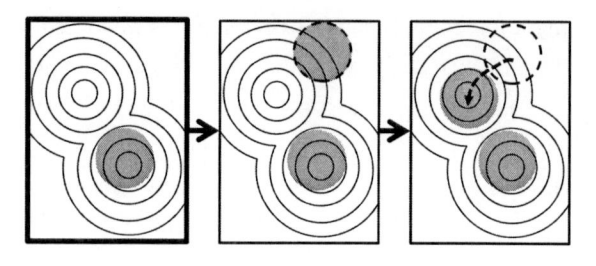

translocations between continents. In this situation, the introduced population can be thought of as representing a phenotypic excursion on the new Adaptive Landscape. It is also important to recognize that although the initial excursion is not caused by changes in the Adaptive Landscape, it will immediately then influence the Adaptive Landscape through changes in phenotypic variation (Whitlock, 1995) and density/frequency-dependence (Rueffler et al. 2006, Chapter 7).

12.4 Human impacts

Having outlined a conceptual framework, we now put it to use in considering the evolutionary effects of different human-caused environmental changes. We start with invasive species, which provide an opportunity to outline, discuss, and exemplify many of the ways in which evolution on Adaptive Landscapes can be altered (henceforth "effects"). We then consider other types of disturbance, including hunting/harvesting, climate change, habitat loss and fragmentation, and changes in habitat quality. Among these influences, hunting/harvesting involves humans directly influencing selection on natural populations, such as through the selective removal of large individuals. For the other influences, the effects of humans on selection are more typically indirect: humans cause environmental change that then causes selection. In each case, we do not try to discuss many potential effects but rather focus in more depth on a few that seem particularly relevant and interesting. The various types of disturbance and effects on evolution do interact and grade into each other but their separate categorization aids conceptualization, organization, and exposition.

12.4.1 Invasive species

One major influence that humans have on the environment is the introduction (translocation) of species to locations where they are not native. Some of these introduced species become invasive; spreading over large areas, attaining high abundance, and having dramatic ecological consequences. They can also have a wide variety of evolutionary effects.

Topography—numbers
From the perspective of native species, an invasive species can sometimes represent a brand new fitness peak on the local Adaptive Landscape. If the fitness valley separating the new peak from already occupied peaks is low enough, the new peak might be colonized by an offshoot from an existing population. As introduced earlier, excellent examples come from native herbivorous insects forming new host races on introduced plants, with examples including *Rhagoletis* colonizing apple trees (Bush 1969; Filchak et al. 2000) and Japanese honeysuckle (Schwarz et al. 2005; Schwarz et al. 2007), soapberry bugs colonizing golden rain trees (Carroll and Boyd 1992; Carroll et al. 1997), and weevils colonizing Eurasian watermilfoil (Sheldon and Jones 2001). In many cases, adaptive genetic differences and reproductive barriers are now present between the new and old host races (see the earlier given references). In such cases, humans have had a positive influence on biodiversity by increasing the number of peaks on local Adaptive Landscapes. Of course, invasive species can also remove adaptive peaks for native species by consuming their original food sources or occupying their preferred habitats. For example, the introduction of red squirrels to Newfoundland in 1963 has apparently caused

the extinction of endemic crossbills, which feed on the same resources (Parchman and Benkman 2002).

Topography—gradient
In some cases, invasive species can smooth fitness landscapes to the point that incipient (or at least "young") species fuse into a single variable population or hybrid swarm. This effect has been argued for beak size morphs of Darwin's finches (photo: Fig. 12.2) exposed to new foods in the Galapágos (Hendry et al. 2006; De León et al. 2011), and also for sympatric benthic/limnetic three-spine stickleback species in a lake recently invaded by crayfish (Taylor et al. 2006; Behm et al. 2010). The basic idea is that a valley between formerly distinct fitness peaks has been erased, or at least become shallower, owing to habitat changes that result from the invasive species. These changes relax selection against intermediate forms, which ultimately converts multimodal phenotypic distributions into unimodal distributions. Humans here have a negative effect on biodiversity by smoothing Adaptive Landscapes and causing the fusion of populations or species formerly specialized on distinct fitness peaks. (Grading into the previously-described effect, the smoothing of a landscape can be caused by the addition of a new fitness peak between two existing peaks—thus eliminating a valley.)

Dimensionality
Invasive species sometimes cause selection on native species that acts along a brand new phenotypic axis, and can thereby change the dimensionality of a local Adaptive Landscape. A clear example is the introduction of predators/parasites to oceanic islands that formerly lacked them. The native species are usually maladapted in the face of these new predators/parasites and a frequent outcome is extinction, as in Hawaiian honeycreepers (Atkinson et al. 2000), Pacific island flightless rails (Steadman 1995), and many bird species on Guam (Fritts and Rodda 1998). In each of these cases, maladaptation and selection were so strong along the new dimension (resistance/tolerance to predators/parasites) that adaptive evolution to the new peak was not rapid enough to forestall extinction. (And we know that adaptation is ultimately possible given that related bird species often coexist with the same predators on the mainland.) In other cases, however, native species persist and adapt following the introduction of predators or parasites. As one example, native predatory snakes of Australia have evolved increased resistance to the toxin found in invasive cane toads (Phillips and Shine 2006). Some snake populations have less exposure to toads than do others, and so divergent selection presumably occurs along this new phenotypic axis.

Excursions
Invasive species undertake, promote, or represent phenotypic excursions. First, invasive species sometimes hybridize with native species, leading to traits that increase invasiveness (Schierenbeck and Ellstrand 2009). Second, invasive species can promote hybridization between native species. For example, the introduction of Japanese honeysuckle to North America caused the evolution of a new host race of *Rhagoletis* flies through hybridization of two native *Rhagoletis* species (Schwarz et al. 2005, 2007). Third, translocation to a new place in the world (i.e. a new local Adaptive Landscape) means that the introduced species will often be positioned away from fitness peaks. We might therefore expect maladaptation to be often so high as to prevent success. Consistent with this causal explanation (as well as with other explanations), few introduced species are ever seen in the wild, few of those ever become self-sustaining, and few of those ever become fully invasive (Williamson and Fitter 1996; Sax and Brown 2000). Introductions that are successful might represent those situations where the introduced species was reasonably near an adaptive peak to start with. Indeed, some invasive species appear pre-adapted to resist new enemies in the introduced range—presumably because of previous adaptation to similar enemies in the native range (Strauss et al. 2006; Ricciardi and Ward 2006). It seems unlikely, however, that introduced species will ever fall perfectly on a fitness peak in the new local Adaptive Landscape—and so directional selection and contemporary evolution is always expected. Indeed, many of our best examples of contemporary (rapid) evolution come from

invasive species (Reznick and Ghalambor 2001; Stockwell et al. 2003; Cox 2004). Another type of excursion occurs when a native organism (e.g. an herbivorous insect) recognizes an invasive organism (e.g. plant) as a resource and therefore tries to use it. If the native species can survive on and adapt to the new host, then a new form might result (Bush 1969; Thomas et al. 1987; Carroll and Boyd 1992; Singer et al. 1993), potentially enhancing diversification. In some cases, however, native organisms might not be able to survive on the new resource—even though they don't discriminate against it (Chew 1980). This "evolutionary trap" (Schlaepfer et al. 2005) could potentially reduce fitness of the native species—unless the resulting selection can eliminate genetic variation for attraction to the new resource.

Multifarious effects
As noted earlier, our different categories of effects are not mutually exclusive, and here we have an opportunity for illustration. Consider the situation where continental organisms are introduced to oceanic islands where native flora and fauna are relatively depauperate. This situation can be considered an excursion (as above) coupled with changes in dimensionality (fewer traits under selection), numbers of fitness peaks (fewer types of resources), and the overall gradient (smoother transitions between peaks and valleys owing to fewer competitors). In combination, these effects can lead to relaxed selection on some traits, in the sense that phenotypic variation no longer matters so much for fitness (Lahti et al. 2009). For example, weaverbirds native to Africa often have their nests parasitized by cuckoos. When these weaverbirds are introduced to islands without cuckoos, the selective pressure to reject foreign eggs is removed. Traits that increase the potential for egg rejection then can be lost quickly (if they are costly) or can be retained for long periods (if they are nearly neutral). As a putative example of each, Village Weaverbirds introduced to islands without cuckoos have lost egg coloration patterns that facilitate the recognition of foreign eggs (Lahti 2005), but have not lost the ability to recognize foreign eggs experimentally placed in their nests (Lahti 2006).

12.4.2 Climate change

Past evolution has accomplished dramatic adaptation to different climates: penguins breed in the dead of winter in the Antarctic and on baking lava in the Galapágos! Humans, however, have recently accelerated the pace of climate change, which may hamper the ability of evolution to rescue populations from a maladapted state. Although climate change could have a number of influences on Adaptive Landscapes, we here focus on just one—peak shifts for reproductive timing—so as to enable a detailed discussion of a single particularly likely effect.

Topography—position
Organisms generally breed when conditions are most suitable for reproductive success. Here we focus on altricial birds in temperate environments, which typically time their breeding so that chick rearing coincides with high insect abundance. Peak insect abundance corresponds, in turn, to when leaves flush out on plants (budburst), which is largely determined by local temperature. Warmer temperatures cause earlier budburst, which causes earlier peak insect abundance, which should favor earlier egg laying by birds (Both et al. 2009). Fitting this expectation, the reproductive timing of many organisms has been advancing over the past 10–50 years (Parmesan 2007). Many of these changes are strongly influenced by phenotypic plasticity (individuals reproduce earlier in warmer years), but some organisms are constrained in their ability to mount sufficient plastic responses (Visser 2008; Phillimore et al. 2010). The reproductive timing of these organisms should therefore become maladaptive with climate change and they might therefore decline in abundance. In this case, directional selection will favor earlier reproduction, and adaptive evolution would be expected, and evolutionary rescue might result.

When it comes to the plastic matching of reproductive timing to appropriate conditions, non-migratory organisms should have it the easiest—owing to the local availability of proximate cues, such as temperature or tree phenology (Réale et al. 2003; Charmantier et al. 2008). Migratory organisms, however, have access to such cues only after arriving at the site of reproduction. The time of this arrival is strongly influenced by the time of departure from wintering grounds, which is generally cued by photoperiodic responses that evolved under past selection to produce appropriate timing. Under climate warming, then, migratory organisms might arrive too late to optimally time their reproduction. These possibilities have been studied in detail for pied flycatchers that breed in Europe but winter in Africa. Both et al. (2006) showed that locations where caterpillar dates have advanced the most now have flycatcher populations that are showing the greatest numerical declines. The birds in those populations seem to be arriving too late to take full advantage of the caterpillar bonanza (Both et al. 2006). For flycatchers, then, the fitness peak represented by reproductive time has advanced to the point that evolved departure times from Africa are now maladaptive. Selection should therefore favor earlier departure from Africa, the evolution of which might or might not arrest the population declines.

Although the adaptive evolution of timing events is therefore expected in response to climate change, this has proven very difficult to unequivocally demonstrate in most cases (Bradshaw and Holzapfel 2008). That is, apparently adaptive phenotypic changes are often documented (Parmesan 2007) but their genetic basis has remained uncertain because standard methods (e.g. common garden experiments) are difficult to implement in a temporal context and more recent methods (e.g. animal model comparisons of "breeding values") have proven unreliable (Hadfield et al. 2010). However, a few clear examples are known. For instance, the photoperiodic response of pitcher plant mosquitoes has clearly evolved in response to warming conditions (Bradshaw and Holzapfel 2001).

We focused in detail on reproductive timing shifts in response to warming. However, peak shifts in reproductive timing are also likely as a result of other aspects of climate change and other types of traits might evolve in response to warming. As an example of the first alternative, changes in precipitation that increase droughts have advanced the reproductive timing of some plants (Franks et al. 2007). As an example of the second alternative,

warmer ocean temperatures are selecting for greater temperature tolerance in some fish species (Pörtner and Knust 2007), although evolutionary responses have not yet been examined. In addition, body size is declining in many bird populations experiencing warmer conditions, perhaps because smaller bodies are favored under warmer temperatures (i.e. Bergmann's Rule) (Van Buskirk et al. 2010).

12.4.3 Hunting and harvesting

The acceleration of phenotypic changes in human-disturbed populations (Hendry et al. 2008) is greatest when the humans act as "predators" through hunting or harvesting (Darimont et al. 2009). The likely reason is that hunting and harvesting directly selects on phenotypes rather than acting indirectly through environmental change. Among the numerous potential selective effects of hunting and harvesting, perhaps the most common is that larger individuals are more likely to be removed from the population (Fenberg and Roy 2008). This change in selection could have several evolutionary effects.

Topography—peak position
It is often assumed that natural selection favors large body size (Kingsolver and Pfennig 2004). The data supporting this assumption come largely from studies considering a single component of fitness, such as fecundity or mating success. However, positive selection acting through some fitness components will generally be offset by negative selection acting through other fitness components, such as when faster growing individuals suffer higher mortality, reduced performance, or reduced structural integrity (Arendt 1997; Billerbeck et al. 2001; DiBattista et al. 2007). Thus, average body size in most natural populations is probably in the general vicinity of a fitness peak and is therefore under stabilizing selection owing to offsetting fitness components (Fig. 12.6). What hunting/harvesting often does is to increase the component of selection acting against large size (Carlson et al. 2007; Olsen and Moland 2010) (Fig. 12.6). As expected, then, hunting/harvesting generally causes shifts toward smaller body sizes (Fig. 12.6), with clear examples coming from ungulates (Coltman et al. 2003), plants (Law and Salick 2005), and fish (Sharpe and

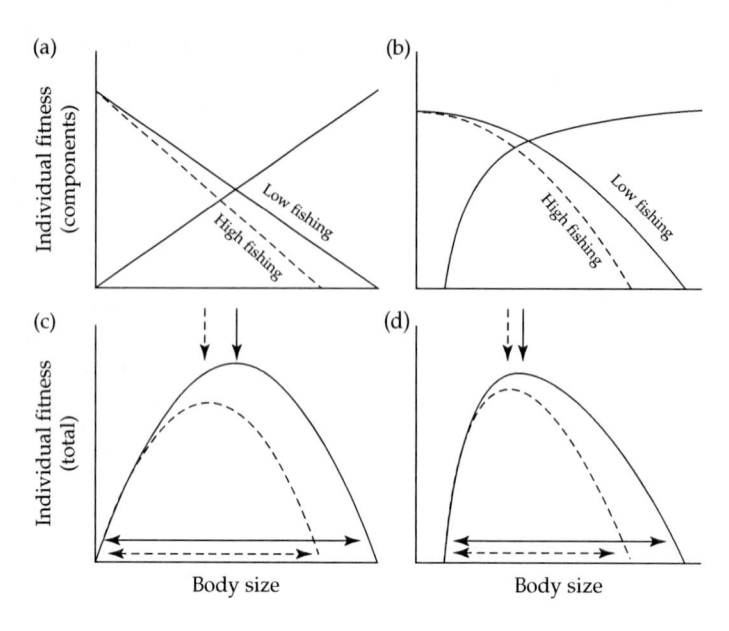

Figure 12.6 Peak movement caused by changes in one of two offsetting fitness components, exemplified here by a consideration of body size evolution with low (solid lines) and with high (dashed lines) fish harvesting. Panel (a) shows one linear fitness component favoring large body size and a second linear fitness component disfavoring large size. Total fitness for an individual with a given trait value is the product of the two functions and is shown in (c). These individual fitness functions will be closely related to the Adaptive Landscape, although the latter will be smoother. Fish harvesting is represented as an increasing strength of selection against large fish (dashed lines) and the outcome is a shift in the fitness peak to a smaller body size (compare vertical arrows) and an increasing strength of stabilizing selection (compare horizontal arrows). Panels (b) and (d) show the same effects but for non-linear fitness functions.

Hendry 2009). These phenotypic shifts can be gradual and incremental, consistent with the expectation of ongoing movement of the fitness peak due to perennial removal of the largest individuals within any given generation. At some point, however, this shrinkage must stop given that organisms cannot forever evolve smaller size. This end of directional evolution could occur through extinction, as has been argued for many prehistoric megafauna that were hunted by humans (Alroy 2001). Or it could occur if hunting/harvesting is only desirable, profitable, or legal when body sizes or abundances are above a certain level. In essence, peak movement here stops because the human induced contribution to selection stabilizes. Alternatively, peak movement might stop if natural selection against smaller body size gets proportionally stronger as average sizes get smaller (Fig. 12.6).

Topography—gradient
Another expected outcome of increased selection against large body size might be a reduction in body size variance. The reason is that an increase in selection will generally sharpen a fitness peak (Fig. 12.6), thereby more assiduously removing individuals that deviate too much from the peak and causing a reduction in population variance. Few studies have tested for such effects, but Olsen et al. (2009) showed that variance in body size decreased over 85 years for Atlantic cod harvested along the Norwegian Skagerrak coast (Fig. 12.7). Although this trend is consistent with the previously presented hypothesis, it could also be the result of other factors (Olsen et al. 2009). Moreover, the data are for juvenile cod and so harvest-based selection would have to be having correlated effects on early-life history stages.

12.4.4 Habitat loss and fragmentation

One frequent influence of humans is to decrease the area of, and connectivity between, patches of suitable habitat. This habitat loss and fragmentation has a number of immediate demographic effects,

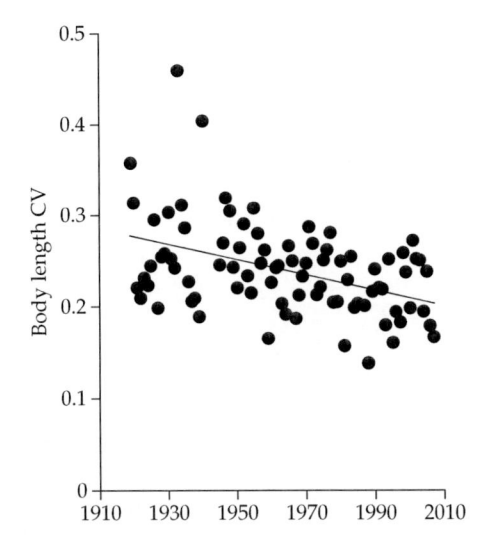

Figure 12.7 Decreases in body size variation in an Atlantic cod population under harvesting. Shown are coefficients of variation for body size in each year, except for 1940–1944, when sample sizes were very small. The study is Olsen et al. (2009) and the values shown here were calculated from data provided by E. Olsen.

such as reducing population sizes and increasing population fluctuations, both of which can increase the chances of extirpation (Andrén 1994; Gonzalez et al. 1998; Fahrig, 2003). Habitat loss and fragmentation also have well known genetic effects, such as reduced genetic variation within populations, increased genetic variation among populations, and inbreeding depression (Young et al. 1996; Keller and Waller 2002; DiBattista 2008). We will here go beyond these effects by considering how habitat loss and fragmentation can influence evolution on Adaptive Landscapes.

Excursion

Habitat loss and fragmentation can influence excursions by altering the connectivity and genetic constitution of populations. As one example, the adaptation of some populations is constrained by gene flow between divergent environments (Hendry and Taylor 2004; Bolnick and Nosil 2007; Garant et al. 2007). Reduced connectivity would reduce this maladaptive gene flow and perhaps allow greater adaptive divergence. As an exemplar, Timema walkingsticks generally adapt to different host plants, but their success in doing so is compromised by gene flow between adjacent populations on different host plants (Nosil and Crespi 2004; Bolnick and Nosil 2007). In a fortuitous "natural" experiment, new road construction separated two formerly contiguous patches of different host plants, which decreased gene flow and resulted in increasing adaptive divergence of the resident walkingstick populations (Nosil 2009). Alternatively, decreased connectivity might sometimes hinder adaptive divergence by reducing the spread of advantageous mutations (Garant et al. 2007). On the flip side, increased connectivity (e.g. translocations or corridors) can increase gene flow and hybridization, initiating their attendant excursionary effects (see earlier).

Topography—peak position

A reduction in the amount of suitable habitat ("habitat area") could alter the position of fitness peaks. Smaller areas, for instance, will usually have fewer total resources, fewer predators, and fewer interspecific competitors, differences repeatedly shown to influence body size evolution. On the time span of millennia, sea-level changes that converted large landmasses into smaller ones were frequently accompanied by the evolution of body size (Millien and Damuth 2004; Millien 2006). We might expect similar effects when humans reduce the habitat area available to contemporary populations. Indeed, rodents introduced from the mainland to small islands often undergo a variety of morphological changes—although the type of change is not easily predictable (Pergams and Ashley 2001). In addition, decreased connectivity and habitat area could select for decreased migration or dispersal if mortality risks or energy consumption increase during the movement phase. Often cited in support of this assertion is the reduction of dispersal ability of organisms that colonize small and remote islands (Cody and Overton 1996). In the context of human influences, at least some plants evolve reduced dispersal in fragmented urban environments (Cheptou et al. 2008). Similarly, butterflies appear to have evolved reduced dispersal when their habitats become increasingly fragmented (Thomas et al. 1998; Hill et al. 1999; Schtickzelle et al. 2006). Interestingly, at least one of these butterfly population also appears to have evolved increased ability to survive during dispersal (Schtickzelle et al.

2006). Despite the intuitive appeal of the above argument, increased fragmentation could select for increased dispersal under some conditions (Heino and Hanski 2001)—and a putative example has been advanced (Taylor and Merriam 1995).

Topography—peak number
Reductions in habitat area could reduce the number of local fitness peaks because smaller areas tend to have less diverse habitats and resources. Consistent with this idea, larger islands tend to hold (and generate) a greater diversity of organisms within a given adaptive radiation (Lack 1947; Losos and Schluter 2000; Losos 2009). Moreover, these effects could generate feedbacks if "diversity begets diversity" (Whittaker 1977; Emerson and Kolm 2005; Forbes et al. 2009). That is, reductions in diversity owing to fewer fitness peaks could, in turn, further reduce the number of fitness peaks.

12.4.5 Habitat quality

Although changes in habitat area (one consequence of habitat loss and fragmentation—see earlier) and quality (considered here) grade into each other to some extent, we keep them separate because habitat quality has some additional interesting properties that warrant separate discussion. For the sake of exposition, we consider habitat area as the spatial spread of a resource, holding constant the amount of resource per unit area, and habitat quality to represent the amount or quality of a resource per unit area. Examples of the latter might include a change in the amount of available nutrients (e.g. fertilizer) or food (e.g. additions or depletions) or the presence of contaminants (e.g. pollution) in a given location.

Topography—elevation 1
Changes in habitat quality could alter the height of fitness peaks, leading to several possible outcomes. Most obviously, decreasing habitat quality (e.g. food) could cause a peak to sink so much as to disappear—instigating any of the aforementioned consequences of peak loss. It is also possible that a peak might sink to some degree without entirely disappearing—and the most likely consequence would be a decrease in population density. This decrease could then have any of the well-known effects of small population size discussed earlier, as well as several other evolutionary effects. First, populations could change from being density regulated to density unregulated, or vice versa, which could cause evolution along the classic r–K (fast–slow) life history continuum. For example, guppies translocated from high-resource to low-resource environments (which also differ in predation intensity) rapidly diverge from r-type life histories toward K-type life histories, such as later age at maturity, larger and fewer offspring, and lower reproductive effort (Reznick et al. 1997; Gordon et al. 2009). Second, the particular factor causing density regulation could change (e.g. nesting sites versus food availability), thus altering selection on traits that improve competitive ability for the different resources.

Dimensionality
New contaminants reduce habitat quality and can expose organisms to brand new selective forces that require brand new adaptations. Many examples come from human pathogens evolving in response to antibiotics or antivirals (Bergstrom and Feldgarden 2007), insect pests evolving in response to chemical or transgenic insecticides (Whalon et al. 2008), and weeds evolving in response to herbicides (Heap 1997). Numerous examples also come from more "natural" situations, such as adaptation to toxic mine tailings by plants (Jain and Bradshaw 1966). More recent examples include adaptation of oligochaete worms to industrial cadmium pollution (Levinton et al. 2003) and Poecilid fish to barbasco toxicants used during indigenous religious rituals (Tobler et al. 2011). Evidence also exists that this increased dimensionality increases the potential for speciation between populations in contaminated versus non-contaminated environments. For example, adaptation has led to the evolution of several reproductive barriers between plants on toxic mine tailings and on those in adjacent pasture (McNeilly and Antonovics 1968).

Topography—elevation 2
All of the previous examples of adaptation to contaminants come from situations where the populations persisted, perhaps because evolution rescued them from a maladaptive state that would have

caused extinction. But evolutionary rescue is not inevitable given that contaminants have certainly caused many extirpations (Bradshaw 1991). Theoretical work suggests that the potential for evolutionary rescue depends on the novelty and severity of the contaminant, initial population size, generation length, and the presence and supply of genetic variation (Bürger and Lynch 1995; Gomulkiewicz and Holt 1995; Orr and Unckless 2008). In the laboratory, Bell and Gonzalez (2009) confirmed that evolutionary rescue in the face of a contaminant (salt) was more likely when the focal populations (yeast) were larger. This effect arose because larger populations had a greater supply of new mutations, increasing their potential for rapid and effective adaptive evolution. More work on the factors influencing evolutionary rescue is clearly needed given that few potential contributors have been examined and these only in the laboratory.

Excursions
Both of our suggested contributors to habitat quality (resource levels and contaminants) can have plastic effects on organisms that represent excursions. For instance, low resource levels or new contaminants could lead to slower individual growth and smaller body sizes. These plastic changes could be adaptive (e.g. smaller individuals require less food) or a simple maladaptive consequence of insufficient food or metabolic problems. In the maladaptive case, a potential consequence is genetic accommodation, where evolution increases body size to compensate for the plastic reduction (Grether 2005) and the original peak is again ascended. Alternatively, plastic excursions could place populations in the domain of attraction of different peaks (Price et al. 2003), with the usual potential for fusion, exclusion, or replacement. Stress created by low food or contaminants could also increase trait variances by inducing developmental instabilities (Hoffmann and Merilä 1999). These variants might then be weeded out by selection, causing phenotypes to contract again around the original peak, or it could cause part of the population to come into the domain of attraction of new fitness peaks. Finally, reduced habitat quality could increase hybridization rates, such as when eutrophication of Lake Victoria degraded the sexual signals that kept species of cichlid fishes separate and thereby lead to a hybrid swarm (Seehausen et al. 1997; Seehausen et al. 2008).

12.5 What have we gained?

Our chapter can serve several purposes. First, we hope it shows how human disturbances can be used to study Adaptive Landscapes. It is normally hard to characterize Adaptive Landscapes in nature because one can't measure fitness for phenotypes that don't normally exist. Given that undisturbed populations are probably reasonably near to adaptive peaks, we have very little understanding of fitness away from those peaks. Human disturbances, however, frequently displace adaptive peaks away from phenotypes or phenotypes away from adaptive peaks. The variety of mismatches that result can help to fill out the space of phenotypic possibility and thereby assess fitness across the Adaptive Landscape. Human disturbances can thus help us to construct and understand Adaptive Landscapes in a level of detail that isn't otherwise possible.

Second, we hope our chapter shows how situations of conservation concern fit into a unifying framework of evolutionary theory. This realization suggests the value of deploying the range of predictive tools based on the Adaptive Landscape to better understand and ameliorate fitness declines in species of concern. For example, measurements of selection acting on populations provide an indication of the shape of the local fitness surface experienced by the population (Arnold et al. 2001; Estes and Arnold 2007). Coupled with information on genetic variances and covariances among traits, these estimates can be used to predict evolutionary trajectories (Chapter 13). These predictions can then be updated as the population moves across the landscape—or the landscape changes.

But does the Adaptive Landscape generate any truly novel perspectives for conservation biology? For instance, a skeptic might argue that the specific examples we presented are already well understood, and so no additional insight was gained by interpreting them in an Adaptive Landscape perspective. Indeed, we chose these examples specifically because they are well understood, making it easier to show how they can fit into an Adaptive

Landscape perspective. The hope was to inspire confidence in the Adaptive Landscape perspective, which could then motivate its use in situations that are not already well understood.

In addition to its potential for conceptual organization, the Adaptive Landscape perspective does force at least one novel consideration into conservation biology. At present, most practitioners recognize that human influences cause mismatches between existing phenotypes and "optimal" phenotypes, and these mismatches then compromise fitness, population size, and persistence. It is also often recognized that these mismatches must be reduced in order to achieve self-sustainability. Attempts are therefore made to determine optimal phenotype under new conditions (e.g. Phillimore et al. 2010), and the goal then becomes to facilitate movement of phenotypes to the new optima (Stockwell et al. 2003).

The additional perspective brought by Adaptive Landscapes is an emphasis on alternative phenotypic trajectories that disturbed populations could follow toward a new fitness peak. For instance, patterns of genetic covariance among traits can cause a curved trajectory through phenotypic spaces toward the new peak (Schluter 2000b; Arnold et al. 2001). In addition, the persistence of a population will depend on trends in its mean fitness as it moves toward the new peak (Gomulkiewicz and Holt 1995; Bell and Gonzalez 2009). A particular trajectory, perhaps the shortest one to the new peak, might lead the population through a valley of very low fitness, potentially compromising conservation efforts. But perhaps the same Adaptive Landscape has a ridge of higher fitness connecting to the new peak along a less direct phenotypic path. Or maybe an intervening valley of low fitness is very narrow, such that the demographic cost of moving through it would not last very long: maybe it is better to jump directly into the chilly lake than to wade in slowly. In all of these cases, characterization of the Adaptive Landscape would aid estimates of the population sizes necessary to transit successfully along various routes.

Armed with information about the Adaptive Landscape, efforts could perhaps be made to facilitate the movement of phenotypes along the optimal route to adaptation (although we won't here go into the various ways this could be done—e.g. Stockwell et al. 2003). An analogy can be found in the consideration of corridors for dispersal or migration. For instance, it has become common to use graph theory (Pinto and Keitt 2009) or circuit theory (McRae et al. 2008) to determine the "least-cost" path(s) for organisms (akin to the above ridge of high fitness), while also considering their behavior tendencies (akin to the bias induced by genetic covariance). With this valuable precedent, perhaps the trajectory-based thinking that derives from Adaptive Landscapes could be rather quickly and smoothly incorporated into conservation biology.

12.6 Summary and conclusions

Humans are currently the world's greatest evolutionary force (Palumbi 2001; Stockwell et al. 2003; Hendry et al. 2008; Darimont et al. 2009)—and the reasons are several. First, humans dramatically alter the environment, such as through urbanization, pollution, translocations, and climate change. Second, humans directly select on some populations, such as through hunting, harvesting, and agriculture. Third, humans sometimes directly induce phenotypic change. Our thesis has been that these various impacts can be organized and evaluated through the lens of the Adaptive Landscape concept.

We first outlined three basic ways in which evolution can be altered on Adaptive Landscapes: through changes in topography, changes in dimensionality, and phenotypic excursions. Changes in topography involve the numbers, positions, gradients, and elevations of surface features on the landscape, such as peaks and valleys. Changes in dimensionality involve the number of at least partially independent traits that are under selection. Excursions typically involve more or less abrupt changes in the phenotypic position of populations on existing Adaptive Landscapes, such as through plasticity, hybridization, or genetic manipulation. These different types of change can be used to generate predictions for changes in selection and alterations in evolution—assuming the population can persist through the disturbance. These different sorts of alterations have different effects on selection and measurements of selection in natural populations might therefore provide some useful initial insights into how Adaptive Landscapes are changing.

We then considered several basic types of human disturbance and how these might influence topography, dimensionality, and excursions. Invasive species can have all of these classes of effects, either for the invasive species itself or for native species. Climate change will most obviously involve a shift in peak position, such as breeding times under warmer temperatures. Hunting/harvesting will also often involve a shift in peak position, particularly toward smaller and slower growing individuals, and it might also decrease phenotypic variance. Habitat loss and fragmentation will influence numbers and positions of adaptive peaks, and can also influence excursions by altering patterns of gene flow in meta-populations. Finally, a decrease in habitat quality can decrease the heights of fitness peaks and cause Adaptive Landscapes to become smoother. It can also change dimensionality, such as through the introduction of a new contaminant.

The specific examples we provided were chosen to best illustrate a particular way in which humans influence evolution on Adaptive Landscapes. As they are examples, however, they are merely an initial illustration of how the Adaptive Landscape metaphor can be used for organizing our thoughts about how humans influence evolution. They are not meant to be exhaustive or exclusive. In addition, we recognize that the various categories of change we have outlined and illustrated grade into each other and are not entirely independent—complex combinatorial responses are likely. Nevertheless, we feel they illustrate the value of viewing human-induced environmental change in the framework of changes to Adaptive Landscapes, whilst also offering new perspectives for future research.

Acknowledgments

A. H. was supported by the Natural Sciences and Engineering Research Council of Canada. A. G. and H. L. were supported by the Canada Research Chair Program and the Natural Sciences and Engineering Research Council of Canada.

References

Alfaro, M. E., Bolnick, D. I. and Wainwright, P. C. (2005). Evolutionary consequences of many- to-one mapping of Jaw morphology to mechanics in labrid fishes. American Naturalist, 165, E140–E154.

Alroy, J. (2001). A multispecies overkill simulation of the end-Pleistocene megafaunal mass extinction. Science, 292, 1893–1896.

Andrén, H. (1994). Effects of habitat fragmentation on birds and mammals in landscapes with different proportions of suitable habitat: a review. Oikos, 71, 355–366.

Arendt, J. D. (1997). Adaptive intrinsic growth rates: an integration across taxa. Quarterly Review of Biology, 72, 149–177.

Arnold, S. J., Pfrender, M. E. and Jones, A. G. (2001). The Adaptive Landscape as a conceptual bridge between micro- and macroevolution. Genetica, 112–113, 9–32.

Atkinson, C. T., Dusek, R. J., Woods, K. L. and Iko, W. M. (2000). Pathogenicity of avian malaria in experimentally-infected Hawaii Amakihi. Journal of Wildlife Diseases, 36, 197–204.

Barbaro, G., Scozzafava, A., Mastrolorenzo, A. and Supuran, C. T. (2005). Highly active antiretroviral therapy: current state of the art, new agents and their pharmacological interactions useful for improving therapeutic outcome. Current Pharmaceutical Design, 11, 1805–1843.

Beckie, H. J. and Reboud, X. (2009). Selecting for weed resistance: herbicide rotation and mixture. Weed Technology, 23, 363–370.

Behm, J. E., Ives, A. R. and Boughman, J. W. (2010). Breakdown in postmating isolation and the collapse of a species pair through hybridization. American Naturalist, 175, 11–26.

Bell, G. and Gonzalez, A. (2009). Evolutionary rescue can prevent extinction following environmental change. Ecology Letters, 12, 942–948.

Benkman, C. W. (2003). Divergent selection drives the adaptive radiation of crossbills. Evolution, 57, 1176–1181.

Bergstrom, C. T. and Feldgarden, M. (2007) The ecology and evolution of antibiotic-resistant bacteria. In S. C. Stearns and J. C. Koella (eds.) Evolution in Health and Disease. Oxford University Press, Oxford, pp. 124–138.

Billerbeck, J. M., Lankford, T. E., Jr. and Conover, D. O. (2001). Evolution of intrinsic growth and energy acquisition rates. I. Trade-offs with swimming performance in Menidia menidia. Evolution, 55, 1863–1872.

Blackburn, T. M., Cassey, P., Duncan, R. P., Evans, K. L. and Gaston, K. J. (2004). Avian extinction and mammalian introductions on oceanic islands. Science, 305, 1955–1958.

Bolnick, D. I. and Nosil, P. (2007). Natural selection in populations subject to a migration load. Evolution, 61, 2229–2243.

Both, C., Bouwhuis, S., Lessells, C. M. and Visser, M. E. (2006). Climate change and population declines in a long-distance migratory bird. Nature, 441, 81–83.

Both, C., van Asch, M., Bijlsma, R. G., van den Burg, A. B. and Visser, M. E. (2009). Climate change and unequal phenological changes across four trophic levels: constraints or adaptations? Journal of Animal Ecology, 78, 73–83.

Bradshaw, A. D. (1991). Genostasis and the limits to evolution. Philosophical Transactions of the Royal Society B: Biological Sciences, 333, 289–305.

Bradshaw, W. E. and Holzapfel, C. M. (2001). Genetic shift in photoperiodic response correlated with global warming. Proceedings of the National Academy of Sciences of the United States of America, 98, 14509–14511.

Bradshaw, W. E. and Holzapfel, C. M. (2008). Genetic response to rapid climate change, it's seasonal timing that matters. Molecular Ecology, 17, 157–166.

Brown, W. L. and Wilson, E. O. (1956). Character displacement. Systematic Zoology, 5, 49–64.

Bürger, R. and Lynch, M. (1995). Evolution and extinction in a changing environment: a quantitative-genetic analysis. Evolution, 49, 151–163.

Bush, G. L. (1969). Sympatric host race formation and speciation in frugivorous flies of Genus Rhagoletis (Diptera, Tephritidae). Evolution, 23, 237–251.

Carlson, S. M., Edeline, E., Vollestad, L. A., Haugen, T. O., Winfield, I. J., Fletcher, J. M., et al. (2007). Four decades of opposing natural and human- induced artificial selection acting on Windermere pike (*Esox lucius*). Ecology Letters, 10, 512–521.

Carroll, S. P. and Boyd, C. (1992). Host race radiation in the soapberry bug: natural history with the history. Evolution 46, 1052–1069.

Carroll, S. P., Dingle, H. and Klassen, S. P. (1997). Genetic differentiation of fitness-associated traits among rapidly evolving populations of the soapberry bug. Evolution, 51, 1182–1188.

Charmantier, A., McCleery, R. H., Cole, L. R., Perrins, C., Kruuk, L. E. B. and Sheldon, B. C. (2008). Adaptive phenotypic plasticity in response to climate change in a wild bird population. Science, 320, 800–803.

Cheptou, P. O., Carrue, O., Rouifed, S. and Cantarel, A. (2008). Rapid evolution of seed dispersal in an urban environment in the weed *Crepis sancta*. Proceedings of the National Academy of Sciences of the United States of America, 105, 3796–3799.

Chevin, L. M., Lande, R. and Mace, G. M. (2010). Adaptation, plasticity, and extinction in a changing environment: towards a predictive theory. PLoS Biology, 8, 1–8.

Chew, F. S. (1980). Food preferences of Pieris caterpillars (Lepidoptera). Oecologia, 46, 347–353.

Cody, M. L. and Overton, J. M. (1996). Short-term evolution of reduced dispersal in island plant populations. Journal of Ecology, 84, 53–61.

Coltman, D. W., O'Donoghue, P., Jorgenson, J. T., Hogg, J. T., Strobeck, C. and Festa-Bianchet, M. (2003). Undesirable evolutionary consequences of trophy hunting. Nature, 426, 655–658.

Cox, G. W. (2004). Alien species and evolution. Island Press, Washington, DC.

Darimont, C. T., Carlson, S. M., Kinnison, M. T., Paquet, P. C., Reimchen, T. E. and Wilmers, C. C. (2009). Human predators outpace other agents of trait change in the wild. Proceedings of the National Academy of Sciences of the United States of America, 106, 952–954.

De León, L.F., Raeymaekers, J.A.M., Bermingham, E., Podos, J., Herrel, A., and Hendry, A.P. (2011). Exploring possible human influences on the evolution of Darwin's finches. Evolution, 65, 2258–2272.

DiBattista, J. D. (2008). Patterns of genetic variation in anthropogenically impacted populations. Conservation Genetics, 9, 141–156.

DiBattista, J. D., Feldheim, K. A., Gruber, S. H. and Hendry, A. P. (2007). When bigger is not better: selection against large size, high condition and fast growth in juvenile lemon sharks. Journal of Evolutionary Biology, 20, 201–212.

Dieckmann, U., Doebeli, M., Metz, J. A. J. and Tautz, D. (eds.) (2004) Adaptive speciation, Cambridge, Cambridge University Press.

Emerson, B. C. and Kolm, N. (2005). Species diversity can drive speciation. Nature, 434, 1015–1017.

Endler, A. J. (1986). Natural selection in the wild. Princeton University Press, Princeton, NJ.

Estes, S. and Arnold, S. J. (2007). Resolving the paradox of stasis: models with stabilizing selection explain evolutionary divergence on all timescales. American Naturalist, 169, 227–244.

Fahrig, L. (2003). Effects of habitat fragmentation on biodiversity. Annual Review of Ecology Evolution and Systematics, 34, 487–515.

Fear, K. K. and Price, T. (1998). The adaptive surface in ecology. Oikos, 82, 440–448.

Fenberg, P. B. and Roy, K. (2008). Ecological and evolutionary consequences of size-selective harvesting: how much do we know? Molecular Ecology, 17, 209–220.

Filchak, K. E., Roethele, J. B. and Feder, J. L. (2000). Natural selection and sympatric divergence in the apple maggot *Rhagoletis pomonella*. Nature, 407, 739–742.

Forbes, A. A., Powell, T. H. Q., Stelinski, L. L., Smith, J. J. and Feder, J. L. (2009). Sequential sympatric speciation across trophic levels. Science, 323, 776–779.

Franks, S. J., Sim, S. and Weis, A. E. (2007). Rapid evolution of flowering time by an annual plant in response to a climate fluctuation. Proceedings of the National Academy of Sciences of the United States of America, 104, 1278–1282.

Fritts, T. H. and Rodda, G. H. (1998). The role of introduced species in the degradation of island ecosystems: a case history of Guam. Annual Review of Ecology and Systematics, 29, 113–140.

Garant, D., Forde, S. E. and Hendry, A. P. (2007). The multifarious effects of dispersal and gene flow on contemporary adaptation. Functional Ecology, 21, 434–443.

Gavrilets, S. (2004). Fitness landscapes and the origin of species. Princeton University Press, Princeton, NJ.

Gomulkiewicz, R. and Holt, R. D. (1995). When does evolution by natural selection prevent extinction. Evolution, 49, 201–207.

Gonzalez, A., Lawton, J. H., Gilbert, F. S., Blackburn, T. M. and Evans-Freke, I. (1998). Metapopulation dynamics, abundance, and distribution in a microecosystem. Science, 281, 2045–2047.

Gordon, S. P., Reznick, D. N., Kinnison, M. T., Bryant, M. J., Weese, D. J., Räsänen, K., et al. (2009). Adaptive changes in life history and survival following a new guppy introduction. American Naturalist, 174, 34–45.

Grether, G. F. (2005). Environmental change, phenotypic plasticity, and genetic compensation. American Naturalist, 166, E115–E123.

Groom, M. J., Meffe, G. K. and Carroll, C. R. (2006) Principles of conservation biology, 3rd edn. Sinauer Associates, Sunderland, MA.

Hadfield, J. D., Wilson, A. J., Garant, D., Sheldon, B. C. and Kruuk, L. E. B. (2010). The misuse of BLUP in ecology and evolution. American Naturalist, 175, 116–125.

Hansen, T. F. (2003). Is modularity necessary for evolvability? Remarks on the relationship. between pleiotropy and evolvability. BioSystems, 69, 83–94.

Hansen, T. F. and Houle, D. (2008). Measuring and comparing evolvability and constraint in multivariate characters. Journal of Evolutionary Biology, 21, 1201–1219.

Heap, I. M. (1997). The occurrence of herbicide-resistant weeds worldwide. Pesticide Science, 51, 235–243.

Heino, M. and Hanski, I. (2001). Evolution of migration rate in a spatially realistic metapopulation model. American Naturalist, 157, 495–511.

Hendry, A. P., Farrugia, T. J. and Kinnison, M. T. (2008). Human influences on rates of phenotypic change in wild animal populations. Molecular Ecology, 17, 20–29.

Hendry, A.P. and Gonzalez, A. (2008). Whither adaptation? Biology and Philosophy, 23, 673–699.

Hendry, A. P., Grant, P. R., Grant, B. R., Ford, H. A., Brewer, M. J. and Podos, J. (2006). Possible human impacts on adaptive radiation, beak size bimodality in Darwin's finches. Philosophical Transactions of the Royal Society B: Biological Sciences, 273, 1887–1894.

Hendry, A. P. and Taylor, E. B. (2004). How much of the variation in adaptive divergence can be explained by gene flow? An evaluation using lake-stream stickleback pairs. Evolution, 58, 2319–2331.

Hill, J. K., Thomas, C. D. and Lewis, O. T. (1999). Flight morphology in fragmented populations of a rare British butterfly, Hesperia comma. Biological Conservation, 87, 277–283.

Hoffmann, A. A. and Merilä, J. (1999). Heritable variation and evolution under favourable and unfavourable conditions. Trends in Ecology & Evolution, 14, 96–101.

Hughes, J. B., Daily, G. C. and Ehrlich, P. R. (1997). Population diversity: its extent and extinction. Science, 278, 689–692.

Jain, S. K. and Bradshaw, A. D. (1966). Evolutionary divergence among adjacent plant populations. 1. Evidence and its theoretical analysis. Heredity, 21, 407–441.

Keller, L. F. and Waller, D. M. (2002). Inbreeding effects in wild populations. Trends in Ecology & Evolution, 17, 230–241.

Kettlewell, B. (1973). Industrial melanism: the study of a recurring necessity; with special reference to industrial melanism in the Lepidoptera. Clarendon Press, Oxford.

Kingsolver, J. G. and Pfennig, D. W. (2004). Individual-level selection as a cause of Cope's rule of phyletic size increase. Evolution, 58, 1608–1612.

Kinnison, M. T. and Hairston Jr., N. G. (2007). Eco-evolutionary conservation biology: contemporary evolution and the dynamics of persistence. Functional Ecology, 21, 444–454.

Lack, D. (1947). Darwin's finches. Cambridge University Press, Cambridge.

Lahti, D. C. (2005). Evolution of bird eggs in the absence of cuckoo parasitism. Proceedings of the National Academy of Sciences of the United States of America, 102, 18057–18062.

Lahti, D. C. (2006). Persistence of egg recognition in the absence of cuckoo brood parasitism: pattern and mechanism. Evolution, 60, 157–168.

Lahti, D. C., Johnson, N. A., Ajie, B. C., Otto, S. P., Hendry, A. P., Blumstein, D. T., et al. (2009). Relaxed selection in the wild. Trends in Ecology & Evolution, 24, 487–496.

Lande, R. (1976). Natural selection and random genetic drift in phenotypic evolution. Evolution, 30, 314–334.

Lande, R. (1979). Quantitative genetic-analysis of multivariate evolution, applied to brain-body size allometry. Evolution, 33, 402–416.

Law, W. and Salick, J. (2005). Human-induced dwarfing of Himalayan snow lotus, *Saussurea laniceps* (Asteraceae). Proceedings of the National Academy of Sciences of the United States of America, 102, 10218–10220.

Levinton, J. S., Suatoni, E., Wallace, W., Junkins, R., Kelaher, B. and Allen, B. J. (2003). Rapid loss of genetically based resistance to metals after the cleanup of a Superfund site. Proceedings of the National Academy of Sciences of the United States of America, 100, 9889–9891.

Losos, J. B. (2009). Lizards in an evolutionary tree: ecology and adaptive radiation of Anoles. University of California Press, Berkeley, CA.

Losos, J. B. and Schluter, D. (2000). Analysis of an evolutionary species-area relationship. Nature, 408, 847–850.

Mallet, J. (2007). Hybrid speciation. Nature, 446, 279–283.

McNeilly, T. and Antonovics, J. (1968). Evolution in closely adjacent plant populations. 4. Barriers to gene flow. Heredity, 23, 205–218.

McRae, B. H., Dickson, B. G., Keitt, T. H. and Shah, V. B. (2008). Using circuit theory to model connectivity in ecology, evolution, and conservation. Ecology, 89, 2712–2724.

Millien, V. (2006). Morphological evolution is accelerated among island mammals. PLoS Biology, 4, 1863–1868.

Millien, V. and Damuth, J. (2004). Climate change and size evolution in an island rodent species: new perspectives on the island rule. Evolution, 58, 1353–1360.

Nosil, P. (2009). Adaptive population divergence in cryptic color-pattern following a reduction in gene flow. Evolution, 63, 1902–1912.

Nosil, P. and Crespi, B. J. (2004). Does gene flow constrain adaptive divergence or vice versa? A test using ecomorphology and sexual isolation in Timema cristinae walking-sticks. Evolution, 58, 102–112.

Nosil, P., Harmon, L. J. and Seehausen, O. (2009). Ecological explanations for (incomplete) speciation. Trends in Ecology & Evolution, 24, 145–156.

Olsen, E. M., Carlson, S. M., Gjøsæter, J. and Stenseth, N. C. (2009). Nine decades of decreasing phenotypic variability in Atlantic cod. Ecology Letters, 12, 622–631.

Olsen, E. and Moland, E. (2010). Fitness landscape of Atlantic cod shaped by harvest selection and natural selection. Evolutionary Ecology, 25, 695–710.

Orr, H. A. and Unckless, R. L. (2008). Population extinction and the genetics of adaptation. American Naturalist, 172, 160–169.

Palumbi, S. R. (2001). Humans as the world's greatest evolutionary force. Science, 293, 1786–1790.

Parchman, T. L. and Benkman, C. W. (2002). Diversifying coevolution between crossbills and black spruce on Newfoundland. Evolution, 56, 1663–1672.

Parmesan, C. (2007). Influences of species, latitudes and methodologies on estimates of phenological response to global warming. Global Change Biology, 13, 1860–1872.

Pergams, O. R. W. and Ashley, M. V. (2001). Microevolution in island rodents. Genetica, 112–113, 245–256.

Phillimore, A. B., Hadfield, J. D., Jones, O. R. and Smithers, R. J. (2010). Differences in spawning date between populations of common frog reveal local adaptation. Proceedings of the National Academy of Sciences of the United States of America, 107, 8292–8297.

Phillips, B. L. and Shine, R. (2006). An invasive species induces rapid adaptive change in a native predator: cane toads and black snakes in Australia. P Philosophical Transactions of the Royal Society B: Biological Sciences, 273, 1545–1550.

Pimm, S. L., Russell, G. J., Gittleman, J. L. and Brooks, T. M. (1995). The future of biodiversity. Science, 269, 347–350.

Pinto, N. and Keitt, T. H. (2009). Beyond the least-cost path: evaluating corridor redundancy using a graph-theoretic approach. Landscape Ecology, 24, 253–266.

Pörtner, H. O. and Knust, R. (2007). Climate change affects marine fishes through the oxygen limitation of thermal tolerance. Science, 315, 95–97.

Price, T. D., Qvarnström, A. and Irwin, D. E. (2003). The role of phenotypic plasticity in driving genetic evolution. Philosophical Transactions of the Royal Society B: Biological Sciences, 270, 1433–1440.

Réale, D., McAdam, A. G., Boutin, S. and Berteaux, D. (2003). Genetic and plastic responses of a northern mammal to climate change. Philosophical Transactions of the Royal Society B: Biological Sciences, 270, 591–596.

Reznick, D. N. and Ghalambor, C. K. (2001). The population ecology of contemporary adaptations: what empirical studies reveal about the conditions that promote adaptive evolution. Genetica, 112–113, 183–198.

Reznick, D. N., Shaw, F. H., Rodd, F. H. and Shaw, R. G. (1997). Evaluation of the rate of evolution in natural populations of guppies (Poecilia reticulata). Science, 275, 1934–1937.

Ricciardi, A. and Ward, J. M. (2006). Comment on "Opposing effects of native and exotic herbivores on plant invasions." Science, 313, 298a.

Rieseberg, L. H., Archer, M. A. and Wayne, R. K. (1999). Transgressive segregation, adaptation and speciation. Heredity, 83, 363–372.

Rieseberg, L. H., Raymond, O., Rosenthal, D. M., Lai, Z., Livingstone, K., Nakazato, T., et al. (2003). Major ecological transitions in wild sunflowers facilitated by hybridization. Science, 301, 1211–1216.

Roush, R. T. (1998). Two-toxin strategies for management of insecticidal transgenic crops: can pyramiding succeed where pesticide mixtures have not? Philosophical Transactions of the Royal Society B: Biological Sciences, 353, 1777–1786.

Rueffler, C., Van Dooren, T. J. M., Leimar, O. and Abrams, P. A. (2006). Disruptive selection and then what? Trends in Ecology & Evolution, 21, 238–245.

Rundle, H. D. and Nosil, P. (2005). Ecological speciation. Ecology Letters, 8, 336–352.

Sax, D. F. and Brown, J. H. (2000). The paradox of invasion. Global Ecology and Biogeography, 9, 363–371.

Schierenbeck, K. A. and Ellstrand, N. C. (2009). Hybridization and the evolution of invasiveness in plants and other organisms. Biological Invasions, 11, 1093–1105.

Schlaepfer, M. A., Sherman, P. W., Blossey, B. and Runge, M. C. (2005). Introduced species as evolutionary traps. Ecology Letters, 8, 241–246.

Schluter, D. (2000a). Ecological character displacement in adaptive radiation. American Naturalist, 156, S4-S16.

Schluter, D. (2000b). The ecology of adaptive radiation. Oxford University Press, Oxford.

Schluter, D. and Grant, P. R. (1984). Determinants of morphological patterns in communities of Darwin finches. American Naturalist, 123, 175–196.

Schtickzelle, N., Mennechez, G. and Baguette, M. (2006). Dispersal depression with habitat fragmentation in the bog fritillary butterfly. Ecology, 87, 1057–1065.

Schwarz, D., Matta, B. M., Shakir-Botteri, N. L. and McPheron, B. A. (2005). Host shift to an invasive plant triggers rapid animal hybrid speciation. Nature, 436, 546–549.

Schwarz, D., Shoemaker, K. D., Botteri, N. L. and McPheron, B. A. (2007). A novel preference for an invasive plant as a mechanism for animal hybrid speciation. Evolution, 61, 245–256.

Seehausen, O., Takimoto, G., Roy, D. and Jokela, J. (2008). Speciation reversal and biodiversity dynamics with hybridization in changing environments. Molecular Ecology, 17, 30–44.

Seehausen, O., vanAlphen, J. J. M. and Witte, F. (1997). Cichlid fish diversity threatened by eutrophication that curbs sexual selection. Science, 277, 1808–1811.

Sharpe, D. M. T. and Hendry, A. P. (2009). Life history change in commercially exploited fish stocks: an analysis of trends across studies. Evolutionary Applications, 2, 260–275.

Sheldon, S. P. and Jones, K. N. (2001). Restricted gene flow according to host plant in an herbivore feeding on native and exotic watermilfoils (Myriophyllum: Haloragaceae). International Journal of Plant Sciences, 162, 793–799.

Siepielski, A. M., DiBattista, J. D. and Carlson, S. M. (2009). It's about time: the temporal dynamics of phenotypic selection in the wild. Ecology Letters, 12, 1261–1276.

Simpson, G. G. (1944). Tempo and mode in evolution. Columbia University Press, New York.

Simpson, G. G. (1953). The major features of evolution. Columbia University Press, New York.

Singer, M. C., Thomas, C. D. and Parmesan, C. (1993). Rapid human-induced evolution of insect host associations. Nature, 366, 681–683.

Steadman, D. W. (1995). Prehistoric extinctions of Pacific Island birds: biodiversity meets zooarchaeology. Science, 267, 1123–1131.

Stockwell, C. A., Hendry, A. P. and Kinnison, M. T. (2003). Contemporary evolution meets conservation biology. Trends in Ecology & Evolution, 18, 94–101.

Strauss, S. Y., Webb, C. O. and Salamin, N. (2006). Exotic taxa less related to native species are more invasive. Proceedings of the National Academy of Sciences of the United States of America, 103, 5841–5845.

Taylor, E. B., Boughman, J. W., Groenenboom, M., Sniatynski, M., Schluter, D. and Gow, J. L. (2006). Speciation in reverse: morphological and genetic evidence of the collapse of a three-spined stickleback (*Gasterosteus aculeatus*) species pair. Molecular Ecology, 15, 343–355.

Taylor, P. D. and Merriam, G. (1995). Wing morphology of a forest damselfly is related to landscape structure. Oikos, 73, 43–48.

Thomas, C. D., Hill, J. K. and Lewis, O. T. (1998). Evolutionary consequences of habitat fragmentation in a localized butterfly. Journal of Animal Ecology, 67, 485–497.

Thomas, C. D., Ng, D., Singer, M. C., Mallet, J. L. B., Parmesan, C. and Billington, H. L. (1987). Incorporation of a European weed into the diet of a North American herbivore. Evolution, 41, 892–901.

Tobler, M., Culumber, Z. W., Plath, M., Winemiller, K. O. and Rosenthal, G. G. (2011). An indigenous religious ritual selects for resistance to a toxicant in a livebearing fish. Biology Letters, 7, 229–232

Van Buskirk, J., Mulvihill, R. S. and Leberman, R. C. (2010). Declining body sizes in North American birds associated with climate change. Oikos, 119, 1047–1055.

van Klinken, R. D. and Edwards, O. R. (2002). Is host-specificity of weed biological control agents likely to evolve rapidly following establishment? Ecology Letters, 5, 590–596.

Visser, M. E. (2008). Keeping up. with a warming world: assessing the rate of adaptation to climate change. Philosophical Transactions of the Royal Society B: Biological Sciences, 275, 649–659.

Vitousek, P. M., Mooney, H. A., Lubchenco, J. and Melillo, J. M. (1997). Human domination of Earth's ecosystems. Science, 277, 494–499.

Wagner, G. P. and Altenberg, L. (1996). Complex adaptations and the evolution of evolvability. Evolution, 50, 967–976.

Walsh, B. and Blows, M. W. (2009). Abundant genetic variation + strong selection = multivariate genetic constraints: a geometric view of adaptation. Annual Review of Ecology Evolution and Systematics, 40, 41–59.

Whalon, M. E., Mota-Sanchez, D. and Hollingworth, R. M. (2008). Global pesticide resistance in arthropods. CAB International, Oxfordshire.

Whitlock, M. C. (1995). Variance-induced peak shifts. Evolution, 49, 252–259.

Whittaker, R. H. (1977). Evolution of species diversity in land communities. Evolutionary Biology, 10, 1–67.

Williamson, M. and Fitter, A. (1996). The varying success of invaders. Ecology, 77, 1661–1666.

Wright, S. (1932). The roles of mutation, inbreeding, crossbreeding and selection in evolution. Proceedings of the Sixth Annual Congress of Genetics, 1, 356–366.

Young, A., Boyle, T. and Brown, T. (1996). The population genetic consequences of habitat fragmentation for plants. Trends in Ecology & Evolution, 11, 413–418.

PART IV
Speciation and Macroevolution

CHAPTER 13

Adaptive Landscapes and Macroevolutionary Dynamics

Thomas F. Hansen

Adaptive zones, not only the animals occupying them, evolve.

—Simpson (1944, p. 190)

13.1 Introduction

Natural selection is amazingly effective. It is becoming increasingly clear that any measurable strength of selection on any measurably evolvable character can cause large changes within geological time spans. Yet, evolution often appears surprisingly slow in the fossil record. This paradox of stasis is an important theoretical problem in evolutionary biology (Williams 1992; Hansen and Houle 2004; Futuyma 2010). The seemingly obvious solution is that traits evolve rapidly to local adaptive peaks and stay there, but we also know that environments and selection pressures in nature are highly variable both spatially and temporally. Explaining stasis with stabilizing selection around a local adaptive optimum thus only serves to shift the problem to a different level; why are adaptive optima stable in a changing environment? This is the theoretical challenge, and the core problem in linking microevolution to macroevolution.

Solving the problem of peak stability is a part of the more general problem of developing a theory of the dynamics of Adaptive Landscapes. The development of population genetics theory has given us a highly successful mathematical theory of microevolutionary dynamics on defined fitness surfaces. This theory is not limited to static landscapes, as it also includes a detailed description of frequency- and density-dependent selection, which simply means that there is feedback from trait (or gene) dynamics to the fitness landscape itself, and it includes the effects of various forms of plasticity and environmental stochasticity (Chapters 7 and 14). Even with these extensions, the theory is still almost completely microevolutionary in that it is formulated on a generational time scale and contains parameters and assumptions that are not operationally accessible on macroevolutionary time scales.

These remarks refer to phenotypic trait evolution and the Simpsonian Adaptive Landscape. For Wrightian (genotype) landscapes we really do have some well-developed mathematical models pertaining to macroevolutionary time scales. These include the neutral theory of molecular evolution with its many modifications and extensions (Kimura 1983; Gillespie 1991; Lynch 2007). Very little of this theory is applicable to trait evolution, however, because evolution on the physiological, morphological, and behavioral levels is radically different and much less regular than evolution at the levels of the gene and genome structure. For phenotypic traits we only have verbal or crude extrapolative theories of long-term dynamics. This is well illustrated by the important paper of Estes and Arnold (2007), which formulates macroevolutionary extrapolations of explicit microevolutionary models and tests them formally against fossil or other historical data on evolutionary change on different time scales. None of these extrapolations could give a reasonable explanation of the data, either because they failed to fit the pattern, or because they made biologically unreasonable ad hoc assumptions. This is not very surprising, as they are essentially models of how populations should move in a fixed landscape. Instead, what we need are models of how the Adaptive Landscape itself is changing. We need to go from models of evolution *on* the Adaptive Landscape to models of evolution *of* the Adaptive Landscape. This is not a novel idea. A changing Adaptive Landscape was focal in Simpson's (1944, 1953) foundational work on interpreting large-scale evolutionary change, where it was reflected in his ideas of shifting adaptive zones and quantum evolution. More recently, the dynamical landscape was the centerpiece of Arnold's et al. (2001) influential review of the Adaptive Landscape as a bridge between micro- and macroevolution (see also Chapters 4 and 7).

In this chapter I will first outline some pertinent, and often misrepresented, facts about the potential for evolution and selection, and I will review some recent findings on patterns of long-term evolution in quantitative characters. I then evaluate some simple evolutionary process models in the light of these results and argue that no simple extrapolation of microevolution on a fixed Adaptive Landscape can be made consistent with all the facts. I then review ideas about macroevolutionary change in relation both to patterns of evolution and to possible microevolutionary interpretations. As did Simpson (1944), Arnold et al. (2001), and Estes and Arnold (2007), and indeed many contributors to the present book, I argue that a successful theory of macroevolution is likely to be grounded in microevolutionary process models combined with a theory of changes in the Adaptive Landscape. Finally, I make some remarks on the role of constraints in macroevolutionary dynamics.

In this chapter I will use the term Adaptive Landscape to refer to the individual fitness landscape, the mapping from trait to fitness, and not to mappings from gene or trait frequencies or means to mean fitness (overview over different notions of the Adaptive Landscape metaphor can be found in many chapters of this book, e.g. Chapter 2 and 19).

13.2 Patterns of evolution

13.2.1 Potential for evolution: evolvability and strengths of selection

A key event in the operationalization of the Adaptive Landscape came when Lande (1979) and Lande and Arnold (1983) started to write the response to selection as $R = G\beta$, a product between the additive genetic variance (the G-matrix for multivariate characters) and the selection gradient, β, the local slope of the fitness landscape (see Chapter 13 for details). This provided for the first time a quantifiable conceptual distinction between a measure of evolutionary potential on one side and a measure of the shape of the landscape on the other. This distinction between population variation and the fitness landscape is confounded in other descriptions of the response to selection. For example, in the mathematically, but not conceptually, equivalent breeder's equation, selection is measured as a selection differential, the covariance between fitness and trait, which is a mixture of landscape shape and variational potential, and evolutionary potential is measured as heritability, which is seriously confounded by evolutionary irrelevant residual variation (Hereford et al. 2004; Houle et al. 2011; Hansen et al. 2011). In contrast, the "Lande equation" provides a reasonable prediction of the response to selection over a few generations in terms of *independent* measures of evolutionary potential and landscape shape. These measures were also made empirically operational. The additive genetic variance can be estimated from quantitative-genetics breeding experiments or from pedigree information in natural populations (e.g. Lynch and Walsh 1998) and the selection gradient can be estimated as the regression slope of relative fitness on the trait (Lande and Arnold 1983; Chapter 9). I start by reviewing what we know about these two factors.

Following earlier work (Hansen et al. 2003a; Hansen and Houle 2008), I will define the evolvability of a trait as its expected response to a unit strength of selection, where a unit strength of selection is the strength of selection on fitness itself (i.e. a selection gradient of unity). Note that this only makes sense if trait change is measured in units of its own mean (i.e. as proportional change). It is a fundamental fact of selection mathematics (e.g. Price 1970) that (Wrightian) fitness needs to be standardized with its own mean to correctly predict changes in frequencies of types (e.g. allele frequencies). Hence, the selection gradient on a trait can only be compared to unit selection if the trait is standardized by its own mean. This leads to the mean-standardized selection gradient, β_μ, which measures strength of selection as the proportional change in fitness with a proportional change in the trait (Van Tienderen 2000; Hereford et al. 2004). The corresponding measure of evolvability is e_μ, the expected proportional response in the trait when $\beta_\mu = 1$ (Hansen et al. 2003a). The Lande equation tells us that e_μ equals the mean-scaled additive genetic variance (i.e. additive variance divided by the square of the trait mean). I believe the use of mean-scaled, i.e. proportional, measures of evolvabilities, selection strengths, and evolution rates is essential to understand the relative roles of selection and variation in evolutionary change, because the alternative standardization with phenotypic variation conflates the separation of landscape shape and variational potential that was achieved with the Lande–Arnold setup. With variation-based standardization, the steepness of the fitness landscape is measured in units of variation (standard deviation), and this introduces a negative correlation between measures of selection and measures of variational potential that makes them problematic to interpret in isolation. Furthermore, it is a remarkable fact that the heritability, the variance-scaled additive variance, is almost completely uncorrelated with the mean-scaled evolvability, and thus practically void of information about evolutionary potential (Houle 1992; Hansen et al. 2011). Hence, I will report and discuss all evolvabilities and rates of evolution in per cent of the mean.

In Fig. 13.1 I show the distribution of 394 published estimates of mean-scaled additive variances for linear measures of morphological traits. These estimates are gathered from a survey of the journals *Evolution* and *Journal of Evolutionary Biology* in the years 1992–2009 (Hansen et al. 2011). The overall median of the estimates is close to $e_\mu = 0.1\%$, meaning that the predicted per cent change in the trait mean per generation under unit selection is a tenth of a per cent. In an earlier review, Houle (1992) found comparable figures. This seems small, and such a change is unlikely to be detectable over a single generation, but if we consider thousands of generations, this level of evolvability is enough to make dramatic changes. If we assume that the evolvability and the slope of the Adaptive Landscape stay constant at $e_\mu = 0.1\%$ and $\beta_\mu = 1$, then the trait will

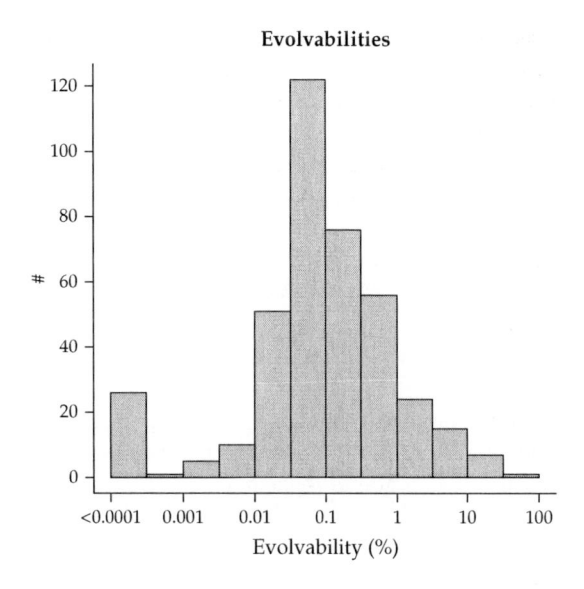

Figure 13.1 Histogram of 394 evolvability estimates, e_μ, for linear size traits gathered from the journals *Evolution* and *Journal of Evolutionary Biology* from 1992–2009. On the x-axis is additive variance divided by the square of the trait mean expressed as a per cent (interpretable as per cent response per generation for a trait under unit selection as predicted by the Lande equation). The median of this sample is 0.085%.

change with a factor $(1 + e_\mu \beta_\mu)^t$ over t generations, which with these values imply that doubling the mean size of a trait will take 693 generations. This is much shorter than the resolution of most fossil time series (but see Bell et al. 2006). Linear morphological measures, as plotted in Fig. 13.1, are also traits that tend to have relatively low evolvabilities. Most other trait categories have higher median evolvabilities. In this database, life-history traits, as well as volume and weight measurements have median evolvabilities around 1%, while physiological traits are around 0.5% and behavioral traits approach 2%. With an evolvability of 1%, unit selection can double the trait in 69 generations.

Unit selection is strong selection. Few biologists would expect that traits are often experiencing selection of a strength comparable to the strength of selection on fitness itself. But empirical estimates of the strength of directional selection in nature are often astonishingly high. In a literature survey of published selection gradients, Hereford et al. (2004) found that the median absolute value of univariate estimates of β_μ for morphological traits was 0.9

(after correcting for the bias that results from averaging absolute values); almost the same strength of selection as on fitness itself. There are many issues with this number, but it is still true that our best estimates to date indicate that traits in natural populations are often under strong directional selection. Note, however, that this does not necessarily imply that direct selection for adaptation of the trait is also very strong. Univariate selection gradients include all sources of selection on the trait, including indirect selection due to correlations with other traits under selection. The direct selection caused by the trait itself is likely to be much weaker than this, and indeed, the median mean-scaled gradient from multitrait selection studies in the Hereford et al. (2004) database was $\beta_\mu = 0.28$, a little less than a third of the strength of selection on fitness. This is still strong selection, but few studies are likely to include a set of traits that includes all, or even most, relevant dimensions of selection. We can thus not rule out that the direct selection is usually much weaker.

The picture that emerges is that it is common, even typical, for traits to be under very strong directional selection. This is consistent with the conclusions of Endler (1986) and Bell (2010), but contrary to the conclusion of a major review of field-based selection gradients by Kingsolver et al. (2001), who analyzed the gradients on a variance-standardized scale. Detailed criticism of the Kingsolver interpretation can be found in Hereford et al. (2004) and Houle et al. (2011), and I take it as tentatively established that strong selection is commonplace. It is, however, quite possible that the observed selection is largely indirect and not causally related to the focal trait. As there are many correlated traits that may be influenced by a variety of selective factors, such indirect selection may not be very consistent in strength or direction. Indeed, recent reviews by Siepielski et al. (2009) and Bell (2010) conclude that there is considerable temporal variation in the strength, shape and direction of selection, although it remains unclear how much of this could be due to estimation error in selection gradients, which is often substantial. Spatial variation in selection is also commonly found (e.g. Schluter 2000a; Gosden and Svensson 2008; Hereford 2009). Taken at face value these facts imply that Adaptive Landscapes are quite dynamic even on short temporal and spatial scales (see also Chapters 7, 9, and 10).

A difficulty with this scenario is that additive genetic variation eventually gets depleted by fixation of alleles under persistent directional selection. We then need to consider how fast new additive genetic variance can be generated by mutation. Assuming an additive genotype-phenotype map, new mutational variance can simply be added to the existing additive variance. Estimates of the new mutational variance that arise each generation are surprisingly high (Houle et al. 1996; Houle 1998; Lynch et al. 1999; Houle and Kondrashov 2006). The mean-scaled mutational variance per generation can be taken as an estimate of mutational evolvability, the evolvability due to new mutations only. Squaring the median coefficients of mutational variance reported by Houle and Kondrashov (2006) gives a mutational evolvability of $e_m = 0.00058\%$ for morphological traits, and for life-history traits it gives a whopping $e_m = 0.0216\%$. If combined with a selection gradient of $\beta_\mu = 0.3$, the former number would lead to a doubling of the trait in about 400,000 generations. This suggests that genetic constraints could be important for adaptive evolution of morphological traits at least, but this is still rather rapid evolution by the standards of the fossil record. Another way of looking at the same thing is to consider the ratio between mutational and additive genetic variance, which tells us how many generations it would take to replenish typical levels of additive genetic variance. In *Drosophila*, Houle (1998) found this to be on the order of a few hundred generations for morphological traits and typically less than hundred generations for life-history traits.

Genetic correlations are often strong, and pleiotropic constraint is a further factor to consider when assessing evolvabilities (Kirkpatrick 2009; Blows and Walsh 2009; Walsh and Blows 2009). High levels of variation in individual characters may give a misleading picture of evolvability if it is due to genes with pleiotropic effects on other characters. My collaborators and I have argued that the degree of pleiotropic constraint on evolvability can be quantified through the concept of conditional evolvability, defined as the predicted response to unit directional selection when the directional selection has come to an equilibrium with stabilizing selection along all other directions in morphospace (Hansen 2003; Hansen et al. 2003b; Hansen and Houle 2008). This concept must be measured relative to a defined set of constraining characters, but is still a useful tool for assessing the quantitative effects of constraints. Given a set of characters, Hansen and Houle (2008) suggested that comparing the average conditional evolvability over all possible directions of evolutionary change, $\bar{c}$, to the average (unconditional) evolvability over those same directions, $\bar{e}$, may be a measure of the degree of constraint inherent in a given genetic variance matrix (G-matrix). For example, for a G-matrix of 20 wing dimensions in *Drosophila melanogaster*, we found that $\bar{c}$ was only 7% of $\bar{e}$. Strikingly, both evolvabilities and conditional evolvabilities were much higher along directions of species differences in *Drosophila* indicating that there is relationship between rates of species diversification and evolvability. McGuigan and Blows (2010) found a different result for wing shape in *D. bunnanda*, where the evolvability of trait landmarks were only reduced by 28% by conditioning on nine other trait landmarks. They did, however, find very small conditional evolvabilities for individual cuticular hydrocarbon frequencies (9% of unconditional evolvabilities). In fact, moderate to substantial reductions in evolvability after conditioning on a few traits have been found in most studies using this approach (Hansen et al 2003b; Jensen et al 2003; Parker and Garant 2004; Rolff et al 2005; Rønning et al. 2007).

Although there are thousands of evolvability estimates in the literature, we know little about the costs of evolutionary change and the longer-term evolution of evolvability itself. In consequence, the current understanding of the potential for long-term evolutionary change is poor, and indeed, the common assumption of unrestricted evolvability has been questioned in several recent reviews (Hansen and Houle 2004; Blows and Hoffmann 2005; Hansen 2006; Willi et al. 2006; Kirkpatrick 2009; Walsh and Blows 2009).

13.2.2 The rate of evolution

There is a huge literature on the rates of phenotypic evolution in different traits and time scales stemming from the historical and fossil record (reviewed in Hendry and Kinnison 1999; Kinnison and Hendry 2001; Gingerich 2001, 2009; Hunt 2006, 2007a; Svensson and Gosden 2007; Hendry et al. 2008; Chapter 15) or phylogenetic comparative studies (Harmon et al. 2010; Uyeda et al 2011). Here I will focus on morphological traits only. One fundamental fact to start with is that there is a strong time-scale effect on evolutionary rates. The longer the time span of the measurement, the lower the estimated rate (Gingerich 1983, 2009). This pattern was dramatically illustrated, first by Gingerich (1983), and then by Estes and Arnold (2007), who found that size of observed evolutionary changes in historical and fossil time series were essentially independent of the time span over which they were measured. This is an astonishing finding, as the time spans they considered ranged from one generation to millions of generations. This is also a pattern that contradicts a number of other findings in evolutionary biology, including the common finding of strong phylogenetic signal in a large number of phylogenetic comparative studies (see e.g. Gittleman et al. 1996; Freckleton et al. 2002; Blomberg et al. 2003; Ashton 2004 for reviews). This contradiction arises because if there is no relationship between the time span and the amount of change, then there can be no relationship between phylogenetic and morphological distance.

For this reason, Uyeda et al. (2011) decided to combine the time-series data of Estes and Arnold (originally from Gingerich 2001) with data from comparative phylogenetic studies (as well as additional time-series data). The results of this excercise are shown in Fig. 13.2, which is a plot of change in (log) morphological size against the time interval they are measured over. As in Estes and Arnold's study there is hardly any tendency for the changes to increase over time spans up to about a million years, but at about a million years we start to see a regular increase in the amount of change. This is largely driven by phylogenetic comparative data, which commonly involve time scales from one to several tens of millions of years (the time scales

Evolutionary rates

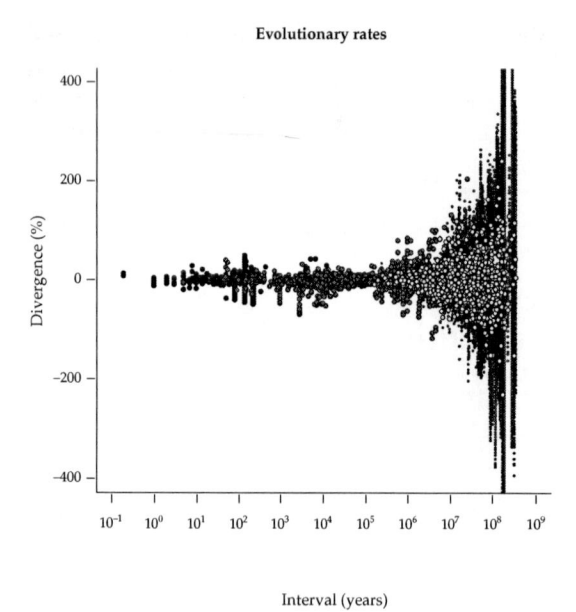

Figure 13.2 Plot of evolutionary divergence in size-related traits expressed as an approximate percent change against the length of the time interval over which the change took place. Dark-grey points refer to field studies and historical data (allochronic and synchronic, n = 315 + 1916), medium-grey points are from fossil time series (n = 3138), and light-grey points are from phylogenetic comparative data based on node-averaged contrasts (n = 2627). The small light-grey points are pairwise contrasts between extant species from the phylogenetic comparative data, and was not used to fit models, but added to give a better visual sense of the degree of divergence (n = 6340). The divergence is only approximately equal to percent change, as it is measured as $(\ln(z_1) - \ln(z_2))/k$, where z_1 and z_2 are the phenotypic means of two samples to be compared and k is the dimensionality of the trait (division by k is to make the data comparable to the linear size measures as in Fig 13.1). Data are from Uyeda et al. (2011), which can be consulted for more detail.

that typically separate species in a genus). It may be tempting to interpret this as a simple contradiction between fossil time-series data and phylogenetic comparative data, but statistical modeling shows that the difference is better explained by the time scale than by the type of data. There is after all a substantial overlap between the time scales of time series and comparative data, and in the region of overlap they do not differ. We can also see hints of the beginning trend in figure 1 of Estes and Arnold (2007), but they had too little data above a million-year time spans to reject a stationary process over the entire time span.

Uyeda et al. (2011) found that the best model to explain this pattern is a combination of a white-noise process (i.e. rapid, uncorrelated, non-cumulative changes) with a point process where large normally-distributed changes happen with long intervals in between. It is important to realize that this is consistent with the common finding of a component of Brownian-motion-like evolution in comparative data (e.g. Freckleton et al. 2002), since the among-species (co)variance structure of this point process will converge on that of Brownian motion on long time scales (Hansen and Martins 1996). I suggest the point process be interpreted as rare semi-permanent changes in the Adaptive Landscape, and the white noise as frequent ephemeral changes causing rapid evolution on very short time scales, but without long-term trends. This pattern is also consistent with the common observation of strong and temporally-varying directional selection as discussed earlier. The evolutionary changes estimated from the variance of the short-term fluctuations had a standard deviation of 9.6% (of the mean on the original scale). I suggest this number should be interpreted as an estimate of the change over single episodes of ephemeral environmental changes, which may take anything from one to maybe a hundred generations. Under a Poisson model the average waiting time to the events that lead to cumulative change was estimated to be as much as 25 million years, and the estimated standard deviation of change indicates a 27.2% change in the trait mean per event. The chances of experiencing such an event over periods less than 1 million years is then about 4%, thus explaining the near stationary pattern of evolution in fossil time series found by Estes and Arnold (2007). Nevertheless, there is also evidence of such phase shifts in fossil time series (e.g. Hunt 2006, 2008; Hunt et al. 2008).

These results raise questions about the biological cause of these rare cumulative changes in the Adaptive Landscape. I will return to this below, but before leaving the description of the patterns I want to review a few recent studies that have looked at correlates between evolutionary rates and other factors. The first factor to consider is speciation. The theory of punctuated equilibrium (Eldredge and Gould 1972; Gould and Eldredge 1993; Gould 2002; Eldredge et al. 2005) postulates that most macroevolutionary change should be associated with speciation events. Attempts at testing this with fossil data have been inconclusive, mostly due to the difficulties of inferring speciation from the fossil record (e.g. Hoffman 1989), but some recent studies have used phylogenetic comparative data to throw light on this question. For example, Mattila and Bokma (2008; Bokma 2008) studied body-size evolution of 2143 extant mammal species. They decomposed among-species body-size variance into one component that relates to phylogenetic branch length and one that relates to an estimated number of speciation events leading to the species. They found that 65–80% of the variance could be explained by speciation events. I suspect this is an overestimate, because the variance assigned to speciation events must also include variance due to measurement error and due to anagenetic changes that do not scale linearly with branch length such as stationary fluctuations and other forms of constrained evolutionary change. Their estimates also exclude clades in which some species have undergone large changes, and in these cases speciation explained much less variation. Nevertheless, this study shows that phylogenetic splits may be an important factor influencing macroevolutionary divergence. Adams et al. (2009), however, found little correlation between speciation and morphological diversification in a study of 190 species of Plethodontid salamanders.

The comparative data on mammalian body sizes were also studied by Cooper and Purvis (2010) and Harmon et al. (2010) with somewhat different results. Both studies fitted the same three models, a Brownian-motion process, an Ornstein–Uhlenbeck process, which contains a deterministic pull towards a central state, and an "early-burst model," which is Brownian motion with an exponentially decreasing rate parameter. Cooper and Purvis (2010) found support for the early-burst model in a phylogeny of 3473 species of mammals, indicating that evolutionary rates were high early in the radiation of mammals and then slowed down, but within specific orders there were cases of support for all three models. Harmon et al. (2010) found little support for the early-burst model in various subclades of the mammals or in many other comparative body-size data sets. In their analysis, an Ornstein–Uhlenbeck process often fitted well indicating that body-size evolution is somewhat constrained and tends towards an intermediate size. Their exception was the birds, for which the early-burst model showed the best fit. Hence, there are indications of elevated rates of body size evolution early in the radiation of birds and mammals, but less so within subclades. Cooper and Purvis also compared the rate parameter in the Brownian-motion model with various ecological factors and found that high evolutionary rates were associated with cold, low-lying, species-poor, high-energy, mainland ecoregions, thus indicating that evolutionary diversification of body size is broadly associated with ecological, climate, and community factors.

13.2.3 The rate of adaptation

Evolution is not adaptation. Many discussions of macroevolutionary change seem to presuppose a simplistic split between neutral evolution influenced by genetic drift on the one hand and adaptive evolution driven by selection on the other. I will argue below that genetic drift can be safely rejected as an explanation for almost all phenotypic change, but this does not mean that all trait evolution is adaptive. Adaptation is not the same as natural selection. The observed selection on a trait can be direct, in the sense of being causally related to the trait itself, but also indirect, due to correlations with other traits under selection (Sober 1994). The redness of vertebrate blood is certainly a selected trait, but being an indirect response to selection for binding oxygen, it is not an adaptation. Trait correlations are common, and on short time scales trait changes must be strongly influenced by stochastic-like responses to selection on other unobserved traits. This is supported by the observation that multivariate selection gradients tend to be much smaller than univariate selection gradients (Hereford et al. 2004).

Since evolution and adaptation have been treated as more-or-less synonymous there have been few attempts at estimating rates of adaptation as distinct from rates of evolution. What we have are some mostly qualitative reviews of limits to adaptation (e.g. Bradshaw 1991; Williams 1992; Crespi 2000; Hansen and Houle 2004; Blows and Hoffmann 2005; Hansen et al. 2006a; Labandeira 2007; Hereford 2009; Futuyma 2010). These reviews show that maladaptation is not uncommon. One general example concerns species ranges (Hoffman and

Parsons 1997; Wiens and Graham 2005; Kellerman et al. 2009; Futuyma 2010). Many species have geographic ranges that are not limited by abrupt changes of habitat, and have thus failed to adapt to some more or less continuous ecological gradient.

Some quantitative estimates of the rate of adaptation are also available. Labandeira (2007) estimated the time it took to establish different forms of herbivory during the colonization of land by arthropods. He found that the first signs of herbivory on roots, leaves, seeds, stems, and sporangia appeared 98, 76, 54, 4, and 0 millions of years, respectively, after they first could have appeared in the fossil record. He also found that the first signs of wood boring in fungi and plants took 20 and 62 millions of years, respectively, to appear. Hence, major ecological adaptations often take long to establish. Indeed, any biologist can think of many examples were large phylogenetic groups have failed to acquire some likely beneficial qualitative adaptation. The near absence of marine insects (Ruxton and Humpheries 2008) and Vermeij's (2010) intriguing observation that no mollusk has ever evolved the ability to communicate with sound are cases in point.

The adaptation of simple quantitative characters could be expected to be quicker, however. Such estimates are available from use of comparative methods explicitly designed to study adaptation towards a niche-dependent optimal state (Hansen 1997; Butler and King 2004; Hansen et al. 2008; Labra et al. 2009). In these models the rate of adaptation can be expressed as a phylogenetic half-life, the time it takes for a species evolving in a new niche to evolve on average half the distance from its ancestral state to its estimated (primary) optimum in the new niche. Explicit estimates of such half-lives are given in Table 13.1. This table reveals that adaptation of quantitative characters is highly variable and often slow enough to be detectable on

Table 13.1 Rates of adaptation: the table shows estimates of phylogenetic half-life, $t_{1/2}$, from different studies using the adaptation-inertia model (Hansen 1997; Butler and King 2004; Hansen et al. 2008). The model estimates primary optima in relation to different niches and the half-life gives the time it takes to evolve half the distance from the ancestral state to a primary optimum. Half-lives are given either in years (yrs), millions of years (myr), or in % of tree height (th), the distance from root of the phylogeny to the most remote tip. In all these studies the tree height will be several millions of years. In making this table I have left out many studies and traits using this approach, either because estimates of half-lives were not reported or because the hypothesized adaptations were not well supported in the sense of not explaining much variation in the trait

Species	Trait	Adaptation	$t_{1/2}$	Reference
Horses	Teeth morp. (hypsodonty)	Browsing → Grazing	2.8 myr	Hansen (1997)
	+ Size		1.3 myr	
Anolis lizards	Size	Habitat	28% th	Butler and King (2004)
	Sexual size dimorphism		48% th	
Rainforest birds	Coloration	Habitat	0–17% th	Gomez and Thery (2007)
Maples	Leaf size	Mating system	60,000 yrs	Verdu and Gleiser (2006)
	Infl. size		88,000 yrs	
Sedge grasses	Chromosome#	Clade specific	~ 0 yrs	Hipp (2007)
Mammals	Recombination rate	Clade specific	~ 0 yrs	Dumont and Payseur (2008)
Primates	Sexual size dimorphism	female size	8.1 myr	Hansen et al. (2008)
Centrarchid fishes	Jaw morphology	Feeding mode	13.1 myr	Collar et al. (2009)
Lacertid lizards	Muscle phys.	Predator escape strategy	33%th	Scales et al. (2009)
Varanid lizards	Trunk length	Locomotory strategy	10,000yrs	Schuett et al. (2009)
	Femur length		120,000yrs	
	Tail length		Infinity	
Frogs	Tadpole shape	Habitat	10–13 myr	van Buskirk (2009)
Liolaemus lizards	Preferred body temperature	Climate, temperature	13%th, 0.5%th	Labra et al. (2009)
	Critical thermal minimum	Size	18%th	
Anolis lizards	Display traits	Environmental noise	0–Infinity (th)	Ord et al. (2010)

a macroevolutionary time scale. One such example is the estimated rate of adaptation for the evolution of high-crowned (hyposodont) teeth in horses when adapting to grassland (Hansen 1997). Here, I found that the phylogenetic half-life for the height to length ratio of molars was in the range of 1.3–6.5 millions of years with a best estimate of 2.8 million years (this dropped to 1.3 million years when body size was included in the model). The primary optimum for the molar ratio in the grassland adaptive zone was 2.35, and grazing horses evolved toward this from an ancestral (browsing) primary optimum with a molar ratio of 0.57. These estimates are also consistent with the paleobiological analysis of Stromberg (2006), who found that the appearance of full hypsodonty and grassland adaptation in horses most likely took several million years (see also MacFadden 1992 and Chapter 15).

Note that the half-lives in the table are in relation to specific adaptive hypotheses where a primary adaptive optimum has been shown to vary meaningfully with some ecological predictor variable. This is different from a half-life computed from a simple Ornstein–Uhlenbeck process around single central state, which can be taken as a measure of phylogenetic signal since it measures the rate of decay of the correlation with ancestral states. Even if rates of adaptation are often slow in the sense of taking tens of thousands to millions of generations to complete, they are usually more rapid than indicated by the phylogenetic signal in the trait. For example, Labra et al. (2009) found rapid adaptation of lizard preferred body temperatures to climate and environmental temperature, but when these environmental factors were not included in the model, the half-life approached infinity, indicating Brownian-motion-like evolution with no decay of ancestral effects. The interpretation is that the phylogenetic signal in preferred body temperatures was caused by Brownian-motion-like changes in the thermal habitats to which the temperature preferences were adapted. This shows that a phylogenetic signal is not necessarily evidence for a lag in adaptation. Instead, a discrepancy between rate of adaptation and phylogenetic signal can be taken as evidence for niche tracking. See also Kozak and Wiens (2010) for a similar result on salamanders.

13.3 Processes of evolution

My review of the patterns of and potential for evolution can be summarized in the following points:

1. Evolvabilities computed from standing genetic variation or mutation rates in individual traits are highly variable, but usually high enough to produce nearly instantaneous change under directional selection on macroevolutionary time scales.
2. Evolvabilities are likely to be constrained by correlations with other traits under stabilizing selection, but it is unclear how strong these constraints can be.
3. Directional selection in nature is often extremely strong, and fluctuates rapidly in strength on short time scales.
4. Indirect selection due to correlations with other traits may be a major component of observed univariate selection.
5. Evolutionary divergence in quantitative traits can be divided into two components: one component of rapid non-cumulative changes on short time scales and one component of rare cumulative changes happening on a million-year time scale.
6. Rates of adaptation are highly variable. Complex functional adaptations are often slow and erratic, and we find considerable maladaptation on macroevolutionary time scales, but adaptation of quantitative traits can also be fast, and phylogenetic signals may often be caused by slow changes in niches or adaptive zones more than by inertia in adaptation to these niches.

In this section I review and evaluate, in relation to these patterns, models of evolution that have been proposed to explain macroevolutionary changes in quantitative characters. I divide the models into those that describe evolution on an Adaptive Landscape and those that describe evolution of the Adaptive Landscape itself. I do not consider models of qualitative changes such as the emergence of novel characters.

13.3.1 Evolution on the Adaptive Landscape

The simplest Adaptive Landscape is a completely flat one. If we assume constant evolvability and a normal distribution of breeding values, the change in the population mean from one generation to the next on a flat landscape will be a normally distributed variable with mean zero and variance equal to the additive genetic variance, V_A, divided by the effective population size, N_e, (Lande 1976). This predicts that the trait mean will follow Brownian motion with a constant rate parameter, such that the variance in trait means after t generations equals $V_A t/N_e$. From this it follows that the expected per cent change in the trait would be $\sqrt{2e_\mu t/\pi N_e}$. With our median evolvability of $e_\mu = 0.1\%$, a 10% change in the trait (approximately the estimated standard deviation of short-term fluctuations from Uyeda et al. 2011) is expected to happen in about $16 N_e$ generations. This would require unrealistically small N_e to explain evolution on time scales of a few generations, and unrealistically large N_e to explain evolution on time scales on hundreds of thousands of generations or more (see Estes and Arnold 2007 for a similar argument). The drift model also fails to explain the non-cumulative nature of the evolutionary changes, as we expect a gradual accumulation of change under Brownian motion.

Lynch (1990, 1993) developed a more realistic version of this model by assuming that the additive genetic variance is maintained in mutation-drift equilibrium. This model predicts that the variance in trait means after t generations will equal $2V_m t$, where V_m is the mutational variance. Thus there is no dependence on the effective population size. In my notation this model predicts that the expected per cent change in the trait after t generations will be $\sqrt{4e_m t/\pi}$. With our median mutational evolvability of $e_m = 0.00058\%$, this predicts that a 10% increase can be expected in 1 354 generations and a doubling of the trait in 135,413 generations. Hence, this model is also too slow to explain the fluctuations on short time scales and much too fast and gradual to be consistent with evolution on million-year time scales. The later point was demonstrated by Lynch (1990) in relation to the fossil record of mammalian skeletal traits.

Neutral models of phenotypic evolution are thus unable to explain observed patterns of evolutionary change. They are also inconsistent with the common observations of strong natural selection and at least some degree of adaptation, and for most quantitative traits it is biologically implausible that they would be able to evolve to extreme states without any effect on fitness. Even if we could find traits with little causal connection to fitness, the ubiquity of pleiotropy makes it implausible that they would also be immune to indirect selection. In conclusion, fitness landscapes are not flat.

Constant directional selection (i.e. a tilted monotonic Adaptive Landscape) is equally implausible. It is incompatible with the common observations of frequent reversals in selection gradients (Siepielski et al. 2009; Bell 2010), and, as I have shown earlier, it predicts much too high rates of evolution on macroevolutionary time scales when combined with even minuscule evolvabilities. More realistically, we could imagine a slow trend being produced by evolution moving along a narrow corridor of high fitness with steep edges (Burger 1986; Wagner 1988; Zeng 1988; Armbruster and Schwaegerle 1996). The rate of evolution is then limited by the conditional evolvability along the corridor (Hansen 2003). A cordillera shape with many small fluctuating peaks could account for fluctuating directional selection in this model. More-or-less steady trends are, however, inconsistent with all the phenomenological models, such as Brownian motions, Ornstein–Uhlenbeck models or point processes, that usually show a good fit to evolution on long time scales.

Lande (1976) developed a simple model of evolutionary change on an Adaptive Landscape with a single peak. Assuming constant evolvability and population size, and a fitness function on the form $W(z) \approx 1 - s(z-\theta)^2$, where z is the trait, this model predicts that the population mean will fluctuate around the optimum, θ, due to genetic drift in a way that can be approximated by an Ornstein–Uhlenbeck process, and it could thus be a possible explanation for frequent observation of rapidly fluctuating directional selection and the non-cumulative nature of evolutionary change. In this model the (absolute) expected change (in % of trait mean) is equal to $2s_\mu e_\mu (1-\theta_\mu)$, where s_μ is the curvature of the Adaptive Landscape scaled with the trait mean (i.e. multiplied with the mean squared), and θ_μ is the optimum scaled with the trait mean such that $1-\theta_\mu$ is the relative distance from mean to optimum. There is also a variance in this change due to genetic drift equal to e_μ/N_e. This model has two obvious difficulties in accounting for the observed patterns of evolution. The first is that it requires extremely weak selection to produce any level of phylogenetic inertia or lag in adaptation. The estimated phylogenetic half life predicted from this model is $t_{1/2} = \ln2/2s_\mu e_\mu$ generations. With $e_\mu = 0.1\%$, a half life of 1 million generations would then require that $s_\mu = 0.00035$, which is extremely weak stabilizing selection. There are few direct estimates of mean-scaled quadratic selection gradients in the literature, but this number implies that a mean-scaled selection gradient would increase with 0.035% for each 1% change in the trait away from the optimum, surely not an empirically detectable effect. In fact, $\beta_\mu/s_\mu = 2(1 - \theta_\mu)$ in this model, and to produce a "typical" directional selection gradient of $\beta_\mu = 0.3$ the mean would need to differ from the optimum with a factor of 432, which is clearly unrealistic (and way outside the bounds of the model). Hence, this model can not simultaneously explain phylogenetic inertia on longer time scales and strong directional selection on short time scales. The model also fails to account for short-term fluctuations in isolation. The predicted stationary variance in the mean in stochastic equilibrium between stabilizing selection and drift is $1/(2s_\mu N_e)$, which implies that the expected deviance of the population mean from the optimum will be $\sqrt{1/\pi s_\mu N_e}$, and at this value of the mean, the observed linear selection gradient would be $|\beta_\mu| = 2\sqrt{s_\mu/\pi N_e}$. To make $\beta_\mu = 0.3$, we would need $s_\mu/N_e = 0.0707$, and even if the population was as small as $N_e = 100$, we would need $s_\mu \approx 7$, which would mean that the gradient would increase sevenfold for every per cent the mean moves from the optimum. This seems unrealistically strong stabilizing selection, and genetic drift around a single fitness peak is not a feasible explanation for observed levels of fluctuating directional selection unless the effective population size is extremely small (and would surely imply demographic extinction on relevant time scales).

Lande (1985, 1986) also considered genetic drift on a multipeaked Adaptive Landscape. Is it possible that the rare evolutionary events could be due to peak shifts caused by genetic drift? Lande's (1985) main result was that the expected time to a peak shift would be proportional to P^{2N_e}, where P is the ratio between the maximal fitness at the ancestral peak and the minimum fitness in the valley separating the peaks. This makes it practically impossible for a large population to cross an adaptive valley unless this valley is extremely shallow. Lande (1985) suggested that a population of size $N_e = 100$ could be expected to cross a fitness valley with a 5% reduction in fitness in about 1 million generations, but with $N_e = 200$ this estimate raises to more than 10 billion generations (see also Estes and Arnold 2007; Chapter 15).

While this rules out demographic peak shifts in large populations as a relevant mode of macroevolution, it does not rule out a shifting-balance pattern of evolution (Wright 1931,1932) in which peak shifts could happen in small isolated populations, and later spread to the metapopulation as a whole. This would, however, require a directed pattern of migration from the isolated populations (see e.g. Coyne et al. 1997; Wade and Goodnight 1998; Chapter 4 for review and criticisms). Finally, genetic drift is not the only source of stochasticity that could induce a peak shift. Peak shifts in a focal character can be induced by temporal and spatial environmental stochasticity (e.g. Kirkpatrick 1982; Schluter 2000a), and by correlated responses to selection on other characters (Price et al. 1993). In my view, the data on selection gradients and the ubiquity of genetic correlations indicate that most traits must be subject to frequently shifting indirect selection that would appear largely stochastic from the perspective of the focal trait, and such stochastic "background selection" is likely to swamp the stochastic effects of demographic genetic drift. However, the conditions for such stochasticity to induce shifts on the right time scale remain to be investigated. The hypothesis remains open, but in any case we have now moved on to models of changes in the landscape itself.

13.3.2 Evolution of the Adaptive Landscape

The bottom line is that none of the simple process models of evolution on an Adaptive Landscape proposed in the literature can account for the known patterns of evolution. It would of course be possible to imagine specific models of complex Adaptive Landscapes that could fit with some of the patterns, e.g. by postulating particular changes in local population sizes or genetic structure that would allow peak shifts at the right time, but I find it hard to see how a reasonably general model of evolution based on a stationary Adaptive Landscape could account for all the general patterns of evolution outlined earlier. Instead, a successful theory of macroevolution must be sought in models of changing Adaptive Landscapes. Such a theory must be consistent with population-genetics theory, but the real challenge is to model processes that affect Adaptive Landscapes on a broad range of time scales.

There are many models of macroevolutionary change. These come in two categories. One is a category of explicit, phenomenological models than can fitted to data. This includes simple stochastic models such as the Brownian motions, Ornstein–Uhlenbeck processes and Markov chains that are used in comparative methods and fossil time-series analyses. Each of these models are open to different interpretations of what evolutionary processes may have generated the patterns they describe, and the challenge is to derive them from evolutionary first principles (see Hansen and Martins 1996; Arnold et al. 2001; Estes and Arnold 2007 for review). The other category contains qualitative, largely verbal theories of macroevolutionary change such as punctuated equilibrium (Eldredge and Gould 1972), ephemeral divergence (Futuyma 1987, 1988, 2010), correlated progression (Kemp 2006, 2007), adaptive radiation (Schluter 2000a), Red Queen (Van Valen 1973; Stenseth and Maynard Smith 1984; Vermeij 1987), geographic mosaic (Thompson 2004), niche tracking (Eldredge 1989; Wiens 2004), evolution driven by stress in extreme environments (Hoffmann and Parsons 1991), and many other accounts. Here the challenge is to sharpen the theories into operational models that can be evaluated both against data and against population-genetics theory.

Rather than evaluating every proposed scenario, I will restrict myself to some selected comments on what I regard to be the three main explanatory challenges: (1) strong selection and rapid evolution on short time scales; (2) the non-cumulative nature of these changes resulting in long-term stationarity; (3) the slow, episodic nature of adaptive divergence.

The most obvious explanation for the frequent observations of strong and fluctuating selection gradients is that local adaptive peaks are quite volatile, but for this to be consistent with macroevolutionary stationarity, the peak movements must be constrained to a limited range. I will briefly outline three non-exclusive hypotheses to explain this constraint. The first is simply that the environment presents a limited set of possibilities. A flower adapted to attract pollinators may be under rapidly shifting selection pressures due to changes in the local abundance of pollinators, but if there are only a few possible candidates for effective pollination then there are only so many states of the environment and consequently a limited range of optimal states. Evidence for this hypothesis can be found in observations of repeated and convergent evolution, as seen in the repeated radiation of Anoles on islands (e.g., Losos et al. 1998; Harmon et al. 2005) or of sticklebacks in postglacial lakes (e.g., Colosimo et al. 2005; Berner et al. 2010). A second hypothesis is that the constraints stem from gene flow from a larger metapopulation. Although the environment may change frequently in any local population, these changes average out in the larger metapopulation (e.g. Lieberman and Dudgeon 1996). This scenario predicts that local populations are often maladapted due to recent changes in the local environment and due to gene flow from other populations. This prediction seems consistent with observations (Hereford 2009), and we can imagine frequent changes in selection gradients both due to local environmental changes, and due to local changes in migration rates between populations that alter the balance between selection and gene flow. What remains to be explained under this hypothesis is why the average optimum in a metapopulation should be stable. A third hypothesis is that traits are kept in a balance between causal direct selection and more or less stochastic indirect selection due to a large number of factors. Our focal trait may have a stable adaptive function, but may be constantly pulled from its optimal state by changes in any one of a large number of correlated traits. Since these changes have many different causes they may appear stochastic to an observer of the focal trait in isolation. Due to the increasing strength of direct selection when the trait is pulled farther from its functional (primary) optimum, these changes become constrained to a limited range.

None of these hypotheses explain long-term stationarity or stasis, because they simply assume that the range of environments or the causal basis for adaptations is constant, and assumptions are not explanations. Hypotheses for long-term stasis have received considerable attention (reviewed in Sheldon 1996; Jablonski 2000; Gould 2002; Hansen and Houle 2004; Eldredge et al. 2005; Futuyma 2010). Still, I regard the problem as largely unsolved. Hypotheses to explain stasis can be divided into those based on selective "constraints" and those based on variational constraints. I will return to variational constraints later in the chapter and briefly consider selective constraints here. The selective-constraint hypotheses all boil down to stabilizing selection in a constant environment on some level. For example, stasis may be caused by constant habitat preferences that make organisms track the niche to which they are adapted (e.g. Eldredge 1989) or it may be caused by gene flow from a source population in a hyperstable niche (e.g. Futuyma 1987; Williams 1992), but such explanations beg the question (Hansen and Houle 2004); why would the habitat preferences be under stabilizing selection, and why would there be hyperstable niches in the first place? A satisfactory selection account of stasis needs delve into this type of questions. For example Lieberman et al. (2007; Lieberman and Dudgeon 1996) claim that directional selection is improbable in widespread species distributed over a wide range of environments, but they do not provide a population-ecological basis for this claim. It should not be ruled out, but without a microevolutionary mechanism, the hypothesis has little explanatory value.

It is also difficult to come up with a good account of episodes of cumulative adaptive changes without a good account of stasis. It seems likely that some form of selective or variational constraints must be breaking down during such episodes, but without a firm understanding of the nature of the constraints it is also difficult to predict how they would break down. In the case of selective constraints, we can imagine two main classes of explanation. One is based on the breakdown of gene flow into a local population, which then becomes free to evolve a stronger degree of local adaptation. This is the essence of the ephemeral-divergence hypothesis and some newer versions of punctuated equilibrium (Futuyma 1987, 2010; Gould 2002; Lieberman et al. 2007). According to the punctuated-equilibrium hypotheses gene flow is broken by the rapid evolution of reproductive isolation. I find it hard to see this as an exclusive mechanism, however, because a reduction of population size in the general metapopulation, e.g. due to disappearance or change of its main niche, would have the same effects. It cannot possibly matter whether gene flow disappears due to reproductive isolation or due to disappearance or diminishment of a constraining source population.

The other class of explanation is based on some form of permanent change in the Adaptive Landscape for the species. The main challenge here is to understand why such changes would occur episodically on relatively long time scales. One possibility is to search for regular or irregular environmental changes happening on a similar time scale. While not denying the importance of large-scale environmental events in the history of life, I find it implausible as a general explanation of the relatively small-scale adaptive changes in quantitative traits that I am seeking to explain. I find it more likely that episodic adaptive shifts are relatively sudden responses to what might be more or less gradual change in the availability of alternative niches or adaptive zones. Consider the evolution of horses again. Species of early Miocene horses may have been well adapted to various browsing niches, but as grasslands started to spread, an increasing fraction of individuals would become exposed to a new adaptive zone (Simpson 1944; MacFadden 1992, 2000; Stromberg 2006). Morphological adaptations to this niche would initially have been constrained by gene flow from more abundant and productive browsing niches to which the species are well adapted, but eventually grasslands became sufficiently abundant to exert strong selective pressures on the whole or a part of the species. This would start to shift morphological optima toward the grazing niche, which in itself could increase the usage and productivity of this niche, further shifting the balance of selection. We can then imagine a relatively sudden phase shift from one dominant primary optimum to another, as in Simpson's (1944) quantum-evolution model. This scenario has theoretical support in the asymmetry that exists in the selective balance between core and marginal niches. Selection in an abundant niche to which the species is well adapted will dominate selection in a less abundant or marginal niche simply because of the relative reproductive output of the niches (Holt and Gaines 1992; Kawecki 1995, 2008; Holt 1996; Holt et al. 2003). This sets up a positive feedback system in which sudden phase shifts are likely. Such a shift may or may not be aided by population fragmentation and speciation, and it may or may not involve a prolonged period of refinement of adaptation to the niche as indicated by estimates of phylogenetic half lives in Table 13.1. It is tempting to speculate that the relatively strong phylogenetic inertia estimated for the evolution of hypsodonty in horses may be due to the need for correlated evolution of many other adaptations to grasslands including larger bodies, longer legs, and altered digestive systems, as in the correlated-progression scenario of Kemp (2006, 2007).

In Hansen (1997) I built a method for comparative analysis around this scenario. Here, the goal was to estimate primary optima associated with hypothesized niches or adaptive zones, such as grazing and browsing, mapped onto a phylogeny. This method presupposes relatively sudden shifts from one selective regime to another, and also allows for shorter-term stochastic changes in the optima. Central to the method is the concept of a primary optimum defined as a hypothetical average optimum of species well adapted to the niche. Hence, each species may have a different optimum due to differences in any one of a suite of secondary factors, but if they are all living in the same general adaptive zone they will all experience some of the same selection pressures that will keep their optima within a similar range, as outlined in the third, "balance" scenario for restricted optima discussed earlier.

Recently, we expanded the model to allow more continuous, Brownian-motion type evolution for the primary optima (Hansen et al. 2008; Labra et al. 2009). Since the phylogenetic variances and covariances predicted by Brownian motion can be derived from episodic point processes as long as events are not too rare within the time scale of the phylogeny (Hansen and Martins 1996), we can view this model as applicable to somewhat longer time scales where many adaptive shifts have taken place.

In this scenario, adaptive change is driven by environmental change, and there is no special role for speciation and species fragmentation. This does not mean that the scenario is inconsistent with an influence of speciation on evolutionary change. Changes in patterns of gene flow may play a role in reaching the tipping points between alternative primary niches, and speciation may also by itself constitute an important change in the selective environment, for example through character displacement associated with sympatric or parapatric speciation (Gavrilets 2004) or through secondary contact after allopatric speciation. Indeed, there is evidence that character displacement is common among congeneric species (Schluter 2000a,b). Still, competition with closely related species is only one out of many ways a new niche can appear or become favorable. The appearance of new predators, diseases, mutualists, sources of food or unrelated competitors is probably more important, and there are many more subtle or indirect ways in which species could influence each other. There is also no doubt that many evolutionary changes are driven by mechanisms that are largely intrinsic to the species, such as many forms of sexual selection. And finally, abiotic changes must at least sometimes be important. Thus, I favor a view in which a variety of biotic and abiotic changes can form the basis of shifts in primary niches. The episodic nature of these shifts may reflect episodic environmental changes or mechanisms, phase shifts caused by mechanisms favoring adaptation to current niches, or possibly constraints on evolvability that are only rarely overcome (e.g. require a key adaptation to appear). Bell (2010) has recently outlined a similar view of macroevolutionary change.

This view is consistent with adaptive radiations that happen when a whole range of niches become available, as for example when a species appears in a new place, acquires a key adaptation, or in the aftermath of a mass extinction. There are a fair number of specific cases that seem to fit the adaptive-radiation scenario quite well (Schluter 2000a, 2000b;

Yoder et al. 2010), but as discussed previoously, the fit of the early-burst model to diversification on phylogenies has been mixed (Agrawal et al. 2009; Cooper and Purvis 2010; Harmon et al. 2010), and Harmon et al. (2010) explicitly reject adaptive radiation as a common mode of evolution.

13.3.3 A role for constraints?

Estimates of short-term evolvability show that variational constraints are unlikely to be qualitatively important for microevolutionary changes in individual characters, but our understanding of the limitations imposed by pleiotropy and the evolutionary dynamics of the evolvability itself are limited, and variational constraints remain a viable hypothesis for explaining larger-scale stasis and maladaptation.

Pleiotropy and genetic correlations are widespread and pleiotropic constraints are almost surely quantitatively important (Walsh and Blows 2009). As previously discussed, studies of conditional evolvability show that stabilizing selection on even a handful of other traits may often constitute a significant constraint that could easily reduce evolvabilities with an order of magnitude after the first few generations of response (i.e. until directional and stabilizing selection equilibrate, Hansen 2003). Indeed, an increasing number of studies conclude that patterns of species divergence tend to correlate with directions of high evolvability (e.g. Schluter 1996; Blows and Higgie 2003; Hansen et al. 2003a,b; Marroig and Cheverud 2005; Hunt 2007b; Hansen and Houle 2008; Hohenlohe and Arnold 2008; Chenoweth et al. 2010; Grabowski et al. 2011). If a few traits have such dramatic effects, what would then be the total effect of pleiotropy? One hypothesis is that a large fraction of genetic variation, and particularly mutational variation, in any given character is due to effects of more or less unconditionally deleterious alleles that may fuel short-term evolutionary changes in the character, but only at the price of a physiological deterioration in the organism at large. In this case, long-term evolvability could be severely limited. An alternative hypothesis is that there is sufficient variation in pleiotropy across genes that it will be possible to find compensatory changes, and that pleiotropic constraints only act as temporary breaks on evolution. For example, reduced pelvic spines have evolved repeatedly in sticklebacks during freshwater radiation over time scales of a few thousand years based on changes in a pelvic-specific enhancer in the *Pitx1* gene (Shapiro et al. 2006; Bell 2009; Chan et al. 2010). This shows that quasi-independent evolution of the stickleback pelvis is possible on a relatively short time scale. The generality of such examples is unclear, however, and the *Pitx1* example also hints at the possibility that macroevolution may be based on a very small subset of normally observed quantitative genetic variation, and hence that macroevolutionary evolvability may be restricted.

The variational properties of a character are fundamentally determined by the genotype–phenotype map, and to understand long-term constraints we need to understand how the genotype–phenotype map evolves (Wagner and Altenberg 1996; Houle 2001; Hansen 2006; Polly 2008). I will not attempt to review this vast topic here except for reiterating my view that, although evolutionary changes in genetic architecture and evolvability are to be expected, adaptive changes in evolvability may not, because evolution of the genotype–phenotype map is likely to be dominated by correlated responses to selection on the traits themselves, and these correlated responses may be as likely to increase as to decrease evolvability (Hansen 2006, 2011). From the point of view of evolvability, the crucial question is how the additive effects, i.e. effects of single allele substitutions or new mutations, are likely to evolve. For traits under directional selection, it has been shown that this depends on the type of epistasis, or equivalently, the curvature of the genotype–phenotype map (Chapter 18), such that we would expect increasing evolvability under positive directional epistasis in the direction of selection, but canalization and reduced evolvability under negative directional epistasis (Carter et al. 2005; Hansen et al. 2006b). It is a simple extension to see that patterns of multivariate directional epistasis will determine the evolution of pleiotropy, and that this may often lead to changes that do not favor evolvability (Hansen 2011; but see e.g. Wagner et al. 2007 or Lee and Gelembiuk 2008 for a different view). For traits under stabilizing selection there is a general, but weak, tendency for canalization, but how far this could proceed will depend on genetic details, because the canalization of alternative loci may conflict (Hermisson et al. 2003). A weak canalizing effect is also easily swamped by inevitable fluctuating indirect selection pressures.

Hence, we cannot exclude the possibility that some traits have been trapped in regions of the genotype–phenotype map where they are strongly canalized with respect to change in at least some directions. An epistatic constraint could explain long-term stasis, and negative epistasis could act to restrict the range of evolutionary change and hence be consistent with stationary fluctuations in traits. This scenario predicts that we should find highly canalized traits that have stalled under directional selection, and that may have directionally asymmetric evolvabilities. While this seems inconsistent with most single-trait estimates of evolvability, we have to remember that it is very hard to demonstrate that additive variance is absent, that few studies have looked at asymmetries, which are often found in artificial-selection experiments (e.g. Frankham 1990), and that observed additive variances and responses to artificial selection may be tied to deleterious pleiotropic effects to the point where the conditional evolvability is much reduced.

13.4 Final remarks

A quantitative theory of macroevolutionary change is at its infancy. Although we have a detailed mathematical theory of microevolutionary change, the salient features of macroevolutionary change are not explained by an extrapolation of this theory. While any macroevolutionary model must be consistent with standard population-genetics models, we also need to include new explanatory elements. I believe these elements are to be found in two areas. One element is an understanding of the dynamics of the Adaptive Landscape. While some microevolutionary changes can be understood as populations evolving on a fixed Adaptive Landscape, macroevolutionary changes must be associated with changes in the landscape itself, as argued by Simpson (1944, 1953), Arnold et al. (2001), and indeed many contributors to this book. The other element is an improved understanding of constraints. What are the properties of the variation that can be made available to selection, and how do these properties evolve? This involves dynamical models for the evolution of the genotype–phenotype map, and in particular the evolutionary dynamics of additive effects, pleiotropy, and epistasis.

A crucial aspect of building a theory of macroevolutionary dynamics is to distinguish selection and adaptation. Selection is ubiquitous, but adaptation for any given function must be balanced against other selective pressures on the trait and on correlated traits. Such background selection needs to be taken into account. In fact, it gives promise to the search for a general model. Adaptation is extremely diverse, and not likely to yield to anything but specific models for specific situations. The background selection, however, may be more regular in a stochastic sense, as it may usually be due to a large number of different factors. The success of the neutral theory in molecular evolution is based on modeling regular stochasticity. A similar neutral theory of phenotypic evolution has not been successful, but we can imagine a stochastic theory based on modeling regularities in background selection. Indeed such a theory has been proposed for molecular evolution. Hahn (2008) argued that the neutral model should be replaced by a background-selection model as the standard of inference in molecular evolution. A similar background-selection model for phenotypic evolution remains to be developed, but a manageable model of this sort is much more realistic than a general model of adaptive evolution.

An operational theory of macroevolution would consist of a set of mathematical models well-grounded in population-genetics first principles with empirical measurable variables and parameters. Such a theory would be enormously useful in analyzing and interpreting the increasing amount of detailed macroevolutionary data that is being produced. This includes paleobiological data and phylogenetic comparative data. Simple stochastic models such as Brownian motions, Ornstein–Uhlenbeck processes, and Markov chains are frequently used to model phylogenetic correlations in such data, but they are short on satisfactory interpretations in terms of relevant macroevolutionary processes such as adaptive evolution (e.g. Hansen 1997; Price 1997; Hansen and Orzack 2005). For example, Ornstein–Uhlenbeck processes are frequently interpreted in terms of Lande's (1976) model of drift and stabilizing selection, but this makes little sense in most comparative studies, because, as I argued above, this model does not predict any phylogenetic correlation on the time scales usually involved in comparative analyses, and it can not account for typical levels of variation across species. To accommodate phylogenetic correlations over millions of generations, we need to invoke processes taking place on relevant time scales, and this requires a return to Simpson's focus on the evolving Adaptive Landscape

Acknowledgments

I thank Michael Bell, Andrew Hendry, David Houle, Christophe Pélabon, Josef Uyeda, an anonymous reviewer, and the editors for helpful discussions and/or comments on the manuscript.

References

Adams, D. C., Berns C. M., Kozak K. H. and Wiens J. J. (2009). Are rates of species diversification correlated with rates of morphological evolution? Proceedings of the Royal Society B: Biological Sciences, 276, 2729–2738.

Agrawal, A. A., Fishbein, M., Hlitschke, R., Hastings, A. P. and Rabosky, D. L. (2009). Evidence for adaptive radiation from a phylogenetic study of plant defences. Proceedings of the National Academy of Sciences of the United States of America, 106, 18067–18072.

Armbruster, W. S. and Schwaegerle, K. E. (1996). Causes of covariation of phenotypic traits among populations. Journal of Evolutionary Biology, 9, 261–276.

Arnold, S. J., Pfrender M. E. and Jones A. G. (2001). The Adaptive Landscape as a conceptual bridge between micro- and macroevolution. Genetica, 112/113, 9–32.

Ashton, K. G. (2004). Comparing phylogenetic signal in intraspecific and interspecific body size data sets. Journal of Evolutionary Biology, 17, 1157–1161.

Bell, G. (2010). Fluctuating selection: the perpetual renewal of adaptation in variable environments. Philosophical Transactions of the Royal Society B: Biological Sciences, 365, 87–97.

Bell, M. A. (2009). Implications of a fossil stickleback assemblage for Darwinian gradualism. Journal of Fish Biology 75, 1977–1999.

Bell, M. A., Travis, M. P. and Blouw, D. M. (2006). Inferring natural selection in a fossil threespine stickleback. Paleobiology 32, 562–577.

Berner, D., Stutz, W. E. and Bolnick, D. I. (2010). Foraging trait (co)variances in stickleback evolve deterministically and do not predict trajectories of adaptive diversification. Evolution, 64, 2265–2277.

Blomberg, S. P., Garland T. Jr. and Ives, A. R. (2003). Testing for phylogenetic signal in comparative data: behavioral traits are more labile. Evolution, 57, 717–745.

Blows, M. W. and Higgie, M. (2003). Genetic constraints on the evolution of mate recognition under natural selection. American Naturalist, 161, 240–253.

Blows, M. W. and Hoffmann, A. A. (2005). A reassessment of genetic limits to evolutionary change. Ecology 86, 1371–1384.

Blows, M. and Walsh, B. (2009). Spherical cows grazing in flatland: constraints to selection and adaptation. In J. Vanderwerf, H. U. Graser, R. Frankham, and C. Gondro (eds.) Adaptation and fitness in animal populations—evolutionary and breeding perspectives on genetic resource management. Springer, Dordrecht, pp. 83–101.

Bokma, F. (2008). Detection of "punctuated equilibrium" by Bayesian estimation of speciation and extinction rates, ancestral character states, and rates of anagenetic and cladogenetic evolution on a molecular phylogeny. Evolution, 62, 2718–2726.

Bradshaw, A. D. (1991). Genostasis and the limits to evolution. Philosophical Transactions of the Royal Society B: Biological Sciences, 333, 289–305.

Butler, M. A. and King, A. A. (2004). Phylogenetic comparative analysis: A modeling approach for adaptive evolution. American Naturalist, 164, 683–695.

Bürger, R. (1986). Constraints for the evolution of functionally coupled characters: A nonlinear analysis of a phenotypic model. Evolution, 40,182–193.

Carter A. J. R., Hermisson, J. and Hansen, T. F. (2005). The role of epistatic gene interactions in the response to selection and the evolution of evolvability. Theoretical Population Biology, 68, 179–196.

Chan, Y. F., Marks, M. E., Jones, F. C., Villarreal, Jr., G., Shapiro, M. D., Brady, S. D., et al. (2010). Adaptive evolution of pelvic reduction in sticklebacks by recurrent deletion of a *Pitx1* enhancer. Science, 327, 302–305.

Chenoweth S. F., Rundle H. D. and Blows M. W. (2010). The contribution of selection and genetic constraints to phenotypic divergence. American Naturalist, 175, 186–196.

Collar D. C., O'Meara, B. C., Wainwright, P. C., Near, T. J. (2009). Piscivory limits diversification of feeding morphology in centrarchid fishes. Evolution, 63, 1557–1573.

Colosimo, P.F., Balabhadra, S., Villarreal, G., Dickson, M., Grimwood, J., Schmutz, J., et al. (2005). Widespread parallel evolution in sticklebacks by repeated fixation of Ectodysplasin alleles. Science, 307, 1928–1933.

Cooper N. and Purvis, A. (2010). Body size evolution in mammals: Complexity in tempo and mode. American Naturalist, 175, 727–738.

Coyne, J. A., Barton N. H. and Turelli. M. (1997). A critique of Sewall Wright's shifting balance theory of evolution. Evolution, 51, 643–671.

Crespi, B. J. (2000). The evolution of maladaptation. Heredity 84, 623–639.

Dumont, B. L. and Payseur, B. A. (2008). Evolution of the genomic rate of recombination in mammals. Evolution, 62, 276–294.

Eldredge, N. (1989). Macroevolutionary patterns and evolutionary dynamics: Species, niches and adaptive peaks. McGraw-Hill, New York.

Eldredge, N. and Gould S. J. (1972). Punctuated equilibria: an alternative to phyletic gradualism. In T. J. M Schopf (ed.) Models in Paleobiology. Freeman, San Francisco, CA, pp. 82–115.

Eldredge, N., Thompson, J. N., Brakefield, P. M., Gavrilets, S., Jablonski, D., Jackson, J. B. C., et al. (2005). The dynamics of evolutionary stasis. Paleobiology 31, 133–145.

Endler, J. A. (1986). Natural Selection in the Wild. Monographs in population biology 21. Princeton University Press, Princeton, NJ.

Estes, S. and Arnold, S. J. (2007). Resolving the paradox of stasis: models with stabilizing selection explain evolutionary divergence on all timescales. American Naturalist, 169, 227–244.

Frankham, R. (1990). Are responses to artificial selection for reproductive fitness characters consistently asymmetrical? Genetical Research, 56, 35–42.

Freckleton R. P., Harvey, P. H. and Pagel, M. (2002). Phylogenetic analysis and comparative data: A test and review of evidence. American Naturalist, 160, 712–726.

Futuyma, D. J. (1987). On the role of species in anagenesis. American Naturalist, 130, 465–473.

Futuyma, D. J. (1988). Macroevolutionary consequences of speciations: Inferences from phygophagus insects. In D. Otte and J. A. Endler (eds.) Speciation and its consequences. Sinauer, Sunderland, MA

Futuyma, D. J. (2010). Evolutionary constraint and ecolgoicial consequences. Evolution, 64, 1865–1884.

Gavrilets, S. (2004). Fitness Landscapes and the Origin of Species. Monographs in population biology. Levin, S. A. and Horn, H. (eds). Princeton University Press, Princeton, NJ.

Gillespie, J. H. (1991). The Causes of Molecular Evolution. Oxford University Press, Oxford.

Gingerich, P. D. (1983). Rates of evolution: Effects of time and temporal scaling. Science 222,159–161.

Gingerich, P. D. (1993). Quantification and comparison of evolutionary rates. American Journal of Science, 293-A: 453–478.

Gingerich, P. D. (2001). Rates of evolution on the time scale of the evolutionary process. Genetica, 112/113, 127–144.

Gingerich, P. D. (2009). Rates of evolution. Annual Review of Ecology, Evolution and Systematics 40, 657–675.

Gittleman, J. L., Anderson, C. G., Kot M. and Luh H.-K. (1996). Phylogenetic lability and rates of evolution: a comparision of behavioral, morphological and life history traits. In E. P. Martins (ed.) Phylogenies and the comparative method in aminal behavior. Oxford University Press, Oxford, pp. 166–205.

Gomez, D. and Thery M. (2007). Simultaneous crypsis and conspicuousness in color patterns: Comparative analysis of a neotropical rainforest bird community. American Naturalist, 169, S42-S61.

Gosden, T.P. and Svensson, E.I. (2008). Spatial and temporal dynamics in a sexual selection mosaic. Evolution, 62, 845–856.

Gould, S. J. (2002). The structure of evolutionary theory. Belknap, Cambridge, MA.

Gould, S. J. and Eldredge, N. (1993). Punctuated equilibrium comes of age. Nature, 366, 223–227.

Grabowski, M. W., Polk J. D. and Roseman, C. C. (2011). Divergent patterns of integration and reduced constraint in the human hip and the origins of bipedalism. Evolution, 65, 1336–1356.

Hahn, M. W. (2008). Toward a selection theory of molecular evolution. Evolution, 62, 255–265.

Hansen, T. F. (1997). Stabilizing selection and the comparative analysis of adaptation. Evolution, 51, 1341–1351.

Hansen, T. F. (2003). Is modularity necessary for evolvability? Remarks on the relationship between pleiotropy and evolvability. Biosystems, 69, 83–94.

Hansen, T. F. (2006). The evolution of genetic architecture. Annual Review of Ecology, Evolution, and Systematics, 37, 123–157.

Hansen, T. F. (2011). Epigenetics: Adaptation or contingency? In Hallgrimsson, B. and B. K. Hall (eds.) Epigenetics: Linking genotype and phenotype in development and evolution. University of California Press, Berkeley, CA, pp. 354–373.

Hansen, T. F., Alvarez-Castro J. M., Carter A. J. R., Hermisson J. and Wagner G. P. (2006b). Evolution of genetic architecture under directional selection. Evolution, 60, 1523–1536.

Hansen T. F., Armbruster, W. S., Carlson, M. L., and Pélabon, C. (2003b). Evolvability and genetic constraint in *Dalechampia* blossoms: Genetic correlations and conditional evolvability. Journal of Experimental Zoology, 296B, 23–39.

Hansen, T. F., Carter A. J. R. and Pélabon, C. (2006a). On adaptive accuracy and precision in natural populations. American Naturalist, 168, 168–181.

Hansen, T. F. and Houle, D. (2004). Evolvability, stabilizing selection, and the problem of stasis. In M. Pigliucci and K. Preston (eds.) Phenotypic integration: Studying the ecology and evolution of complex phenotypes. Oxford University Press, Oxford, pp. 130–150.

Hansen, T. F. and Houle, D. (2008). Measuring and comparing evolvability and constraint in multivariate characters. Journal of Evolutionary Biology, 21, 1201–1219.

Hansen, T. F. and Martins, E. P. (1996). Translating between microevolutionary process and macroevolutionary patterns: the correlation structure of interspecific data. Evolution, 50, 1404–1417.

Hansen, T. F. and Orzack, S. H. (2005). Assessing current adaptation and phylogenetic inertia as explanations of trait evolution: The need for controlled comparisons. Evolution, 59, 2063–2072.

Hansen T. F., Pélabon, C., Armbruster, W. S. and Carlson, M. L. (2003a). Evolvability and genetic constraint in *Dalechampia* blossoms: Components of variance and measures of evolvability. Journal of Evolutionary Biology, 16, 754–765.

Hansen, T. F., Pélabon, C. and D. Houle. (2011). Heritability is not evolvability. Evolutionary Biology, 38, 258–277.

Hansen, T. F., Pienaar, J. and Orzack, S. H. (2008). A comparative method for studying adaptation to a randomly evolving environment. Evolution, 62, 1965–1977.

Harmon, L. J., Kolbe, J. J., Cheverud, J. M., and Losos, J. B. (2005). Convergence and the multidimensional niche. Evolution 59, 409–421.

Harmon, L. J., Losos, J. B., Davies, T. J., Gillespie, R. G., Gittleman, J. L., Jennings, W. B., et al. (2010). Early bursts of body size and shape evolution are rare in comparative data. Evolution, 64, 2385–2396.

Hendry, A. P., Farrugia T. J. and Kinnison, M. T. (2008). Human influences on rates of phenotypic change in wild animal populations. Molecular Ecology, 17, 20–29.

Hendry, A. P. and Kinnison M. T. (1999). Perspective: the pace of modern life: measuring rates of contemporary microevolution. Evolution, 53, 1637–1653.

Hereford, J. (2009). A quantitative survey of local adaptation and fitness trade-offs. American Naturalist, 173, 579–588.

Hereford J., Hansen, T. F., Houle, D. (2004). Comparing strengths of directional selection: How strong is strong? Evolution, 58, 2133–2143.

Hermisson, J., Hansen T. F. and Wagner G. P. (2003). Epistasis in polygenic traits and the evolution of genetic architecture under stabilizing selection. American Naturalist, 161, 708–734.

Hipp, A. L. (2007). Nonuniform processes of chromosome evolution in sedges (*Carex: Cyperaceae*). Evolution, 61, 2175–2194.

Hohenlohe, P. A. and Arnold, S. A. (2008). MiPoD: A hypothesis-testing framework for microevolutionary inference from patterns of divergence. American Naturalist, 171, 366–385.

Hoffman, A. (1989). Arguments on evolution: a paleontologist's perspective. Oxford University Press, Oxford.

Hoffmann, A. A. and Parsons, P. A. (1997). Extreme environmental change and evolution. Cambridge University Press, Cambridge.

Holt, R. D. (1996). Demographic constraints in evolution: towards unifying the evolutionary theories of senescence and niche conservativism. Evolutionary Ecology, 10, 1–11.

Holt, R. D. and Gaines, M. S. (1992). Analysis of adaptation in heterogeneous landscapes: Implications for the evolution of fundamental niches. Evolutionary Ecology, 6, 433–447.

Holt, R. D., Gomulkiewicz R. and Barfield, M. (2003). The phenomenology of niche evolution via quantitative traits in a "black-hole" sink. Proceedings of the Royal Society B: Biological Sciences, 270, 215–224.

Houle, D. (1992). Comparing evolvability and variability of quantitative traits. Genetics 130, 195–204.

Houle, D. (1998). How should we explain variation in the genetic variance of traits? Genetica, 102/103, 241–253.

Houle, D. (2001). Characters as the units of evolutionary change. In G. P. Wagner (ed.) The character concept in evolutionary biology. Academic Press, New York, pp. 109–140.

Houle, D. and Kondrashov, A. (2006). Mutation. In C. W. Fox and J. B. Wolf (eds.) Evolutionary Genetics: Concepts and case studies. Oxford University Press, Oxford, pp. 32–48.

Houle, D., B. Morikawa and M. Lynch. (1996). Comparing mutational variabilities. Genetics 143, 1467–1483.

Houle, D., Pélabon, C., Wagner, G. P. and Hansen, T. F. (2011). Measurement and meaning in biology. Quarterly Review of Biology, 86, 3–34.

Hunt, G. (2006). Fitting and comparing models of phyletic evolution: random walks and beyond. Paleobiology, 32, 578–601.

Hunt, G. (2007a). The relative importance of directional change, random walks, and stasis in the evolution of fossil lineages. Proceedings of the National Academy of Sciences of the United States of America, 104, 18404–18408.

Hunt, G. (2007b). Evolutionary divergence in directions of high phenotypic variance in the Ostracode genus *Poseidonamicus*. Evolution, 61, 1560–1576.

Hunt, G. (2008). Gradual or pulsed evolution: when should punctuational explanations be preferred? Paleobiology 34, 360–377.

Hunt, G., Bell, M. A. and Travis, M. (2008). Evolution toward a new adaptive optimum: Phenotypic evolution in a fossil stickleback lineage. Evolution, 62, 700–710.

Jablonski, D. (2000). Micro- and macroevolution: scale and hiearchy in evolutionary biology and paleobiology. Paleobiology 26(supplement), 15–52.

Jensen, H., Sæther, B.-E., Ringsby, T. H., Tufto, J., Griffith, S. G., and Ellegren, H. (2003). Sexual variation in heritability and genetic correlations of morphological traits in house sparrow (*Passer domesticus*). Journal of Evolutionary Biology, 16, 1296–1307.

Kawecki, T. J. (1995). Demography of source-sink populations and the evolution of ecological niches. Evolutionary Ecology, 9, 38–44.

Kawecki, T. J. (2008). Adaptation to marginal habitats. Annual Review of Ecology, Evolution, and Systematics, 39, 321–342.

Kellermann, V., van Heerwaarden, B., Sgro, C. M. and Hoffmann, A. A. (2009). Fundamental evolutionary limits in ecological traits drive *Drosophila* species distributions. Science, 325, 1244–1246.

Kemp, T. S. (2006). The origin of mammalian endothermy: a paradigm for the evolution of complex biological structure. Zoological Journal of the Linnean Society, 147, 473–488.

Kemp, T. S. (2007). The origin of higher taxa: macroevolutionary processes, and the case of the mammals. Acta Zoologica, 88, 3–22.

Kimura, M. (1983). The neutral theory of molecular evolution. Cambridge University Press, Cambridge.

Kingsolver, J. G., Hoekstra, H. E., Hoekstra, J.M., Berrigan, D., Vignnieri, S.N., Hill, C. E., et al. (2001). The strength of phenotypic selection in natural populations. American Naturalist, 157, 245–261.

Kinnison, M. T. and A. P. Hendry. (2001). The pace of modern life II: from rates of contemporary microevolution to pattern and process. Genetica, 112/113, 145–164.

Kirkpatrick, M. (1982). Quantum evolution and punctuated equilibria in continuous genetic characters. American Naturalist, 119, 833–848.

Kirkpatrick, M. (2009). Patterns of quantitative genetic variation in multiple dimensions. Genetica, 136, 271–284.

Kozak, K. H. and Wiens, J. J. (2010). Niche conservatism drives elevational diversity patterns in Appalachian salamanders. American Naturalist, 176, 40–54.

Labandeira, C. (2007). The origin of herbivory on land: Initial patterns of plant tissue consumption by arthropods. Insect Science, 14, 259–275.

Labra A., Pienaar J. and Hansen, T. F. (2009). Evolution of thermal physiology in *Liolaemus* lizards: Adaptation, phylogenetic inertia, and niche tracking. American Naturalist, 174, 204–220.

Lande, R. (1976). Natural selection and random genetic drift in phenotypic evolution. Evolution, 30, 314–334.

Lande, R. (1979). Quantitative genetic analysis of multivariate evolution, applied to brain:body size allometry. Evolution, 33, 402–416.

Lande, R. (1980). Microevolution in relation to macroevolution. Paleobiology, 6, 235–238.

Lande, R. (1985). Expected time for random genetic drift of a population between stable phenotypic states. Proceedings of the National Academy of Sciences of the United States of America, 82, 7641–7645.

Lande, R. (1986). The dynamics of peak shifts and the pattern of morphological evolution. Paleobiology, 12, 343–354.

Lande, R. and Arnold, S. J. (1983). The measurement of selection on correlated characters. Evolution, 37, 1210–1226.

Lee, C. E. and Gelembiuk, G. W. (2008). Evolutionary origins of invasive populations. Evolutionary Applications, 1, 427–448.

Lieberman, B. S. and Dudgeon, S. (1996). An evaluation of stabilizing selection as a mechanism for stasis. Paleaeogeography, Paleaeoclimatology, Palaeoecology, 127, 229–238.

Lieberman B. S., Miller W. III. and Eldredge, N. (2007). Paleontological patterns, macroecological dynamics and the evolutionary process. Evolutionary Biology, 34, 28–48.

Losos, J. B., Jackman, T. R., Larson, A., de Queiroz, K. and Rodriguez-Schettino, L. (1998). Contingency and determinism in replicated adaptive radiations of island lizards. Science, 279, 2115–2118.

Lynch, M. (1990). The rate of morphological evolution in mammals from the standpoint of the neutral expectation. American Naturalist, 136, 727–741.

Lynch, M. (1993). Neutral models of phenotypic evolution. In L. Real (ed). Ecological genetics. Princeton University Press, Princeton, NJ, pp. 86–108.

Lynch, M. (2007). The origins of genome architecture. Sinauer, Sunderland, MA.

Lynch, M., Blanchard, J., Houle, D., Kibota, T., Schultz, S., Vassilieva, L., Willis, J. (1999). Perspective: Spontaneous deleterious mutation. Evolution, 53, 645–663.

Lynch, M. and Walsh, B. (1998). Genetics and analysis of quantitative characters. Sinauer.

MacFadden, B. J. (1992). Fossil horses: Systematics, Paleobiology and Evolution of the Family Equidae. Cambridge University Press, Cambridge.

MacFadden, B. J. (2000). Cenozoic mammalian herbivores from the americas: reconstructing ancient diets and terrestrial communities. Annual Review of Ecology and Systematics, 31, 33–60.

Mattila, T. M. and Bokma, F. (2008). Extant mammal body masses suggest punctuated equilibrium. Proceedings of the Royal Society B: Biological Sciences, 275, 2195–2199.

Marroig, G. and Cheverud, J. M. (2005). Size as a line of least evolutionary resistance: Diet and adaptive morphological radiation in New World monkeys. Evolution, 59, 1128–1142.

McGuigan, K. and Blows, M. W. (2010). Evolvability of individual traits in a multivariate context: Patitioning the additive genetic variance into common and specific components. Evolution, 64, 1899–1911.

Ord, T. J., Stamps J. A. and Losos, J. B. (2010). Adaptation and plasticity of animal communications in fluctuating environments. Evolution, 64, 3134–3138.

Parker, T. H. and Garant, D. (2004). Quantitative genetics of sexually dimorphic traits and capture of genetic variance by a sexually-selected condition-dependent ornament in red junglefowl (*Gallus gallus*). Journal of Evolutionary Biology, 17, 1277–1285.

Polly, D. (2008). Developmental dynamics and G-matrices: Can morphometric spaces be used to model phenotypic evolution? Evolutionary Biology, 35, 83–96.

Price G. R. (1970). Selection and covariance. Nature, 227, 520–521.

Price T. (1997). Correlated evolution and independent contrasts. Philosophical Transactions of the Royal Society B: Biological Sciences, 352, 519–529.

Price, T., Turelli, M. and Slatkin, M. (1993). Peak shifts produced by correlated response to selection. Evolution, 47, 280–290.

Rolff J., Armitage, S. A. O., and Coltman, D. W. (2005). Genetic constraints and sexual dimorphism in immune defence. Evolution, 59, 1844–1850.

Rønning, B., Jensen, H., Moe, B., and Bech, C. (2007). Basal metabolic rate: Heritability and genetic correlations with morphological traits in the zebra finch. Journal of Evolutionary Biology, 20, 1815–1822.

Roopnarine, P. D. (2003). Analysis of rates of morphologic evolution. Annual Review of Ecology, Evolution and Systematics, 34, 605–632.

Ruxton, G. D. and Humphries, S. (2008). Can ecological and evolutionary arguments solve the riddle of the missing marine insects? Marine Ecology 29, 1–4.

Scales, J. A., King, A. A. and Butler, M. A. (2009). Running for your life or running for your dinner: What drives fiber-type evolution in lizard locomotor muscles? American Naturalist, 173, 543–553.

Schluter, D. (1996). Adaptive radiation along genetic lines of least resistance. Evolution, 50, 1766–1774.

Schluter, D. (2000a). The ecology of adaptive radiation. Oxford University Press, Oxford.

Schluter, D. (2000b). The role of ecological character displacement in adaptive radiation. American Naturalist, 156, S4–S16.

Schuett G. W., Reiserer, R. S. and Early, R. L. (2009). The evolution of bipedal postures in varanoid lizards. Biological Journal of the Linnean Society, 97, 652–663.

Shapiro, M. D., Bell, M. A. and Kingsley D. M. (2006). Parallel genetic origins of pelvic reduction in vertebrates. Proceedings of the National Academy of Sciences of the United States of America, 103, 13753–13758.

Sheldon, P. R. (1996). Plus ca change—a model for stasis and evolution in different environments. Paleaeogeography, Paleaeoclimatology, Palaeoecology, 127, 209–227.

Siepielski, A. M., DiBattista J. D. and Carlson, S. M. (2009). It's about time: The temporal dynamics of phenotypic selection in the wild. Ecology Letters, 12, 1261–1276.

Simpson, G. G. (1944). Tempo and mode in Evolution. Colombia University Press, New York.

Simpson, G. G. (1953). The major features of Evolution. Columbia University Press, New York.

Sober, E. (1984). The Nature of Selection: Evolutionary Theory in Philosophical Focus. Bradford books.

Stenseth, N. C. and Maynard Smith, J. (1984). Coevolution in ecosystems: red queen evolution or stasis? Evolution, 38, 870–880.

Stromberg, C. A. E. (2006). Evolution of hypsodonty in equids: testing a hypothesis of adaptation. Paleobiology, 32, 236–258.

Svensson, E. I. and Gosden, T. P. (2007). Contemporary evolution of secondary sexual traits in the wild. Functional Ecology, 21, 422–433.

Thompson J. N. (1994). The coevolutionary process. University of Chicago Press, Chicago, IL.

Uyeda, J. C, Hansen, T. F., Arnold, S. J., and Pienaar, J. (2011). The million-year wait for macroevolutionary bursts. Proceedings of the National Academy of Sciences of the United States of America, 108, 15908–15913.

Van Buskirk, J. (2009). Getting in shape: adaptation and phylogenetic inertia in morphology of Australian anuran larvae. Journal of Evolutionary Biology, 22, 1326–1337.

Van Tienderen, P. M. (2000). Elasticities and the link between demographic and evolutionary dynamics. Ecology 81, 666–679.

Van Valen, L. (1973). A new evolutionary law. Evolutionary Theory, 1, 1–30.

Verdu, M. and Gleiser, G. (2006). Adaptive evolution of reproductive and vegetatitve traits driven by breeding system. New Phytologist, 169, 409–417.

Vermeij, G. J. (1987). Evolution and escalation: an ecological history of life. Princeton University Press, Princeton, NJ.

Vermeij, G. J. (2010). Sound reasons for silence: why do molluscs not communicate acoustically? Biological Journal of the Linnean Society, 100, 485–493.

Wade, M. J. and Goodnight, C. J. (1998). Perspective: the theories of Fisher and Wright in the context of metapopulations: when nature does many small experiments. Evolution, 52, 1537–1553.

Wagner, G. P. (1988). The influence of variation and of developmental constraints on the rate of multivariate phenotypic evolution. Journal of Evolutionary Biology, 1, 45–66.

Wagner, G. P. and Altenberg, L. (1996). Complex adaptations and evolution of evolvability. Evolution, 50, 967–976.

Wagner G. P., Pavlicev M. and Cheverud, J. M. (2007). The road to modularity. Nature Reviews Genetics, 8, 921–931.

Walsh, B. and Blows, M. W. (2009). Abundant genetic variation + strong selection = multivariate genetic constraints: a gemometric view of adaptation. Annual Review of Ecology, Evolution and Systematics, 40, 41–59.

Wiens, J. J. (2004). Speciation and ecology revisited: Phylogenetic niche conservativism and the origin of species. Evolution, 58, 193–197.

Wiens, J. J. and Graham, C. H. (2005). Niche conservativism: Integrating evolution, ecology, and conservation biology. Annual Review of Ecology, Evolution and Systematics, 36, 519–539.

Willi, Y., van Buskirk J. and Hoffmann, A. A. (2006). Limits to the adaptive potential of small populations. Annual Review of Ecology, Evolution and Systematics, 37, 433–458.

Williams, G. C. (1992). Natural Selection: Domains, Levels, and Challenges. Oxford University Press, Oxford.

Wright, S. (1931). Evolution in Mendelian populations. Genetics 16, 97–159.

Wright, S. (1932). The roles of mutation, inbreeding, crossbreeding and selection in evolution. Proceedings of the Sixth Annual Congress of Genetics, 1, 356–366.

Yoder, J. B., E. Clancey, S. Des Roches, J. M. Eastman, L. Gentry, W. Godsoe, et al. (2010). Ecological opportunity and the origin of adaptive radiations. Journal of Evolutionary Biology, 23, 1581–1596.

Zeng, Z. B. (1988). Long-term correlated response, interpopulation covariation, and interspecific allometry. Evolution, 42, 363–374.

CHAPTER 14

Adaptive Dynamics: A Framework for Modeling the Long-Term Evolutionary Dynamics of Quantitative Traits

Michael Doebeli

14.1 Introduction

In evolutionary models based on Wright's classic fitness landscape metaphor (Wright 1932), fitness is usually a static quantity, and selection is determined by factors that are extrinsic to an evolving population (see e.g. Chapters 2 and 4, this volume). In particular, the fitness of a given type does not depend on the distribution of other types present in the population. In this perspective, evolution becomes an optimization process that unfolds on a given fixed fitness landscape. Locally, i.e. in the neighborhood of a single peak of the landscape, this optimization process is described by Fisher's Fundamental Theorem (Fisher 1930). (Strictly speaking, the local process of climbing fitness peaks is described by the first part of the fundamental theorem, see Chapter 4 in this volume for an enlightening discussion.)

Accordingly, in deterministic, i.e. non-stochastic, evolutionary dynamics, populations are assumed to climb the nearest fitness peak, that is, evolution is assumed to produce the type that is best able to cope with the extrinsic factors, and hence is best adapted. This perspective already poses difficult problems when extrinsic environments are temporally varying, and when life histories or genetic architectures are complex. Moreover, it sets the stage for Wright's classical shifting balance problem Wright (1931), which refers to the evolutionary mechanisms that can move populations from one local peak to another. A large body of theory has been developed in population genetics and life history theory to deal with these complexities (e.g. Fisher 1930; Wright 1984; Metz and Diekmann 1986; Tuljapurkar 1990; Caswell 1989; Bürger 2000; Rice 2004; Gavrilets 2004).

It has of course long been recognized, and is indeed integral to evolutionary thinking, that the fitness of a given type may not only be determined by extrinsic factors, but also by the distribution of other types present in an evolving population, or in co-evolving populations of a different species. In other words, fitness may be frequency-dependent. Frequency-dependence has, for example, already been studied extensively in Fisher's sex ratio theory (Fisher 1930). The assumption of frequency-dependent fitness is perhaps made most explicitly in evolutionary game theory (Maynard Smith and Price 1973, Maynard Smith 1982), where one assumes that payoffs from interactions between individuals depend on the strategies employed by the interacting individuals. But much of evolutionary game theory is concerned with understanding the frequency dynamics of a finite and discrete set of strategies, and typically, game theoretical models lack an ecological embedding (and hence often assume fixed population sizes). In particular, evolutionary game theory was not conceived for determining dynamics in continuous phenotype spaces (although it can be extended in that direction, see, e.g. Sandholm (2010)).

Frequency-dependence implies that the fitness landscapes determining selection pressures change themselves dynamically as populations respond to selection, i.e. as frequency distributions change. To describe evolutionary dynamics one would therefore like to have a general theory that incorporates the feedback between evolutionary change and changes in the fitness landscape, and hence incorporates the dynamics of fitness landscapes. In fact, such a connection between ecology and evolution is already implicitly contained in Fisher's fundamental theorem. As explained in Steve Frank's chapter in this volume (Chapter 4), Fisher partitioned the total change in fitness into change caused by natural selection (which is always ≥ 0) and change caused by the deterioration in the environment. This second component essentially incorporates the ecological feedback that is due to evolutionary change. In fact, Fisher argued that the two components generally tend to cancel each other, so that the overall level of adaptation remains constant, precisely because of this ecological feedback (see Chapter 4).

When invoking the fundamental theorem this second component of the theorem is typically ignored (Chapter 4). Yet when considering frequency-dependent selection, it is exactly inclusion of this second component that is required for arriving at a description of evolutionary dynamics, because the second component describes how selection pressures change due to evolutionary change. In other words, the second component is required for a closed description of evolutionary dynamics that incorporates the feedback between selection pressures and evolutionary change, and hence contains a description of the dynamics of the fitness landscape.

To study evolutionary dynamics in continuous phenotype spaces, several modeling frameworks are in principle available for incorporating this feedback, and hence for studying evolution under frequency-dependent selection. Chief among them are the quantitive genetics models based on the approach developed by Lande (Lande 1979), the fitness generating functions developed by Vincent and Brown (Vincent and Brown 2005), and the framework of adaptive dynamics developed by Metz and others (Dieckmann and Law 1996; Geritz et al. 1998; Metz et al. 1992). Of these approaches, adaptive dynamics has two distinctive advantages. First, adaptive dynamics can be derived from first principles governing individual-based ecological processes that give rise to a stochastic birth–death process. This process can be described by a stochastic master equation, from which the adaptive dynamics framework can be derived by making a number of simplifying assumptions, as has been demonstrated in the seminal paper Dieckmann and Law (1996) (see also Champagnat et al. 2006). Thus, in contrast to frequency-dependent quantitative genetic models and models based on fitness generating functions, adaptive dynamics is based on an underlying individual-based process for frequency-dependent ecological interactions.

Second, adaptive dynamics has the advantage of being based on a fitness definition that not only has a precise ecological meaning, but is model-independent. In particular, and in contrast to other frameworks, adaptive dynamics explicitly contains the ecological feedback, and hence a description of the dynamics of fitness landscapes. This description is based on the notion of invasion fitness (Metz et al. 1992), which refers to the ecological per capita growth rate of rare mutant types in an environment that is determined by monomorphic resident populations, and in particular by the ecological dynamics of such resident populations. The invasion fitness can be extracted from the individual-based master process (Dieckmann and Law 1996) and serves as a unifying notion that allows for a model-independent classification of major regimes of adaptive dynamics. As in classical dynamical systems theory, this delimits the possibilities for any specific model one might want to study, and can thus serve as a useful investigative guide.

Two different stability concepts that are both derived from the invasion fitness function are of crucial importance in such a classification: convergence stability and evolutionary stability. In this article, I will review some fundamental definitions and properties of adaptive dynamics, including the definitions of convergence and evolutionary stability. In sections 14.2 and 14.3 I will present examples that illustrate the definition of invasion fitness and the paradigmatic dynamic regimes that can occur when bifurcations lead to the loss of either convergence or evolutionary stability. In section 14.4 I will illustrate the importance of an ecologically based fitness definition by briefly describing the phenomenon of evolutionary suicide. The basic theory presented in sections 14.2 and 14.3 is an abridged version of the Appendix in Doebeli (2011) (see also Diekmann (2004) and Dercole and Rinaldi (2008) for extensive treatments of the principles of adaptive dynamics). The examples are taken from the existing literature on the evolution of quantitative traits determining competitive and predator–prey interactions.

14.2 Basics of adaptive dynamics: convergence stability

The general problem in adaptive dynamics is to follow the co-evolutionary dynamics of real-valued traits x_i, $i = 1, \ldots, m$ in m different species, where the traits $x_i = (x_{i1}, \ldots, x_{in_i}) \in \mathbb{R}^{n_i}$ are themselves in general given by a multidimensional vector of dimension n_i. This is done by first assuming that at any point in time, there is a resident community in which all m species are monomorphic for some trait value x_i, $i = 1, \ldots, m$, i.e. that all individuals currently present in species i have the same trait value x_i. This resident community determines the ecological environment for any newly appearing mutants, and in each of the m species, the invasion fitness function $f_i(x, y_i)$ is then defined as the long term per capita growth rate of a rare mutant type $y_i = (y_{i1}, \ldots, y_{in_i})$ of species i, appearing in the resident community in which each species is monomorphic for the resident type x_i. (Technically, if n is the population density of the mutant type at time t, the invasion fitness is calculated as an average of either dn/n in continuous time, or n_{t+1}/n_t in discrete time, over a suitable time interval in which the resident population traces out its ecological attractor, and under the assumption that n is small; see Metz et al. (1992) for details.)

The invasion fitness functions f_i may or may not be amenable to analytical calculation. For example, if the ecological dynamics of the resident community exhibits a stable equilibrium, then the invasion fitness function is simply the per capita growth rate of a mutant type evaluated at this equilibrium (see later examples). However, if the ecological dynamics of the resident community is complicated, e.g. when the m resident populations exhibit chaotic fluctuations, then the per capita growth rate of rare mutants must be evaluated at all population densities on the attractor of the resident population dynamics, and then averaged in the limit as time goes to infinity (Metz et al. 1992). This makes it impossible in general to calculate the invasion fitness function explicitly, but many features of adaptive dynamics can be derived without such explicit calculations.

For example, we must always have $f_i(x, x_i) = 0$ for all i. This is because the long-term growth rate of a mutant that has the same trait value as the resident must have the same long term ecological growth rate as the resident, and the resident is assumed to persist ecologically (i.e. to neither go extinct nor to increase indefinitely), and hence its long-term growth rate must be zero (assuming time is continuous; in discrete time long-term persistence is equivalent to a long-term expected number of offspring of one per individual and per generation, and the corresponding consistency condition for the invasion fitness function is then $f_i(x, x_i) = 1$ for all i, which can be translated into $f_i(x, x_i) = 0$ for all i by taking logarithms).

Given that $f_i(x, x_i) = 0$, one now restricts attention to small mutational deviations, i.e. to mutants y_i that are close to the resident value x_i, and one would like to know whether such small mutations would lead to per capita growth rates that are smaller or larger than the growth rate of the resident (i.e. than 0). Therefore, one considers the partial derivatives of the invasion fitness functions evaluated at the resident value, i.e. the selection gradients

$$s_i(x_1, \ldots, x_m) = \left.\frac{\partial f_i(x, y_i)}{\partial y_i}\right|_{y_i = x_i}. \quad (14.1)$$

Note that in general, i.e. when traits are multidimensional, the selection gradients are themselves multidimensional vectors:

$$s_i(x_1, \ldots, x_m) = (s_{i1}, \ldots, s_{in_i}), \quad (14.2)$$

where

$$s_{ij} = \left.\frac{\partial f_i(x, y_i)}{\partial y_{ij}}\right|_{y_i = x_i} \quad (14.3)$$

is the partial derivative of f_i with respect to the jth component of the mutant trait y_i, evaluated at the resident trait $y_{ij} = x_{ij}$.

The selection gradients indicate the directions in phenotype space in which invasion of mutants that are phenotypically close to the resident is most likely. Therefore, the selection gradients are an essential ingredient to determine the adaptive dynamics of the traits x_i. However, in general one also has to take into account how the mutational process in the various species warps the phenotype space by making certain directions more or less likely to be reached by mutations. The mutational process in species i is described by the mutational variance-covariance matrix $B_i(x)$, which characterizes the distribution of mutations in species i, as well as the rate at which mutations occur. Specifically, this matrix reflects the expected size of mutations in each phenotypic direction, as well as the correlations between mutations in different phenotypic directions (given as non-zero off-diagonal entries in $B_i(x)$). It is typically assumed that the mutation distribution is symmetric around the resident phenotype, and hence that the $B_i(x)$ are symmetric matrices. The matrix $B_i(x)$ also reflects the rate of occurrence of new mutations, which may for example vary due to changes in population size. Therefore, the matrix $B_i(x)$ depends in general on the current resident trait values, as these determine the ecology, and hence the population sizes of the various interacting species.

The adaptive dynamics of the trait vectors x_i are determined by the selection gradients s_i and by the mutational matrices B_i and take the form of a system of coupled differential equations:

$$\frac{dx_i}{dt} = B_i(x) \cdot s_i(x) \quad i = 1, ..., m, \quad (14.4)$$

where $s_i(x)$ is the (column) vector (equation 14.2) of selection gradients on each phenotypic component in species i, and where $x = (x_1, \ldots, x_m) = \in \mathbb{R}^{\sum_{i=1}^{m} n_i}$ is the vector of all trait values.

Equation 14.4 is known as the canonical equation of adaptive dynamics. It can be derived from an underlying stochastic process as described in Dieckmann and Law (1996), to which I refer for more details. The general goal of adaptive dynamics is to find the solutions $x(t) = (x_1(t), ..., x_m(t))$ of the system of differential equations (equation 14.4) and determine their behaviour. As with all dynamical systems, equilibrium points are useful starting points, and in adaptive dynamics, a point $x^* = (x_1^*, \ldots, x_m^*)$ in trait space satisfying $B_i(x^*) \cdot s_i(x^*) = 0$ for all $i = 1, \ldots, m$ is called a singular point of the adaptive dynamics (equation 14.4), or an evolutionary singularity.

As long as all the mutational matrices $B_i(x)$ have non-zero determinants, which is usually assumed, singular points are given as solutions of the m equations $s_i(x^*) = 0, i = 1, \ldots, m$. Evolutionary singularities are a central focus of attention in adaptive dynamics, and the most basic question is whether a singular point is an attractor for the adaptive dynamics for which it is a stationary point. A common approach to answering this question is to consider the linearization of the adaptive dynamical system (equation 14.4) around the singular point, i.e. to consider the Jacobian matrix $J(x^*)$ at the singular point. The Jacobian matrix is the matrix of partial derivatives of the functions defining the differential equations (equation 14.4). In other words, the Jacobian matrix $J(x^*)$ is given by

$$J(x^*) = \begin{pmatrix} \frac{\partial [B_1(x) \cdot s_1(x)]}{\partial x_1}\Big|_{x=x^*} & \cdots & \frac{\partial [B_1(x) \cdot s_1(x)]}{\partial x_m}\Big|_{x=x^*} \\ \vdots & \cdots & \vdots \\ \frac{\partial [B_m(x) \cdot s_m(x)]}{\partial x_1}\Big|_{x=x^*} & \cdots & \frac{\partial [B_m(x) \cdot s_m(x)]}{\partial x_m}\Big|_{x=x^*} \end{pmatrix}$$

(14.5)

Note that all the partial derivatives occurring in the expression for $J(x^*)$ are themselves vectors of partial derivatives; for example, $\frac{\partial [B_1(x) \cdot s_1(x)]}{\partial x_1} = \left(\frac{\partial [B_1(x) \cdot s_1(x)]}{\partial x_{11}}, \ldots, \frac{\partial [B_1(x) \cdot s_1(x)]}{\partial x_{1n_1}} \right)$.

It is well known from dynamical systems theory that the singular point x^* is locally stable for the dynamics (equation 14.4) if all eigenvalues of the Jacobian matrix have negative real parts. Accordingly, a singular point x^* for an adaptive dynamical system is called convergent stable if all the eigenvalues of the Jacobian $J(x^*)$ at the singular point have negative real parts. In this case, if the initial conditions of the adaptive dynamics (equation 14.4) are sufficiently close to x^*, then the adaptive dynamics will converge to the singular point, and x^* is thus an attractor for the adaptive dynamics.

Even though the selection gradients s_i, and hence the invasion fitness functions f_i, play a crucial role in determining convergence stability of singular points, it is important to note that the Jacobian matrix also depends on the mutational matrices B_i. Therefore, convergence stability is in general not a property of the selection gradients alone. For example, as we will see later, it can happen that for a given set of selection gradients, all eigenvalues of the Jacobian matrix at a singular point have negative real parts for some choices of the mutational quantities $B(x_i^*)$, but not for other choices. However, there are certain circumstances in which this difficulty does not arise, and hence in which the selection gradients contain all the information about the convergence stability of a singular point. This situation has been captured in the definition of strong convergence stability in Leimar (2001, 2005, 2009), to which I refer for an in depth discussion of this and related topics (including the concept of absolute convergence stability).

The notion of convergence stability sets the stage for evolutionary cycling, which can for instance occur in adaptive dynamical systems of dimensions higher than 1 when a singular point loses stability through a Hopf bifurcation. This is illustrated by the following example of co-evolution in a predator–prey system, which is due to Dieckmann et al. (1995) and Marrow et al. (1996) (see also chapter 5 in Doebeli (2011)).

We assume that prey individuals are variable for a one-dimensional trait x, and predator individuals are variable for a one-dimensional trait y (i.e. $x, y \in \mathbb{R}$). The ecological dynamics setting the stage for determining the invasion fitness functions in the prey and the predator is assumed to be described by a classical Lotka–Volterra predator–prey model with linear functional response, in which the prey trait x determines the carrying capacity of the prey, and both the prey trait x and the predator trait y determine the search efficiency of the predator. More precisely, let $N(x)$ be the density of a prey population that is monomorphic for trait value x, and let $P(y)$ be the density of a predator population that is monomorphic for trait value y. Then the ecological dynamics of $N(x)$ and $P(y)$ are given by

$$\frac{dN(x)}{dt} = rN(x)\left(1 - \frac{N(x)}{K(x)}\right) - \beta(x, y)N(x)P(y) \tag{14.6}$$

$$\frac{dP(y)}{dt} = c\beta(x, y)N(x)P(y) - dP(y). \tag{14.7}$$

Here $K(x)$ is the carrying capacity of a prey population that is monomorphic for x, i.e. the equilibrium population density of such a population in the absence of predators. (Note that even though the carrying capacity is a population property, it is straightforward to interpret the parameter $K(x)$ as a component of per capita death rates, and hence as a property of individuals; see section 14.4 and chapter 3 in Doebeli (2011) for a discussion of this.) Here we assume the carrying capacity function to be Gaussian:

$$K(x) = K_0 \exp\left[\frac{-x^2}{2\sigma_K^2}\right]. \tag{14.8}$$

However, the precise form of the carrying capacity is not important for the results reported later in this section, as long as it is a unimodal differentiable function. The parameter $r > 0$ in equation 14.6 is the intrinsic per capita growth rate of the prey (which is assumed to be independent of the trait x). The interaction between the prey and the predator is described by a linear functional response, in which the efficiency $\beta(x, y)$ with which a predator of type y attacks and kills prey of type x depends on the difference in trait values, $x - y$. We assume that this function is also Gaussian:

$$\beta(x, y) = b \exp\left[\frac{-(x - y)^2}{2\sigma_\beta^2}\right]. \tag{14.9}$$

(Biologically, this implies that the prey and predator traits can be appropriately scaled in such a way that for any given prey trait x, the predator with trait $y = x$ has maximal attack rate.) Finally, the parameter c in equation 14.7 is the conversion efficiency of the predator (i.e. the rate at which captured prey is converted into predator offspring), and the parameter d is the per capita death rate in the predator population. Both these parameter are assumed to be independent of the predator phenotype y.

It is well known (e.g. Kot 2001) that the ecological model given by equations 14.6 and 14.7 has a stable equilibrium $(\hat{N}(x, y), \hat{P}(x, y))$ at which both species persist if and only if $d < c\beta(x, y)K(x)$. In that case,

$$(\hat{N}(x, y), \hat{P}(x, y))$$
$$= \left(\frac{d}{c\beta(x, y)}, \frac{r}{\beta(x, y)}\left(1 - \frac{d}{c\beta(x, y)K(x)}\right)\right), \tag{14.10}$$

and all ecological trajectories starting out with positive prey and predator densities converge to the equilibrium $(\hat{N}(x, y), \hat{P}(x, y))$ (which can be a stable node or a stable focus, see Kot (2001)). Note that the equilibrium densities of monomorphic populations are functions of the two trait values x and y.

To derive the adaptive dynamics of the traits x and y, we assume that these traits are confined to regions in trait space in which the conditions for the existence of the stable equilibrium $(\hat{N}(x, y), \hat{P}(x, y))$ are satisfied (i.e. we assume that $d < c\beta(x, y)K(x)$ always holds). It should be mentioned, however, that there are interesting evolutionary scenarios in which the trait values evolve to the boundary of the coexistence region, thus leading to the extinction of the predator (Dieckmann et al. 1995).

To determine the invasion fitness functions for mutant traits in both prey and predator, we assume monomorphic resident populations that are at their ecological equilibrium $(\hat{N}(x, y), \hat{P}(x, y))$, and remain so during an invasion attempt of rare mutants. Then the per capita growth rate of a rare mutant type x' in the prey is

$$f_{prey}(x, y, x') = r\left(1 - \frac{\hat{N}(x, y)}{K(x')}\right) - \beta(x', y)\hat{P}(x, y).$$

(14.11)

The first term of the right-hand side of this equation reflects intrinsic growth of the mutant as well as intraspecific competition that the mutant experiences from the resident prey type (but not from itself, because the mutant is assumed to be rare). The second term on the right-hand side of equation 14.11 reflects per capita predation pressure of the resident predator population density $\hat{P}(x, y)$ on the mutant, on which the resident predator has an attack rate $\beta(x', y)$.

Assuming the same resident (x, y), the per capita growth rate of a rare mutant type y' in the predator is

$$f_{pred}(x, y, y') = c\beta(x, y')\hat{N}(x, y) - d.$$

(14.12)

The first term on the right-hand side is the per capita production of mutant offspring as a consequence of predation on the resident prey, while the second term is the phenotype-independent per capita death rate.

Given the invasion fitness functions, the selection gradients in the prey and the predator are determined as

$$s_{prey}(x, y) = \left.\frac{\partial f_{prey}(x, y, x')}{\partial x'}\right|_{x'=x}$$

$$= \frac{r\hat{N}(x, y)K'(x)}{K(x)^2} - \frac{\partial \beta}{\partial x}(x, y)\hat{P}(x, y)$$

(14.13)

$$s_{pred}(x, y) = \left.\frac{\partial f_{pred}(x, y, y')}{\partial y'}\right|_{y'=y} = c\frac{\partial \beta}{\partial y}(x, y)\hat{N}(x, y).$$

(14.14)

Finally, the adaptive dynamics of the traits x and y are then given by

$$\frac{dx}{dt} = B_{prey}(x, y) \cdot s_{prey}(x, y)$$

(14.15)

$$\frac{dy}{dt} = B_{pred}(x, y) \cdot s_{pred}(x, y).$$

(14.16)

Here the mutational matrices $B_{prey}(x, y) = m_{prey}\hat{N}(x, y)$ and $B_{pred} = m_{pred}\hat{P}(x, y)$ are 1×1 matrices, i.e. real-valued functions of the trait values x and y. Because mutation production is determined by the current population size, these functions are proportional to the respective resident population size, with the coefficients of proportionality m_{prey} and m_{pred} reflecting the rate and distributions of mutations produced per individual (Dieckmann and Law 1996).

Equilibria of the adaptive dynamics equations 14.5 and 14.6, i.e. singular points, are obtained by setting the right hand side of these equations equal to 0. Since we assume coexistence of prey and predator at all times, this is equivalent to requiring that the fitness gradients in both species vanish. Thus, equilibria (x^*, y^*) of the adaptive dynamics are given as solutions to the equations

$$\frac{r\hat{N}(x^*, y^*)K'(x^*)}{K(x^*)^2} = \frac{\partial \beta}{\partial x}(x^*, y^*)\hat{P}(x^*, y^*) \quad (14.17)$$

$$c\frac{\partial \beta}{\partial y}(x^*, y^*)\hat{N}(x^*, y^*) = 0. \quad (14.18)$$

Given the functional form of $\beta(x, y)$, equation 14.9, it is immediately clear from the second of these equations that an equilibrium (x^*, y^*) must satisfy $y^* = x^*$. But then the right-hand side of the first equation is zero, and hence the prey equilibrium must satisfy $K'(x^*) = 0$ as well. Given the unimodal function $K(x)$, eq. 14.8, it follows that $x^* = 0$, i.e. the prey equilibrium trait must be the trait value maximizing the prey carrying capacity. Thus, the adaptive dynamics has a unique singular point at $(x^*, y^*) = (0, 0)$.

The Jacobian matrix $J(x, y)$ of the adaptive dynamics in equations 14.15 and 14.6 at a point (x, y) is

$$J(x, y) = \begin{pmatrix} \frac{\partial}{\partial x}\left[m_{prey}\hat{N}(x, y)s_{prey}(x, y)\right] & \frac{\partial}{\partial y}\left[m_{prey}\hat{N}(x, y)s_{prey}(x, y)\right] \\ \frac{\partial}{\partial x}\left[m_{pred}\hat{P}(x, y)s_{pred}(x, y)\right] & \frac{\partial}{\partial y}\left[m_{pred}\hat{P}(x, y)s_{pred}(x, y)\right] \end{pmatrix}.$$
(14.19)

Evaluating at $(0, 0)$, and taking into account that the selection gradients vanish at the singular point, one finds

$$J(0, 0) = \begin{pmatrix} m_{prey}\hat{N}(0, 0)\frac{\partial s_{prey}}{\partial x}(0, 0) & m_{prey}\hat{N}(0, 0)\frac{\partial s_{prey}}{\partial y}(0, 0) \\ m_{pred}\hat{P}(0, 0)\frac{\partial s_{pred}}{\partial x}(0, 0) & m_{pred}\hat{P}(0, 0)\frac{\partial s_{pred}}{\partial y}(0, 0) \end{pmatrix}.$$
(14.20)

It is well known that the two eigenvalues of J have negative real parts, and hence the singular point is convergence stable, if and only if the determinant of J is positive and the trace of J (i.e. the sum of its diagonal elements) is negative. Using the functions $K(x)$ and $\beta(x, y)$ given by equations 14.8 and 14.9, as well as expression 14.10 for the equilibrium densities $\hat{N}(0, 0)$ and $\hat{P}(0, 0)$ and expressions 14.13 and 14.14 for the selection gradients, one can calculate that the determinant of J is a positive multiple of $cbK_0 - d$ (where $b = \beta(0, 0)$ and $K_0 = K(0)$). Because we always assume $d - cbK_0 < 0$, the determinant of J is therefore always positive. Also, the trace of J is a positive multiple of the expression $(cm_{pred} - m_{prey})\sigma_K^2(d - cbK_0) - dm_{prey}\sigma_\beta^2$. It follows that if $cm_{pred} < m_{prey}$, the trace will be positive, and hence the singular point unstable, if σ_K is large enough.

Since c is the conversion efficiency of captured prey to predator offspring, the condition $cm_{pred} < m_{prey}$ can be interpreted as requiring that production of mutant predators per (captured) prey individual is slower than production of mutant prey, or that predators are evolutionarily less plastic, i.e. have a smaller mutational variance. If in addition, the stabilizing selection in the prey due to the carrying capacity $K(x)$ is weak enough, i.e. if σ_K is large enough, then, intuitively speaking, there is enough evolutionary freedom for the prey to escape the singular point. Moreover, this escape (i.e. instability of the singular point) can lead to sustained evolutionary oscillations. Indeed, using σ_K as a bifurcation parameter, it is easy to come up with scenarios in which the singular point $(0, 0)$ changes from a stable to an unstable focus in a Hopf bifurcation, which occurs as the parameter σ_K is increased above a critical value. As a consequence, the adaptive dynamics changes from a regime of damped oscillations converging to the singular point, to a regime of sustained evolutionary oscillations. This is illustrated in Fig. 14.1.

The oscillating adaptive dynamics correspond to an evolutionary arms race between the prey and the predator. It is important to note that it is really the joint effects of mutational and ecological parameters that determine whether such arms races occur. For example, the singular point is always stable if the mutation rate in the predator is high enough (i.e. if $cm_{pred} > m_{prey}$), but if the mutation rate in the predator is low, stability of the singular point depends on the ecological parameters (i.e. on c, σ_K, d, b, K_0 and σ_β). The derivation of the adaptive dynamics from the invasion fitness functions clearly reveals these relationships.

The oscillatory regimes of the adaptive dynamics (equations 14.15 and 14.16) resulting from convergence instability of the singular point are discussed more fully in Dieckmann et al. (1995) and Marrow et al. (1996), and Dercole et al. (2006) describe more complex dynamic regimes in the adaptive dynamics of predator–prey coevolution. However, even if the singular point $(0, 0)$ is convergent stable, it need not be the end point of the evolutionary process. To see this, one needs to consider a second stability concept, namely evolutionary stability. Evolutionary stability is historically a more familiar notion than convergence stability, and in the context of adaptive dynamics it is again defined in terms of the invasion fitness functions. Evolutionary stability, as well as potential consequences of evolutionary instability, will be considered in the next section using a simpler class of models. For the adaptive dynamics of predator-prey interactions, evolutionary stability and instability of singular points has been explored in Doebeli

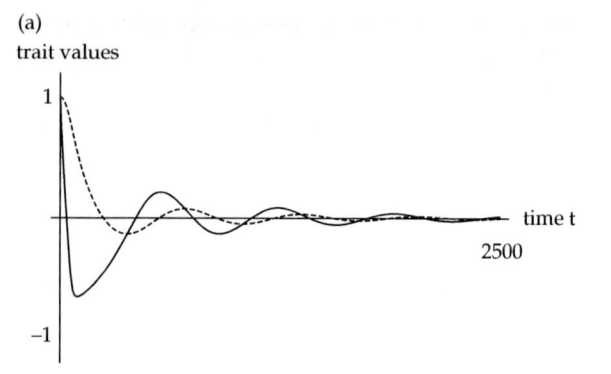

Figure 14.1 Coevolutionary adaptive dynamics (equation 14.15) of the prey trait (continuous lines) and (equation 14.16) of the predator trait (dashed lines). In (a), the adaptive dynamics exhibits damped oscillations and convergence to the singular point (0, 0), which is a stable focus because the Jacobian matrix (equation 14.20) has two complex conjugate eigenvalues with negative real parts and non-zero imaginary parts. In (b), the real parts of the Jacobian matrix (equation 14.20) have become positive due to an increase in the width of the carrying capacity function, and hence the singular point is an unstable focus, and the adaptive dynamics exhibits cyclic behaviour, in which the predator trait continually chases the escaping prey trait. Parameter values were $r = 1$, $K_0 = 1$, $b = 1$, $\sigma_\beta = 2$, $c = 1$, $d = 0.1$, $m_{prey} = 1$, $m_{pred} = 0.2$ for both panels, and $\sigma_K = 0.7$ for panel (a), and $\sigma_K = 1.2$ for panel (b).

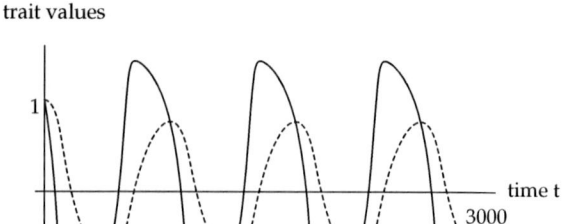

and Dieckmann (2000) and more fully in Doebeli (2011).

14.3 Basics of adaptive dynamics: evolutionary stability and evolutionary branching

By definition, singular points x^* of an adaptive dynamical system are points in phenotype space at which the selection gradients are 0, i.e., at which the first derivative of the invasion fitness functions with respect to mutant trait values vanish. For one-dimensional traits, this leaves generically only two possibilities: as a function of mutant traits y, the invasion fitness function $f(x^*, y)$ either has a maximum or a minimum at $y = x^*$. In particular, if the singular point is a fitness maximum, no nearby mutants can invade, in which case the singular point is called evolutionarily stable. The extension of this notion to higher dimensions is straightforward: a singular point is evolutionarily stable if, in each of the interacting species, the singular point is a fitness maximum for the invasion fitness functions $f_i(x^*, y)$. If there is only one species involved, and if one is considering the evolution of a multi-dimensional trait $x = (x_1, \ldots, x_n)$ in a single species determined by the invasion fitness function $f(x, y)$, the condition for evolutionary stability is that the Hessian matrix of second derivatives of the invasion fitness function with respect to mutant trait values at the singular point is negative definite. In other words, all eigenvalues of the matrix

$$H(x^*) = \begin{pmatrix} \frac{\partial^2 f}{\partial y_1 \partial y_1}(x^*, y)\Big|_{y=x^*} & \cdots & \frac{\partial^2 f}{\partial y_1 \partial y_n}(x^*, y)\Big|_{y=x^*} \\ \vdots & \ddots & \vdots \\ \frac{\partial^2 f}{\partial y_n \partial y_1}(x^*, y)\Big|_{y=x^*} & \cdots & \frac{\partial^2 f}{\partial y_n \partial y_n}(x^*, y)\Big|_{y=x^*} \end{pmatrix}$$

(14.21)

must be negative. In the case of multiple coevolving species, a singular point is evolutionarily stable if the Hessian of the invasion fitness functions in each of the interacting species is negative definite.

Note that in contrast to the definition of convergence stability given above, the definition of evolutionary stability only involves the invasion fitness functions, but not the mutational matrix B. This notion of evolutionary stability conforms with the classical definition of an evolutionary stable strategy (ESS) (Maynard Smith and Price 1973): if almost all individuals in coevolving populations have phenotypes specified by an evolutionary stable singular point, then no (nearby) mutant has a positive invasion fitness, and hence no nearby mutant can invade this population. Thus, once such a strategy becomes established in a population, no further evolutionary change through small mutations occurs deterministically.

Also, it is very important to note that even though both convergence stability and evolutionary stability are derived from second derivatives of the invasion fitness functions, the two notions do in general not coincide mathematically. For example, a singular point may be evolutionarily stable, but not convergent stable, a situation which has been called a Garden-of-Eden configuration by Nowak and Sigmund (1990). If a population is at such a singularity, it will stay there forever, because no mutant can invade, but if the population starts its adaptive dynamics at a different trait value, it will never reach that particular singular point, because the singular point is not an attractor for the adaptive dynamics.

More importantly, singular points that are both convergent stable and evolutionarily stable are classically called continuously stable strategies (CSS) (Eshel 1983). Such singular points represent endpoints of the evolutionary process. In stark contrast to this are singular points that are convergent stable, but evolutionarily unstable. Such singular points may be the starting point for adaptive diversification. For example, consider the one-dimensional adaptive dynamics of a scalar trait x in a single species (i.e. $m = 1$ and $n_1 = 1$ in equations 14.1 and 14.2). Then the mutational matrix $B(x)$ is a real number that scales time, and for simplicity we assume $B(x) = 1$ at all times. The adaptive dynamics is then given by

$$\frac{dx}{dt} = s(x) = \left.\frac{\partial f(x, y)}{\partial y}\right|_{y=x}, \quad (14.22)$$

where $f(x, y)$ is the invasion fitness function. If x^* is a singular point, the condition for convergence stability is

$$\left.\frac{ds}{dx}\right|_{x=x^*} = \left.\frac{\partial^2 f(x, y)}{\partial x \partial y}\right|_{y=x=x^*} + \left.\frac{\partial^2 f(x, y)}{\partial y^2}\right|_{y=x=x^*} < 0, \quad (14.23)$$

whereas the condition for evolutionary stability is

$$\left.\frac{\partial^2 f(x^*, y)}{\partial y^2}\right|_{y=x^*} < 0. \quad (14.24)$$

Clearly, if $\left.\frac{\partial^2 f(x,y)}{\partial x \partial y}\right|_{y=x=x^*} < 0$, it is possible that the first condition is satisfied, but the second is not, and hence that the singular point is convergent stable, but evolutionarily unstable. Once the trait value in such an adaptive system reaches the basin of attraction of the singular point, the trait will converge to x^*. Since x^* is also evolutionarily unstable, this system will therefore exhibit convergence towards a fitness minimum, i.e. convergence to a point in trait space at which the resident population can be invaded by all nearby mutants. This seemingly paradoxical evolutionary process has been found in many adaptive dynamics models (Doebeli 2011). In fact, convergence to fitness minima is a generic feature of adaptive dynamics, as is explained in Fig. 14.2. Convergence to a fitness minimum is possible because of the feedback between ecology and evolution: as phenotypes evolve, the ecological environment that they generate changes, and hence the fitness landscapes change (Fig. 14.2). Here I will give an example of generic convergence to fitness minima by means of a classical model for the adaptive dynamics of a scalar trait determining competitive interactions.

Let's assume that the scalar trait x determines per capita birth and death rates of individuals such that the dynamics of a population that is monomorphic for trait value x is given by the logistic equation

$$\frac{dN(x)}{dt} = N(x)(b - c(x)N(x)), \quad (14.25)$$

where it is assumed that the per capita growth rate b is independent of x, and the per capita death rate of individuals is density-dependent and given

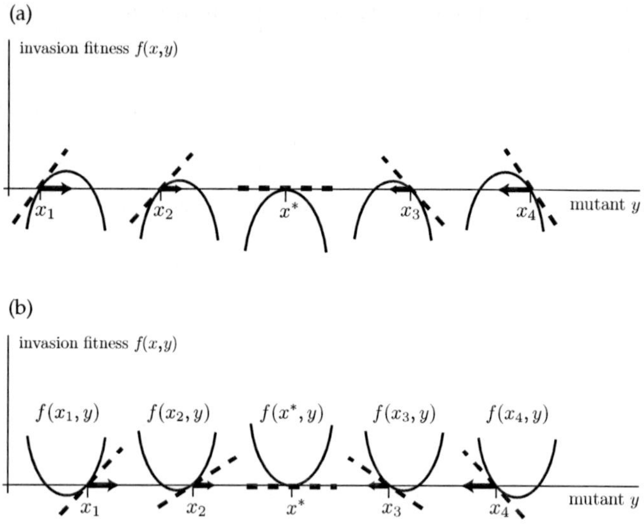

Figure 14.2 The two generic scenarios for convergence to a singular point in one-dimensional trait spaces. The various continuous line segments show the invasion fitness of mutants y for different resident trait values x_i. The dashed lines indicate the selection gradients, i.e. the slopes of the invasion fitness function at the resident values $y = x_i$. When this slope is negative, lower trait values than the current resident are favored, as indicated by the arrows, and vice versa when the selection gradient is positive. In both panels, the selection gradients generate convergence to the singular point x^*. When considering invasion fitness functions up to second order, the two convergence scenarios shown are the only ones that occur generically. In panel (a), convergence to the singular point is generated by invasion fitness functions with a negative curvature, so that the invasion fitness function $f(x^*, y)$ with the singular value x^* as the resident has a maximum at x^* (the selection gradient is by definition 0 at the singular point x^*). Conversely, in panel (b) convergence to the singular point is generated by invasion fitness functions with a positive curvature, so that the invasion fitness function $f(x^*, y)$ has a minimum at x^*, and hence even though x^* is convergent stable, it can be invaded by all nearby mutants. Reproduced from Doebeli, M. (2011). Adaptive Diversification. Princeton University Press, Princeton, NJ.

by $c(x)N(x)$, where $c(x)$ is a function that determines how well phenotype x copes with competition. (Equation 14.25) can be reformulated in the more familiar form

$$\frac{dN(x)}{dt} = bN(x)\left(1 - \frac{bN(x)}{K(x)}\right), \quad (14.26)$$

where $K(x) = b/c(x)$ is the carrying capacity of a population that is monomorphic for x (and hence $K(x)$ can be seen as an individual property, as it is derived from the individual property $c(x)$).

To determine the invasion fitness function $f(x, y)$, we assume the effective density experienced by the mutant y when the resident is at its monomorphic equilibrium $K(x)$ is $a(y, x)K(x)$, where the competition kernel $a(y, x)$ describes the strength of competition that phenotype x exerts on phenotype y. Furthermore, we assume that $a(y, x) = a(x, y) = a(x - y)$ is a function of the difference in trait values only, and that $a(x - y)$ is unimodal with a maximum of 1 at $x - y = 0$ (so that competitive effects are symmetric and largest between individuals of the same phenotype. We also assume that the carrying capacity $K(x)$ is unimodal with a maximum at $x = 0$. In the resulting model, the competition kernel a is the source of negative frequency-dependent selection, whereas as the carrying capacity function K is the source of stabilizing selection (see e.g. chapter 3 in Doebeli (2011)).

With these assumptions, the invasion fitness $f(x, y)$, i.e. the per capita birth rate of a rare mutant y in the resident x, is then given by

$$f(x, y) = b - \frac{ba(y, x)K(x)}{K(y)}. \quad (14.27)$$

Assuming as before that the mutational (scalar) matrix is scaled to 1 at all times, the adaptive dynamics of the trait x is therefore

$$\frac{dx}{dt} = s(x) = \left.\frac{\partial f(x, y)}{\partial y}\right|_{y=x} = b\frac{K'(x)}{K(x)}. \quad (14.28)$$

(The last equality is obtained by observing that $\left.\frac{\partial a(y,x)}{\partial y}\right|_{y=x} = 0$ by assumption.) It immediately follows that the only singular point is the unique maximum $x^* = 0$ of $K(x)$, and that $ds/dx(x^*) < 0$, i.e., that the singular point is convergent stable.

On the other hand, the second derivative of $f(x, y)$ at the singular point is

$$\left.\frac{\partial^2 f(x^*, y)}{\partial y^2}\right|_{y=x^*} = b\left[\frac{K''(x^*)}{K(x^*)} - \left.\frac{\partial^2 a(y, x^*)}{\partial y^2}\right|_{y=x^*}\right]. \quad (14.29)$$

It follows that the singular point is a fitness minimum if

$$\left.\frac{\partial^2 a(y, x^*)}{\partial y^2}\right|_{y=x^*} < \frac{K''(x^*)}{K(x^*)}. \qquad (14.30)$$

Thus, the maximum of the carrying capacity is always an evolutionary attractor, but whether or not this attractor is evolutionarily stable depends on the relative curvature of the competition kernel and the carrying capacity, i.e. on the relative strengths of frequency-dependent and stabilizing selection. Intuitively, the singular point is evolutionarily unstable if for a mutant y appearing in a resident at the singular point, the advantage of experiencing less competition due to being different from the resident is larger than the disadvantage of having a smaller carrying capacity. In particular, convergence to an evolutionarily unstable singular points occurs generically in this model (i.e. for an open set of parameter space).

But if x^* is a fitness minimum, what happens after convergence to x^*? Since all nearby mutants can now invade the resident, an immediate question is whether such nearby mutants can invade each other and hence coexist, thus forming a resident dimorphism. If a resident dimorphism is formed, one can derive the corresponding two-dimensional adaptive dynamics in the two coexisting residents, and it is then important to know whether this adaptive dynamics leads to gradual divergence of the two resident strains, eventually resulting in two clearly distinct and coexisting phenotypic clusters. Such a process of convergence to the singular point and subsequent divergence of coexisting phenotypes is called evolutionary branching. Accordingly, in general adaptive dynamical systems, a singular point x^* is called an evolutionary branching point if the following four conditions are satisfied:

1. x^* is convergent stable.
2. x^* is evolutionarily unstable in at least one direction of trait space.
3. Mutual invasibility (i.e. coexistence) holds around x^* in the directions of evolutionary instability.
4. The coexisting phenotypic branches around x^* diverge evolutionarily from x^*.

Condition 1 is necessary for the system to be attracted to the singularity at which branching is to take place (note again that in general, convergence stability may depend on the mutational variance-covariance matrix). Condition 2 says that nearby mutants are able to invade populations that are monomorphic for the singular trait values specified by x^*. Condition 3 is necessary for nearby mutants to be able to coexist, and condition 4 guarantees that the different branches constituting a polymorphism around the singular point according to 3 diverge from each other over evolutionary time.

One can show that for one-dimensional adaptive dynamics, i.e. for scalar traits, conditions 1 and 2 imply conditions 3 and 4. For example, for the adaptive dynamics given by equation 14.28, the singular point x^* is an evolutionary branching point whenever it is a fitness minimum, and it is straightforward to extend the adaptive dynamics to two coexisting phenotypes (Doebeli 2011). This adaptive dynamics will in general again converge to a singular point in two-dimensional phenotype space, i.e. to a singular coalition of two coexisting resident strains, in both of which the selection gradients are 0. One can then ask again whether this convergent stable singular coalition is evolutionarily stable, or whether the 'adaptive tree' (Geritz et al. 1998) continues to grow through repeated evolutionary branching. For example, in the competition model considered previously the number of successive branching events generally depends on the functional form of the competition kernel and the carrying capacity (Doebeli 2011). Fig. 14.3 shows an example of the adaptive dynamics of evolutionary branching in individual-based versions of the competition model considered here.

In general, the more complicated case of evolutionary branching in multidimensional adaptive dynamics is not yet very well understood. Conditions 1–3 in the definition for evolutionary branching are in general independent requirements in such systems. For systems of coevolving scalar traits, condition 4 follows from conditions 1–3, but it is not known whether conditions 1–3 imply condition 4 in the adaptive dynamics of multidimensional traits. Some results for evolutionary branching in multi-dimensional traits in a single

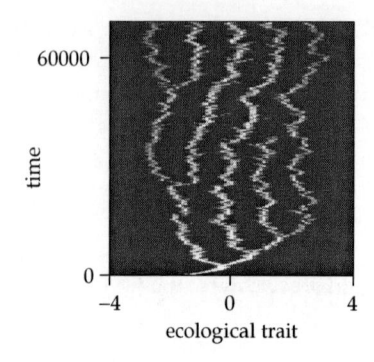

Figure 14.3 Example of repeated evolutionary branching in individual-based simulations based on competition models with Gaussian carrying capacity and competition kernel: $K(x) = K_0 \exp[-x^2/(2\sigma_K^2)]$ and $\alpha(x, y) = \exp[-(x - y)^2/(2\sigma_\alpha^2)]$. The plots shows the frequency distribution of the evolving trait x over time, with brighter areas indicating higher frequencies. Parameter values were chosen to satisfy the branching condition (equation 14.30): $\sigma_K = 2$, $\sigma_\alpha = 0.75$, $K_0 = 200$, $b = 1$. Reproduced from Doebeli, M. (2011). Adaptive Diversification. Princeton University Press, Princeton, NJ.

species have been obtained by Doebeli and Ispolatov (2010).

14.4 The importance of ecological dynamics

Consideration of ecological dynamics is essential for adaptive dynamics, and to conclude, I will briefly consider the phenomenon of evolutionary suicide (Gyllenberg and Parvinen 2001; Webb 2003; Parvinen 2005), which makes the importance of ecological dynamics for evolutionary processes particularly clear. Gyllenberg and Parvinen (2001) considered the following ecological model for asymmetric competition between two phenotypes x and y with an Allee effect:

$$\frac{dN(x)}{dt} = N(x)\left(\rho(x)\frac{a(N(x) + N(y))}{1 + N(x) + N(y)} - b - c(N(x) + \alpha(x - y)N(y))\right) \quad (14.31)$$

$$\frac{dN(y)}{dt} = N(y)\left(\rho(y)\frac{a(N(x) + N(y))}{1 + N(x) + N(y)} - b - c(N(y) + \alpha(y - x)N(x))\right). \quad (14.32)$$

Here x and y are scalar trait values, $N(x)$ and $N(y)$ are the population densities of all x- and y-individuals, respectively, and b and c are parameters influencing individual death rates. The function $\rho(z)$ describes how the intrinsic per capita birth rate depends on the phenotype, and is taken to be

$$\rho(z) = \exp[-\frac{(z - z_0)^2}{2\sigma_\rho^2}]. \quad (14.33)$$

Note that this function reflects stabilizing selection for phenotype z_0. The function $\alpha(z)$ is a competition kernel reflecting asymmetric frequency-dependent competition and taken to be

$$\alpha(z) = \exp[-\beta z - \frac{z^2}{2\sigma_\alpha^2}]. \quad (14.34)$$

Finally, the term $a(N(x) + N(y))/(1 + N(x) + N(y))$ is assumed to reflect an Allee effect.

By setting $N(y) = 0$ in (equation 14.31), one obtains the ecological dynamics of a resident population that is monomorphic for phenotype x. It is easy to see that because of the Allee effect, the following holds. There are two trait values x_{min} and x_{max} such that for $x < x_{min}$ and $x > x_{max}$, the extinction state 0 is a global attractor for the ecological dynamics of the resident x. In contrast, for intermediate $x_{min} < x < x_{max}$, the ecological dynamics of the resident x has two locally attracting equilibria, one at a positive density and the other at 0 (i.e. extinction). These equilibria are separated by an unstable equilibrium (and which one of the stable equilibria is reached depends on initial conditions). To consider the adaptive dynamics of the trait x in the interval (x_{min}, x_{max}), we assume that the resident is at the positive equilibrium, denoted by $\hat{N}$.

By setting $N(x) = \hat{N}$ in (equation 14.32) and letting $N(y) \to 0$, we obtain the per capita growth rate of rare mutants y in the resident x, i.e. the invasion fitness $f(x, y)$:

$$f(x, y) = \rho(y)\frac{a\hat{N}}{1 + \hat{N}} - b - c\alpha(y - x)\hat{N}. \quad (14.35)$$

Asymmetric competition generates an evolutionary trend towards higher trait values, and it is easy to come up with scenarios in which the selection gradient $s(x) = \partial f/\partial y|_{y=x} > 0$ for all $x \in (x_{min}, x_{max})$ in the feasible trait interval. Therefore, selection drives the trait to higher and higher values, until it reaches the boundary of feasible values, x_{max}. Once the current resident x is close enough to x_{max}, a small mutation will occur to a value $y > x_{max}$.

Invasion of this mutant then leads to the collapse and extinction of the whole population, because the ecological dynamics for y does not have a positive equilibrium. It is important to note that this form of evolutionary suicide does not occur due to a gradual decline to 0 of resident population sizes. Rather, extinction occurs once mutations take the evolving population across a bifurcation point at which an unstable and a stable equilibrium of the ecological resident dynamics collide (at a non-zero value of the density), leaving the extinction equilibrium as the only attractor of the ecological dynamics. Gyllenberg and Parvinen (2001) and Parvinen (2005) discuss the general conditions necessary for this type of evolutionary suicide in more detail.

Switching of population dynamical attractors can also lead to interesting evolutionary dynamics in other scenarios, for example, in the evolution of dispersal rates in metapopulation models consisting of two habitat patches that are coupled by dispersal. When the local populations in each patch exhibit non-equilibrium dynamics, a resident metapopulation can exhibit non-synchronous population dynamics for low dispersal rates (i.e. dynamics in which the population densities are different in the two patches at all times), but synchronous population dynamics for high dispersal rates (i.e. dynamics in which the population densities are the same in the two patches at all times). It is known that asynchronous temporal fluctuations in population density can select for higher dispersal rates (Holt and McPeek 1996; Doebeli and Ruxton 1997), while if there is a cost to dispersal, synchronous fluctuations select for lower dispersal rates. This generates an interesting eco-evolutionary feedback that is based on attractor switching: at low dispersal rates, asynchronous ecological dynamics select for higher dispersal rates; as a consequence, dispersal rates evolve to higher values, for which the ecological dynamics become synchronous, which in turn induces selection for lower dispersal rates. As a result, the dispersal rate exhibits fluctuating adaptive dynamics (Doebeli and Ruxton 1997). The same mechanism can lead to the "resident-strikes-back" phenomenon (Doebeli 1998; Mylius and Diekmann 2001), in which invasion of a rare mutant is successful when the resident is on a particular population dynamical attractor, but due to the corresponding increase in the frequency of the mutant the ecological dynamics switches to another attractor on which the mutant is at a disadvantage, and hence the mutant goes extinct despite successful initial invasion. As with evolutionary suicide, it is clear that such phenomena can only be captured by evolutionary frameworks that take ecological dynamics explicitly into account.

14.5 Conclusions

Adaptive dynamics is a gradient dynamics on temporally varying fitness landscapes given by the invasion fitness function $f(x, y)$. Viewed as functions of mutant trait values y, fitness landscapes change over time because they are determined by the resident trait value x, which itself changes dynamically due to the adaptive dynamics generated by the invasion fitness function. Gradient dynamics has of course a long tradition in evolutionary theory, starting with Fisher's fundamental theorem (Fisher 1930). As Frank points out (Chapter 4), the gradient dynamics is in fact only one half of the fundamental theorem, describing how evolution follows the fitness gradient. The other half, often omitted, implicitly describes how the gradient itself changes over time. One virtue of adaptive dynamics is that it takes this second component, i.e. the dynamics of the fitness landscape, explicitly into account, which is made possible by the ecological definition of (invasion) fitness. Thus, in some sense adaptive dynamics can be viewed as one extension of Fisher's theory in which the role of ecology is made precise.

Frequency dependence has also been incorporated in other models for gradient dynamics, e.g. in Russ Lande's framework for quantitative genetics (Lande 1979), which has been prominently used in ecological contexts by Peter Abrams and colleagues (e.g. Abrams et al. 1993; Abrams 2001), as well as by others (e.g. Kirkpatrick and Barton 1997). Compared to quantitative genetic models, as well as to models based on the framework of fitness generating functions (Vincent and Brown 2005), adaptive dynamics has the general conceptual advantage of being derived as a limiting case from an underlying individual-based stochastic process

(Dieckmann and Law 1996), which in principle always allows one to compare results derived for deterministic adaptive dynamics models with corresponding individual-based, stochastic simulations (an example of which is shown in Fig. 14.3). More importantly, compared to other frameworks, adaptive dynamics has the advantage of being derived from a fitness function whose definition is completely model-independent, yet has a precise ecological meaning: the long-term growth rate of rare mutant types in the ecological environment determined by a resident moving on an attractor of its population dynamics. It is important to note that this fitness definition applies to a large class of ecological dynamics of the resident, including e.g. chaotic dynamics (Metz et al. 1992). In fact, the importance of an ecologically based fitness definition becomes apparent when considering evolution in systems in which the resident population dynamics can be complicated. This is exemplified by the phenomenon of evolutionary suicide described in section 14.4, which can only occur deterministically when the resident population dynamics has more than one locally stable attractor (Parvinen 2005).

The model-independent definition of fitness serves as a unifying core concept that allows one to derive general properties of adaptive dynamical systems without having to specify the underlying ecological model. One particular strength of the adaptive dynamics approach is that it allows for a clear and explicit distinction between two crucial stability concepts: convergence stability and evolutionary stability. It is conceptually important to distinguish between the two forms of stability, because based on these two notions of stability, there are two fundamentally different ways in which an evolutionary equilibrium, i.e. a singular point, can become unstable and give rise to an evolutionary bifurcation. The first type of bifurcation occurs when singular points lose their convergence stability, leading to temporally fluctuating adaptive dynamics, an example of which was described in section 14.2.

The second type of evolutionary bifurcation occurs when the selection gradients cease to satisfy the ESS condition at a convergent stable singular point, i.e. when a singular point becomes evolutionarily unstable, and hence an evolutionary branching point, as described in section 14.3. For adaptive dynamics in one-dimensional trait spaces, the concepts of convergence and evolutionary stability allow for a complete classification: apart from scenarios in which traits evolve to ever bigger or ever smaller values (and in the absence of external fluctuations in environmental parameters), the only long-term dynamical behaviours possible for a scalar trait in a single species are convergence to an evolutionarily stable singular point, representing a CSS and hence a final evolutionary stop, or convergence to a fitness minimum, i.e. to an evolutionary branching point (see Fig. 14.2). Geritz et al. (1998) have given a more fine-grained classification of one-dimensional adaptive dynamics, but for many purposes the distinction between CSS and evolutionary branching points suffices. In particular, this classification shows that evolutionary branching is a generic and robust feature of one-dimensional adaptive dynamics. While evolution to fitness minima has been observed before the advent of adaptive dynamics (e.g. Christiansen 1991; Brown and Pavlovic 1992; Abrams et al. 1993), it is probably fair to say that only with adaptive dynamics was it possible to see this phenomenon as a general feature of evolutionary dynamics under frequency-dependent selection. Indeed, the generic existence of evolutionary bifurcations generating evolutionary branching points in adaptive dynamics greatly invigorated the investigation of processes of adaptive diversification and speciation in many different ecological contexts (Doebeli 2011).

Like any modeling approach, adaptive dynamics makes a number of simplifying assumptions. Chief among those is the assumption that the evolutionary process is mutation limited, and hence that there is very limited standing genetic variation. Also, as formulated by equation 14.4 adaptive dynamics is a deterministic theory, and hence does not account for genetic drift and the fixation of mutations with negative effects on fitness. It should be noted, however, that in particular the assumption of monomorphic resident population allows to clearly identify the conditions for evolutionary branching, i.e. for the actual emergence of polymorphism. Moreover, after evolutionary branching adaptive dynamics can be used to describe the evolutionary dynamics of the resulting polymorphism, i.e. of coexisting (monomorphic) resident strains. In addition, because there is an explicit connection between adaptive dynamics and underlying stochastic models, results derived from adaptive dynamics can be verified by individual-based simulations of finite populations, which are polymorphic in general and in which genetic drift is possible. Overall, it turns out that results from individual-based models are generally very compatible with results derived from adaptive dynamics (e.g. Doebeli 2011). Finally, results from adaptive dynamics are also generally compatible with results derived from deterministic models for the dynamics of phenotype distributions (Doebeli et al. 2007, Doebeli 2011). Such models are formulated as partial differential equations and are "maximally polymorphic" in the sense that they describe the simultaneous dynamics of a continuous range of phenotypes. For example, when an adaptive dynamics model predicts evolutionary branching, the corresponding partial differential equation models typically predicts pattern formation in the phenotype distribution in the form of multiple phenotypic modes.

Another fundamental limiting assumption of adaptive dynamics is that of asexual reproduction. This assumption is most critical for the process of evolutionary branching, which can unfold in clonal population unhindered by recombination after evolutionary convergence to the branching point. With sexual reproduction, populations will still converge to evolutionary branching points (due to the convergence stability of such points), but the subsequent diversification is not possible without some form of assortative mating with respect to the trait that is under disruptive selection at the branching point. Thus, in sexual populations evolutionary branching requires assortative mating, and when it does occur, evolutionary branching leads to adaptive speciation (Dieckmann et al. 2004). There is a substantial body of theoretical literature investigating the conditions under which various forms of assortment, as well as the evolution of assortment, can lead to evolutionary branching in sexual populations, and hence to adaptive speciation (Dieckmann and Doebeli 1999; Dieckmann et al. 2004; Doebeli 2011).

Overall, due to the universality of the underlying fitness concept, adaptive dynamics has proven to be a fruitful modeling technique for addressing many different evolutionary questions, as is exemplified by the list of papers that can be found on Eva Kisdi's website (http://mathstat.helsinki.fi/kisdi/addyn.htm). Like evolutionary game theory did four decades ago, adaptive dynamics has opened up new dimensions for modeling evolutionary dynamics.

References

Abrams, P. A. (2001). Modelling the adaptive dynamics of traits involved in inter- and intraspecific interactions: An assessment of three methods. Ecology Letters, 4(2), 166–175.

Abrams, P. A., Matsuda, H., and Harada, Y. (1993). Evolutionarily unstable fitness maxima and stable fitness minima of continuous traits. Evolutionary Ecology, 7(5), 465–487.

Brown, J. S. and Pavlovic, N. B. (1992). Evolution in heterogeneous environments: effects of migration on habitat specilaization. Evolutionary Ecology, 6, 360–382.

Bürger, R. (2000). The Mathematical Theory of Selection, Recombination, and Mutation. Wiley, New York.

Caswell, H. (1989). Matrix Population Models. Sinauer Associates, Sunderland, MA.

Champagnat, N., Ferriére, R., and Meleard, S. (2006). Unifying evolutionary dynamics: From individual stochastic processes to macroscopic models. Theoretical Population Biology, 69(3), 297–321.

Christiansen, F. B. (1991). On conditions for evolutionary stability for a continuously varying character. American Naturalist, 138(1), 37–50.

Dercole, F., Ferriére, R., Gragnani, A., and Rinaldi, S. (2006). Coevolution of slow-fast populations: evolutionary sliding, evolutionary pseudo-equilibria and complex red queen dynamics. Proceedings of the Royal Society B-Biological Sciences, 273(1589), 983–990.

Dercole, F. & Rinaldi, S. (2008). Analysis of Evolutionary Processes. Princeton University Press, Princeton, NJ.

Dieckmann, U. & Doebeli, M. (1999). On the origin of species by sympatric speciation. Nature, 400(6742), 354–357.

Dieckmann, U., Doebeli, M., Metz, J. A. J., & Tautz, D. (eds.) (2004). Adaptive Speciation. [Cambridge Studies in Adaptive Dynamics] Cambridge University Press, Cambridge.

Dieckmann, U. and Law, R. (1996). The dynamical theory of coevolution: A derivation from stochastic ecological processes. Journal of Mathematical Biology, 34(5–6), 579–612.

Dieckmann, U., Marrow, P., and Law, R. (1995). Evolutionary cycling in predator-prey interactions- population-dynamics and the red queen. Journal of Theoretical Biology, 176(1), 91–102.

Dieckmann, O. (2004). A beginner's guide to adaptive dynamics. Banach Center Publications, 63, 47–86.

Doebeli, M. (1998). Invasion of rare mutants does not imply their evolutionary success: a counterexample from metapopulation theory. Journal of Evolutionary Biology, 11, 389–401.

Doebeli, M. (2011). Adaptive Diversification. Princeton University Press, Princeton, NJ.

Doebeli, M., Blok, H. J., Leimar, O., and Dieckmann, U. (2007). Multimodal pattern formation in phenotype distributions of sexual populations. Proceedings of the Royal Society B-Biological Sciences, 274(1608), 347–357.

Doebeli, M. and Dieckmann, U. (2000). Evolutionary branching and sympatric speciation caused by different types of ecological interactions. American Naturalist, 156, S77–S101. S.

Doebeli, M. & Ispolatov, Y. (2010). Continuously stable strategies as evolutionary branching points. Submitted.

Doebeli, M. & Ruxton, G. D. (1997). Evolution of dispersal rates in metapopulation models: Branching and cyclic dynamics in phenotype space. Evolution, 51(6), 1730–1741.

Eshel, I. (1983). Evolutionary and continuous stability. Journal of Theoretical Biology, 103, 99–111.

Fisher, R. A. (1930). The Genetical Theory of Natural Selection. Clarendon, Oxford.

Gavrilets, S. (2004). Fitness Landscapes and the Origin of Species. Princeton University Press, Princeton, NJ.

Geritz, S. A. H., Kisdi, E., Meszéna, G., and Metz, J. A. J. (1998) Evolutionarily singular strategies and the adaptive growth and branching of the evolutionary tree. Evolutionary Ecology, 12(1), 35–57.

Gyllenberg, M. & Parvinen, K. (2001). Necessary and sufficient conditions for evolutionary suicide. Bulletin of Mathematical Biology, 63(5), 981–993.

Holt, R. D. & McPeek, M. A. (1996). Chaotic population dynamics favors the evolution of dispersal. American Naturalist, 148(4), 709–718.

Kirkpatrick, M. & Barton, N. H. (1997). Evolution of a species' range. American Naturalist, 150(1), 1–23.

Kot, M. (2001). Elements of Mathematical Ecology. Cambridge University Press, Cambridge.

Lande, R. (1979). Quantitative genetic-analysis of mulitvariate evolution, applied to brain- body size allometry. Evolution, 33(1), 402–416.

Leimar, O. (2001). Evolutionary change and darwinian demons. Selection, 2, 65–72.

Leimar, O. (2005). The evolution of phenotypic polymorphism: Randomized strategies versus evolutionary branching. American Naturalist, 165(6), 669–681.

Leimar, O. (2009). Multidimensional convergence stability. Evolutionary Ecology Research, 11, 191–208.

Marrow, P., Dieckmann, U., and Law, R. (1996). Evolutionary dynamics of predator-prey systems: An ecological perspective. Journal of Mathematical Biology, 34(5-6), 556–578.

Maynard Smith, J. (1982). Evolution and the Theory of Games. Cambridge University Press, Cambridge.

Maynard Smith, J. and Price, G. R. (1973). Logic of animal conflict. Nature, 246(5427), 15–18.

Metz, J. A. J. & Diekmann, O. (1986). The Dynamics of Physiologically Structured Populations. Springer, Berlin.

Metz, J. A. J., Nisbet, R. M., and Geritz, S. A. H. (1992). How should we define fitness for general ecological scenarios. Trends in Ecology & Evolution, 7(6), 198–202.

Mylius, S. D. and Diekmann, O. (2001). The resident strikes back: Invader-induced switching of resident attractor. Journal of Theoretical Biology, 211, 297–311.

Nowak, M. and Sigmund, K. (1990). The evolution of stochastic strategies in the prisoners-dilemma. Acta Applicandae Mathematicae, 20(3), 247–265.

Parvinen, K. (2005). Evolutionary suicide. Acta Biotheoretica, 53, 241–264.

Rice, S. H. (2004). Evolutionary Theory: Mathematical and Conceptual Foundations. Sinauer Associates, Sunderland, MA.

Sandholm, W. S. (2010). Population Games and Evolutionary Dynamics. MIT Press, Cambridge, MA.

Tuljapurkar, S. (1990). Population dynamics in Variable Environments. Springer, Berlin.

Vincent, T. L. and Brown, J. L. (2005). Evolutionary Game Theory, Natural Selection, and Darwinian Dynamics. Cambridge University Press, Cambridge.

Webb, C. (2003). A complete classification of darwinian extinction in ecological interactions. American Naturalist, 161, 181–205.

Wright, S. (1931). Evolution in mendelian populations. Genetics, 16, 97–159.

Wright, S. (1932). The roles of mutation, inbreeding, crossbreeding and selection in evolution. Proceedings of the Sixth Annual Congress of Genetics, 1, 356–366.

Wright, S. (1984). Evolution and the Genetics of Populations: Genetics and Biometric Foundations v. 2 (Theory of Gene Frequencies); New Edition. University of Chicago Press, Chicago, IL.

CHAPTER 15

Adaptive Landscapes, Evolution, and the Fossil Record

Michael A. Bell

15.1 Evolutionary mechanism and paleontological pattern

Reconciliation of microevolutionary mechanisms with paleontological patterns has been a nagging problem since the time of Darwin and Lyell. The 150th anniversary of *The Origin of Species* in 2009 prompted assessment of Darwin's views on this problem (e.g. Bell 2009; chapters in Bell et al. 2010; Hunt 2010). In Darwin's Chapter X, "On the Geological Succession of Organic Beings," he emphasized the orderly appearance of higher taxa in the fossil record to support evolution as an historical fact. Nevertheless, Chapter IX, "On the Imperfection of the Geological Record," was devoted to explaining the absence of gradual transitions between fossil species, a logical expectation from natural selection on small differences. He concluded that the fossil record is intrinsically fragmentary and poorly sampled. Darwin's (1859) doubts about the potential for the fossil record to provide direct evidence on the role of natural selection in evolution have largely been borne out.

Simpson (1944) was the leading paleontologist among the founders of the evolutionary synthesis. He transformed Wright's (1931, 1932) genotypic fitness landscape metaphor into a phenotypic Adaptive Landscape and popularized it among evolutionary biologists. But Simpson's (1944) landscape had little impact on paleontological practice (Laporte 2000). Mechanistic inferences in paleontology continue to emphasize analysis of higher taxa (e.g. see Wagner 2010; Foote 2010). In this review, I explain the limited appeal of Adaptive Landscapes to paleontologists. I briefly describe the history (see Chapter 1) and conceptual difficulties (Chapter 2 and 3) of landscape metaphors in evolutionary biology. Poor temporal resolution in the fossil record is the most serious impediment to using biostratigraphic sequences to study Adaptive Landscapes, but lack of suitable statistical methods has also limited their value. A more generally applicable approach uses the locations or rates of movement of species on Adaptive Landscapes to infer evolutionary mechanisms. I turn first to the history of fitness landscapes and Adaptive Landscapes.

15.2 Sewall Wright's fitness (genotypic) landscape

Seven decades after the publication of *The Origin of Species*, Sewall Wright (1931, 1932) proposed the fitness landscape metaphor to illustrate the relationship between stepwise, multilocus genotypic change, genetic drift, and Darwinian fitness. Although it was not the first use of a landscape to describe the action of natural selection, subsequent landscape models in evolutionary biology can be traced back to Wright's (1931, 1932) concept (Chapter 1). Peaks in the landscape represent genotypic combinations that confer high fitness under prevailing conditions, and valleys represent genotypes with low fitness. Wright's (1931, 1932) fitness landscapes focused on individuals and populations of a single species. He (Wright 1932; Fig. 15.1, top) originally represented them as contour maps with isolines of equal fitness on a surface viewed from above in the space of two genotypic axes. They are also represented as two-dimensional graphs (i.e. one genotypic and one fitness axis) viewed from the side (Fig. 15.1, bottom) or three-dimensional graphs (i.e. two genotypic and one

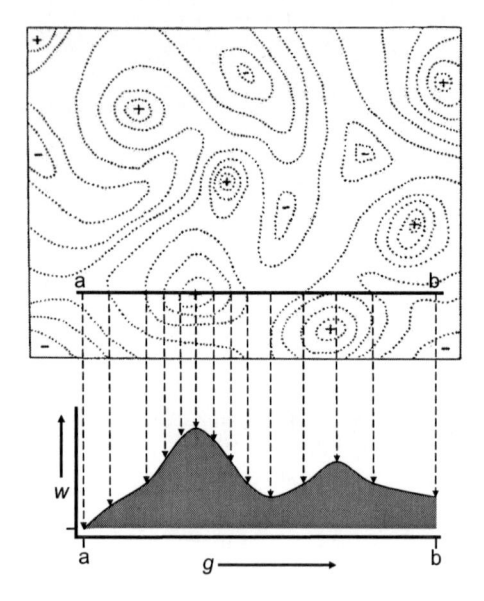

Figure 15.1 Wright's (1932, fig. 1b) original Adaptive Landscape (Top) in contour-map format. Dotted lines represent points of equal fitness, pluses (+) are adaptive peaks, and minuses (−) are adaptive valleys, but the axes are undefined. (Bottom) Two-dimensional plot of fitness values (w) along line a-b on the original landscape, which forms genotypic axis g of the bottom landscape. Vertical dashed arrows connect corresponding fitness values from the upper to the lower landscape. (Figure adapted from Wright 1932.)

fitness axis) viewed in perspective. They provide an intuitively appealing image that relates microevolutionary dynamics to change along genotypic axes. While the actual meaning of distance along genotypic axes is obscure and may be misleading without considering high-dimensionality of genotypes (Pigliucci 2008, Chapter 3; Chapter 17), the landscape concept is among the most effective metaphors in evolutionary biology (e.g. Arnold et al. 2001; McGhee 2007; Chapter 2).

15.3 G. G. Simpson's Adaptive (phenotypic) Landscape

Wright's fitness landscape achieved its greatest impact in George Gaylord Simpson's (1944) hands (Chapter 3). The independent variables, genotypic axes in Wright's (1932) fitness landscape, became phenotypic axes in Simpson's (1944) Adaptive Landscape (Chapter 1). Like Wright (1932), Simpson (1944) rendered the Adaptive Landscape as a contour map, with concentric isolines surrounding adaptive peaks and valleys (Fig. 15.2). In contrast to Wright's (1932) fitness landscape, which represented individuals or populations of a single species, Simpson's (1944) Adaptive Landscape represented divergence among species and higher taxa.

Simpson was strongly influenced by Dobzhansky's (1937) book, *Genetics and the Origin of Species*, which had re-established the role of natural selection on minor, heritable, phenotypic variants in evolution. This role had been repudiated by the mutationists after rediscovery of Mendelian genetics in 1900 and had always been questioned by paleontologists, many of whom favored orthogenesis and similar intrinsic evolutionary forces (Simpson 1944, 1952; Bowler 1983). Dobzhansky (1937) reasserted the importance of natural selection of small heritable differences for microevolution, and it was left for Simpson to explain the potential for this process to produce differences between species and higher taxa (i.e. macroevolution). Thus, in the light of mid-twentieth century genetics and a much improved fossil record, Simpson (1944) reasserted Darwin's (1859) claim that natural selection on minor heritable differences can account for the appearance of new species and higher taxa over geological time.

Simpson's (1944) account of Adaptive Landscapes began by explaining the meaning of peaks and basins and the importance of slopes as fitness gradients (Fig. 15.2). He emphasized that Adaptive Landscapes are dynamic and change as the environment and genetic variation change; they are more like the surface of a choppy sea than like hills and valleys on dry land (see Chapters 7 and 13). He illustrated his concept with the evolution of Tertiary horses (Equidae; Fig. 15.3). Earlier horses (Eocene Hyracotheriinae) occupied an adaptive peak for browsing on leaves. Increasing body size caused the browsing peak to approach the unoccupied peak for grazing (Early to Late Oligocene Anchitheriinae). (Hansen's [1997] subsequent analysis indicated that size did not influence adaptation for grazing.) Simpson (1944) described the Late Oligocene browsing peak as a broad, low peak separated by a shallow valley from the higher, growing, but vacant grazing peak. Modal browsing phenotypes remained on the browsing peak, but outliers among the browsers spilled onto the

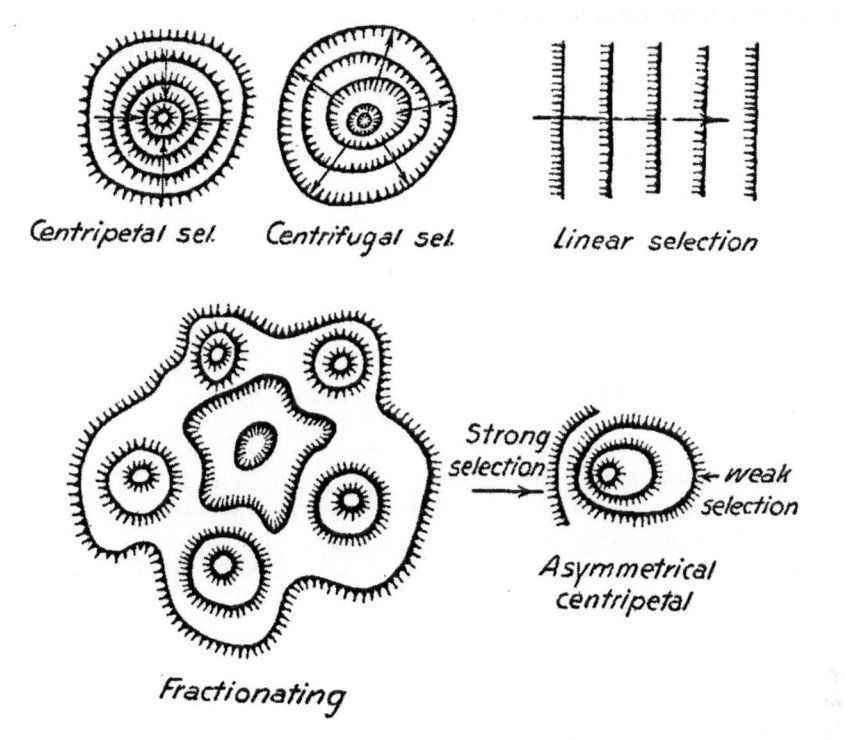

Figure 15.2 Simpson's (1944, fig. 11) diagrams of modes of selection on adaptive (selective) landscapes in contour-map format. Hatchures indicate the direction of reduced fitness relative to the adjacent contour line, and arrows indicate the direction of selection, but the axes are undefined. (From *Tempo and Mode in Evolution*, by G. G. Simpson. Copyright© 1984.)

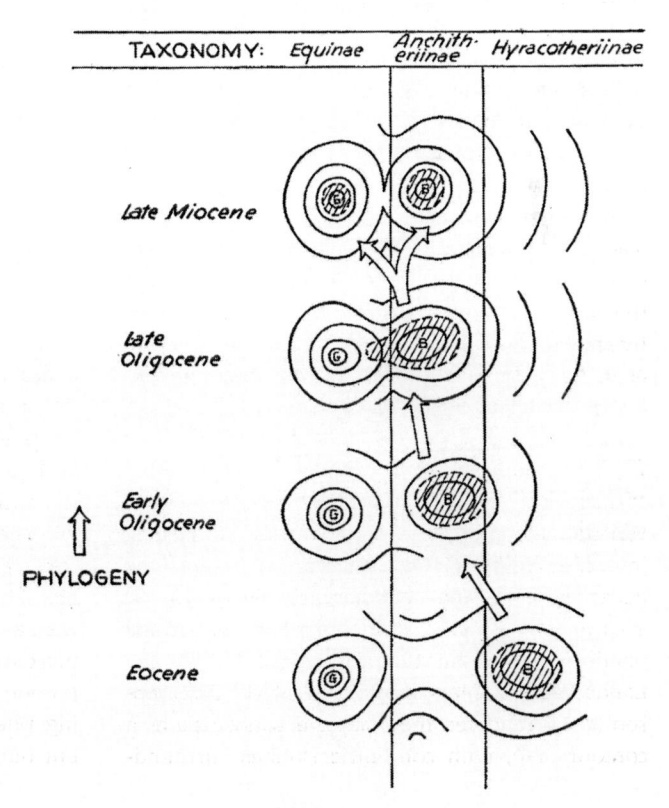

Figure 15.3 Simpson's (1944, fig. 13) illustration of Adaptive Landscapes for the transition in Tertiary horses (Equidae) from browsing ancestors (Hyracotheriinae) prior in the Eocene to the previously vacant grazing adaptive peak of the Equinae in the Late Miocene. The landscapes are represented by contour maps, and the distribution of phenotypes is indicated by diagonal hatchures within dotted lines. (From *Tempo and Mode in Evolution*, by G. G. Simpson. Copyright© 1984.)

base of the grazing peak (Late Oligocene) and were eventually (Late Miocene Equinae) driven by directional selection onto its summit (see MacFadden 1992 for update). While this scenario invokes unknown properties of ancient horses (Laporte 2000), its use of Adaptive Landscape imagery to explain a classic paleontological case successfully popularized the Adaptive Landscape metaphor.

15.4 Paleontology and Adaptive Landscapes

Tempo and Mode in Evolution (Simpson 1944) received a more enthusiast reception from evolutionary biologists who study modern organisms (i.e. "neontologists") than from paleontologists (Laporte 2000). In retrospect, Mayr (1988) commented that empirical research in paleontology, including Simpson's, did not concern evolutionary processes within species (i.e. microevolution). Thus, Simpson (1944) integrated paleontology into the evolutionary synthesis, but paleontologists did not generally apply the synthesis to their research. There is probably a good reason for this failure, and this brings us back to Simpson's rendition of Wright's fitness landscape.

Wright's (1932) fitness landscape concerns genetic change within populations in response to fitness differences associated with allelic combinations at multiple loci. As recognized by evolutionary biologists ranging from Darwin (1859) to Simpson (1944) and Eldredge and Gould (1972), however, fossil samples are too coarsely distributed through time to observe the phenotypic consequences of natural selection within populations (e.g. Gingerich 2001; Kidwell and Holland 2002). Consequently, Simpson's (1944) rendition of Adaptive Landscapes was an appealing metaphor for the evolution of transitions between fossil species, but it has rarely been operational to explain such transitions because they are largely invisible (Pigliucci 2008, Chapter 3). Although the Adaptive Landscape metaphor is closely associated with Simpson, a paleontologist, it has had limited application to empirical paleontology.

McGhee's (2007) proposal to integrate Adaptive Landscapes and theoretical morphospaces is a conspicuous exception. He applied Adaptive Landscapes to phenomena such as convergent evolution and cladogenesis, but these applications were still heuristic. He concluded that previous applications of Adaptive Landscapes had usually been heuristic but that they could be combined with theoretical morphospaces for analytical purposes.

15.5 Temporal resolution in paleontology and evolutionary rates in populations

Eldredge and Gould's (1972) punctuated equilibria paper stimulated analyses of the limits of temporal resolution in the stratigraphic record and computer simulations to estimate plausible evolutionary rates. Recent interest in the evolution of contemporary populations has added independent insights into plausible evolutionary rates. Together, these studies indicate that differences between species may evolve too rapidly to be resolved in the fossil record, limiting application of Adaptive Landscape models to analysis of evolution in biostratigraphic sequences.

15.5.1 The limits of temporal resolution in the stratigraphic and fossil records

The rate at which sediment accumulates and its fate after deposition place limits on temporal resolution in the stratigraphic (i.e. rock) record and can distort patterns of evolutionary change (Schindel 1982; Dingus and Sadler 1982; Behrensmeyer et al. 2000; Kidwell and Holland 2002). Three important properties of stratigraphic sequences, temporal scope, microstratigraphic acuity, and stratigraphic completeness, define the limits of temporal resolution in transitions between fossil species. Temporal scope is the amount of time spanned by a stratigraphic sequence. The greater scope is, the more likely it is that evolution will occur within a geological section and that enough samples can be collected to characterize the pattern of evolutionary change. Microstratigraphic acuity is the time between the earliest and latest specimens within individual fossil samples, and poor acuity limits temporal resolution of adjacent samples (Behrensmeyer et al. 2000; Kowalewski and Bambach 2003). Fossils that accumulated during periods of no or

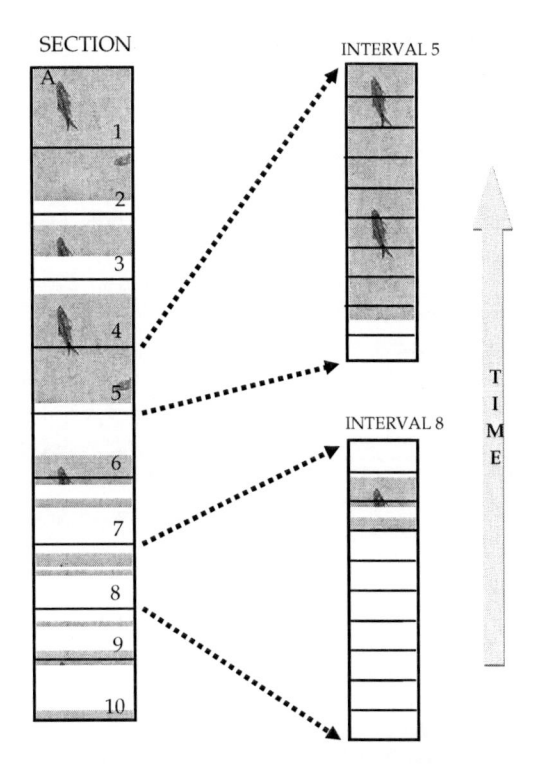

Figure 15.4 Stratigraphic completeness of a hypothetical stratigraphic section. Section A (left) is divided into ten intervals of equal duration, and intervals 5 and 8 are each divided into ten subunits of equal duration. The shaded portions of each section represent the time during which sediment accumulated, and sediment is absent (not deposited or eroded) from the unshaded portions. Section A is 100% complete at the scale of the numbered intervals because each interval contains sediment. However, dividing Section A into ten finer subunits (right) produces shorter intervals with lower completeness. Interval 5 is 90% complete because nine of its ten subunits contain sediment, and interval 8 is only 20% complete. Also note that resolution is limited to a coarser scale in the lower half of Section A by low stratigraphic completeness at a finer scale.

low sedimentation will occur within a narrow rock section, and long time intervals will be represented by very short stratigraphic distances. Microstratigraphic acuity is increased by faster sedimentation but reduced by low or discontinuous sedimentation or by postdepositional mixing (i.e. reworking, bioturbation) of deposited sediment. Most paleontological time series involve shelly metazoan or unicellular species from near-shore and continental shelf deposits and vertebrates from fluvial deposits (Erwin and Anstey 1995). These depositional environments produce microstratigraphic acuity of decades to hundreds of thousands of years (Behrensmeyer et al. 2000). Stratigraphic completeness (Fig. 15.4) is the fraction of time intervals at a specific temporal scale that are represented by sediment in a geological section. This property depends on the time scale over which completeness is measured (Sadler 1981; Schindel 1982). As a stratigraphic sequence is divided into finer time intervals, the fraction of intervals that contain at least some sediment tends to decrease, reducing completeness. Thus, the per cent stratigraphic completeness for a single stratigraphic section tends to get lower as the length of the intervals at which completeness is measured gets smaller. Discontinuities in sedimentation distort apparent rates and patterns of evolution (Schindel 1982; Kidwell and Holland 2002). Stratigraphic sections with high completeness at a fine time scale will portray the time between samples most accurately. Unfortunately, highly complete sequences with fine resolution are unlikely to accumulate for long before a basin fills and sedimentation shifts elsewhere, limiting temporal scope (Kowalewski and Bambach 2003). Furthermore, they will be very thick and unlikely to be included within individual exposures, complicating sampling. Thus, it is unlikely that all three of these properties, temporal scope, microstratigraphic acuity, and stratigraphic completeness, will be optimal to study transitions between species in any single stratigraphic sequence (Schindel 1982; Kidwell and Holland 2002).

The fossils themselves create additional problems. Those at lower levels in a sequence may not be ancestral to later fossils (Smith 1994). Low stratigraphic completeness may indicate environmental changes that could have caused extinction of one lineage and its replacement by another (Schindel 1982). Evolution during a period of poor microstratigraphic acuity may inflate sample variances if individuals are pooled over a period during which there was substantial evolution (Bell et al. 1987, but see Hunt 2004), reducing evolutionary rates in Haldanes (which are scaled in variance units), and the potential to detect differences between sample means. It also creates trait covariances within fossil samples that are not due to covariance among individuals within populations (Bell et al. 1989). Fossils may be absent from some sedimentary horizons, reducing temporal resolution. Temporal resolution in fossil sequences can be no better and will often be much worse than resolution in the sedimentary record, and it is rarely better than thousands to tens of thousands of years (Behrensmeyer et al. 2000; Kidwell and Holland 2002; Erwin 2006).

Even fossil sequences with the most favorable properties may still be misleading. For example, Bell and Haglund (1982) sampled a sequence with abundant, well-preserved fossils, microstratigraphic acuity of one year, and nearly 100% stratigraphic completeness at the time scale of one year. They sampled at irregular intervals averaging 20,000 years, which is shorter than possible in most fossil sequences, but missed dramatic morphological change that became conspicuous at finer intervals (Bell et al. 1985a, 2006a). Distortions imposed by limited temporal resolution on analysis of a population's ascent of an adaptive peak depend on the irregularity and rate of ascent. Unfortunately, few fossil sequences include a sufficient number of large samples to resolve morphological change in fossil sequences into a series of small steps (Kidwell and Holland 2002; Hunt 2010), limiting the application of Adaptive Landscapes to analysis of change within individual fossil lineages.

15.5.2 Simulated rates of species-level morphological evolution

Punctuated equilibria (Eldredge and Gould 1972) was originally proposed as the manifestation of Mayr's (1963) peripatric speciation in the fossil record, but Gould (e.g. 1980) and others subsequently advocated unconventional evolutionary mechanisms to explain punctuations (Gould 2002). In the 1980s, simulations were performed to test the hypothesis that directional selection on small heritable differences, a conventional Neo-Darwinian mechanism, could produce morphological differences between related species that are too fast to resolve in the fossil record. The details of these simulations will be deferred until section 15.6.1, but they invariably indicated that directional selection on small phenotypic differences could produce species-level morphological evolution within a few hundred generations, a shorter interval than can be resolved in most biostratigraphic sequences.

15.5.3 Rates of phenotypic evolution in contemporary populations

Industrial melanism is the classic case of contemporary evolution (Majerus 1998), and numerous pest and pathogen species have rapidly evolved resistance to chronically applied toxins (Gould 2010). Nevertheless, until recently, it has generally been assumed that morphological phenotypes evolve too slowly under natural conditions to be observed in contemporary populations. However, Hendry and Kinnison (1999) stimulated research on contemporary evolution of traits that would be interpreted as species differences in the fossil record (e.g. Carroll et al. 2007). For example, after a sea-run population of threespine stickleback fish (*Gasterosteus aculeatus*) colonized a lake, it evolved enough within a decade to become virtually indistinguishable from other lake populations (Bell et al. 2004; Arif et al. 2009; Aguirre and Bell, in press). Comparable differences have been used to diagnose nominal stickleback species (Bell 1995) and to distinguish biological species of threespine sticklebacks (e.g. McPhail 1994).

Both simulation studies and empirical analyses of contemporary evolution indicate that the gradual transitions between fossil species that Darwin (1859) so fervently desired to observe and Simpson (1944) modeled in Adaptive Landscapes can be completed within tens to hundreds of generations, almost always too fast to be resolved in the fossil record. Even species with long generation times (e.g. tetrapods frequently studied by vertebrate paleontologists) could evolve too rapidly to resolve species transitions—the ascent of a slope on an Adaptive Landscape—in typical fossil sequences.

15.6 Application of Adaptive Landscapes in paleontology

15.6.1 Evolution of simulated individual lineages on Adaptive Landscapes

The evolutionary response to directional natural selection represents a population's ascent of an adaptive peak. Simulation studies in the 1980s were explicit rebuttals of claims epitomized by Gould (1980) that punctuations imply the importance in evolution of mutations with large effects (e.g.

Gould 1980). Although Gould (2002) eventually abandoned this claim, it led to several simulation studies aimed at showing that directional selection on heritable traits with small differences could produce species-level morphological change faster than could be resolved (i.e. punctuations) in the fossil record. These simulations assumed reasonable values for heritability (h^2), phenotypic variance (σ^2), and selection (s). They began with the ancestral population on an ancestral adaptive peak and presented scenarios for it to cross an adjacent valley to reach the base of a new adaptive peak. Changes in the topography of the Adaptive Landscape, which result from either genetic or environmental change (Chapter 7), initiate evolution up a new adaptive peak.

Petry (1982) envisioned the shift of a population's mean phenotype (z_t or $\bar{z}$) from one adaptive peak to another due to a new mutation that created a key innovation and made a new adaptive peak accessible (see Agrawal et al. 2001). He defined an evolutionary change of the mean phenotype by 5 σ (standard deviation units) as formation of a new species, and this much change took only 700 generations. A similar analysis by Kirkpatrick (1982) envisioned a phenotypic shift to a new peak through either environmental change or perturbation of the developmental system to place some individuals on the slope of an adjacent adaptive peak, much like Simpson's (1944) Late Oligocene Adaptive Landscape for horses (Fig. 15.3). This simulation produced phenotypic change of 2.4 σ to 15.3 σ within 15 to 400 generations. Milligan (1986) assumed a bimodal resource distribution and developed an ecological scenario in which a new adaptive peak arose from environmental change. Phenotypic differences associated with the ecological shift took only about 100 generations, but phenotypic means were stable for extended periods before and after the peak shift, as predicted by punctuated equilibria.

Unlike the other studies, Lande (1985, 1986) assumed a static landscape unaltered by favorable mutation or environmental change and modeled movement of a population off of its ancestral peak, across an adaptive valley, and onto an adjacent peak. He also stipulated the effective population size (N_e), that the adaptive valley between peaks was shallow, and that fitness slopes near peaks and valleys were low. Otherwise, his assumptions resembled those of the other studies. Even under these stringent conditions, populations tended to drift near the summit of their ancestral peaks (i.e. evolutionary stasis or equilibria) for several million (> 10^6, except at small N_e) generations, but if they happened to drift rapidly to the bottom of the adaptive valley, selection could either drive them back to their original peak or to the other peak within a few hundred generations. Although Lande's (1985, 1986) model is concerned with populations, it also resembles Simpson's (1944) scenario for the evolution of grazing in horses. Newman et al. (1985) performed a similar simulation and reached similar conclusions.

These simulations consider landscapes with only a single phenotypic or genetic dimension, but Lande (1986) also modeled the effect of correlated traits on evolutionary trajectories on Adaptive Landscapes. Unless genetic correlations were very high, they did not affect the dynamics of single traits. However, Gavrilets (2004) argued that more realistic fitness (i.e. genotypic) landscapes with more than two trait dimensions contain holes through which a population can evolve onto peaks that appear to be isolated in two- or three-dimension landscapes. Similarly, Wagner (Chapter 17, see also Maynard Smith 1970) argued that high-dimensional fitness landscapes contain long ridges that represent alternative metabolic pathways of equivalent fitness along which populations can evolve great distances. Although Wagner (Chapter 17) focused on metabolic evolution, similar (though less analytically tractable) ridges exist for morphological adaptations (Wainwright 2007) that could be observed in the fossil record. Thus, residence time of a population on an adaptive peak (i.e. stasis or equilibria) may be shorter than three-dimensional drift-based models suggest. However, trait correlations may reduce the dimensionality of fitness landscapes, reducing the rate of the evolutionary response to selection (e.g. Walsh and Blows 2009; Kirkpatrick 2010). Further research will be required to investigate these theoretical results, but effective landscape dimensionality clearly has important effects on evolutionary dynamics that are not captured in two- and three-dimensional landscapes.

These simulations used similar approaches to see if the pattern of punctuated equilibria (Eldredge and Gould 1972) could be produced by conventional evolutionary mechanisms. Combined with the poor temporal resolution of most stratigraphic sequences, these studies show that morphological change during a peak shift is unlikely to be resolved into a sequence of small, observable, stepwise changes in the fossil record but will appear to be punctuated. Subsequent research on contemporary evolution also indicates that evolution can be too fast to resolve as a trend in the fossil record. Thus, although Simpson (1944) modeled macroevolution as the behavior of species on Adaptive Landscapes, actual lineages should generally experience peak shifts that are too fast to resolve in the fossil record (i.e. punctuations) and spend most of their time atop adaptive peaks (i.e. equilibria).

15.6.2 Ascent of adaptive peaks by an individual fossil lineage

Many changes in the fossil record appear to be adaptive. For example, Devonian tetrapods became adapted to land, Mesozoic birds to flight, and Miocene horses to spreading grasslands, but temporal resolution of these events is too coarse to investigate the role of natural selection. Additionally, statistical analyses to infer directional selection (i.e. adaptive peak climbing) in the fossil record generally depend on rejection of the null hypothesis of pure genetic drift manifested as an unbiased random walk through geological time. Statistical analyses can be performed by comparing consistency of the direction of change or of the rate or magnitude of change to expectations based on an unbiased random walk. Relatively few analyses using these criteria have been performed (e.g. Raup and Crick 1981; Bookstein 1987, 1988; Lynch 1990; Cheetham and Jackson 1995). They almost invariably fail to exclude a random walk, which would implicate natural selection, because a mixture of directional selection plus random components (i.e. random environmental change, genetic drift, sample error due to finite fossil sample sizes) is likely to be indistinguishable (i.e. p >0.05) from a purely unbiased random walk (Lande 1986; Bell 1994; Sheets and Mitchell 2001; Bell et al. 2006a; Hunt 2006, 2010).

To address this problem, Hunt (2006) developed a maximum likelihood method that compares the potential for alternative models to account for biostratigraphic patterns. The fundamental variables for all such models, mean change between successive samples (i.e. step size, $\mu_{step'}$) and step size variance ($\sigma^2_{step'}$), were incorporated into models for different patterns of change through time. For example, a distribution of steps with equal probabilities of morphological increase or decrease and a mean step size of zero produces an unbiased random walk. A model that differs only for the probability of step frequency or mean step sizes in different morphological directions could produce a biased random walk in the direction of the bias. The second model incorporates an additional parameter to impose the bias for change. Since more complex models are more likely than simpler ones to explain a pattern, Hunt (2006) employed the Akaike Information Criterion, which takes model complexity into account to evaluate competing models.

Hunt et al. (2008) applied this method to a sequence of fossil stickleback fish (*Gasterosteus doryssus*) with excellent properties to study high-resolution patterns of change in the fossil record. More than 80 large ($\bar{n} > 36$) samples comprising specimens from consecutive 250 year intervals were used, and conventional methods had failed to reject the null hypothesis of an unbiased random walk (Bell et al. 2006a). Extensive evidence from the evolutionary ecology of closely related modern stickleback, the paleoecology of *G. doryssus*, and congruent change of three armor traits indicated that change in the fossils was a response to directional selection, and that failure to reject the null hypothesis of an unbiased random walk was a Type II error (see also Sheets and Mitchell 2001).

Hunt et al. (2008) reanalyzed these data by contrasting an unbiased random walk to a model for an Ornstein–Uhlenbeck process. The unbiased random walk represents a population drifting across a perfectly flat Adaptive Landscape. The Ornstein–Uhlenbeck process represents a population's ascent of an adaptive peak starting at low fitness and a substantial distance from the summit of the peak and coming more strongly under the influence of genetic drift as the fitness slope declines near the summit. Adaptive Landscape geometry is at the core of this analysis. In contrast to analyses using an unbiased random walk as the null hypothesis, this method strongly favored (all p ≤ 0.003) the Ornstein–Uhlenbeck model and directional selection.

15.6.3 Punctuated equilibria and natural selection in sets of fossil lineages

Eldredge and Gould (1972) emphasized the ubiquity of stasis (equilibria) in the fossil record, and subsequent research has supported their contention (Erwin and Anstey 1995; Gould 2002; Eldredge et al. 2005). Several mechanisms have been proposed for stasis, including persistent stabilizing selection, genetic and developmental constraints, habitat selection, gene flow, and local extinction and replacement (see Estes and Arnold 2007 for references). Two studies concerning evolutionary stasis in the fossil record explicitly used the Adaptive Landscape metaphor, and earlier studies concerning evolutionary stasis can be reinterpreted from this perspective.

Stanley and Yang (1987) defined evolutionary stasis operationally for 24 traits in 19 Neogene bivalve species. They compared variation of the individual traits and a multivariate measure of those traits in the fossils to their range of variation in extant, conspecific populations. The range of variation in the extant populations may be viewed as a broad peak or set of adjacent peaks in a combined Adaptive Landscape. They concluded that all 19 lineages had been in stasis for millions of years because their morphology never significantly exceeded the morphological limits set by extant, conspecific populations. This simple approach has rarely been used, though it is broadly applicable, and analyses that are rooted more explicitly in Adaptive Landscapes have been developed.

Hansen (1997) proposed a general method to use the locations of species from a clade on a restricted region of an Adaptive Landscape to tease apart the contributions of common ancestry and stabilizing selection to stasis. He acknowledged that other effects on a species (i.e. selection on correlated traits, mutation, drift, environmental fluctuation) would be confounding factors but felt that they would be reduced to random noise in comparisons of multiple species. Hansen (1997) also assumed an Ornstein–Uhlenbeck process and found that it would rapidly bring species to the summit of adaptive peaks, where they would spend most of their time.

He used hypsodonty (i.e. possession of high-crowned premolars and molars) in horses to illustrate this method. Hypsodonty is favored in grazing horses (Equinae) over the ancestral (plesiomorphic) brachydont condition of browsing horses (Hyracotheriinae and Anchitheriinae) because grasses are abrasive and cause rapid tooth wear. Hansen (1997) analyzed persistence of hypsodonty of the first upper molar in equines. He incorporated taxa ≤18 million years old, starting with *Mesohippus*, when the number of grazing horse species and the extent of grasslands increased (MacFadden 1992; MacFadden and Cerling 1994). His analysis suggested that increasing grassland availability favored maintenance of hypsodonty in equines. This method exploits retention of ancestral states in relation to a shared environmental factor instead of focusing on rare independent origin of derived (homoplastic) states to infer selection. The equines appear to have occupied peaks on a reasonably stable Adaptive Landscape in which hypsodonty was imposed by stabilizing selection.

Estes and Arnold (2007) used quantitative genetic models to account for the distribution of evolutionary rates (Haldanes) from Gingerich (2001) for diverse traits from numerous taxa that were measured at ≥2 time intervals separated by 10^0 to 10^7 generations. They assumed a reasonable range of values for effective population size (N_e, and hence genetic drift), trait heritability (h^2), and the strength of stabilizing selection (ω^2) on phenotypic traits (z). In their models, the adaptive optimum or peak (θ) changed position with a variance of σ_θ^2, and the mean phenotype ($\bar{z}$) after adaptation had a mean of θ and variance of σ_θ^2 plus the variance (σ^2) due to finite effective population size. They compared the distribution of evolutionary rates to expectations from models for the dynamics of z, including selective neutrality ($s \approx 0$), slight displacement of $\bar{z}$ from θ, θ moving directionally at a constant rate, white-noise motion of θ about a stable value,

Brownian motion of θ, and movement of $\bar{z}$ from an ancestral θ, across an adaptive valley, and onto a second θ (peak shift, Lande 1985, 1986). Except in the peak-shift model, $\bar{z}$ remained within the domain of attraction of the original θ. Observed rates were lower than observed under artificial selection, indicating that natural selection is typically opposed by some constraint (e.g. genetic correlation, ecological or functional tradeoffs). Most models produced change that is too fast or too slow compared to observed rates at long and short time intervals and cannot be rescued by manipulating parameter values (e.g. h^2, N_e) within reasonable ranges.

The best model for the distribution of the evolutionary rates over a wide range of time scales was the displaced optimum model. Under this model, $\bar{z}$ rapidly approached θ, which varied by σ_θ^2, and subsequently varied with a distribution of σ^2 from θ. Estes and Arnold (2007) concluded that directional and stabilizing selection can account for the distribution of evolutionary rates in contemporary, historical, and fossil sequences provided that individual values of z remain within the domain of attraction of θ. Hansen (1997) and Estes and Arnold (2007) concluded that adaptive peaks tend to wobble within a limited range that corresponds to Simpson's (1944, 1953) adaptive zones and that retention of ancestral (plesiomorphic) states is likely to indicate stabilizing selection (i.e. not require unconventional evolutionary mechanisms).

15.7 Adaptive Landscapes, evolution, and the fossil record

Wright's (1931, 1932) fitness landscapes had their greatest impact after Simpson (1944) recast them as Adaptive Landscapes with phenotypic axes. Simpson's (1944) Adaptive Landscape linked microevolution to macroevolution and paleontology and contributed greatly to the evolutionary synthesis. Arnold et al. (2001) and McGhee (2007) argued that Adaptive Landscapes can play a significant role in the analysis of temporal variation in the fossil record, and several theoretical and empirical analyses have exploited the Adaptive Landscape metaphor for such analyses.

The distinction between fitness landscapes and Adaptive Landscapes is important (e.g. Pigliucci 2008, Chapters 3, 13, and 19). It is related to the difference between genotypes and phenotypes and between natural selection and the evolutionary response to selection. Natural selection acts on phenotypes, not directly on genotypes. Since phenotypes must be heritable for selection to produce evolution, the evolutionary response to selection will be shaped by the way phenotypes map to genotypes. Potentially complex mapping of phenotypes onto genotypes defines the relationship between Wright's (1931, 1932) and Simpson's (1944) superficially similar landscapes.

This mapping process may produce unexpected trajectories on Adaptive Landscapes and even create unexpected peaks. For example, vestigial pelvic phenotypes may be favored by selection in the threespine stickleback fish (*G. aculeatus*). Remarkably, the vestiges are usually larger on the left side (Bell et al. 1985b), and based solely on selection, directional asymmetry represents an unexpected adaptive peak (Bell et al. 1993). *Pitx1* is usually the major gene for pelvic reduction in this species (Shapiro et al. 2004; Bell et al. 2006b), and homozygosity of recessive *Pitx1* alleles causes both pelvic reduction (the target of selection) and greater size of the left pelvic vestige (Shapiro et al. 2004). Recessiveness of the *Pitx1* allele for pelvic reduction was invoked by Hunt et al. (2008) to explain the 2750-year delay in the response of pelvic structure to directional selection in a fossil stickleback. Thus, genetic architecture influenced both the location of the adaptive peak (i.e. directional pelvic asymmetry instead of fluctuating asymmetry or bilaterally symmetry) and the trajectory (i.e. 2750-year delay) taken by a fossil lineage to surmount it. Similarly, genetic correlations between phenotypes can deflect their evolutionary paths away from the maximum slope of ascent up adaptive peaks (e.g. Schluter 1996). Although Wright's (1931, 1932) fitness landscapes present logical difficulties (Provine 1986; Pigliucci 2008, Chapter 3), strictly phenotypic Adaptive Landscapes may fail to be predictive because they ignore the intricate mapping of phenotypes (which are selected) on genotypes (which evolve).

However, Hansen's (1997) and Estes and Arnold's (2007) results indicate that Adaptive Landscapes combined with more detailed modeling can predict evolutionary change well enough to test hypothesis using fossil data. Most of the methods reviewed so far required extensive modeling with additional parameters and a narrow range of parameter values to either reconcile theory with paleontological observation or to distinguish competing hypotheses. For instance, Lande's (1985, 1986) demonstration that populations could drift into an adaptive valley and ascend an adjacent peak fast enough to explain punctuations in the fossil record required shallow valleys and similar curvatures of the peaks and valleys. Similarly, Hunt et al. (2008) modeled evolution as an Ornstein–Uhlenbeck process to allow comparison of the explanatory power of natural selection and genetic drift. Thus, while the Adaptive Landscape metaphor is valuable to model change in the fossil record, it must be augmented with more specific model features to provide testable hypotheses.

Adaptive Landscapes were important to clarify some issues surrounding punctuated equilibria (Eldredge and Gould 1972), the major issue of late twentieth century paleontology. Several simulations that modeled evolution as movement across an Adaptive Landscape showed that it was unnecessary to propose unconventional evolutionary mechanisms to account for presumably rapid evolution during punctuations in the fossil record. Using reasonable assumptions and standard population genetic models, these simulations showed that ascent of an adaptive peak and even shifts between peaks across a valley, which would appear as punctuations, should be too fast to observe as a sequence of small biased steps in most fossil sequences. Adaptive Landscapes were valuable to show that punctuated equilibria could be produced by natural selection on conventional genetic variation.

More recently, Adaptive Landscapes have been used to analyze paleontological data. Hansen's (1997) and Estes and Arnold's (2007) analyses of multiple lineages imply Adaptive Landscapes on which species quickly ascend peaks and spend most of their time tracking wobbly summits. Thus, directional and stabilizing selection readily explain punctuated equilibria without invoking unconventional mechanisms. Hunt's (2006) method is applicable to individual lineages, and it can accommodate a variety of Adaptive Landscape topographies. Simpson (1944) originally used Adaptive Landscapes to demonstrate the compatibility of population genetics with patterns in the fossil record, and recent applications of Adaptive Landscapes have facilitated tests of the explanatory value of alternative population genetic (microevolutionary) models for patterns of change in the fossil record.

As Estes and Arnold (2007) and McGhee (2007) argued, additional modeling effort will be required to fully exploit Adaptive Landscapes for paleontology. However, application of Hunt's (2006) method is not limited by modeling effort but by the availability of biostratigraphic sequences for individual lineages with fine temporal resolution. Fine resolution is crucial both to approach the time scale at which natural selection operates (Gingerich 2001) and to increase confidence that observed sequences actually represent individual lineages. While analyses of aggregated data for multiple species represent valuable and broadly applicable methods, fine-scale records for individual lineages provide a crucial perspective on tempo and mode in the fossil record.

15.8 Conclusions

Darwin (1859) emphasized the value of the fossil record to validate evolution as an historical fact, but he questioned its utility to study evolutionary mechanisms. One hundred and fifty years of paleontological research has confirmed his pessimism. Thus, although Simpson (1944) was a paleontologist, and he popularized the Adaptive Landscape metaphor, it is not surprising that paleontologists rarely use it in their research because temporal resolution in fossil sequences is almost always too coarse to resolve evolutionary change into long time series composed of numerous, large samples that can be used to infer evolutionary mechanisms from fossil sequences.

However, intraspecific fossil sequences sometimes are suitable for analyses based on Adaptive Landscape models. Identifying these sequences and using the finest possible temporal resolution to study them should be given a high priority by paleontologists. The criteria for selection of fossil sequences to study evolutionary mechanisms include: (1) long stratigraphic sections deposited at high rates without interruption or disturbance (to infer the time between samples), (2) presence of abundant, well-preserved, and continuously distributed fossils (to reliably estimate the statistical properties of evolutionary time series), (3) multiple, easily measurable traits (to study multivariate evolution), and (4) close relationship of the fossil form to living species in which intra- and interpopulation variation, ecology, and genetics have been studied (to help interpret paleontological patterns).

An additional important obstacle to inference of the causes for change in fossil sequences has been lack of effective statistical methods. The null hypothesis of genetic drift (i.e. a flat Adaptive Landscape) is almost impossible to reject, but models based on Adaptive Landscapes can provide alternative predictions that can be compared on an equal footing with genetic drift (Hunt 2006).

Analysis of sets of species distributed upon an Adaptive Landscape is a more generally applicable method because it does not depend on fine temporal resolution; conventional fossil samples can be used. Analyses of this kind can be more broadly applied without acquisition of new fossil material to investigate mechanisms that have influenced the history of life.

Acknowledgments

I thank D. Erwin, G. Hunt, J. S. Levinton, and M. Pigliucci, for enlightening email messages and M. Pigliucci for allowing me to read his manuscript for this volume in advance. I thank T. D. Connelly of the Genetics Society of America for permission to use Fig. 15.1 and J. Simpson Burns for permission to use Figs. 15.2 and 15.3 from her father's book. C. Noto, Pélabon, J. Rollins, and another reviewer generously made constructive suggestions to improve this chapter. This is contribution 1205 from Ecology and Evolution at Stony Brook University and is based in part on research supported by DEB0322818. This chapter is dedicated to the memory of my friend and colleague, George C. Williams, who died the day I finished the first draft.

References

Agrawal, A. F., Brodie, E. D., III and Rieseberg, L. H. (2001). Possible consequences of genes of major effect: transient changes in the G-matrix. Genetica, 112–113, 33–43.

Aguirre, W. E. and M. A. Bell. In press. Twenty years of body shape evolution in a threespine stickleback population adapting to a lake environment. Biological Journal of the Linnean Society.

Arif, S., Aguirre, W. E. and Bell, M. A. (2009). Evolutionary diversification of opercle shape in Cook Inlet threespine stickleback. Biological Journal of the Linnean Society, 97, 832–844.

Arnold, S. J., Pfrender, M. E. and Jones, A. G. (2001). The Adaptive Landscape as a conceptual bridge between micro- and macroevolution. Genetica, 112–113, 9–32.

Behrensmeyer, A. K., Kidwell, S. M. and Gastaldo, R. A. (2000). Taphonomy and paleobiology. Paleobiology, 26, 103–147.

Bell, M. A. (1994). Paleobiology and evolution of threespine stickleback. In M. A. Bell and S. A. Foster (eds.) The evolutionary biology of the threespine stickleback. Oxford University Press, Oxford, pp. 439–471.

Bell, M. A. (1995). Intraspecific systematic of *Gasterosteus aculeatus* populations: implications for behavioral ecology. Behaviour, 132, 1131–1152.

Bell, M. A. (2009). Implications of a fossil stickleback assemblage for Darwinian gradualism. Journal of Fish Biology, 75, 1977–1999.

Bell, M. A., Aguire, W. E. and Buck, N. J. (2004). Twelve years of contemporary armor evolution in a threespine stickleback population. Evolution, 58, 814–824.

Bell, M. A., Baumgartner, J. V. and Olson, E. C. (1985a). Patterns of temporal change in single morphological characters of a Miocene stickleback fish. Paleobiology, 11, 258–271.

Bell, M. A., Francis, R. C. and Havens, A. C. (1985b). Pelvic reduction and its directional asymmetry in threespine sticklebacks from the Cook Inlet region, Alaska. Copeia, 1985, 437–444.

Bell, M. A., Futuyma, D. J., Eanes, W. F. and Levinton, J. S. (eds.) (2010). Evolution since Darwin: the first 150 years. Sinauer, Sunderland, MA.

Bell, M. A. and Haglund, T. R. (1982). Fine-scale temporal variation in the Miocene stickleback, *Gasterosteus doryssus*. Paleobiology, 8, 282–292.

Bell, M. A., Khalef, V. and Travis, M. P. (2006b). Directional asymmetry of pelvic vestiges in threespine stickleback. Journal of Experimental Zoology (Molecular and Developmental Evolution), 306B, 189–199.

Bell, M. A., Ortí, G., Walker, J. A. and Koenings, J. P. (1993). Evolution of pelvic reduction in threespine stickleback fish: a test of competing hypotheses. Evolution, 47, 906–914.

Bell, M. A., Sadagursky, M. S. and Baumgartner, J. V. (1987). Utility of lacustrine deposits for the study of variation within fossil samples. Palaios, 2, 455–466.

Bell, M. A., Travis, M. P. and Blouw, D. M. (2006a). Inferring natural selection in a fossil threespine stickleback. Paleobiology, 32, 562–577.

Bell, M. A., Wells, C.E. and Marshall, J. A. (1989). Mass-mortality layers of fossil stickleback fish: catastrophic kills of polymorphic schools. Evolution, 43, 607–619.

Bookstein, F. L. (1987). Random walk and the existence of evolutionary rates. Paleobiology, 13, 446–464.

Bookstein, F. L. (1988). Random walk and the biometrics of morphological characters. Evolutionary Biology, 9, 369–398.

Bowler, P. J. (1983). The eclipse of Darwinism: anti-Darwinian evolution theories in the decades around 1900. Johns Hopkins University Press, Baltimore, MA.

Carroll, S. P., Hendry, A. P., Reznick, D. N. and Fox, C. W. (2007). Evolution on ecological time-scales. Functional Ecology, 21, 387–393.

Cheetham, A. H. and Jackson, J. B. C. (1995). Process from pattern: tests for selection versus random change in punctuated bryozoan speciation. In D. H. Erwin and R. L. Anstey (eds.) New approaches to speciation in the fossil record. Columbia University Press, New York, pp. 184–207.

Darwin, C. R. (1859). The origin of species by means of natural selection or preservation of favoured races in the struggle for life. [Avenal 1979 reprint of the first edition]. Avenal Books, New York.

Dingus, L. and Sadler, P. M. (1982). The effects of stratigraphic completeness on estimates of evolutionary rates. Systematic Zoology, 31, 400–412.

Dobzhansky, T. (1937). Genetics and the origin of species. Columbia University Press, New York.

Eldredge, N. and Gould, S. J. (1972). Punctuated equilibria: an alternative to phyletic gradualism. In T. J. M. Schopf, ed. Models in paleobiology. Freeman Cooper, San Francisco, pp. 82–115.

Eldredge, N., Thompson, J. N., Brakefield, P. M., Gavrilets, S., Jablonski, D., Jackson, J. B. C., et al. (2005). The dynamics of evolutionary stasis. Paleobiology, 31, 133–145.

Erwin, D. H. (2006). Dates and rates: temporal resolution in the deep time stratigraphic record. Annual Reviews of Earth and Planetary Sciences, 34, 569–590.

Erwin, D. H. and Anstey, R. L. (1995). Speciation in the fossil record. In D. H. Erwin and R. L. Anstey (eds.) New approaches to speciation in the fossil record. Columbia University Press, New York.

Estes, S. and Arnold, S. J. (2007). Resolving the paradox of stasis: models with stabilizing selection explain evolutionary divergence at all timescales. American Naturalist, 169, 227–244.

Foote, M. (2010). The geological history of biodiversity. In: M. A. Bell, D. J. Futuyma, W. F. Eanes, and J. S. Levinton (eds.) Evolution since Darwin: the first 150 years. Sinauer, Sunderland, MA, pp. 479–510.

Gavrilets, S. (2004). Fitness landscapes and the origin of species. Princeton University Press, Princeton, NJ.

Gingerich, P. D. (2001). Rates of evolution on the time scale of the evolutionary process. Genetica, 112, 127–144.

Gould, F. (2010). In M. A. Bell, D. J. Futuyma, W. F. Eanes, and J. S. Levinton (eds.) Evolution since Darwin: the first 150 years, Sinauer, Sunderland, MA, pp. 591–621.

Gould, S. J. (1980). Is a new and general theory of evolution emerging? Paleobiology, 6, 119–130.

Gould, S. J. (2002). The structure of evolutionary theory. Belknap Press, Cambridge, MA.

Hansen, T. (1997). Stabilizing selection and the comparative analysis of adaptation. Evolution 51, 1341–1351.

Hendry, A. P. and Kinnison, M. T. (1999). The pace of modern life: measuring rates of contemporary microevolution. Evolution, 53, 1637–1653.

Hunt, G. (2004). Phenotypic variation in fossil samples: modeling the consequences of time-averaging. Paleobiology, 30, 426–443.

Hunt, G. (2006). Fitting and comparing models of phyletic evolution: random walks and beyond. Paleobiology, 32, 578–601.

Hunt, G. (2010). Evolution in fossil lineages: paleontology and the origin of species. American Naturalist 176 (Suppl. 1), S61–S76.

Hunt, G., Bell, M. A., and Travis, M. P. (2008). Evolution toward a new adaptive optimum: phenotypic evolution in a fossil stickleback lineage. Evolution, 62, 700–710.

Kidwell, S. M. and Holland, S. M. (2002). The quality of the fossil record: implications for evolutionary analysis. Annual Review of Ecology and Systematics, 33, 561–588.

Kirkpatrick, M. (1982). Quantum evolution and punctuated equilibria in continuous genetic characters. American Naturalist, 119, 833–848.

Kirkpatrick, M. (2010). Rates of adaptation: why is Darwin's machine so slow? In M. A. Bell, D. J. Futuyma, W. F. Eanes, and J. S. Levinton (eds.) Evolution since Darwin: the first 150 years, pp. 177–195. Sinauer, Sunderland, MA.

Kowalewski, M. and Bambach, R. K. (2003). The limits of paleontological resolution. In P. J. Harries (ed.) High-resolution approaches in stratigraphic paleontology. Kluwer, New York, pp. 1–48.

Lande, R. (1985). Expected time for random genetic drift of a population between stable phenotypic states. Proceedings of the National Academy of Sciences of the United States of America, 82, 7641–7645.

Lande, R. (1986). The dynamics of peak shifts and the pattern of morphological evolution. Paleobiology, 12, 343–354.

Laporte, L. F. (2000). George Gaylord Simpson: paleontologists and evolutionist. Columbia University Press, New York.

Lynch, M. (1990). The rate of morphological evolution in mammals from the standpoint of the neutral expectation. American Naturalist, 136, 727–741.

MacFadden, B. J. (1992). Fossil horses. Cambridge University Press, Cambridge.

MacFadden, B. J. and Cerling, T. E. (1994). Fossil horses, carbon isotopes and global change. Trends in Ecology and Evoution, 9, 481–486.

Majerus, M. E. N. (1998). Melanism: evolution in action. Oxford University Press, Oxford.

Maynard Smith, J. (1970). Natural selection and the concept of protein space. Nature, 225, 563–564.

Mayr, E. (1963). Animal species and evolution. Harvard University Press, Cambridge, MA.

Mayr, E. (1988). Toward a new philosophy of biology. Harvard University Press, Cambridge, MA.

McGhee, G. (2007). The geometry of evolution. Cambridge University Press, New York.

McPhail, J. D. (1994). Speciation and the evolution of reproductive isolation in the sticklebacks (*Gasterosteus*) of south-western British Columbia. In Bell, M. A., and Foster, S. A. (eds.) The evolutionary biology of the threespine stickleback. Oxford University Press, Oxford, pp. 399–437.

Milligan, B. G. (1986). Punctuated evolution induced by ecological change. American Naturalist, 127, 522–532.

Newman, C. M., J. E. Cohen and C. Kipnis. (1985). Neo-darwinian evolution implies punctuated equilibria. Nature, 315, 400–401.

Petry, D. (1982). The pattern of phyletic speciation. Paleobiology, 8, 56–66.

Pigliucci, M. (2008). Sewall Wright's adaptive landscapes: 1932 vs. 1988. Biological Philosophy, 23, 591–603.

Provine, W. B. (1986). Sewall Wright and evolutionary biology. University of Chicago Press, Chicago, IL.

Raup, D. M. and Crick R. E. (1981). Evolution of single characters in the Jurassic ammonite *Kosmoceras*. Paleobiology, 7, 200–215.

Sadler, P. M. (1981). Sediment accumulation rates and the completeness of stratigraphic sections. Journal of Geology, 89, 569–584.

Schindel, D. E. (1982). Resolution analysis: a new approach to the gaps in the fossil record. Paleobiology, 8, 340–353.

Schluter, D. (1996). Adaptive radiation along genetic lines of least resistance. Evolution, 50, 1766–1774.

Sheets, H. D. and Mitchell, C. E. 2001. Why the null matters: statistical tests, random walks and evolution. Genetica, 112–113, 105–125.

Shapiro, M. D., Marks, M. E., Peichel, C. L., Blackman, B. K., Nereng, K. S., Jónsson, B., et al. (2004). Genetic and developmental basis of evolutionary pelvic reduction in threespine sticklebacks. Nature, 428, 717–723.

Simpson, G. G. (1944). Tempo and mode in evolution. Columbia University Press, New York. [1984 reprint.]

Simpson, G. G. (1952). The meaning of evolution. Yale University Press, New Haven, CT.

Simpson, G. G. (1953). The major features of evolution, Columbia University Press, New York.

Simpson, G. G. (1984). Tempo and mode in evolution. Columbia University Press, New York. [Reprint of the 1944 edition with a new introduction by Simpson.]

Smith, A. B. (1994). Systematics and the fossil record: documenting evolutionary patterns. Blackwell, Boston, MA.

Stanley SM, and Yang X (1987). Approximate evolutionary stasis for bivalve morphology over millions of years: a multivariate, multilineage study. Paleobiology, 13, 113–139.

Wagner, P. J. (2010). Paleontological perspectives on morphological evolution. In: M. A. Bell, D. J. Futuyma, W. F. Eanes and J. S. Levinton (eds.) Evolution since Darwin: the first 150 years, pp. 451–478. Sinauer, Sunderland, MA.

Wainwright, P. C. (2007). Functional versus morphological diversity in macroevolution. Annual Review of Ecology, Evolution, and Systematics, 38, 381–401.

Walsh, B. and Blows, M. W. (2009). Abundant genetic variation + strong selection = multivariate genetic constraints: a geometryc view of adaptation. Annual Review of Ecology, Evolution, and Systematics, 40, 41–59.

Wright, S. (1931). Evolution in Mendelian populations. Genetics, 16, 97–159.

Wright, S. (1932). The roles of mutation, inbreeding, crossbreeding and selection in evolution, Proceedings of the sixth International Congress of Genetics, 1, 356–366.

PART V
Development, Form, and Function

CHAPTER 16

Mimicry, Saltational Evolution, and the Crossing of Fitness Valleys

Olof Leimar, Birgitta S. Tullberg, and James Mallet

16.1 Introduction

The relative contribution of gradual and saltational change to evolution has been debated ever since Darwin (1859) emphasized gradualism in his theory of evolution by natural selection. The phenomenon of mimicry was an important example in this debate. In mimicry evolution, members of a population or species become similar in appearance to an aposematic model species and thereby gain increased protection from predation. The number of steps in the approach to mimicry could be few or many and their sizes either large or small. In 1915, Punnett published an influential book on mimicry in butterflies, in which he summed up his strong opposition to gradualistic accounts of mimicry evolution. He dismissed previous suggestions by Poulton (e.g. Poulton 1912, 1913) that mimicry could emerge in a sequence of steps, beginning with the appearance of a rough likeness of the would-be mimic to its model, followed by further improvement in resemblance. Punnett's main argument was representative of the thinking of the early Mendelians, who often pointed to a lack of intermediates between existing variants, inferring that the variants had originated as mutants in a single step, as opposed to being molded by natural selection through successive replacements of intermediate forms. As supporting evidence, Punnett used examples of female-limited polymorphic mimicry, for instance the one found in the butterfly *Papilio polytes*, where the developmental switching between female morphs was known to be controlled by a small number of Mendelian factors, later shown to be alleles at an autosomal locus (Clarke and Sheppard 1972). Punnett's position was thus a radical

saltationism. In terms of Adaptive Landscapes, if the mimic-to-be resides on one adaptive peak and the model on another, mimicry evolves in a single mutational leap, and the issue of natural selection only enters through the constraint that the peak jumped to should be higher than the starting peak. This would apply both to Müllerian mimicry, where the starting point is an aposematic species, and to Batesian mimicry, where the starting adaptive peak of the palatable mimic-to-be is determined by functions other than aposematism, for instance crypsis or other protective coloration, like flash coloration (Cott 1940; Ruxton et al. 2004), or partner choice.

In response to claims like those by Punnett (1915), and as part of his efforts to unify gradualism and Mendelian genetics, Fisher (1927, 1930) presented a fully gradualistic alternative. He envisaged a genetically variable population of mimics-to-be and proposed that individuals with trait values deviating from the population mean in the direction of the traits of the model species would be slightly favored over deviations in the opposite direction, because of a slightly higher probability of being mistaken for the model by predators. The outcome could be a gradual shifting of the mean trait values in the direction of improved resemblance. It appears that Fisher intended his gradualistic scenario to apply both to Batesian and Müllerian relations, but it gained more attention in the latter case. In terms of (frequency-dependent) Adaptive Landscapes of Müllerian mimicry evolution, Fisher's process would be a gradual shifting of two peaks, until they overlap and, approximately, merge to a single peak.

Fisher's (1927, 1930) proposition was not generally accepted. The strongest opposition came from

Goldschmidt (1945a,b), who ended his examination of the issue by coming to the conclusion that "Punnett's interpretation of polymorphism by mutation (saltation) agrees better with the facts than Fisher's neo-Darwinian theory." Although not subscribing to the saltationism of Punnett and Goldschmidt, even Fisher's close associates (for instance E. B. Ford and P. H. Sheppard), who were preoccupied with the problem of Batesian mimicry evolution, came to deviate in their views from Fisher's original proposition (see Turner 1985 for an overview). Instead the so-called two-step process, where a large mutation first achieves approximate similarity to the model, after which smaller changes can improve the likeness, became accepted as describing Batesian mimicry evolution. The idea is often credited to Nicholson (1927), although Poulton (1912) had already suggested it. Over time, the two-step process became accepted also in the context of Müllerian mimicry (Turner 1984; Sheppard et al. 1985). In terms of Adaptive Landscapes, the process entails a mutational leap from the adaptive peak of the mimic-to-be, protected as it is by predator learning of that phenotype, to somewhere on the slope of the higher, more protective peak of the model, thus crossing a fitness valley, followed by a series of modifications climbing the higher peak. Because mimicry often involves several traits, which at least initially can be genetically independent, this first mutational leap is clearly a less demanding assumption than a saltation as argued for by Punnett and Goldschmidt. Even so, the assumption needs to be backed up by arguments or observations making it likely that predators in fact would avoid attacking the first, quite imperfect, mutant mimic.

A different kind of ingredient in explanations of mimicry evolution is that, possibly only in a particular region or period of time, evolutionary forces other than mimicry may modify the appearances of mimics-to-be, fortuitously bringing about sufficient resemblance to a model to start off mimicry evolution. Examples could be selection in relation to mate choice or thermoregulation (Mallet and Singer 1987). Random genetic drift in small populations is another general category of this kind. In aposematism, with learnt attack avoidance by predators, selection tends to operate on deviations of appearances from the current population mean. The selective peaks could be constrained, and in part formed by the aesthetics of predator learning, to particular regions of phenotypic space, as well as being influenced by selection towards the current mean. If the population mean appearance changes to explore new aesthetic combinations, predator learning need not always act to bring the mean back to a previous value. This implies that the mean may perform a random walk over evolutionary time, exploring parts of the relevant trait space, and possibly different peaks of an Adaptive Landscape, in a shifting balance process (Wright 1977; Mallet and Singer 1987; Coyne et al. 1997; Mallet and Joron 1999; Mallet 2010; see also Chapters 2 and 4–6). This would correspond to a random walk to a new adaptive peak. The phenomenon will be more pronounced in small local populations, and/or in populations with limited predation pressure, where selection towards the current mean will be weakened. The process could be responsible both for mimicry evolution and for the rapid diversification of novel aposematic signals that occurs in many aposematic groups (Mallet and Joron 1999; Mallet 2010).

In the following, we extend this brief review of the history of ideas, outlining some recent work on mimicry evolution. The concept of an Adaptive Landscape will be central in the presentation, in particular the question of how a transition from one adaptive peak to another can come about. The landscapes we consider depict fitness as a function of the phenotypic traits of individuals, rather than as a function of genotype or allele frequencies (see Chapters 1–3, 5, 7, 18 and 19 for expositions and discussions on different types of Adaptive Landscapes). It should be kept in mind that, for aposematism and mimicry, Adaptive Landscapes are strongly density- and frequency-dependent, in the sense that their shapes depend on the traits and population sizes that are present in a prey community (this is also true for many other types of Adaptive Landscapes; see Chapter 7). The reason for the frequency dependence is that predator behavior is influenced by learning and generalization about the properties of the community.

16.2 Transitions between adaptive peaks

Predators learn about prey, for instance to avoid attacking those with particular appearances if they have been found to be distasteful or otherwise unprofitable in previous attacks. Generalization is another crucial aspect of predator psychology: if a predator encounters prey with a different appearance than previous prey, the reaction to the new prey may be a generalized version of the learnt reaction to similar-looking prey. From assumptions about learning and generalization by predators in a predator–prey community, an Adaptive Landscape of prey survival as a function of phenotype can be defined. Fig. 16.1 shows a generalization function and an Adaptive Landscape for a one-dimensional trait space. It is convenient to conceptualize the landscape as the survival of a prey individual with a mutant phenotype, so that the survival in principle is defined for any point in the trait space, whether or not the point is near the traits of the resident prey populations. The landscape in Fig. 16.1b is computed from an individual-based simulation, using assumptions similar to those in Balogh and Leimar (2005) and Ruxton et al. (2008). Briefly, the assumption about generalization along a one-dimensional stimulus space, illustrated in Fig. 16.1a, is among the most commonly used and empirically substantiated in animal psychology (Ghirlanda and Enquist 2003). Similarly, the assumptions about predator learning follow broadly accepted ideas of associative learning (Rescorla and Wagner 1972).

Restricting attention to a single trait, a two-step process of Müllerian mimicry evolution is illustrated in Fig. 16.1b. Because large-effect mutations are known to occur, at least for certain traits, for instance for the hue or intensity of pigmentation (Socha and Nemec 1996), and because large-effect alleles have been found in Müllerian mimicry systems (Joron et al. 206b; Baxter et al. 2009), it follows that a two-step process involving a single-locus mutation could well produce shifts between adaptive peaks in one-dimensional trait spaces.

16.3 Peak shift in multidimensional trait spaces

To gain an appreciation of the possible constraints from predator psychology on the two-step process in a space of complex, multitrait appearances, it is instructive to quote Punnett's (1915, p. 140) remarks on what is required of predators as selective agents in butterfly mimicry, of which birds are believed to be the most important:

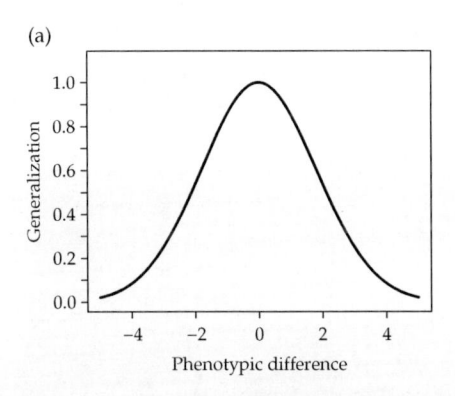

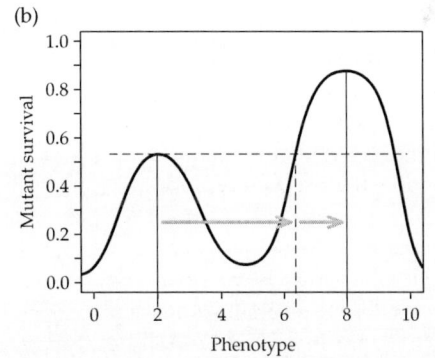

Figure 16.1 Predator generalization function (a) and Adaptive Landscape (b) for Müllerian mimicry evolution in a one-dimensional phenotype space. The generalization function indicates the strength of the tendency of the predator to generalize learning about one prey phenotype to other phenotypes, which depends on the phenotypic difference. The Adaptive Landscape shows the survival over a season of a mutant as a function of its phenotype. The longer gray arrow indicates the size of the mutant change needed to jump from the smaller peak, across the valley and onto the slope of the bigger peak. Further mutant changes (smaller arrow) can then lead to the top of the bigger peak, in accordance with the classical two-step process. There is one resident population of 1000 individuals, each of which has phenotype $x = 2$ and a bigger resident population of 5000 individuals, each of which has phenotype $x = 8$, and the generalization function is a Gaussian with standard deviation 1.8. All individuals are equally distasteful.

In the first place, they must confuse an incipient or 'rough' mimic with a model sufficiently often to give it an advantage over those which have not varied in the direction of the model. In other words, they must be easily taken in. Secondly, they are expected to bring about those marvelously close resemblances that sometimes occur by confusing the exact mimicking pattern with the model, while at the same time eliminating those which vary ever so little from it. In other words, they must be endowed with most remarkably acute powers of discrimination.

Punnett writes that "we must suppose that it is done by different species," and at least in some cases such a conclusion could be warranted. Fig. 16.2 illustrates the basic problem for mimicry evolution in multidimensional trait spaces, for a case of two genetically independent traits, in the sense of an absence of pleiotropy. For a mutant to reach from the smaller peak to a higher point on the slope of the bigger peak, to initiate a climb up that peak, both traits must mutate (Fig. 16.2), and for the double mutant not to break down through subsequent recombination, the genes for the traits must be linked. These are very severe constraints, and for more than two traits they become prohibitive. One should of course keep in mind the possibility of pleiotropic mutants affecting multiple phenotypic traits. Nevertheless, without further assumptions like pleiotropy, a two-step transition between adaptive peaks that are clearly separated along more than one trait becomes more unlikely the more trait dimensions there are.

16.4 Feature-by-feature saltation

The Adaptive Landscape in Fig. 16.2 is built on an assumption that generalization over a multidimensional stimulus space works essentially in the same way as for a single dimension. This possibility is taken into account in animal psychology, but there are important alternatives, involving other psychological mechanisms, relating to the formation of categories (Pearce 2008). Psychological theories of categorization propose that objects are represented as collections of features. By comparing common and distinctive features, individuals categorize objects as similar or dissimilar (Tversky 1977; Treisman and Gelade 1980). Experiments show that animals often use one or a few features when discriminating among stimuli (Troje et al. 1999; Marsh and MacDonald 2008). There are also studies indicating that such categorization occurs in predators discriminating suitable from unsuitable prey (Schmidt 1958; Aronsson and Gamberale-Stille 2008). A related strategy of similarity judgment is to encode stimuli hierarchically at two levels of detail: category level information is used first to sort stimuli into crude categories, whereupon fine-grain information completes the judgment (Huttenlocher et al. 2000; Crawford et al. 2006).

The two-step process in multidimensional trait spaces would be less constrained if there is sequential or hierarchical stimulus processing by predators, such that first a single feature is used for crude

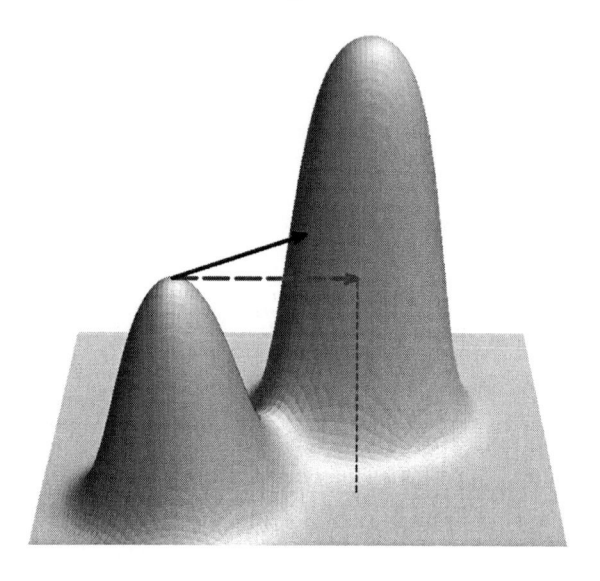

Figure 16.2 Illustration of the difficulty of Müllerian mimicry evolution in multidimensional phenotype spaces. The Adaptive Landscape shows the survival over a season of a mutant as a function of its two-dimensional phenotype. Jumping from the smaller peak, across the valley and onto the slope of the bigger peak requires a simultaneous change in two traits (black arrow). A change in just one trait cannot reach the bigger peak (dashed gray arrow). There is one resident population of 1000 individuals with phenotype at the centre of the smaller peak and another resident population of 5000 individuals with phenotype at the centre of the bigger peak. Predator generalization is given by a bivariate Gaussian with equal standard deviation along each phenotype dimension. All individuals are equally distasteful.

Figure 16.3 Müllerian mimicry evolution occurs more readily if one phenotype dimension is used by predators as a feature to categorize prey. The Adaptive Landscape shows the survival over a season of a mutant as a function of its two-dimensional phenotype. The feature dimension runs along the near-far direction of the figure. For mutant traits in two different intervals (shown as black line segments) of the feature dimension, predators classify prey as belonging to two different categories. These categories correspond to the two resident populations: a smaller population of 1000 individuals with phenotype at the centre of the smaller ridge and another resident population of 5000 individual with phenotype at the centre of the bigger ridge. In comparison with the situation in Fig. 16.2, a mutant change in only the feature trait can reach from the top of the smaller ridge, across the fitness valley and onto the slope of the bigger ridge (arrow). From this point, gradual or stepwise changes in each of the two traits can result in a climb to the top of the ridge.

categorization, followed by a comparison of all perceived prey traits (Balogh et al. 2010), or a comparison of additional features. The overall idea is that a trait functions as a feature by aiding the efficient classification of prey (Chittka and Osorio 2007). The initial mutation in the two-step process could then cause prey to acquire a trait that is used by predators as a feature to categorize potential prey as unsuitable. The significance for mimicry evolution is that, if predators have the tendency to generalize broadly between prey types that share a feature, there may be sufficient advantage for new, imperfect mimics. This possibility of a transition between adaptive peaks through an initial single-feature saltation is illustrated in Fig. 16.3. Fine-grained judgments can then favor subsequent improvement of the mimicry (Balogh et al. 2010). More generally, mimicry evolution could be a sequence of feature mutations, combined with gradual adjustment of the mimetic appearance. Given that different species of predators might use different features, or that evolutionary changes in the composition of the prey community could change the categorizations used, the range of evolutionary changes involving feature mutations is expanded. Even so, feature saltations that initiate mimicry evolution from a non-mimetic starting point will have particular significance.

A possible example of a feature is found in *Heliconius*. A number of species in this genus on the West coast of Ecuador have white hindwing fringes, including *H. erato*, *H. melpomene*, *H. cydno*, *H. sapho*, and *H. sara*. However these five species belong to three very differently patterned Müllerian mimicry rings. Because such white hindwing fringes are virtually absent elsewhere, Sheppard et al. (1985, p. 597) suggested that mimicry of the white hindwing feature allows some generalization by predators, even though much of the rest of the pattern remains very different.

16.5 Fisherian peak shifting

We should also examine the applicability of Fisher's (1927, 1930) idea—that mimicry evolution is gradual and driven by occasional predator "mistakes"—to multidimensional trait spaces. Considering a protected species, Fisher took as a starting point that variation is equally frequent in either of two directions around the mean appearance. Deviations in both directions could be expected to lose protection equally, but with another protected species present, variation in the direction towards that appearance might benefit from the increased similarity. Selection would thus favor variation in that direction, which might lead to a gradual change in the mean trait values of the species, in the direction towards each other. A number of aspects of Fisher's proposal have been studied using theoretical modeling (Balogh and Leimar 2005; Franks and Sherratt 2007; Ruxton et al. 2008). A conclusion from these studies is that Fisher's proposal can work, also in multidimensional trait spaces, if there is wide enough predator generalization, creating a noticeable generalization overlap between the traits of the species already at the start of the process, together with ample amounts of standing genetic variation in prey traits.

The effect of generalization overlap on an Adaptive Landscape can be seen in Fig. 16.1b. The left-hand, smaller peak is asymmetric, so that survival falls off more slowly from its maximum for mutants in the right-hand compared to the left-hand direction (there is a similar but smaller asymmetry of the larger peak). Even though there is a fitness valley between the peaks (Fig. 16.1b), the asymmetric peak shapes promote Fisher's process. Given sufficient genetic variation, the peak positions can approach each other, although when there is initially only a small generalization overlap, the approach will be very slow (Ruxton et al. 2008).

In general, for Müllerian mimicry, the process leads to convergence on a mimetic phenotype in trait space somewhere between the original appearances. However, if one species has higher population density, or is more distasteful than the other, the outcome is that one species (the mimic) approaches the approximately unchanging appearance of the other (the model), which is consistent with a seeming absence of coevolution in empirical examples of Müllerian mimicry (Mallet 1999). This corresponds to an Adaptive Landscape where a smaller, initially more asymmetric peak (Fig. 16.1b) gradually approaches and blends into a bigger, less asymmetric peak, rather than vice versa. In this way peak movement overcomes the fitness valley that was present originally (this is an example of how frequency-dependence changes the shapes of Adaptive Landscapes during mimicry evolution). An interesting property of the process is that it can operate also if only a relatively small proportion of the predator community generalizes broadly (Balogh and Leimar 2005), thus to a degree satisfying Punnett's (1915) requirements mentioned earlier. Even so, if there is little or no generalization overlap to begin with, Fisher's process will not work, or will be too slow to make a difference in comparison with other conceivable weak fitness effects.

16.6 Wrightian shifts in aposematic coloration

A basic idea about the adaptive function of aposematic coloration is that there is an advantage to resembling other members of a population, so that aposematism in itself promotes uniformity rather than diversity in coloration. There is nevertheless substantial geographic diversity in aposematic species (Mallet and Joron 1999), as well as a notable diversity among closely related species (Papageorgis 1975). Of the explanations that have been proposed, an important one is that diverging Müllerian mimicry can give rise to geographic variation in appearance, because of geographic variation in model species. However, mimicry alone is unlikely to generate diversity in aposematic coloration over the long term because no novel phenotypes would be produced, while more and more species converge to fewer and fewer Müllerian models (Mallet and Singer 1987; Turner and Mallet 1996; Mallet and Joron 1999; Baxter et al. 2009; Mallet 2010). The shifting balance process (Wright 1977) has been proposed as another general mechanism that could result in geographic variation, such that warning coloration phenotypes come to occupy different adaptive peaks, representing different efficient solutions to the problem of signaling unprofitability to predators (Mallet and Joron 1999; Mallet 2010).

Granted that a shifting balance process could result in an exploration of different adaptive peaks in a trait space of aposematic appearances, the phenomenon could also be of importance for Müllerian mimicry evolution. If different aposematic species diversify over a limited number of adaptive peaks, there is a chance that they occasionally will become similar enough for Müllerian mimicry to evolve. This follows because Fisherian convergence of aposematic phenotypes is likely to happen only if species are already similar enough to be generalized by at least some predators, which would also make the two-step process more likely to occur.

The possible importance of the shifting balance process in setting the stage for mimicry evolution exemplifies the general principle that mimicry is likely to evolve only under restricted circumstances, including restrictions on the starting points from which mimicry is likely to evolve.

For instance, as emphasized by Nicholson (1927), mimicry is more likely to evolve between related species, partly because of an already existing similarity in appearance, but also because of a similarity in the mutants that may be produced. This in effect acts as a developmental genetic constraint on the evolution of mimicry, along the lines of the ideas of Goldschmidt (1945a, 1945b), who suggested that similar saltational genetic pathways would be re-used by model and mimic. In general, if mimicry evolution corresponds to a transition between adaptive peaks, an important prerequisite is a sufficient closeness of the starting positions of the peaks in trait space.

16.7 Well-studied cases

A presentation of a few much studied instances of mimicry evolution illustrates the concepts we have discussed. One example concerns the variable burnet moth, *Zygaena ephialtes*, which has a wide distribution in Europe and exhibits geographic variation in coloration. North of the Alps it resembles other zygaenids, including the abundant *Zygaena filipendulae*, which has red forewing spots and red hindwing patches on a black background. The northern variant of *Z. ephialtes* also has a red abdominal band and is referred to as the red peucedanoid form (Turner 1971). In some parts of southern Europe, the species instead occurs as the yellow ephialtoid form. This southern variant of *Z. ephialtes* resembles the co-occurring *Syntomis* (*Amata*) *phegea* (Arctiidae), which is black with white wing spots and a yellow abdominal band, and lacks hindwing patches. That *Z. ephialtes* mimics *S. phegea* in this region, and not the other way around, is supported by the fact that *S. phegea* and other *Syntomis* species have white and yellow in their coloration also when they occur alone (without *Z. ephialtes*), whereas *Z. ephialtes* is white and yellow only when it co-occurs with *S. phegea*. In addition, the population densities in the areas of sympatry are much higher for *S. phegea* (Sbordoni et al. 1979). There are also regions in Europe where an intermediate variant of *Z. ephialtes* is found—the red ephialtoid form—which lacks hindwing patches and has white wing spots, apart from two red basal forewing spots, and has a red abdominal band

(Fig. 16.4). The intermediate form occurs either as a local monomorphism or in a polymorphism with one or both of the other two variants (Turner 1971). The difference in appearance between the red peucedanoid form and the intermediate form is determined by a single locus and the difference between the intermediate and the yellow ephialtoid form is determined by another, unlinked locus. This implies that the evolutionary transition between red peucedanoid and yellow ephialtoid (Fig. 16.4) must have involved at least two (and probably more than two) steps. One of these steps might represent a feature saltation, for instance the shift from red to white wing spots, or the shift from red to yellow pigmentation.

The switch from the red peucedanoid form to the yellow ephialtoid form of *Z. ephialtes* has been used in evolutionary genetics to exemplify transitions between two adaptive peaks (Coyne et al. 1997; Gavrilets 1997; Barton et al. 2007). The suggestion is that a transition from the original red peucedanoid form to the intermediate form occurred first, being favored in situations where both the models *Z. filipendulae* and *S. phegea* are present, and was followed by a change from red to yellow of the abdominal band, leading to the yellow ephialtoid form (Fig. 16.4; an illustration of suggested Adaptive Landscapes are found in fig. 24.4 of Barton et al. 2007). The scenario is in agreement with the argument by Sbordoni et al. (1979), to the effect that the intermediate form is favored in situations where *S. phegea* is abundant early in the season but only *Z. filipendulae* is present later in the season.

The transition from the original to the intermediate form of *Z. ephialtes* could be a single feature saltation (Balogh et al. 2010), but reality is likely to be more complex. The difference in appearance between the red peucedanoid and the intermediate form involves changing two different traits: the color of the wing spots and the degree of melanism in the hindwings (Fig. 16.4). The locus controlling these traits seems to be a supergene with two closely linked components (Sbordoni et al. 1979). For the degree of melanism, a number of variably melanic forms occur in certain Mediterranean regions (Hofmann 2003; Hofmann et al. 2009), suggesting that a change in hindwing melanism could have preceded a change from red to white wing

Figure 16.4 (See also Plate 3.) Burnet moth mimicry. The top row depicts mimetic forms: red peucedanoid (left), red ephialtoid (middle), and yellow ephialtoid (right) of the variable burnet moth *Zygaena ephialtes*. In the bottom row are two models; the six-spot burnet *Zygaena filipendulae* (left) and the nine-spotted moth *Syntomis phegea* (right) are presumed models for the red peucedanoid and the yellow ephialtoid mimetic forms, respectively. The red ephialtoid form might have been an intermediate in an evolutionary transition from red peucedanoid to yellow ephialtoid; the form might have an advantage if both *Z. filipendulae* and *S. phegea* are present. Images derive from photos by Clas-Ove Strandberg of hand painted illustrations in Boisduval (1834) and Hübner (1805), obtained with permission from the Library of the Royal Swedish Academy of Sciences, deposited in Stockholm University Library. Original illustrations are (top left) nr. 8, (top middle) nr. 5, (top right) nr. 6, (bottom left) nr. 10 on Plate 55 of Boisduval (1834), and (bottom right) nr. 100 on Plate 20 of Hübner (1805). Some original species names and identities vary from those in our illustration.

spots in *Z. ephialtes*, in which case the intermediate form would have evolved in two or more steps.

An alternative scenario is that the appearance of hindwing melanism was followed by a transition from red to yellow coloration, and only after this event the white wing spots appeared. In such a case, the red ephialtoid form would have arisen at a later time, perhaps as a consequence of hybridization of red peucedanoid and yellow ephialtoid populations (Hofmann 2003), and would not have been an intermediate step in mimicry evolution. Thus, in the alternative scenario of the evolutionary transition in *Z. ephialtes*, hindwing melanism first appears, possibly for thermoregulatory reasons, setting the stage for an abrupt transition from red to yellow pigmentation (which might be a feature saltation), followed by a whitening of the wing spots (Hofmann 2003).

Heliconius butterflies have been extensively studied with respect to geographic variation and Müllerian mimicry (Turner 1971; Mallet and Gilbert 1995). Common colourations are the "rayed" (orange-rayed hind wings) and "postman" (red and yellow bands and bars; Sheppard et al. 1985; Joron et al. 2006a) appearances, and one of these may be ancestral; (recent molecular data indicate that red forewing bands are ancestral in *H. erato*; Hines et al. 2011) it has been difficult to reconstruct ancestral wing patterns in *Heliconius* because of their rapid diversification (Joron et al. 2006a). A less common *Heliconius* appearance is the "tiger" coloration (orange and yellow stripes and blotches on a black ground; Turner 1971; Sheppard et al. 1985) that involves Müllerian mimicry with Ithomiinae butterflies. The coloration occurs in several species within one, probably monophyletic, group of *Heliconius* species (Beltrán et al. 2007), the so-called silvaniform group, and is probably a derived character. The central orange and black in the "tiger" pattern is a candidate for a feature that predators use in prey categorization. There is considerable variation in the "tiger" coloration of different silvaniform butterflies, a possible result of evolutionary fine-tuning towards more accurate mimicry of different Ithomiinae species. In general, mimicry evolution in *Heliconius* appears to be a complex process, possibly with a strong influence of introgression of wing pattern genes through hybridization between different species (Gilbert 2003). *Heliconius* is a likely candidate for the operation of a shifting balance process (Mallet 2010), acting to diversify aposematic patterns, and this process perhaps also plays a role in mimicry evolution. Furthermore, two-step scenarios, including instances of feature saltation, could well apply to *Heliconius*, if one takes into account spatial and temporal variation in predator and model communities.

The genetics of mimicry in *Heliconius* has been mapped, and the genomic regions that include mimicry genes have recently been cloned in a number of species. The impressive Müllerian mimicry between multiple, divergent geographic races of *Heliconius erato* and *H. melpomene* are now known to employ major-effect loci from the very same regions of the genome in both species. Surprisingly, the silvaniform *Heliconius numata* also employs one of these same genomic regions to effect switches among phenotypes of a very different "tiger" pattern mimicry of unrelated ithomiine butterflies (Joron et al. 2006b). Not only are major-effect alleles involved, as expected from the saltational two-step model, but also the number of loci developmentally capable of mimicry appear constrained in an almost Goldschmidtian manner (Baxter et al. 2009).

There are also well-studied cases of Batesian mimicry that can illustrate transitions between adaptive peaks. The spectacular polymorphic mimicry in the Mocker Swallowtail butterfly, *Papilio dardanus*, is one example where scenarios of mimicry evolution involving big-effect mutations have been proposed (Clarke and Sheppard 1960; Turner 1984; Nijhout 2003; Clark et al. 2008; Gamberale-Stille et al. 2012). Finally, the feasibility of gradual mimetic evolution has recently been proposed in the context of Batesian coral snake mimicry (Kikuchi and Pfennig 2010), the claim being that, in situations of high model abundance, intermediates between mimics-to-be and accurate mimics are not attacked more frequently than the original appearance, resulting in an Adaptive Landscape without a valley. The suggested explanation is that a highly toxic and abundant model gives rise to wide predator generalization, overlapping the intermediate phenotypes (Kikuchi and Pfennig 2010), turning Batesian mimicry evolution into a straightforward hill-climbing process.

16.8 Concluding comments

Mimicry and aposematism are phenomena for which the concept of an Adaptive Landscape has proven helpful. Turner (1984) may have been the first to present heuristic illustrations of adaptive peaks of prey appearance, with evolutionary transitions between peaks. Results of theoretical modeling have also been presented as multipeaked Adaptive Landscapes (e.g. Figs. 16.1–16.3; Leimar et al 1986; Balogh and Leimar 2005; Franks and Sherratt 2007; Ruxton et al. 2008). Furthermore, there are field experiments demonstrating adaptive peaks with valleys between them (Kapan 2001), so for these phenomena the concept has empirical support and goes beyond the role of just an attractive heuristic (see Chapters 2, 3, and 19 for different viewpoints in this discussion).

A reason for the great attention given to mimicry evolution, over a period of more than a century, could be that the nature of the selection acting on multitrait phenotypes—to achieve visual resemblance to a model—is readily understood. Even so, the different elements needed to clarify how mimicry actually evolved in particular cases have proven difficult to come by. A good understanding requires a combination of experimentation on predator psychology, field studies on predator prey communities, and studies on the developmental and evolutionary genetics of mimetic phenotypes, and all of this should be put into a biogeographic and phylogenetic perspective. While there has been recent progress in several of these aspects, not least in the genetics of mimicry in butterflies (e.g. Joron et al. 2006a; Clark et al. 2008; Baxter et al. 2010; Hines et al. 2011; Reed et al. 2011), the magnitude of the task still seems challenging, even in the face of the substantial and ingenious efforts of the students of mimicry. Still, it seems possible that empirically well supported examples of transitions between adaptive peaks soon will emerge from the field.

Acknowledgments

This work was supported by grants to the Swedish Research Council to O.L. (2007–5614) and B.S.T. (2007–5877), and from the Biotechnology and Biological Sciences Research Council to J.M.

References

Aronsson, M. and Gamberale-Stille, G. (2008). Domestic chicks primarily attend to colour, not pattern, when learning an aposematic coloration. Animal Behaviour, 75, 417–423.

Balogh, A. C. V. and Leimar, O. (2005). Müllerian mimicry: an examination of Fisher's theory of gradual evolutionary change. Proceedings of the Royal Society B: Biological Sciences, 272, 2269–2275.

Balogh, A. C. V., Gamberale-Stille, G., Tullberg, B. S. and Leimar, O. (2010). Feature theory and the two-step hypothesis of Müllerian mimicry evolution. Evolution, 64, 810–822.

Barton, N. H., Briggs, D. E. G., Eisen, J. A., Goldstein, D. B. and Patel, N. H. (2007). Evolution, Cold Spring Harbor Laboratory Press, Cold Spring Harbor.

Baxter, S. W., Johnston, S. E. and Jiggins, C. D. (2009). Butterfly speciation and the distribution of gene effect sizes fixed during adaptation. Heredity, 102, 57–65.

Baxter, S. W., Nadeau, N. J., Maroja, L. S., Wilkinson, P., Counterman, B. A., Dawson, A., et al. (2010). Genomic hotspots for adaptation: the population genetics of Müllerian mimicry in the *Heliconius melpomene* clade. PLoS Genetics, 6, e1000794.

Beltrán, M., Jiggins, C. D., Brower, A. V. Z., Bermingham, E. and Mallet, J. (2007). Do pollen feeding, pupal-mating and larval gregariousness have a single origin in *Heliconius* butterflies? Inferences from multilocus DNA sequence data. Biological Journal of the Linnean Society, 92, 221–239.

Boisduval, J. A. (1834). Icones historique des lépidoptères nouveaux ou peu connus. Tome II. Roret, Paris.

Chittka, L. and Osorio, D. (2007). Cognitive dimensions of predator responses to imperfect mimicry? PLoS Biology, 5, 2754–2758.

Clarke, C. A. and Sheppard, P. M. (1960). Super-genes and mimicry. Heredity, 14, 175–185.

Clarke, C. A. and Sheppard, P. M. (1972). The genetics of the mimetic butterfly *Papilio polytes* L. Philosophical Transactions of the Royal Society of London Series B-Biological Sciences, 263, 431–458.

Clark, R., Brown, S. M., Collins, S. C., Jiggins, C. D., Heckel, D. G. and Vogler, A. P. (2008). Colour pattern specification in the mocker swallowtail *Papilio dardanus*: the transcription factor invected is a candidate for the mimicry locus H. Proceedings of the Royal Society B: Biological Sciences, 275, 1181–1188.

Coyne, J. A., Barton, N. H. and Turelli, M. (1997). Perspective: A critique of Sewall Wright's shifting balance theory of evolution. Evolution, 51, 643–671.

Cott, H. B. (1940). Adaptive coloration in animals. Methuen, London.

Crawford, L. E., Huttenlocher, J. and Hedges, L. V. (2006). Within-category feature correlations and Bayesian adjustment strategies. Psychonomic Bulletin & Review, 13, 245–250.

Darwin, C. (1859). On the origin of species by means of natural selection, Murray, London.

Fisher, R. A. (1927). On some objections to mimicry theory; statistical and genetic. Transactions of the Royal Entomological Society of London, 75, 269–278.

Fisher, R. A. (1930). The genetical theory of natural selection. A complete variorum edition. Oxford University Press, Oxford.

Franks D. W. and Sherratt, T. N. (2007). The evolution of multicomponent mimicry. Journal of Theoretical Biology, 244, 631–639.

Gamberale-Stille, G., Balogh, A. C. V., Tullberg, B. S. and Leimar, O. (2012). Feature saltation and the evolution of mimicry. Evolution, doi:10.1111/j.1558-5646.2011.01482.x.

Gavrilets, S. (1997). Evolution and speciation on holey adaptive landscapes. Trends in Ecology & Evolution, 12, 307–312.

Ghirlanda, S. and Enquist, M. (2003). A century of generalisation. Animal Behaviour, 66, 15–36.

Gilbert, L. E. (2003). Adaptive novelty through introgression in *Heliconius* wing patterns: Evidence for a shared genetic "toolbox" from synthetic hybrid zones and a theory of diversification. In C. L. Boggs, W. B. Watt, and P. L. Ehrlich (eds.) Butterflies: ecology and evolution taking flight. The University of Chicago Press, Chicago, IL, pp. 281–318.

Goldschmidt, R. B. (1945a). Mimetic polymorphism, a controversial chapter of Darwinism. Quarterly Review of Biology, 20, 147–164.

Goldschmidt, R. B. (1945b). Mimetic polymorphism, a controversial chapter of Darwinism (concluded). Quarterly Review of Biology, 20, 205–230.

Hines, H. M., Counterman, B. A., Papa, R., de Moura, P. A., Cardoso, M. Z., Linares, M., et al. (2011). Wing patterning gene redefines the mimetic history of Heliconius butterflies. Proceedings of the National Academy of Sciences of the United States of America 108, 19666–19671.

Hofmann, A. (2003). *Zygaena (Zygaena) ephialtes* (Linnaeus, 1767) im südlichen Balkan nebst Anmerkungen zur Entstehung von Polymorphismus sowie melanistischer *Zygaena*-Formen im Mittelmeerraum (Lepidoptera:

Zyganeidae). Entomologische Zeitschrift, 113, 50–54; 75–86; 108–120.

Hofmann, A., Kia-Hofmann, T., Tremewan, W. G. and Turner, J. R. G. (2009). *Zygaena dorycnii* Ochsenheimer, 1808, morph *araratica* Staudinger, 1871 (Lepidoptera: Zyganeidae): its Mendelian genetics, sex ratios, hybridisation and speciation. Entomologist's Gazette, 60, 3–23.

Hübner, J. (1805). Sammlung europäischer Schmetterlinge. Band 2. Augsburg.

Huttenlocher, J., Hedges, L. V. and Vevea, J. L. (2000). Why do categories affect stimulus judgment? Journal of Experimental Psychology – General, 129, 220–241.

Joron, M., Jiggins, C. D., Papanicolaou, A. and McMillan, W. O. (2006a). *Heliconius* wing patterns: an evo-devo model for understanding phenotypic diversity. Heredity, 97, 157–167.

Joron, M., Papa, R., Beltrán, M., Chamberlain, N., Mavárez, J., Baxter, S., et al. (2006b). A conserved supergene locus controls colour pattern diversity in *Heliconius* butterflies. PLoS Biology, 4, e303.

Kapan, D. D. (2001) Three-butterfly system provides a field test of müllerian mimicry. Nature, 409, 338–340.

Kikuchi, D. W. and Pfennig, D. W. (2010). High-model abundance may permit the gradual evolution of Batesian mimicry: an experimental test. Proceedings of the Royal Society B: Biological Sciences, 277, 1041–1048.

Leimar, O., Enquist, M. and Sillén-Tullberg, B. (1986). Evolutionary stability of aposematic coloration and prey unprofitability: a theoretical analysis. American Naturalist 128, 469–490.

Mallet, J. (1999). Causes and consequences of a lack of coevolution in mullerian mimicry. Evolutionary Ecology, 13, 777–806.

Mallet, J. (2010). Shift happens! Shifting balance and the evolution of diversity in warning colour and mimicry. Ecological Entomology, 35, 90–104.

Mallet, J. and Gilbert, L. E. (1995). Why are there so many mimicry rings? Correlations between habitat, behaviour and mimicry in *Heliconius* butterflies. Biological Journal of the Linnean Society, 55, 159–180.

Mallet, J. and Joron, M. (1999). Evolution of diversity in warning color and mimicry: Polymorphisms, shifting balance, and speciation. Annual Review of Ecology and Systematics, 30, 201–233.

Mallet, J. and Singer, M. C. (1987). Individual selection, kin selection, and the shifting balance in the evolution of warning colors: the evidence from butterflies. Biological Journal of the Linnean Society, 32, 337–350.

Marsh, H. L. and MacDonald, S. E. (2008). The use of perceptual features in categorization by orangutans (*Pongo abelii*). Animal Cognition, 11, 569–585.

Nicholson, A. J. (1927). A new theory of mimicry. Australian Zoologist, 5, 10–104.

Nijhout, H. F. (2003). Polymorphic mimicry in *Papilio dardanus*: mosaic dominance, big effects, and origins. Evolution & Development, 5, 579–592.

Papageorgis, C. (1975). Mimicry in neotropical butterflies. American Scientist, 63, 522–532.

Pearce, J. M. (2008). Animal learning & cognition, 3rd edn. Psychology Press, Hove.

Poulton, E. B. (1912). Darwin and Bergson on the interpretation of evolution. Bedrock, 1, 48–65.

Poulton, E. B. (1913). Mimicry, mutation and Mendelism. Bedrock, 2, 42–56.

Punnett, R. C. (1915). Mimicry in butterflies, Cambridge University Press, Cambridge.

Reed, R. D., Papa, R., Martin, A., Hines, H. M., Counterman, B. A., Pardo-Diaz, et al. (2011). optix drives the repeated convergent evolution of butterfly wing pattern mimicry. Science, 333, 1137–1141.

Rescorla, R. A. and Wagner, A. R. A. (1972). A theory of Pavlovian conditioning. Variations in the effectiveness of reinforcement and non- reinforcement. In A. H. Black and W. F. Prokasy (eds.) Classical conditioning II: current research and theory. Appleton-Century-Crofts, New York, pp. 64–99.

Ruxton, G. D., Sherratt, T. N. and Speed, M. P. (2004). Avoiding attack: the evolutionary ecology of crypsis, warning signals and mimicry, Oxford University Press, New York.

Ruxton, G. D., Franks, D. W., Balogh, A. C. V. and Leimar, O. (2008). Evolutionary implications of the form of predator generalization for aposematic signals and mimicry in prey. Evolution, 62, 2913–2921.

Sbordoni, V., Bullini, L., Scarpelli, G., Forestiero, S. and Rampini, M. (1979). Mimicry in the Burnet moth *Zygaena ephialtes*: population studies and evidence of a Batsian-Müllerian situation. Ecological Entomology, 4, 83–93.

Schmidt, R. S. (1958). Behavioural evidence on the evolution of Batesian mimicry. Animal Behaviour, 6, 129–138.

Sheppard, P. M., Turner, J. R. G., Brown, K. S., Benson, W. W. and Singer, M. C. (1985). Genetics and the evolution of Muellerian mimicry in *Heliconius* butterflies. Philosophical Transactions of the Royal Society of London Series B: Biological Sciences, 308, 433–610.

Socha, R. and Nemec, V. (1996). Coloration and pteridine pattern in a new, yolk body mutant of *Pyrrhocoris apterus* (Heteroptera: Pyrrhocoridae). European Journal of Entomology, 93, 525–534.

Treisman, A. M. and Gelade, G. (1980). A feature-integration theory of attention. Cognitive Psychology, 12, 97–136.

Troje, N. F., Huber, L., Loidolt, M., Aust, U. and Fieder, M. (1999). Categorical learning in pigeons: the role of texture and shape in complex static stimuli. Vision Research, 39, 353–366.

Turner, J. R. G. (1971). Studies of Müllerian mimicry and its evolution in burnet moths and heliconid butterflies. In R. Creed (ed.) Ecological genetics and evolution. Blackwell Scientific Publications, Oxford, pp. 224–260.

Turner, J. R. G. (1984). The palatability spectrum and its consequences. In R. I. Vane-Wright and P. R. Ackery (eds.) The biology of butterflies. Academic Press, London, pp. 141–161.

Turner, J. R. G. (1985). Fisher's evolutionary faith and the challenge of mimicry. Oxford Surveys in Evolutionary Biology 2, 159–196.

Turner, J. R. G., and Mallet, J. L. B. (1996). Did forest islands drive the diversity of warningly coloured butterflies? Biotic drift and the shifting balance. Philosophical Transactions of the Royal Society B: Biological Sciences, 351, 835–845.

Tversky, A. (1977). Features of similarity. Psychological Review, 84, 327–352.

Wright, S. (1977.) Evolution and the Genetics of Populations. Volume 3. Experimental Results and Evolutionary Deductions. University of Chicago Press, Chicago, IL.

CHAPTER 17

High-Dimensional Adaptive Landscapes Facilitate Evolutionary Innovation

Andreas Wagner

17.1 Introduction

The Adaptive Landscape is one of the most influential concepts in evolutionary biology (Wright 1932, Futuyma 1998, see also Chapter 2 of this volume). It is commonly visualized as a surface of rolling hills or rugged mountains in a three-dimensional space. Two of the three dimensions represent allele frequencies or—relevant for my purpose—genotypes. The third dimension represents fitness. The landscape's peaks represent adaptive trait combinations or genotypes. The Adaptive Landscape concept has been highly successful, so much so that it has spawned multiple variants in the hands of different authors, including holey landscapes (Gavrilets 1997) and phenotype landscapes. The variant I emphasize here can be viewed as a phenotype landscape (see also Chapter 18 of this volume for different kinds of phenotype landscapes).

I view the Adaptive Landscape as a metaphor for the evolutionary process. It is an abstraction derived from an immensely complex reality. Such abstraction is necessary for all human understanding of the world around us. Yet like all abstractions, it also has limitations. Several of these limitations are caused by the high dimensionality of genotype spaces. This high dimensionality has already been appreciated by Sewall Wright, the creator of the Adaptive Landscape concept (Wright 1932). Its consequences have been studied at least since the late 1980s (Kauffman and Levin 1987; Conrad 1990; Gavrilets 1997). One important such consequence regards a basic geometric intuition we derive from three-dimensional space. To get from one peak to the next, one needs to cross a maladaptive valley—the valley of death, if you will—with no detour around this valley. This changes in higher dimensional landscapes, where, counterintuitively, "extradimensional bypasses" around maladaptive valleys exist (Conrad 1990). Through such bypasses, one can travel from peak to peak while avoiding valleys of death. Gavrilets pointed out that this principle has implications for the evolution of reproductive isolation (Gavrilets 1997, 2005). I here explain that it also has implications for how biological systems innovate.

Innovations in evolving biological systems are qualitatively new and beneficial new phenotypes. High-dimensional Adaptive Landscapes can facilitate such innovations. In the next section I discuss evidence for this assertion for three classes of systems important for evolutionary innovation. These are metabolic networks, gene regulation circuits, as well as protein and RNA macromolecules. Detailed studies of the high-dimensional genotype spaces of these systems have demonstrated the existence of vast genotype networks. These are connected sets of genotypes with the same phenotype that extend far through genotype space. Genotype networks are important for the ability of biological systems to explore many novel phenotypes. Genotype networks can be viewed as a consequence of the high-dimensionality of genotype spaces. I will begin illustrating this feature in some detail with metabolic networks, and then discuss the other system classes more briefly. Table 17.1 summarizes some of the concepts I will introduce.

Table 17.1 Important concepts for the three system classes discussed in this chapter.

	Genotype	Neighbors	Genotype space	Phenotype
Metabolic Network	DNA encoding all metabolic enzymes/reactions in a metabolism	Networks that differ in one enzyme (reaction)	All possible metabolic networks	Ability to synthesize biomass or other important molecules from a given spectrum of nutrients
Regulatory circuit	DNA encoding regulatory interactions among molecules	Circuits that differ in one regulatory interaction	All possible regulatory circuits	Gene expression or molecular activity of all circuit molecules
Macromolecule (protein/RNA)	Amino acid or nucleotide polymers	Polymers that differ in a single monomer	All possible amino acid/nucleotide polymers	Tertiary structure (fold) and biochemical activity

17.2 Metabolic network space

Metabolic networks are comprised of hundreds to thousands of chemical reactions—catalyzed by enzymes that are encoded by genes—which synthesize all small molecules in biomass from environmental nutrients. In addition, they produce energy and many important secondary metabolites. Such networks are involved in many innovations, from microbes to higher organisms. Examples include the ability of microbes to grow on synthetic antibiotics or other toxic xenobiotic compounds, such as polychlorinated biphenyls, chlorobenzenes, or pentachlorophenol, in some cases using them as their sole carbon source (Cline et al. 1989; van der Meer 1995; van der Meer et al. 1998; Copley 2000; Dantas et al. 2008; Rehmann and Daugulis 2008). They also include the urea cycle, a metabolic innovation of land-living animals that allows them to convert toxic ammonia into urea for excretion. Novel metabolic traits often involve new combinations of chemical reactions (enzymes) in an organism that already exist separately elsewhere. For example, a novel metabolic pathway to degrade pentachlorophenol involves four steps that its host organism assembled—probably through horizontal gene transfer—from enzymes processing naturally occurring chlorinated chemicals, as well as from an enzyme involved in tyrosine metabolism (Copley 2000). Similarly, the urea cycle arose when four widespread enzymatic reactions involved in arginine biosynthesis combined with arginase, a reaction involved in arginine degradation (Takiguchi et al. 1989).

To study innovation in metabolic networks systematically, one needs to be able to represent all possible metabolic networks and their biosynthetic abilities. To do so, it is necessary to define a space of possible metabolic genotypes. The metabolic genotype of any one organism is the part of the organism's genome that encodes the enzymes which catalyze metabolic reactions. Although this genotype is a string of DNA, it is useful to represent it in a more compact way as follows. Consider the known "universe" of enzyme-catalyzed chemical reactions. This universe currently comprises more than 5000 reactions and the associated enzymes. Write the names of these reactions or their stoichiometric equations as a list. For any one organism, write a one next to the reaction in this list if its genome encodes an enzyme catalyzing this reaction, and a zero if it does not. The result will be a long string of ones and zeroes that can represent the metabolic genotype of the organism (Rodrigues and Wagner 2009).

In this representation, a metabolic genotype can be viewed as a single point in a vast metabolic *genotype space*, a space of possible metabolic networks. This space contains all possible presence/absence combinations of enzyme-catalyzed reactions. Each such combination constitutes a metabolic network. For a universe of 5000 reactions, there are 2^{5000} such metabolic networks, a hyperastronomical number, many more than could be realized in organisms on earth. Metabolic genotype space can also be viewed as the set of all binary strings of length 5000. Yet another, geometric representation is that of a 5000-dimensional hypercube graph. This is a graph whose vertices—the vertices of an n-dimensional cube—are the individual genotypes. Two genotypes are connected by an edge if they are (1-mutant) neighbors, that is, if they differ in a single chemical reaction.

This genotype space is a prototypical example of a high-dimensional space. Each of its axes (reactions) corresponds to one dimension. In contrast to our low-dimensional Euclidean space, it is a discrete and not a continuous space, containing an enumerable number of elements. It is thus quite different from Euclidean space. However, it also shares some similarities with this space. The most important of them is that in genotype space intuitive measures of the *distance* between two metabolic genotypes exists. One such measure is simply the fraction of metabolic reactions in which two genotypes differ. In mathematical terms, metabolic genotype space is thus a *metric* space (Searcoid 2007).

17.3 From metabolic genotype to phenotype

A free-living organism such as *Escherichia coli* or yeast needs to synthesize some 50 essential biomass molecules to grow and divide (Forster et al. 2003; Feist et al. 2007). These molecules include all 20 proteinaceous amino acids, RNA and DNA nucleotides, lipids, and enzyme co-factors.

To sustain life, the metabolic network of a heterotrophic organism needs to generate energy and synthesize all these molecules from a limited number of chemicals in the environment. Recent advances in computational methods have made it possible to compute whether a metabolic network can do so, including the rate at which it can synthesize each compound (Schilling et al. 1999; Feist and Palsson 2008; Feist et al. 2009). For this computation, one needs two kinds of information. These are the stoichiometric equations for each of the chemical reactions in the network, and the rate at which an organism can import necessary chemicals from the environment. Given this approach, one can compute metabolic phenotypes from metabolic genotypes.

To study metabolic innovation is to study how qualitatively novel metabolic phenotypes arise. This requires a definition of phenotype that is suitable for a systematic analysis, and suitable to compare phenotypes. There are many ways to define a metabolic phenotype. For example, one could list the essential biomass molecules that a metabolic network is able to synthesize in any one chemical environment. However, unless a metabolic network can synthesize *all* essential molecules, it may not be able to sustain life. This definition is therefore of limited use. To study metabolic innovations, and especially innovations that allow survival on novel sources of energy and carbon, the following definition is more useful. This definition focuses on carbon, because carbon is a central element in life, and because innovations involving carbon metabolism are thus especially important. However, what I describe also applies to other metabolic innovations (Rodrigues and Wagner 2011).

Consider a minimal chemical environment that contains a small number of molecules which can serve as a source of all necessary elements except carbon. For instance, in the case of *E. coli*, this environment comprises only six different kinds of molecules. Now make a list of many different potential sources of carbon and energy, such as glucose, ethanol, glycerol, and so forth. For the sake of the argument, let us consider 100 such sources. For each of these sources, when provided in an otherwise minimal environment as the *sole* carbon source, determine whether a given metabolic network can synthesize all essential biomass metabolites. If so, write a one next to the list of carbon sources. If not, write a zero. Define the resulting string as the *metabolic phenotype* of this metabolic network. It represents the set of carbon sources on which the metabolic network can sustain life, on which it is *viable* (Rodrigues and Wagner 2009).

This definition of a metabolic phenotype is well-suited for a systematic analysis and comparison of phenotypes, including metabolic innovations. First, it encapsulates an astronomical number of different phenotypes (2^{100} for 100 carbon sources). Second, this notion of phenotype makes it easy to compare different phenotypes by comparing their associated binary strings. Third, an evolutionary innovation—viability on new carbon sources—simply corresponds to a phenotype string where one or more zeroes are converted to ones. It fits the definition of an innovation as a new phenotype that can make a qualitative difference to survival in the right environments, namely environments where this carbon source is the only available source. Such a novel phenotype can arise by adding reactions to a metabolic network, for example by adding enzyme-coding genes to a genome via horizontal gene transfer.

17.4 Exploring metabolic genotype space

To characterize metabolic genotype space exhaustively is impossible, but one can obtain much insight into the organization of this space by carefully designed random sampling. One relevant class of approach is Markov chain Monte Carlo sampling, which involves random walks through genotype space (Rodrigues and Wagner 2009; Samal et al. 2010). Such random walks modify a starting network in a series of steps, each of which either eliminates a reaction (such as might occur through a loss-of-function mutation in an enzyme coding gene) or adds a reaction (such as might occur through horizontal gene transfer). It is useful to require that each step of such a random walk preserves the phenotype. During this random walk, one can also determine all metabolic genotypes in the immediate neighborhood of the random walking network, determine their metabolic phenotypes, and ask which of them are novel metabolic phenotypes that allow survival on new carbon sources. This neighborhood is of special interest, because it contains all novel metabolic phenotypes that are easily accessible from a network through changes in a single reaction.

We have carried out such random walks from different starting points, with very different starting metabolic phenotypes, and explored innovations in the utilization of carbon and other elements (Rodrigues and Wagner 2009; Samal et al. 2010). Together, these analyses have revealed some strikingly simple principles of metabolic genotype space organization.

First, individual metabolic genotypes typically have many neighbors with the same metabolic phenotype. In other words, the metabolic phenotypes of these metabolic networks are to some extent *robust* to mutations that involve changes in individual reactions.

Second, genotypes with the same phenotype form vast connected genotype networks that reach far through genotype space. This means that one can step from one genotype to its neighbor, to the neighbor's neighbor, and so on, without ever changing a phenotype. A genotype network can be viewed as a network of metabolic networks in genotype space. Two genotypes that are far apart on this network have the same phenotype but may share fewer than 25% of their chemical reactions (Rodrigues and Wagner 2009).

Third, the neighborhood of any two genotypes on the same genotype network contains very different novel phenotypes (Fig. 17.1). Together, these properties facilitate the evolution of novel phenotypes through exploration of genotype space. They allow a population to keep its phenotype unchanged while exploring different regions of genotype space and many novel phenotypes therein.

Genotype networks are not peculiarities of carbon metabolism. They also occur in the metabolism of other elements (Rodrigues and Wagner 2011). I note in passing that they have very similar properties if one requires that each random walk preserves the rate at which a network can synthesize biomass, and not just the mere ability to synthesize biomass (Samal et al. 2010).

17.5 Regulatory circuits and novel gene expression patterns

The second system class I will briefly discuss here are regulatory circuits. They are systems of interacting gene products that influence each other's biological activity. Their phenotypes are gene expression phenotypes or, more generally, molecular activities of gene products with important biological functions. Such circuits are involved whenever cells and tissues communicate, and whenever gene expression is regulated (Gilbert 1997; Carroll et al. 2001). Both processes are indispensable for the development of any multicellular organism, and thus for the formation of all macroscopic phenotypes. The most important kinds of circuits are transcriptional regulation circuits, because transcriptional regulation provides a regulatory backbone of all organismal life, and because such circuits drive many pattern formation processes in

Figure 17.1 Genotype networks in genotype space.
The large rectangle schematically represents a genotype space. Open circles represent genotypes with the same (hypothetical) phenotype; neighboring genotypes are connected by straight lines. The figure shows that the open circles form a large connected network that extends far through genotype space. The figure also contains symbols of various shapes and shading. Each such symbol represents a genotype with a different new phenotype, where this genotype is adjacent to the genotype network. That is, it can be reached through a single, small genetic change from some genotype on this network. The figure illustrates that many different novel phenotypes can be accessed from a connected genotype network that spreads far through genotype space. The usual caveat that two-dimensional images poorly represent high-dimensional spaces applies. For example, each genotype typically has hundreds to thousands of neighbors, many more than can be shown here. Also, each genotype that is not on the focal genotype network (symbols of different shape and shading) is also part of a large genotype network that is not shown here. Figure from Wagner (2011), used with permission from Oxford University Press.

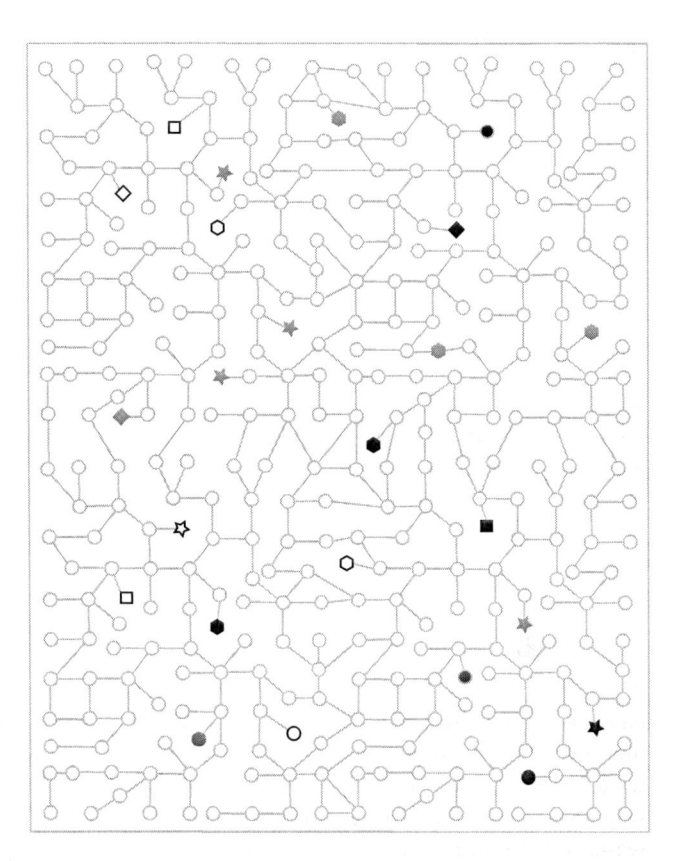

embryonic development. Among the best known examples are Hox gene circuits involved in patterning limbs and many other body structures in animals, as well as circuits involving MADS-box genes important in patterning flowers (Hughes and Kaufman 2002; Irish 2003; Wagner et al. 2003; Causier et al. 2005; Lemons and McGinnis 2006; Hueber and Lohmann 2008).

Regulatory change in transcriptional regulation circuits and other circuits is involved in forming many new novel macroscopic phenotypes. For example, the formation of dissected leaves, an innovation of some plants that may aid thermoregulation, is driven by overexpression of KNOX (KNOTTED1-like homeobox) transcription factors in developing leaves (Bharathan et al. 2002). The predator-deterring eye-spots of butterflies form where the transcription factor *Distal-less* is overexpressed. In these and many other examples (Carroll et al. 2001), changes in the regulation of existing molecules (that also serve other, ancestral functions) are associated with the formation of novel phenotypes.

Because transcriptional regulation circuits are central to embryonic development and to regulatory innovations, many such circuits are well-studied individually (Carroll et al. 2001; Davidson and Erwin 2006). However, a systematic understanding of novel regulatory phenotypes requires analysis of not one circuit, but a systematic analysis of thousands of circuits in the *genotype space* of such circuits. For this purpose, computational models of such circuits are currently indispensable (von Dassow et al. 2000; Albert and Othmer 2003; MacCarthy et al. 2003; Jaeger et al. 2004; Sanchez et al. 2008).

Models that lend themselves to an exploration of a circuit space represent the topology of such a circuit, that is, the pattern of activating and inhibiting regulatory interactions, in a systematic way (MacCarthy et al. 2003; Wagner 2005a; Ciliberti et al. 2007a, 2007b). One can think of a circuit's pattern of regulatory interactions—mediated by the DNA that encodes transcription factors and *cis*-regulatory regions—as the *regulatory genotype* of such a circuit (Fig. 17.1). For a transcriptional regulation circuit, this genotype describes which genes activate or repress each other's expression, and how strongly they do so. Each circuit genotype exists in a space of possible such genotypes. This space contains circuits with all possible patterns of regulatory interactions between circuit genes. Two circuits are neighbors in this space if they differ in exactly one regulatory interaction. The distance between two circuits is the number or fraction of regulatory interactions in which they differ. Two circuits are maximally different, if they have no regulatory interactions in common. The regulatory interactions specified in a regulatory genotype determine the circuit's *phenotype*. This phenotype reflects the activity or expression level of each gene in any one cell, which can be represented either continuously or discretely ("on" or "off"). This highly simplified discrete representation can facilitate enumeration and comparison of different circuit phenotypes (Ciliberti et al. 2007a, 2007b).

Just as in the case of metabolic systems, exploration of the genotype space of circuits shows features very similar to metabolic network space (MacCarthy et al. 2003; Wagner 2005a; Ciliberti et al. 2007a, 2007b; Giurumescu et al. 2009). First, circuits typically have many neighbors in circuit space with the same expression phenotype. Second, circuits with the same phenotype form vast connected genotype networks (MacCarthy et al. 2003; Wagner 2005a; Ciliberti et al. 2007a). Third, the neighborhoods of different circuits contain very different novel phenotypes (Ciliberti et al. 2007b; Giurumescu et al. 2009).

17.6 Macromolecules

The final system class are macromolecules—protein and RNA. They form enzymes, exchange chemicals between cells and their environment, give structural support to cells, are central to locomotion and transport, and perform many other essential functions. Not surprisingly then, many adaptive phenotypic changes are directly traceable to changes in macromolecules. One example regards the ability of some animals to survive temperatures where normal body fluids would freeze. This ability is caused by antifreeze proteins, evolutionary innovations that arose rapidly, multiple times independently, and from different ancestors in arctic and Antarctic fish (Chen et al. 1997; Cheng 1998). Another example involves the ability of the bar-headed goose (*Anser indicus*) to migrate over the Himalayas at altitudes exceeding 10 kilometers (Golding and Dean 1998; Liang et al. 2001). In this bird, one of the subunits of hemoglobin experienced a single proline to alanine substitution. This substitution increases the protein's affinity to oxygen, and thus allows it to transport oxygen at lower oxygen concentrations. It is one of several changes that make the bar-headed goose one of the highest flying birds.

The genotype space of macromolecules is the space of all possible nucleotide or amino acid sequences (Table 17.1). Its structure has been studied for many years (Lipman and Wilbur 1991; Schuster et al. 1994), and shows the same three features I discussed earlier for other genotype spaces. First, individual genotypes typically have many neighbors with the same phenotype (Reidys et al. 1997; Sumedha et al. 2007). For example, random mutagenesis studies of different proteins showed that a large fraction of amino acid changes do not affect protein function (Kleina and Miller 1990; Rennell et al. 1991; Huang et al. 1996).

Second, genotype networks exist in these spaces. This has first been demonstrated for lattice protein models of protein folding, and later for real proteins (Babajide et al. 1997; Todd et al. 1999, 2001; Bastolla et al. 2003; Wagner 2005b). It has also been shown for secondary structure phenotypes of RNA, where genotype networks have been called neutral networks and are extensively characterized (Schuster et al. 1994; Schuster 2006). (I note parenthetically that there are good reason not to use the term neutral network in this context, because evolution along a genotype network need not be neutral in the molecular evolutionist's sense (Wagner 2008).)

Genotype networks in macromolecules typically also extend far through genotype space. The differences between proteins with the same tertiary structure phenotype and common ancestry, for example, can be dramatic. Such proteins may share only a few per cent of their amino acids (Goodman et al.

1988; Rost 1997; Thornton et al. 1999; Todd et al. 1999; Copley and Bork 2000, Bastolla et al. 2003).

Third, different neighborhoods of a genotype network harbor different novel phenotypes. This has first been shown for secondary structure phenotypes of RNA and more recently for proteins and their enzymatic function phenotypes (Schuster et al. 1994; Huynen et al. 1996; Sumedha et al. 2007; Ferrada and Wagner 2010).

17.7 Genotype networks as a consequence of high dimensionality

In sum, three very different classes of biological systems, all of them central for evolutionary innovation, show very similar organizations of their genotype spaces. First, in all three system classes, genotypes have many neighbors with the same phenotype. Specifically, between 10% and more than 50% of a genotype's neighbors typically have the same phenotype, depending on system class, system size, genotype, and phenotype (Wagner 2011). In other words, these systems are to some extent robust to genetic change. Second, genotypes with the same phenotype form vast genotype networks that reach far through genotype space. They typically span between 70–100% of the diameter of this space (Wagner 2011). I note that even though genotype networks are usually astronomically large (10^{50} genotypes in a genotype network are not unusual), and even though they reach far through genotypes space, any one genotype network typically occupies only a vanishing fraction of genotype space. That these properties do not contradict each other results from the fact that genotype space has many dimensions and that it is vast—it has room for myriad genotype networks that are tightly interwoven (Wagner 2011). Third, different neighborhoods of a genotype network generally contain different novel phenotypes. Fig; 17.1 shows a schematic of one such network (open circles) and some novel phenotypes in its neighborhood (symbols of various shapes).

The second and third feature *jointly* facilitate the exploration of novel phenotypes. Specifically, they solve a major problem that innovation poses to living systems: organisms need to preserve old, adaptive phenotypes while exploring innumerable new phenotypes, only few of which may be improvements over the old. Envision a population of organisms that preserves its existing phenotype (through stabilizing selection) while being exposed to mutational change. The existence of genotype networks means that the population can gradually change its genotype while preserving its phenotype. In doing so, it can explore different regions of genotype space. The immediate neighborhood of the population will contain very different novel phenotypes, depending on where its members are located in genotype space. The existence of genotype networks, combined with the diversity of their neighborhoods thus allows exploration of a myriad novel phenotypes.

I note that the discretization of genotypes and phenotypes I used here serves to develop necessary concepts clearly. It also facilitates computational analysis of complex phenotypes and their organization in genotype space. However, a small but growing body of research hints that these concepts can be transferred to systems with continuous genotypes and phenotypes to study how new phenotypes arise in such systems (Wagner 2005a; Giurumescu et al. 2009; Hafner et al. 2009; Raman and Wagner 2011).

The genotype–phenotype relationships I discussed here can be viewed as functions from high-dimensional genotype spaces to high-dimensional spaces of phenotypes. In the case of metabolic networks, the genotypes reflect the presence or absence of metabolic reactions from a reaction universe in any one metabolic network, and the phenotypes reflect the set of carbon sources (or sources of other elements) on which the metabolic network can sustain life. For regulatory circuits, genotypes are the DNA sequences that encode a circuit's regulatory interactions, and phenotypes are activity or concentration patterns of molecules. For macromolecules, genotypes are amino acids and nucleotide sequences, and phenotypes correspond to complex three-dimensional folds of these molecules and their biochemical function (Table 17.1).

I also note that for the systems I consider here, there will generally be more genotypes than phenotypes. For example, for proteins that are N amino acids long, there is an astronomical number of 20^N

genotypes even for moderately large N. In contrast, the number of protein tertiary *structure* phenotypes (protein folds) is of the order 10^4 (Levitt 2009), and for enzymes, the most prominent class of proteins, the number of known *function* phenotypes—the number biochemical reactions they catalyze—is of the same order of magnitude (Ogata et al. 1999). Even if these estimates were to be too low by a factor 100 or 1000, the total number of protein phenotypes would be minute compared to the number of genotypes. In regulatory circuits and metabolism, different arguments lead to the same conclusion (Wagner 2011). For example, in regulatory circuits involving N molecules, the number of regulatory genotypes scales with the number of possible interactions between molecules, and thus with N^2, whereas the number of possible activity states scales with the number of molecules N. There will therefore be more regulatory genotypes than phenotypes. I finally note that if the number of genotypes did not exceed the number of phenotypes, be they protein structures, gene expression phenotypes, metabolic phenotypes, or visible macroscopic traits of organisms, then neutral mutations would not exist, contrary to what empirical genetic data suggest (Eyre-Walker and Keightley 2007).

Any systematic analysis of innovation requires phenotypes that are complex, multidimensional objects, and not just simple scalars, as in many population genetic models of evolution. The phenotypes in all three systems I discuss here meet this criterion. For example, metabolic phenotypes can be represented as vectors in many dimensions, as can the spatial coordinates of a folded protein's amino acids.

The concepts I discuss here share one commonality with holey Adaptive Landscapes used to study speciation (Gavrilets 1997), but they differ in more important ways. The commonality is that genetic change can occur along ridgelines in a high-dimensional space.

The first major difference regards the multidimensional nature of the phenotypes I consider. To study speciation and reproductive isolation, it is may be adequate to consider scalar-valued fitness—or, as in holey landscapes, binary fitness of values 0 and 1—as the only aspect of phenotype. (In this case, the genotypes with fitness zero correspond to the holes in the landscape.) But to study innovation, this will not suffice. It becomes necessary to study phenotypes as complex, multidimensional objects. One could say that for innovation, it is the off-ridge regions in the landscape that are most important (although they are no longer mere holes). They represent genotypes with new phenotypes, some of which may have superior fitness. The second major difference is that to understand innovation, a detailed mechanistic understanding of how phenotypes emerge from genotypes is crucial. The system classes I discuss here meet this requirement. It is not needed to study speciation on holey landscapes, where phenotypes (fitness) are typically assigned *randomly* to genotypes. Finally, I note that innovations also occur in organisms where reproductive isolation, and thus also reproductive isolation by the holey-landscape mechanism, has limited relevance. They include asexual eukaryotes and prokaryotic microbes with unusual forms of sex. Innovation does not generally require reproductive isolation.

How does a map from complex multidimensional genotypes to a multidimensional phenotypes relate to an Adaptive Landscape? To see the connection, we need to simplify this map a bit. Consider the example of metabolic systems, and a specific metabolic phenotype P. For this phenotype, one can define an Adaptive Landscape in metabolic genotype space as a function from this space into the non-negative real numbers. Each genotype's value of this function (its height in the landscape) maps onto the distance of its phenotype from P. Peaks of the landscape correspond to genotypes whose phenotypes are equal to P. Genotypes whose phenotype have a given distance from P occur along the same contour lines (elevation) of the landscape.

Analogous definitions are possible for genotype–phenotype maps in regulatory systems and in macromolecules, because one can define analogous distances among their phenotypes (e.g. Schuster et al. 1994; Ciliberti et al. 2007a). Exactly as for metabolic systems, the height of a point (genotype) corresponds to the distance its phenotype has from some focal phenotype. The genotypes with the focal phenotype are the peaks in this landscape. The more distant a genotype's phenotype is from this focal phenotype, the lower its elevation in this landscape. This simplification renders the maps I consider instances of phenotype landscapes with a scalar phenotype (see also Chapter 18 of this volume).

In Adaptive Landscapes thus defined, the existence of genotype networks clearly violates our geometric intuition derived from low-dimensional spaces. For any one phenotype, a genotype network corresponds to a series of peaks that occur throughout genotype space, and that are all connected to one another. No valleys separate these peaks. In other words, they form an interlaced network of ridges reaching into distant corners of the space. To confound our intuition further, this web of ridges exists for many different focal phenotypes. Each of them has its own genotype network, and all these genotype networks are tightly interwoven with one another (Schuster et al. 1994; Ciliberti et al. 2007b; Rodrigues and Wagner 2009).

Fundamentally, the reason why genotype networks exist is that typical genotypes with some phenotype P have many neighbors with the same phenotype. If this was not the case, and if the genotypes comprising a typical genotype network were otherwise randomly distributed in genotype space, these genotypes would be isolated from one another (Ciliberti et al. 2007a; Wagner 2011). In other words, their robustness to mutation brings forth the vast genotype networks of which they are a part.

The fact that a genotype can have many neighbors at all emerges from the high-dimensionality of genotype space. In a reaction universe of 5000 reactions, each metabolic network has 5000 neighbors. In a transcriptional regulation circuit of 20 genes, there are of the order of 400 possible regulatory interactions; each circuit thus has of the order of 400 neighbors. In a protein of 100 amino acids and 20 possible amino acids, each genotype has $100 \times 19 = 1900$ numbers of neighbors. Because large numbers of neighbors are possible only in a high-dimensional space, so is the possibility to have many neighbors with the same phenotype, and thus the existence of vast genotype networks.

17.8 Outlook

Applying the Adaptive Landscape concept to low-dimensional genotype spaces and to simple, scalar phenotypes limit its utility. This does not mean that we should abandon the concept. There may be philosophical quibbles about it (Provine 1986, discussed also in Chapter 2), but nothing speaks better to its success than its widespread usage almost a century after its conception.

The various incarnations the Adaptive Landscape has taken in the hands of different researchers (e.g., Chapters 2, 3, 9, 13, and 19) show that rather than abandoning the concept, we need to refine it for specific purposes. To study innovation, for example, we can extend it to functions on genotypes whose values are qualitatively different phenotypes in a high-dimensional space. In that case, mathematical or computational analysis needs to replace our geometric intuition. Further refinements will undoubtedly be necessary. For example, some phenotypes and even genotypes are best viewed as objects in a continuous and not in a discrete space. Examples do not only include models from quantitative genetics where genotypes are continuously valued genetic "variables" that underly a uni- or multivariate continuous phenotype. They also include, on the genotypic level, the regulatory genotypes of many signaling circuits, which are defined through parameters such as (continuous) association and dissociation constants, and reaction rates.

On the phenotypic level, they include the ever-changing conformations—defined by the continuous atomic coordinates of amino acids—that one protein can adopt through thermal noise, or the effectively continuous changes of concentrations that gene products can undergo. Where such continuous systems have been studied, phenomena analogous to those in discrete systems seem to exist (Wagner 2005a; Giurumescu et al. 2009; Hafner et al. 2009). However, we lack the theoretical foundation to study the concepts I emphasized here rigorously in such systems. For example, even just defining the analogue to a genotype network in a continuous, high-dimensional genotype space presents challenges. If we meet these challenges, we may discover new worlds in the vast universe of genotype space. There is little doubt that Adaptive Landscapes, despite their limitations, will help us in understanding this universe.

Acknowledgments

I thank Charles Goodnight, the editors, and an anonymous reviewer for suggestions that helped me improve this manuscript. I also acknowledge support through Swiss National Science Foundation grants 315200-116814, 315200-119697, and 315230-129708, as well as through the YeastX project of SystemsX.ch, and the University Priority Research Program in Systems Biology at the University of Zurich.

References

Albert, R. and Othmer, H.G. (2003). The topology of the regulatory interactions predicts the expression pattern of the segment polarity genes in *Drosophila melanogaster*. Journal of Theoretical Biology, 223, 1-18.

Ancel, L.W. and Fontana, W. (2000). Plasticity, evolvability, and modularity in RNA. Journal of Experimental Zoology/Molecular Development and Evolution, 288, 242-283.

Babajide, A., Hofacker, I., Sippl, M., and Stadler, P. (1997). Neutral networks in protein space: a computational study based on knowledge-based potentials of mean force. Folding & Design, 2, 261-269.

Bastolla, U., Porto, M., Roman, H.E., and Vendruscolo, M. (2003). Connectivity of neutral networks, overdispersion, and structural conservation in protein evolution. Journal of Molecular Evolution, 56, 243-254.

Bharathan, G., Goliber, T.E., Moore, C., Kessler, S., Pham, T., and Sinha, N.R. (2002). Homologies in leaf form inferred from KNOXI gene expression during development. Science, 296, 1858-1860.

Carroll, S.B., Grenier, J.K., and Weatherbee, S.D. (2001). From DNA to diversity. Molecular genetics and the evolution of animal design. Blackwell, Malden, MA.

Causier, B., Castillo, R., Zhou, J.L., Ingram, R., Xue, Y., Schwarz-Sommer, Z., et al. (2005). Evolution in action: Following function in duplicated floral homeotic genes. Current Biology, 15, 1508-1512.

Chen, L.B., DeVries, A.L., and Cheng, C.H.C. (1997). Convergent evolution of antifreeze glycoproteins in Antarctic notothenioid fish and Arctic cod. Proceedings of the National Academy of Sciences of the United States of America, 94, 3817-3822.

Cheng, C.C.-H. (1998). Evolution of the diverse antifreeze proteins. Current Opinion in Genetics & Development, 8, 715-720.

Ciliberti, S., Martin, O.C., and Wagner, A. (2007a). Circuit topology and the evolution of robustness in complex regulatory gene networks. PLoS Computational Biology, 3(2), e15.

Ciliberti, S., Martin, O.C., and Wagner, A. (2007b). Innovation and robustness in complex regulatory gene networks. Proceedings of the National Academy of Sciences of the United States of America, 104, 13591-13596.

Cline, R.E., Hill, R.H., Phillips, D.L., and Needham, L.L. (1989). Pentachlorophenol measurements in body-fluids of people in log homes and workplaces. Archives of Environmental Contamination and Toxicology, 18, 475-481.

Conrad, M. (1990). The geometry of evolution. Biosystems, 24, 61-81.

Copley, S.D. (2000). Evolution of a metabolic pathway for degradation of a toxic xenobiotic: the patchwork approach. Trends in Biochemical Sciences, 25, 261-265.

Copley, R.R. and Bork, P. (2000). Homology among (ba)$_8$ barrels: Implications for the evolution of metabolic pathways. Journal of Molecular Biology, 303, 627-640.

Dantas, G., Sommer, M.O.A., Oluwasegun, R.D., and Church, G.M. (2008). Bacteria subsisting on antibiotics. Science, 320, 100-103.

Davidson, E.H. and Erwin, D.H. (2006). Gene regulatory networks and the evolution of animal body plans. Science, 311, 796-800.

Eyre-Walker, A. and Keightley, P.D. (2007). The distribution of fitness effects of new mutations. Nature Reviews Genetics, 8, 610-618.

Feist, A.M., Henry, C.S., Reed, J.L., Krummenacker, M., Joyce, A. R., Karp, P. D., et al. (2007). A genome-scale metabolic reconstruction for *Escherichia coli* K-12 MG1655 that accounts for 1260 ORFs and thermodynamic information. Molecular Systems Biology, 3, 121.

Feist, A.M., Herrgard, M.J., Thiele, I., Reed, J.L., and Palsson, B.O. (2009). Reconstruction of biochemical networks in microorganisms. Nature Reviews Microbiology, 7, 129-143.

Feist, A.M. and Palsson, B.O. (2008). The growing scope of applications of genome-scale metabolic reconstructions using *Escherichia coli*. Nature Biotechnology, 26, 659-667.

Ferrada, E. and Wagner, A. (2010). Evolutionary innovation and the organization of protein functions in sequence space. PLoS ONE, 5(11), e14172.

Forster, J., Famili, I., Fu, P., Palsson, B., and Nielsen, J. (2003). Genome-scale reconstruction of the Saccharomyces cerevisiae metabolic network. Genome Research, 13, 244-253.

Futuyma, D.J. (1998). Evolutionary Biology. Sinauer, Sunderland, MA.

Gavrilets, S. (1997). Evolution and speciation on holey adaptive landscapes. Trends in Ecology & Evolution, 12, 307-312.

Gavrilets, S. (2005). Fitness landscapes and the origin of species. Princeton University Press, Princeton, NJ.

Gilbert, S.F. (1997). Developmental Biology. Sinauer, Sunderland, MA.

Giurumescu, C.A., Sternberg, P.W., and Asthagiri, A.R. (2009). Predicting phenotypic diversity and the underlying quantitative molecular transitions. PLoS Computational Biology, 5, e1000354.

Golding, G.B. and Dean, A.M. (1998). The structural basis of molecular adaptation. Molecular Biology and Evolution, 15, 355–369.

Goodman, M., Pedwaydon, J., Czelusniak, J., Suzuki, T., Gotoh, T., Moens, L., et al. (1988). An evolutionary tree for invertebrate globin sequences. Journal of Molecular Evolution, 27, 236–249.

Hafner, M., Koeppl, H., Hasler, M., and Wagner, A. (2009). "Glocal" robustness in model discrimination for circadian oscillators. PLoS Computational Biology, 5, e1000534.

Huang, W., Petrosino, J., Hirsch, M., Shenkin, P., and Palzkill, T. (1996). Amino acid sequence determinants of beta-lactamase structure and activity. Journal of Molecular Biology, 258, 688–703.

Hueber, S.D. and Lohmann, I. (2008). Shaping segments: Hox gene function in the genomic age. Bioessays, 30, 965–979.

Hughes, C.L. and Kaufman, T.C. (2002). Hox genes and the evolution of the arthropod body plan. Evolution & Development, 4, 459–499.

Huynen, M., Stadler, P., and Fontana, W. (1996). Smoothness within ruggedness: The role of neutrality in adaptation. Proceedings of the National Academy of Sciences of the United States of America, 93, 397–401.

Irish, V.F. (2003). The evolution of floral homeotic gene function. Bioessays, 25, 637–646.

Jaeger, J., Surkova, S., Blagov, M., Janssens, H., Kosman, D., Kozlov, K.N., et al. (2004). Dynamic control of positional information in the early Drosophila embryo. Nature, 430, 368–371.

Kauffman, S. and Levin, S. (1987). Towards a general-theory of adaptive walks on rugged landscapes. Journal of Theoretical Biology, 128, 11–45.

Kleina, L. and Miller, J. (1990). Genetic studies of the lac repressor. 13. Extensive amino-acid replacements generated by the use of natural and synthetic nonsense suppressors. Journal of Molecular Biology, 212, 295–318.

Lemons, D. and McGinnis, W. (2006). Genomic evolution of Hox gene clusters. Science, 313, 1918–1922.

Levitt, M. (2009). Nature of the protein universe. Proceedings of the National Academy of Sciences of the United States of America, 106, 11079–11084.

Liang, Y.H., Hua, Z.Q., Liang, X., Xu, Q., and Lu, G.Y. (2001). The crystal structure of bar-headed goose hemoglobin in deoxy form: The allosteric mechanism of a hemoglobin species with high oxygen affinity. Journal of Molecular Biology, 313, 123–137.

Lipman, D. and Wilbur, W. (1991). Modeling neutral and selective evolution of protein folding. Proceedings of the Royal Society of London Series B, 245, 7–11.

MacCarthy, T., Seymour, R., and Pomiankowski, A. (2003). The evolutionary potential of the Drosophila sex determination gene network. Journal of Theoretical Biology 225, 461–468.

Ogata, H., Goto, S., Sato, K., Fujibuchi, W., Bono, H., and Kanehisa, M. (1999). KEGG: Kyoto Encyclopedia of Genes and Genomes. Nucleic Acids Research, 27, 29–34.

Provine, W.B. (1986). Sewall Wright and evolutionary biology. University of Chicago Press, Chicago, IL.

Raman, K. and Wagner, A. (2011). Evolvability and robustness in a complex signaling circuit Molecular Biosystems, 7, 1081–1092.

Rehmann, L. and Daugulis, A.J. (2008). Enhancement of PCB degradation by Burkholderia xenovorans LB400 in biphasic systems by manipulating culture conditions. Biotechnology and Bioengineering, 99, 521–528.

Reidys, C., Stadler, P., and Schuster, P. (1997). Generic properties of combinatory maps: Neutral networks of RNA secondary structures. Bulletin of Mathematical Biology, 59, 339–397.

Rennell, D., Bouvier, S., Hardy, L., and Poteete, A. (1991). Systematic mutation of bacteriophage T4 lysozyme. Journal of Molecular Biology, 222, 67–87.

Rodrigues, J.F. and Wagner, A. (2009). Evolutionary plasticity and innovations in complex metabolic reaction networks. PLoS Computational Biology, 5, e1000613.

Rodrigues, J.F. and Wagner, A. (2011). Genotype networks innovation, and robustness in sulfur metabolism. BMC Systems Biology, 5, 39.

Rost, B. (1997). Protein structures sustain evolutionary drift. Folding & Design, 2, S19–S24

Samal, A., Rodrigues, J.F.M., Jost, J., Martin, O.C., and Wagner, A. (2010). Genotype networks in metabolic reaction spaces. BMC Systems Biology, 4, 30.

Sanchez, L., Chaouiya, C., and Thieffry, D. (2008). Segmenting the fly embryo: logical analysis of the role of the Segment Polarity cross-regulatory module. International Journal of Developmental Biology, 52, 1059–1075.

Schilling, C.H., Edwards, J.S., and Palsson, B.O. (1999). Toward metabolic phenomics: Analysis of genomic data using flux balances. Biotechnology Progress, 15, 288–295.

Schuster, P. (2006). Prediction of RNA secondary structures: from theory to models and real molecules. Reports on Progress in Physics, 69, 1419–1477.

Schuster, P., Fontana, W., Stadler, P., and Hofacker, I. (1994). From sequences to shapes and back—a case-study in RNA secondary structures. Proceedings of the Royal Society of London Series B, 255, 279–284.

Searcoid, M.O. (2007). Metric spaces. Springer, London.

Sumedha, Martin, O.C., Wagner, A. (2007). New structural variation in evolutionary searches of RNA neutral networks. Biosystems, 90, 475–485.

Takiguchi, M., Matsubasa, T., Amaya, Y., and Mori, M. (1989). Evolutionary aspects of urea cycle enzyme genes. Bioessays, 10, 163–166.

Thornton, J., Orengo, C., Todd, A., and Pearl, F. (1999). Protein folds, functions and evolution. Journal of Molecular Biology, 293, 333–342.

Todd, A., Orengo, C., and Thornton, J. (1999). Evolution of protein function, from a structural perspective. Current Opinion in Chemical Biology, 3, 548–556.

Todd, A., Orengo, C., and Thornton, J. (2001). Evolution of function in protein superfamilies, from a structural perspective. Journal of Molecular Biology, 307, 1113–1143.

van der Meer, J.R. (1995). Evolution of novel metabolic pathways for the degradation of chloroaromatic compounds. Presented at Beijerinck Centennial Symposium on Microbial Physiology and Gene Regulation—Emerging Principles and Applications, The Hague, Netherlands, December.

van der Meer, J.R., Werlen, C., Nishino, S.F., and Spain, J.C. (1998). Evolution of a pathway for chlorobenzene metabolism leads to natural attenuation in contaminated groundwater. Applied and Environmental Microbiology, 64, 4185–4193.

von Dassow, G., Meir, E., Munro, E., and Odell, G. (2000). The segment polarity network is a robust development module. Nature, 406, 188–192.

Wagner, A. (2005a). Circuit topology and the evolution of robustness in two-gene circadian oscillators. Proceedings of the National Academy of Sciences of the United States of America, 102, 11775–11780.

Wagner, A. (2005b). Robustness and evolvability in living systems. Princeton University Press, Princeton, NJ.

Wagner, A. (2008). Neutralism and selectionism: A network-based reconciliation. Nature Reviews Genetics, 9, 965–974.

Wagner, A. (2011). The origins of evolutionary innovations. A theory of transformative change in living systems. Oxford University Press, Oxford.

Wagner, G.P., Amemiya, C., and Ruddle, F. (2003). Hox cluster duplications and the opportunity for evolutionary novelties. Proceedings of the National Academy of Sciences of the United States of America, 100, 14603–14606.

Wright, S. (1932). The roles of mutation, inbreeding, crossbreeding, and selection in evolution. Proceedings of the Sixth International Congress on Genetics, 1, 355–366.

CHAPTER 18

Phenotype Landscapes, Adaptive Landscapes, and the Evolution of Development

Sean H. Rice

18.1 Introduction

The Adaptive Landscape, a function mapping fitness to genotype or phenotype, was originally devised as a visual device that captured the consequences of non-linear (epistatic) interactions between genes (Wright 1932, 1988) (also see Chapter 2 of this volume). A "landscape" is a useful way to visualize a multivariate non-linear function; in order to convert this heuristic device into a mathematical formalism with which to build theories, though, we need a clear understanding of what the axes represent, what kind of fitness is being mapped, and an interpretation of what a particular landscape geometry means biologically.

This last point—the biological interpretation of the form of a landscape—is of particular concern when applying Adaptive Landscape models to questions about the evolution of development. In mapping genetic and developmental traits directly to fitness there are two, very different, sets of biological processes involved: the developmental processes which determine how genetic factors map to phenotype, and the ecological processes that map phenotype to fitness. In order to study developmental evolution, we need to tease apart these different classes of processes.

One way around this problem is to introduce a *phenotype landscape* that explicitly maps genetic and developmental traits to the phenotypic traits that selection acts on. In this chapter we will consider the basic properties of phenotype landscapes—both theoretical and empirically derived—and how we use them, in concert with an Adaptive Landscape, to study the evolution of development. I first give a brief, non-formal, description of phenotype landscapes and how they facilitate the study of developmental evolution. I then cover the formal theory for evolution on phenotype landscapes, discussing how this generalizes the quantitative genetic approaches that are usually applied to Adaptive Landscapes. Next, I discuss how phenotype landscapes can be constructed, and examine some recently published cases in which we can apply the theory to understand the evolution of particular developmental systems. Finally, I show how phenotype landscape theory allows us to study developmental associations between different traits, and discuss the application of this branch of the theory to studying the evolution of genetic covariances and heritability, and to the evolution of novelty.

18.2 Informal theory

A phenotype landscape (sometimes referred to as a "phenotypic landscape" or a "developmental landscape") shows how underlying genetic, epigenetic, and environmental factors map to a phenotypic trait; it thus formalizes and generalizes (through the inclusion of environmental factors) the concept of a "genotype–phenotype map" (Lewontin 1978; Wagner and Zhang 2011). Though phenotype landscapes make no direct reference to selection, they can be used—in combination with a fitness landscape for the trait—to investigate how selection on the trait leads to evolution of the underlying factors.

In this way, we can study the evolution of development while keeping a clear distinction between the developmental processes that build a phenotypic trait and the ecological, physiological, and biomechanical processes through which that phenotype influences fitness.

The phenotype that we consider may be any measurable trait on which we think that selection is acting. The underlying factors can be anything that influences an individual's phenotypic value and to which we can assign a real number. These include explicitly genetic factors, such as the rate of expression of a particular gene; biochemical factors, such as the enzymatic activity of the gene product; or quantitative developmental factors, such as the concentration of a diffusible inducer in a tissue. Because transmission across generations is treated separately from development, the underlying factors need not be heritable. They thus can include environmental factors, such as intensity of sunlight or salinity. In section 18.4, I give some examples of different kinds of underlying factors that can be measured and mapped to phenotype.

For a population, the local geometry of the phenotype landscape determines how the distribution of underlying variation (both genetic and environmental) translates into a distribution of phenotypic variation (Fig. 18.1). For example: in a small region of the landscape, the local slope (captured by the length of the gradient vector) determines how much phenotypic variation will result from a given amount of variation in the underlying factors. Similarly, the local mean curvature (averaged across different directions on the landscape) determines the degree to which a symmetrical distribution of underlying variation translates into an asymmetrical phenotype distribution.

In the case that all underlying factors contribute additively to phenotype (meaning that we can write $\phi = a_1 u_1 + a_2 u_2 + \cdots$, where the a's are numbers), the phenotype landscape is an uncurved plane that looks the same everywhere. Any non-additive contributions of the underlying factors to phenotype are manifest as curvature of the phenotype landscape. In such a case, the local geometry of the landscape changes as a population moves over it.

How a population moves across a phenotype landscape is influenced both by the local geometry of the landscape and by the nature of selection acting on the phenotype. The most familiar kind of selection is *directional* selection, which acts to change the mean of the phenotype distribution and can be captured mathematically as $\frac{\partial w}{\partial \phi}$, the derivative of individual fitness with respect to phenotype (alternately, we can use the regression of fitness on phenotype, which is not identical but plays an analogous role). We can also have *stabilizing* selection (captured by $-\frac{\partial^2 w}{\partial \phi^2}$) or *disruptive* selection (captured

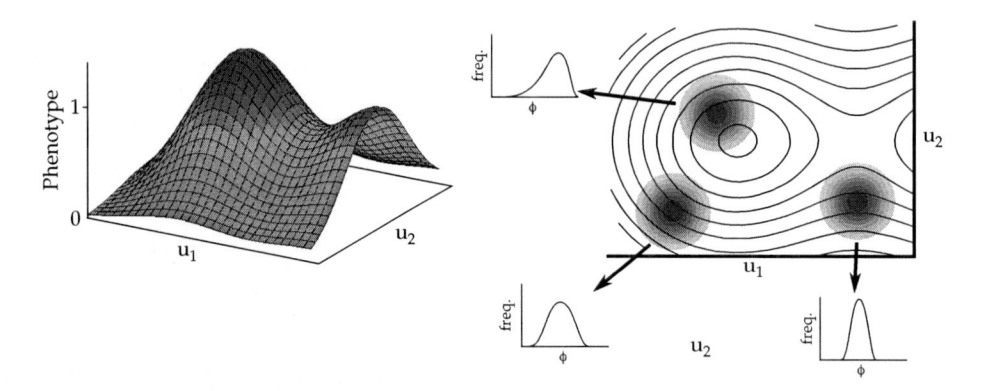

Figure 18.1 A phenotype landscape. The surface on the left is a plot of a phenotypic trait as a function of two underlying factors. On the right is a contour plot of the same landscape. The shaded circular regions represent possible populations (the shading indicating the density of individuals), and the small surrounding graphs show how each distribution of underlying factors translates into a different phenotype distribution; depending on its position on the landscape. Populations positioned on steep parts of the landscape have broader phenotype distributions than those on less steep parts. The population near the peak is in a region of the landscape with high average curvature, which translates into a skewed phenotype distribution.

by $\frac{\partial^2 w}{\partial \phi^2}$) which act to reduce or increase the phenotypic variance, respectively. In fact, selection can act to change any aspect of the phenotype distribution. In the following discussion, we will also encounter cases of asymmetrical selection, captured by the third derivative of the Adaptive Landscape ($\frac{\partial^3 w}{\partial \phi^3}$), which results when fitness drops off more quickly on one side of the optimum than on the other. One of the important results from phenotype landscape theory is that selection that acts to change the shape (rather than the mean) of the phenotype distribution plays at least as important a role in developmental evolution as does directional selection.

If the underlying factors are uncorrelated, and have symmetrical distributions with equal variances, then directional selection should move a population along the gradient (the direction of maximum slope) towards the contour corresponding to the optimum phenotype Rice (1998). Note that this may involve going uphill or downhill on the landscape, depending on whether the optimal phenotypic value is larger or smaller than the current mean.

Once a population is positioned along the contour corresponding to the optimum phenotypic value, $\frac{\partial w}{\partial \phi} = 0$, so *directional* selection on the phenotype is no longer active. It is still the case, though, that most individuals in the population have phenotypic values that differ, in one direction or the other, from the optimum (this is the "imprecision" discussed in detail in Chapter 10). There is thus still *stabilizing* selection (captured by $-\frac{\partial^2 w}{\partial \phi^2}$), and on a curved landscape this will move a population towards regions of minimal slope, where the same amount of underlying variation translates into a narrower phenotype distribution. These processes combined will thus move a population towards points of minimal slope along the contour corresponding to the optimal phenotypic value. We will see later that there is now some direct evidence suggesting that, in some cases, this has happened in natural populations.

Another way to visualize developmental evolution is to graphically combine the phenotype landscape for a trait with the fitness landscape for that same trait, yielding a fitness landscape over the space of underlying factors (Fig. 18.2). Though this approach is generally not as useful as examining the phenotype landscape directly, it does yield one important insight: regardless of the shape of the fitness function for the trait, the fitness function over the space of underlying factors tends to be made up of ridges—meaning that the landscape is "holey" in the sense used by Gavrilets (2003, 2004).

The topology of these ridges may change drastically if selection on the trait changes. Fig. 18.2 shows how a phenotype landscape, combined with a fitness function for the trait, allows us to build the fitness landscape for the set of underlying factors. The key is that a particular value of the trait corresponds to a contour line in the space of underlying factors. An adaptive peak on the trait fitness landscape thus corresponds to an adaptive ridge on the fitness landscape for the underlying factors. Changing the optimal trait value can significantly change the pattern of the network of adaptive ridges.

Fitness landscapes that feature non-linear ridges, such as those in Fig. 18.2, may play an important role in speciation, since movement along an adaptive ridge can easily lead to two populations with the same mean phenotype occupying parts of the space of underlying factors that are separated by "valleys" of low fitness. Hybridization between two such populations is likely to produce hybrids with greatly reduced fitness (Gavrilets 2003). The examples in Fig. 18.2 show that this potential for isolation of different populations on different parts of a ridge is increased if selection on the adult phenotype changes, such that the optimum phenotypic value increases or decreases.

This view of evolution on a phenotype landscape works only if the underlying factors are uncorrelated and have equal variances. If the joint distribution of underlying factors happens to be multivariate normal, then we can always transform the axes to eliminate correlations and equilibrate variances (analogous to a principal components analysis). Otherwise, and especially if there are asymmetries in the distribution of underlying variation, we need to use the formal theory for evolution on phenotype landscapes, which takes into account the joint distribution of the underlying factors.

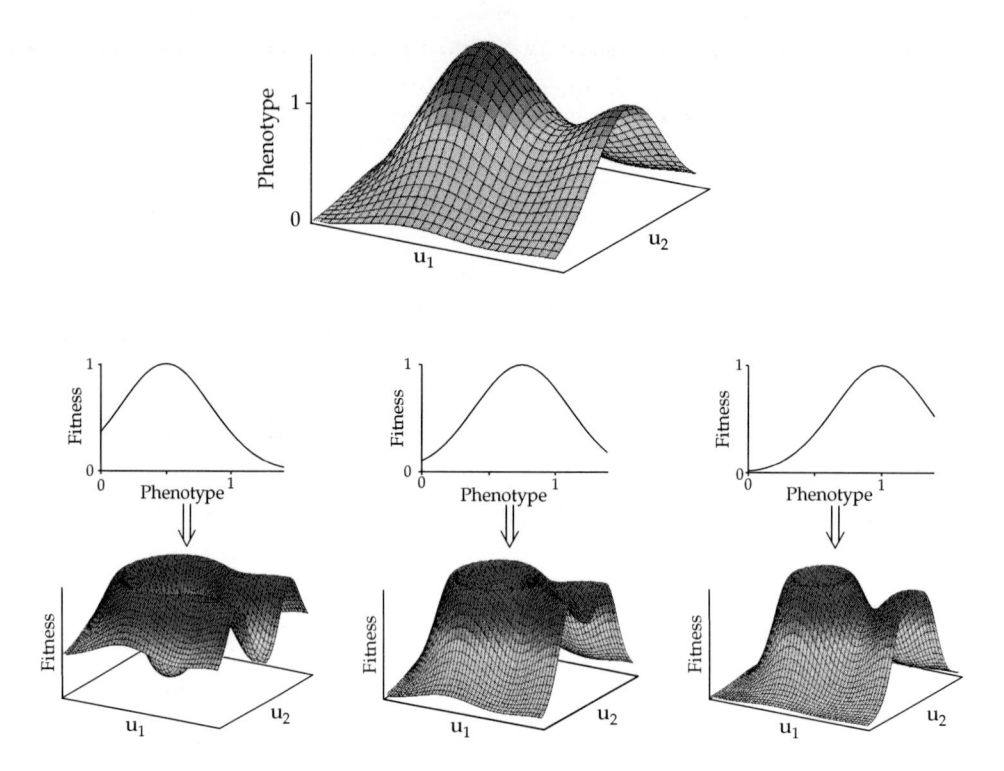

Figure 18.2 Constructing Adaptive Landscapes for underlying factors from a phenotype landscape and different fitness landscapes for phenotype. Each of the fitness functions (plotting fitness as a function of phenotype), when combined with the phenotype landscape at top, produces the corresponding fitness landscape for the underlying factors (bottom figures).

18.3 Formal analysis

The mathematical theory for evolution on a phenotype landscape can be thought of as an extension of quantitative genetics theory in which we relax most of the simplifying assumptions used in quantitative genetics. The full theory is presented in Rice (2004a, 2004b) the following discussion is an outline.

In addition to relaxing the assumptions that underlying factors contribute additively to phenotype, we also will not assume that fitness landscapes are quadratic or Gaussian functions (a common, though not universal (Shaw et al., 2008) assumption of quantitative genetics), and we will not assume that the distribution of variation in a population is multivariate normal.

We will define fitness as the number of descendants that an individual has after one generation (note that we thus treat the individual itself as one of its descendants if it continues to survive for multiple generations). Here we retain one assumption from quantitative genetics: determinism. Specifically, we will treat both fitness and adult and offspring phenotype as deterministic values, rather than as random variables. In reality, both development and evolution are stochastic processes. What this means for fitness is that there is not a single fitness value associated with each phenotype. Rather, prior to reproduction, an individual has a distribution of possible fitness values. Similarly, a particular set of underlying factors should map to a distribution of possible phenotypic values, rather than to a single value. This is a serious issue for Adaptive Landscape models; Rice (2008a) has shown that the evolution of a trait is influenced by the entire distribution of fitness values, not just the mean (see also Gillespie (1977), Proulx (2004), and Lande (2008)). We thus can not simply plot expected fitness as a function of phenotype, as is often done in Adaptive Landscape models. We can still construct a landscape by using expected relative fitness (the expected value of $\frac{w}{\overline{w}}$). However, such a landscape will always be both frequency and density dependent (Rice 2008a; Lande 2008). The reason for this is that mean population fitness, $\overline{w}$, is a function of the phenotype of every individual in the population (since it is the mean of all of their fitness values); $\frac{w}{\overline{w}}$ is thus frequency dependent even when individual fitness (w) is not. The situation for phenotype landscapes is a bit easier. As long as we are studying only the change in mean phenotype, we can consider only the expected phenotype associated with a set of underlying factors—variance around this expectation does not matter.

A consequence of relaxing these simplifying assumptions is that we can not easily write our equations in terms of vectors and matrices alone—we must also consider higher rank tensors. For our purposes, a tensor is just an array of values (there are other criteria for being a tensor, but they are met by all of the things referred to as "tensors" in the following discussion, and will not concern us here). The "rank" of a tensor is just the number of dimensions in the array. So, for example, a vector, being a one-dimensional array, is a rank 1 tensor. The matrices that we use in evolutionary genetics (such as the "G" matrix) are two-dimensional arrays, and as such are rank 2 tensors. A rank 3 tensor is just a three-dimensional array, and so on (see Rice (2004b) for a more extensive discussion of tensors as applied to phenotype landscape theory).

The formal theory of evolution on phenotype landscapes requires that phenotype be a continuous function of the underlying factors. It does not require, though, that the distribution of underlying factors be continuous. This is why we can use gene expression level as an underlying factor; it is reasonable to say that phenotype could be a continuous function of the expression of a particular gene, even if the distribution of that expression level in a population is discrete, corresponding to distinct genotypes or alleles.

We capture the local geometry of the landscape using the various partial derivatives of phenotype with respect to the underlying factors. For example, the slope and orientation of the landscape at a given point is captured by the gradient vector—the vector of first partial derivatives—at that point. We denote the gradient vector as D^1 (the superscript 1 indicating that only first derivatives are involved; this is the same as $\nabla \phi$). On an uncurved landscape, D^1 is all that we need to know, since the local shape of the landscape is the same everywhere. On a curved landscape, though, higher derivatives come into play. The local curvature at a point is captured by the set of second derivatives, measuring curvature in different directions; these are contained in the matrix D^2 (with elements $D^2_{i,j} = \frac{\partial^2 \phi}{\partial u_i \partial u_j}$). We capture higher derivatives in the same way: Just as the first derivatives are arranged into a one-dimensional array (a vector), and the second derivatives into a two dimensional array (a matrix), the nth derivatives are arranged into an n dimensional array—a tensor of rank n.

The shape of the distribution, in a population, of variation of the underlying factors is similarly captured with a set of P tensors, such that the elements of P^n are all of the nth order central moments of the distribution. The matrix (rank 2 tensor) P^2 is just the variance-covariance matrix for the underlying factors. The third moments are captured in a rank three tensor, P^3, with elements defined as:

$$P^3_{i,j,k} = \text{Ave}[(u_i - \overline{u_i})(u_j - \overline{u_j})(u_k - \overline{u_k})] \quad (18.1)$$

The two critical tensor operations that we need are the "inner product" and the "outer product." For two tensors of the same rank, the inner product, denoted $\langle , \rangle$, is just the sum of the products of all of the corresponding elements of each. So, for example, $\langle P^2, D^2 \rangle = \sum_i \sum_j P^2_{i,j} D^2_{i,j}$. If the tensors are of different rank, say P^m and D^n, with $m > n$, then we sum the product of all of the elements of the lower rank tensor with the corresponding elements in the first n indices of the higher rank tensor. The result will be a new tensor of rank $m - n$. For example, $\langle P^3, D^2 \rangle$ is a tensor of rank 1 with elements given by $\langle P^3, D^2 \rangle_k = \sum_i \sum_j P^2_{i,j,k} D^2_{i,j}$.

The other important operation is the outer product (denoted $\otimes$). This combines two tensors of rank m and n into a new tensor of rank $m + n$ that contains all of the pairwise combinations of elements from the initial tensors. So, for example, $D^2 \otimes D^1$ is a tensor of rank 3 in which element i, j, k is the product $D^2_{i,j} D^1_k$.

In the analysis of developmental evolution, P and D tensors never appear in isolation: they are always paired in an inner product that combines a P tensor with one or more D tensors (if there are multiple D tensors, then they are combined with outer products). We thus encounter terms such as $\langle P^2, D^1 \rangle$ or $\langle P^4, D^2 \otimes D^1 \rangle$. This makes sense when we realize that, geometrically, the inner product can be thought of as measuring one tensor in terms of coordinates defined by the other. Thus, the term $\langle P^2, D^1 \rangle$ is simply measuring the gradient vector (D^1) in terms of the variances and covariances between the underlying factors (contained in P^2).

Regardless of the number of dimensions involved, we can identify a point of maximum canalization (or robustness) for a given phenotypic value, ϕ^*, as a point along the ϕ^* contour at which variation in the underlying factors translates into minimum phenotypic variance. Mathematically, such points satisfy $\langle P^4, D^2 \otimes D^1 \rangle = \lambda \langle P^2, D^1 \rangle$, where λ is a constant. For the special case in which the underlying factors are uncorrelated and have equal variances, this is a point along the contour at which the slope of the landscape is locally minimized. (It is also a point at which the vector D^1 is an eigenvector of the matrix D^2.)

A particular combination of P and D tensors describes some aspect of the local shape of the landscape, as "seen by" the population (measured in terms of the distribution of underlying variation in the population). In order to see how this will influence evolution, we must bring in selection, in the form of some derivative of fitness with respect to phenotype (in other words, some measure of the shape of the Adaptive Landscape). If we are interested in pure directional selection, then we are concerned with the first derivative of fitness with respect to phenotype: $\frac{\partial w}{\partial \phi}$. Stabilizing or disruptive selection is captured by the second derivative: $\frac{\partial^2 w}{\partial \phi^2}$ ($\frac{\partial^2 w}{\partial \phi^2} < 0 \Rightarrow$ stabilizing, $\frac{\partial^2 w}{\partial \phi^2} > 0 \Rightarrow$ disruptive). Similarly, the third derivative, $\frac{\partial^3 w}{\partial \phi^3}$, measures selection to skew the phenotype distribution. Though we have no standard name for this sort of selection, we will see below that there are cases where it appears to be happening. Still higher derivatives of fitness with respect to phenotype capture increasingly complex ways that selection can act to change the shape of the phenotype distribution (for example, selection for bimodality will involve $\frac{\partial^4 w}{\partial \phi^4}$).

For a particular aspect of the local shape of the landscape (defined by a particular set of D tensors), everything comes together in the form of a vector, called a Q vector, that shows the component of directional evolution resulting from the specified kind of selection. Once we specify the set of phenotypic derivatives that we are interested in, we can immediately write down the Q vector that corresponds to them. For example, the gradient vector, D^1, shows the fastest rout up or down the phenotype landscape (the length of D^1 is the slope in that direction). The simplest Q vector corresponding to directional selection is thus:

$$Q_1 = \frac{1}{\overline{w}} \frac{\partial w}{\partial \phi} \langle P^2, D^1 \rangle \qquad (18.2)$$

The subscript "1" denotes the fact that D^1 is the only D tensor involved.

Recall that the phenotypic variance is related to the slope of the landscape. The slope along the gradient is given by the length of D^1, and the change in this slope, as we move over the surface, is given by a function that involves the product of second and first derivatives. The simplest Q vector corresponding to selection to change phenotypic variance is thus $Q_{1,2}$, given by:

$$Q_{1,2} = \frac{1}{2\overline{w}} \frac{\partial^2 w}{\partial \phi^2} \langle P^4, D^2 \otimes D^1 \rangle \qquad (18.3)$$

Under stabilizing selection ($\frac{\partial^2 w}{\partial \phi^2} < 0$), the vector $Q_{1,2}$ moves the population towards a point on the landscape at which, locally, the distribution of underlying variation produces minimal phenotypic variation. (The vector $\langle P^4, D^2 \otimes D^1 \rangle$ points in the direction on the landscape in which phenotypic variance would *increase* most quickly, when $\frac{\partial^2 w}{\partial \phi^2}$ is negative, the population evolves in the opposite direction, reducing phenotypic variance). Such a point is a local point of maximal canalization, or phenotypic robustness. Fig. 18.3 illustrates Q_1 and $Q_{1,2}$ and how they relate to the shape of the landscape.

As mentioned earlier, selection might act to skew the distribution of phenotypic variation. There are actually a number of ways that this can happen (Rice 2002), but one case of interest is when, on an

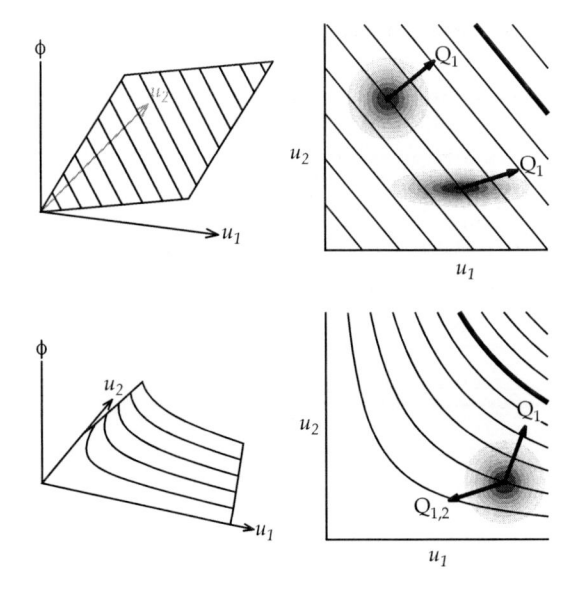

Figure 18.3 Illustration of Q vectors. The landscapes at left correspond to the contour plots on the right. The top landscape is linear, and shows the vector Q_1 for two different phenotype distributions. Note that the vector is a function of both the landscape and the distribution of underlying factors in the population. The bottom landscape shows the vectors Q_1, which points uphill, and $Q_{1,2}$, which points in the direction of maximum decrease in slope of the landscape.

Adaptive Landscape, fitness drops off more quickly on one side of a peak than on the other. In this case, selection should move a population along the optimal phenotype contour to a point at which the phenotype distribution is appropriately skewed. This effect is captured by $Q_{3,1,1}$, given by:

$$Q_{3,1,1} = \frac{1}{12\overline{w}} \frac{\partial^3 w}{\partial \phi^3} \langle P^6, D^3 \otimes D^1 \otimes D^1 \rangle \quad (18.4)$$

Equations 18.2, 18.3, and 18.4 clearly have a similar form. If fact, they are special cases of the general equation for the Q vector corresponding to any set of k derivatives of phenotype with respect to the underlying factors:

$$Q_{a_1 \cdots a_k} = \frac{\gamma}{\overline{w}} \frac{\partial^k w}{\partial \phi^k} \langle P^{1+\Sigma a}, D^{a_1} \otimes \cdots \otimes D^{a_k} \rangle \quad (18.5)$$

The term γ is a scaling factor that is a function of the particular set of phenotype derivatives (Rice 2002, 2004b). The three biologically important parts of equation 18.5 are:

- $D^{a_1} \otimes \cdots \otimes D^{a_k}$—measures some aspect of the local geometry of the phenotype landscape; corresponding to some set of developmental interactions.
- $P^{1+\Sigma a}$—measures the moments of the distribution of underlying variation that are relevant to the developmental interactions captured by the D tensors.
- $\frac{\partial^k w}{\partial \phi^k}$—the kind of selection relevant to the developmental interactions captured by the D tensors. Note that this is determined once we choose a set of D tensors to study.

There is a simple pattern relating all of these parts; the degree of the fitness derivative is equal to the number of D tensors, and the rank of the P tensor is equal to 1 plus the sum of the ranks of the D tensors. In fact, equation 18.5 is for the special case of calculating the change in the *mean* of the distribution of underlying factors. We can find the change in higher moments by changing the rank of the P tensor. For example, substituting $P^{2+\Sigma a}$ gives us a matrix containing the changes in all of the second moments (variances and covariances) of the underlying distribution (Rice 2002).

One striking result from equation 18.5, manifest in equations 18.3 and 18.4, is that we often need to consider more than just the mean and variance of the distribution of underlying factors. For instance, the fourth moments of the distribution of underlying variation are what matter for the evolution of robustness (equation 18.3), and for asymmetrical selection we need to consider the sixth moments (equation 18.4, the same holds for evolution of the curvature of a reaction norm (Rice 2004b, 2008b)). To see the biological significance of this, note that higher moments are increasingly sensitive to outliers. Thus, even relatively small deviations from normality, which would have little effect on the evolution of mean phenotype, can have a significant effect on the evolution of development when selection acts on phenotypic variance or phenotypic plasticity (Rice 2004b, 2008b).

The Q vectors are components of the "selection differential"—the change in mean phenotype resulting only from differential survival and reproduction. In order to get the full response to selection, we need to consider heritability as well, taking into account the fact that the mean phenotype of an individual's offspring may be different from its own phenotype.

To do this, we replace the P tensors, which describe the distribution of variation among the parents, with corresponding C tensors, which describe the joint distribution of offspring and parent phenotypes. Specifically, $C^n_{u_1,\cdots,u_n}$ is just like $P^n_{u_1,\cdots,u_n}$ except that the first value in each element refers to the expected value of an underlying factor in the individual's offspring, rather than in the individual itself. Thus, the elements of C^3 are defined by:

$$C^3_{i,j,k} = \text{Ave}[(u^o_i - \overline{u^o_i})(u_j - \overline{u_j})(u_k - \overline{u_k})] \quad (18.6)$$

where u^o_i is the average value of u_i among an individual's offspring, while $\overline{u^o_i}$ is the average of this value across the entire population (compare with equation 18.1). This puts us on familiar ground, since the second rank C tensor is the offspring–parent covariance matrix which, for our purposes, is the same as the G matrix discussed elsewhere in this volume. The diagonal elements of C thus correspond to the "additive genetic variance" for each trait.

Properly, the matrix C^2 differs from the standard G matrix in two ways (Rice 2004b). First, C^2 need not be symmetrical, meaning that it can be the case that $\text{cov}(\phi^o_1, \phi_2) \neq \text{cov}(\phi^o_2, \phi_1)$. In such a case, the evolutionary change in ϕ_2 resulting from selection on ϕ_1 will be different from the evolutionary change in ϕ_1 resulting from selection on ϕ_2. Second, it is possible (though unlikely) for the diagonal elements of C^2 to be negative. This would correspond to negative heritability—the offspring phenotype being negatively correlated with that of the parents.

In general, equation 18.5 suggests that selection to change mean phenotype may well be less important in developmental evolution than is selection to change other aspects of the phenotype distribution. In order to see if actual developmental systems show the signs of processes like selection for canalization, we need to examine some actual phenotype landscapes.

18.4 Reconstructing phenotype landscapes

Fig. 18.4 shows some examples of phenotype landscapes that have been reconstructed based on various kinds of genetic and developmental data. The fact that all but one of these landscapes is two-dimensional reflects the difficulty of gathering this sort of data—in all cases there are probably many other genetic factors influencing the traits. Nonetheless, these examples show that even with partial data on the development of a trait, a phenotype landscape can be constructed that yields insight into the evolution of the trait. I will briefly consider how these were built, and then discuss their evolutionary implications.

If we know, from prior research, how some set of genes influences a phenotypic trait, we can reconstruct the landscape by fitting a surface to experimental data on how expression of those genes influences the trait. This is how the landscape in Fig. 18.4a, for melanin production in *Daphnia*, was produced by Scoville and Pfrender (2010). In this case, the underlying factors are measures of the expression of known genes. The authors measured the actual values of the underlying factors under three conditions: a) No fish predators and high ultraviolet (UV) radiation (the most common, ancestral, condition for these animals living in high mountain lakes), b) the presence of fish predators—which more readily see darker *Daphnia*—and high UV (the condition for populations in lakes into which fish have been introduces by humans), and c) no predators and low UV radiation (encountered by individuals in protected areas).

Note that all of the occupied points on this landscape lie on or very near points of maximum canalization, with populations in the most likely ancestral environment (Point *a*) being right at the point of minimum slope for that contour (i.e. the combination of underlying factors that produces that level of melanin with minimal phenotypic variation). Scoville and Pfrender (2010) also note that, once fish predators were introduced into some lakes, populations seem to have evolved along the gradient (following Q_1) to a lower level on the landscape (Point *b*). Populations that have not experienced fish

PHENOTYPE LANDSCAPES, ADAPTIVE LANDSCAPES, AND THE EVOLUTION OF DEVELOPMENT   291

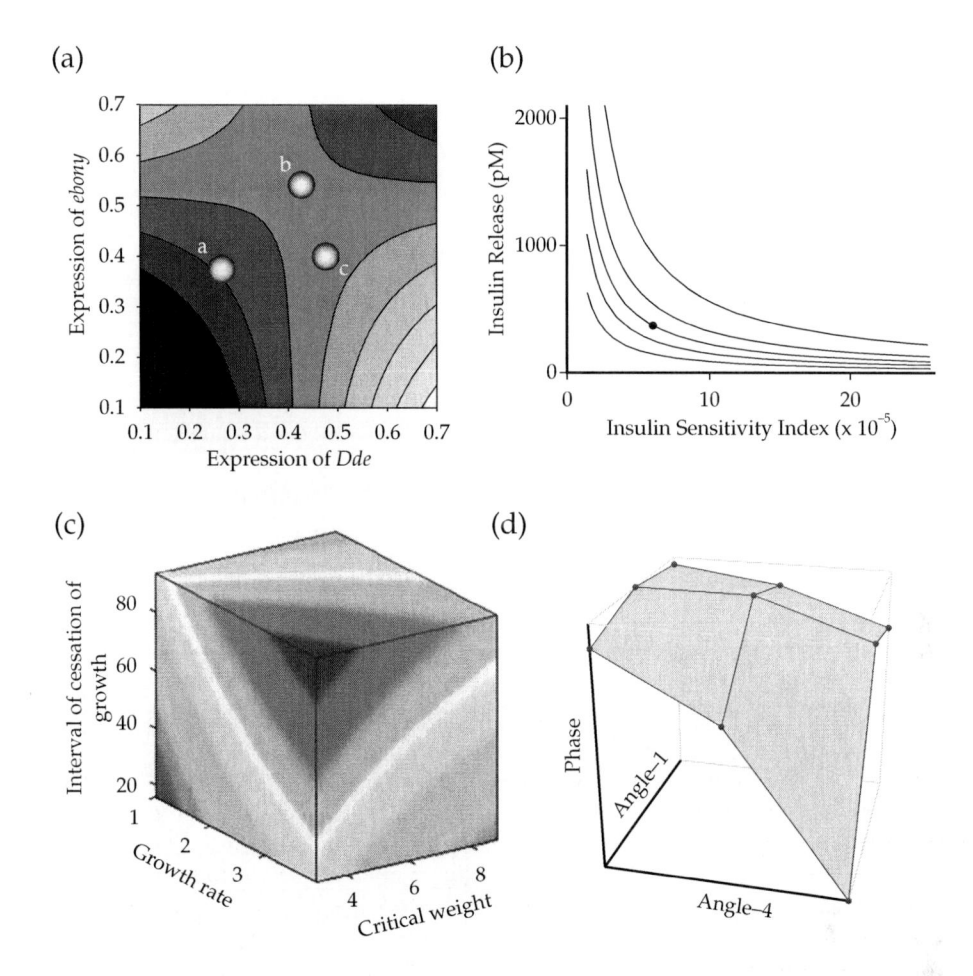

Figure 18.4 Reconstructed phenotype landscapes for a variety of traits. a) Melanin production in *Daphnia* (redrawn and modified from Scoville and Pfrender (2010)). b) Glucose tolerance in humans (redrawn and modified from Kahn (2001)). c) Developmental time in *Manduca*, the three axes are underlying factors, Developmental time is indicated by shade (redrawn and modified from Nijhout et al. (2010)). d) Waking time relative to sundown in mice (redrawn and modified from Rice (2008b), data from Shimomura et al. (2001)).

predation show a plastic response, shifting from *a* to *c* when UV levels are low. It thus appears that the response to selection by predators partially co-opted the plastic response, but also involved evolution at another locus that shifts the population to an even more canalized position.

The landscape in Fig. 18.2b (Kahn et al. 2006; Gibson 2009) was constructed from clinical data on glucose tolerance in humans that suggested that the contours, relating a particular value of glucose tolerance to insulin sensitivity and insulin release, should be hyperbolic (Kahn 2001). This also confirms our intuitive expectation, that the total effect of insulin should be the product of the rate at which it is released and the sensitivity of the tissues that respond to it. (The contours in Fig. 18.2b are exactly what we expect for a landscape of the form $\phi = u_1 u_2$). (Note that, in this case, the contours measure glucose tolerance in percentiles for non diabetic individuals, the spacing between the contours thus does not correspond to a fixed absolute difference in the value of the trait.)

This landscape shows another case in which the population sits close to a point of maximum canalization. Though there is substantial variation in the human population for this trait (including individuals with diabetes, who lie at very low points on the landscape), the median value in the population (solid dot) lies very close to the point of minimum slope along the middle contour in the figure.

Developmental modeling also provides a way to reconstruct phenotype landscapes, though here the underlying factors are generally quantitative traits in development, rather than gene expression rates. The three-dimensional landscape in Fig. 18.4c was constructed by Nijhout and colleagues (Nijhout et al. 2006, 2010) based on experimental work that identified the three main underlying determinants of development time. In this case, the underlying factors are themselves quantitative traits expressed during development. In this case, the region studied does not contain a point of maximum canalization.

Many authors have suspected that selection for canalization or robustness is an important force driving the evolution of development (Wagner et al. 1997; Rice 1998; de Visser et al. 2003; Flatt 2005). It has been difficult, though, to tell just how often, if at all, real developmental systems bear this out. We now can say, at least, that of the relatively few phenotype landscapes that have been reconstructed, the majority show the population at or very near a point of maximum canalization (Rice (2008b) presents another example). This suggests that the expectation that developmental processes are often structured by stabilizing selection on phenotype may be correct. The fact that there are exceptions, though, reminds us that many other processes can also drive developmental evolution. One possible example of a developmental system structured by something other than stabilizing selection alone is shown in Fig. 18.4d.

The landscapes in Fig. 18.4a–c were all built using some prior knowledge of what genes or developmental factors influenced the trait of interest. The landscape in Fig. 18.4d was built without any such prior knowledge. Instead, this landscape was built using quantitative trait locus (QTL) data to identify regions of the genome containing loci that contribute, additively or epistatically, to the trait (the data are from Shimomura et al. (2001), though other QTLs influenced this trait, only these two showed a significant epistatic interaction). Here, the underlying factors are the genotypes at different QTLs;

we thus do not have a clear biological understanding of what the underlying factors really are (and they may not correspond to any continuous variable). Despite this drawback, landscapes based on QTL data have the advantage that they can be constructed even when we do not know the genetic or developmental basis of the trait.

The landscape in Fig. 18.4d is for waking time in mice. Specifically, the trait labeled "Phase" measures the difference between when mice wake up to forage and when it gets dark. Mice are nocturnal foragers, and studies of wild seed-eating rodents suggests that there is a substantial cost to getting a late start on foraging—because others have already reduced the availability of resources (Kotler et al. 1993). Awaking late thus entails a substantial fitness cost. By contrast, awaking early imposes little cost other than some lost sleep, because the mice do not venture out to forage until after dark. We thus expect this to be a case in which the fitness function is asymmetrical around the optimum phenotype.

The likelihood that selection is asymmetrical in this case is significant because the shape of the phenotype landscape for this trait is precisely what we expect for a landscape that produces a skewed phenotype distribution (Rice 2008b). In this case, the distribution of waking times in the population of F2 mice is in fact negatively skewed (i.e. towards earlier waking times (Shimomura et al. 2001)). There were other QTL that contributed additively to this trait, and evolution of those would effectively raise or lower the landscape in Fig. 18.4d. This is important because the optimal value of the trait (phase = 0, meaning wake when it gets dark) is at the plateau on top of the landscape shown. The fact that this system is sitting exactly where it should be, given that the cost of deviating from the optimum in one direction is much greater than it is in the other direction, suggest that stabilizing and skewing selection may have conspired to structure this developmental system.

18.5 Covariance between traits and the evolution of heritability

The discussion so far has focused on the *dynamics* of a population—the evolutionary change over a generation. We can also use phenotype landscapes to study the *statics* of the population—the relationships between different values within a generation.

Of particular interest for evolutionary biology are the phenotypic and genetic variances and covariances for a set of traits. To a first-order approximation, the variance in a phenotypic trait is related to the local (in the vicinity of the population) slope of its phenotype landscape. Wolf et al. (2001) showed that, if the underlying factors are uncorrelated and all have variance 1, then var(ϕ) is equal to the squared magnitude of the gradient vector ($\|D_1\|^2$).

With two traits, ϕ_1 and ϕ_2, there are two different phenotype landscapes, over the space of all underlying factors that contribute to either of them. If both landscapes are additive (or close enough that they can be approximated locally by uncurved surfaces), and the underlying factors are uncorrelated and each have variance 1, then the covariance between ϕ_1 and ϕ_2 is simply the inner product of their gradient vectors: cov(ϕ_1, ϕ_2) = $\langle D_1, D_2 \rangle$. In this case, the correlation between the traits is given by cor(ϕ_1, ϕ_2) = Cos(θ), where θ is the angle between the gradient vectors (Rice 2004a).

In the general case of non-linear landscapes and an arbitrary distribution of underlying variation, the phenotypic covariance between two traits is a function of both development (captured by the D tensors) and the current joint distribution of variation in the underlying factors (captured by the P tensors). Specifically Rice (2004a) showed that:

$$\text{cov}(\phi_1, \phi_2) = \sum_{i=1}^{\infty} \sum_{j=1}^{\infty} \frac{1}{i!j!} \langle P^{i+j} - P^i \otimes P^j, D_1^i \otimes D_2^j \rangle$$

(18.7)

The covariance between two traits is thus a function of both the development of those traits and the joint distribution of variation for underlying factors that influence both traits.

Though equation 18.7 is phrased in terms of the phenotypic covariance between two traits, we can use this same equation to calculate the "additive genetic" variances and covariances for single trait. The trick is to note that we can think of the expected average phenotype of an individual's offspring as another "trait" of that individual. If we know how the underlying factors that influence the trait are transmitted, we can then calculate the "genetic variance" of ϕ as cov(ϕ^o, ϕ). The heritability of ϕ can then be found as $h_\phi^2 = \frac{\text{cov}(\phi^o, \phi)}{\text{var}(\phi)}$.

If all underlying factors, heritable and not, contribute additively to phenotype (i.e. when we can write $\phi = a_1 u_1 + a_2 u_2 + \cdots$, where the a's are numbers), then the offspring–parent covariance and the heritability are not functions of the values of the underlying factors (Rice 2004a). In such cases, heritability and genetic variance will change only as the distribution of underlying variation changes—not as the population moves over the landscape. With non-additive contributions, though, the heritability is a function of the underlying factors, and therefore can (and generally will) evolve, whenever the population moves on the phenotype landscape.

As an example, consider a case in which a trait (ϕ) is influenced by two heritable underlying factors and one non-heritable (environmental) factor. The key is to treat the environment experienced by the parents as a different underlying factor from the environment experienced by their offspring. We now construct separate phenotype landscapes for the parents (ϕ_p) and their offspring (ϕ_o):

$$\phi_p = u_1 u_2 + u_2 u_3$$
$$\phi_o = u_1 u_2 + u_2 u_4 \quad (18.8)$$

Here, we are assuming that the heritable factors u_1 and u_2 are transmitted accurately, so that they have the same values in the offspring as in the parents. The environment is assumed to be non heritable, though, so we treat the parents' environment (u_3) and that of their offspring (u_4) as different underlying factors. (Note that we could make the example more realistic by treating all of the underlying factors for the offspring as different from those of the parents, then capturing variable and incomplete inheritance as covariation between the u's of offspring and parents.) Applying equation 18.7 to the two landscapes in equations 18.8, we can calculate the genetic covariance and heritability for the trait.

Fig. 18.5 shows the phenotype landscape for ϕ_p together with the heritability associated with different values of u_1 and u_2. Note that moving along the phenotype contour corresponding to $\phi = 10$, from upper left to lower right, leads to a change in heritability from $h^2 < 0.2$ to $h^2 > 0.7$.

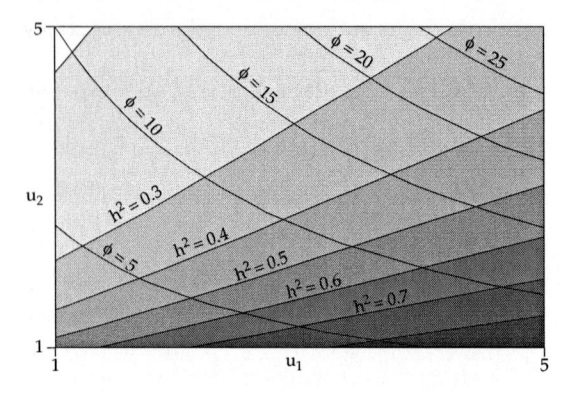

Figure 18.5 Phenotype landscape for parental phenotype, superimposed on the values of heritability (h^2, indicated by shading) at different points on the landscape. In this case, the mean values of u_3 and u_4 are each 1, and their variances are 5 and 1, respectively. A population moving along the $\phi = 10$ contour, from upper left to lower right, would exhibit no change in mean phenotype and only a small change in phenotypic variance, but heritability would change from low (< 0.2) to very high (> 0.7).

Though quantitative genetics models often treat heritability as a constant, there is substantial evidence that it is sometimes a function of the environment in which organisms develop (reviewed in Merila and Sheldon (2001) and Charmantier and Garant (2005)). In fact, analysis of phenotype landscape models shows that heritability will necessarily be a function of the environment whenever environmental and heritable underlying factors interact non-additively in development (Rice 2004a).

The ability to examine the evolution of heritability can also provide insights into a problem that initially seems intractable using Adaptive Landscapes—the evolution of novelty (see also Chapter 16). In order to include a trait in an Adaptive Landscape, we need to know what it is. This seems to preclude using the Adaptive Landscape approach to study the evolution of novel traits, since there are a very large number of possible traits that do not yet exist and we can not anticipate all of them. Furthermore, even if we anticipate that a trait might arise, it is not clear how its appearance would be represented on an Adaptive Landscape.

The analysis of the evolution of heritability discussed earlier suggests another approach. Any function of a set of underlying factors can be thought of as a phenotype, regardless of whether selection is currently acting on it or whether or not it is heritable. In order for such a function to be an evolutionary trait, though, it must be heritable. Thus, from the standpoint of the mechanics of evolution, "novelty" means the appearance of a new heritable function of the genome.

The appearance of a novel trait, in this sense, is something that we could, in principle, observe on a phenotype landscape. Given a distribution, within a population, of a set of heritable and non-heritable underlying factors, we can ask whether or not a particular function of these is heritable. If it is, then that function is currently an evolutionary trait—meaning that it would change if selection acted on it. If the function is not heritable, given the current distribution of the underlying factors, we could use the approach outlined above (and in Rice (2004a)) to ask what changes in the underlying distribution would make it become heritable. At least in principle, we could thus anticipate the conditions under which a novel trait would arise.

Reference

Charmantier, A. and Garant, D. (2005). Environmental quality and evolutionary potential: lessons from wild populations. Proceedings of the Royal Society B: Biological Sciences, 272(1571), 1415–1425.

Flatt, T. (2005). The evolutionary genetics of canalization. Quarterly Review of Biology, 80(3), 287–316.

Gavrilets, S. (2003). Evolution and speciation in a hyperspace: the roles of neutrality, selection, mutation, and random drift. In Towards a Comprehensive Dynamics of Evolution: Exploring the Interplay of Selection, Neutrality, Accident, and Function, J. Crutchfield and P. Schuster (eds.) Oxford University Press, New York, pp. 135–162.

Gavrilets, S. (2004). Fitness Landscapes and the Origin of Species. Princeton University Press, Princeton, NJ.

Gibson, G. (2009). Decanalization and the origin of complex disease. Nature Reviews Genetics, 10, 134–149.

Gillespie, J. H. (1977). Natural selection for variance in offspring numbers: a new evolutionary principle. American Naturalist, 111(981), 1010–1014.

Kahn, S. E. (2001). The Importance of beta-cell failure in the development and progression of type 2 diabetes. Journal of Clinical Endocrinology & Metabolism, 86, 4047–4058.

Kahn, S. E., Hull, R. L. and Utzschneider, K. M. (2006). Mechanisms linking obesity to insulin resistance and type 2 diabetes. Nature, 444, 840–846.

Kotler, B. P., Brown, J. S. and Mitchell, W. A. (1993). Environmental factors affecting patch use in two species of gerbilline rodents. Journal of Mammalogy, 74(3), 614–620.

Lande, R. (2008). Adaptive topography of fluctuating selection in a Mendelian population. Journal of Evolutionary Biology, 21, 1096–1105.

Lewontin, R. C. (1978). Adaptation. Scientific American, 293(3), 213–230.

Merila, J. and Sheldon, B. C. (2001). Avian quantitative genetics. In J. Van Nolan and E. D. Ketterson (eds.), Current Ornithology, Vol 16, Kluwer/Plenum, New York, p. 179.

Nijhout, H., Davidowitz, G. and Roff, D. (2006). A quantitative analysis of the mechanism that controls body size in manduca sexta. Journal of Biology, 5(5), 16.

Nijhout, H. F., Roff, D. A., and Davidowitz, G. (2010). Conflicting processes in the evolution of body size and development time. Philosophical Transactions of the Royal Society B: Biological Sciences, 365(1540), 567–575.

Proulx, S. R. (2004). Sources of stochasticity in models of sex allocation in spatially structured populations. Journal of Evolutionary Biology, 17, 924–930.

Rice, S. H. (1998). The evolution of canalization and the breaking of von Baer's laws: modeling the evolution of development with epistasis. Evolution, 52(3), 647–656.

Rice, S. H. (2002). A general population genetic theory for the evolution of developmental interactions. Proceedings of the National Academy of Sciences, 99, 15518–15523.

Rice, S. H. (2004a). Developmental associations between traits: covariance and beyond. Genetics, 166, 513–526.

Rice, S. H. (2004b). Evolutionary theory: mathematical and conceptual foundations. Sinauer Associates, Sunderland, MA.

Rice, S. H. (2008a). A stochastic version of the Price equation reveals the interplay of deterministic and stochastic processes in evolution. BMC Evolutionary Biology, 8, 262.

Rice, S. H. (2008b). Theoretical approaches to the evolution of development and genetic architecture. Annals of the New York Academy of Sciences, 1133, 67–86.

Scoville, A. G. and Pfrender, M. E. (2010). Phenotypic plasticity facilitates recurrent rapid adaptation to introduced predators. Proceedings of the National Academy of Sciences, 107(9), 4260–4263.

Shaw, R. G., Geyer, C. J., Wagenius, S., Hangelbroek, H. H. and Etterson, J. R. (2008). Unifying life-history analyses for inference of fitness and population growth. American Naturalist, 172(1), E35–E47.

Shimomura, K., Low-Zeddies, S. S., King, D. P., Steeves, T. D., Whiteley, A., Kushla, J., et al. (2001). Genome-wide epistatic interaction analysis reveals complex genetic determinants of circadian behavior in mice. Genome Research, 11(6), 959–980.

de Visser, J., Hermisson, J., Wagner, G. P., Meyers, L. A., Bagheri, H. C., Blanchard, J. L., et al. (2003). Perspective: Evolution and detection of genetic robustness. Evolution, 57(9), 1959–1972.

Wagner, G. P., Booth, G. and Bagheri, H. C. (1997). A population genetic theory of canalization. Evolution, 51(2), 329–347.

Wagner, G. P. and Zhang, J. (2011). The pleiotropic structure of the genotype-phenotype map: the evolvability of complex adaptations. Nature Reviews Genetics, 12, 204–213.

Wolf, J. B., Frankino, W. A., Agrawal, A. F., Brodie, E. D. 3rd, and Moore, A. J. (2001). Developmental interactions and the constituents of quantitative variation. Evolution, 55, 232–245.

Wright, S. (1932). The roles of mutation, inbreeding, crossbreeding and selection in evolution. Proceedings of the 6th International Congress of Genetics, 1, 356–366.

Wright, S. (1988). Surfaces of selective value revisited. American Naturalist, 131(1), 115–123.

PART VI
Concluding Remarks

CHAPTER 19

The Past, the Present, and the Future of the Adaptive Landscape

Erik I. Svensson and Ryan Calsbeek

19.1 Historical context

This volume marks the 80-year anniversary of the publication of Sewall Wright's landmark paper "The roles of mutation, inbreeding, crossbreeding and selection in evolution" (Wright 1932). Since the formulation of the landscape concept in evolutionary biology, it has inspired generations of researchers, but also resulted in deep scientific and philosophical controversies (Chapters 1 and 2). Many of these controversies remain today, and are far from solved (Chapters 3–6). Depending on one's perspective on the scientific process, these controversies and scientific discussions have either been useful and clarifying, or a waste of time and valuable research resources.

Should scientific ideas, models, and metaphors be judged by their absolute truth, their beauty (Fagerström 1987), or how they inspire further research and stimulate scientific discussion? One thing is certain: if one adopts the position that scientific inspiration is important, the Adaptive Landscape idea must be one of the most successful concepts in evolutionary biology. The Adaptive Landscape has inspired research in many different subdisciplines in evolutionary biology, including population genetics (Chapters 4–6), evolutionary ecology (Chapter 7), quantitative genetics (Chapters 8–10), experimental evolution (Chapter 11), conservation biology (Chapter 12), speciation and macroevolutionary dynamics (Chapters 13–15), mimicry, saltational evolution, and behavioral ecology (Chapter 16), molecular biology and protein networks (Chapter 17), and theoretical studies on development (Chapter 18).

Although not all research presented in the chapters in this volume are equally strongly connected to, and critically dependent on the Adaptive Landscape concept, some of these investigations would certainly not have been performed without the conceptual foundations outlined by Sewall Wright in 1932. One might even wonder if certain research fields would have existed at all, since they seem so closely intertwined to the concepts of adaptive peaks and valleys (e.g. bacterial experimental evolution in Chapter 11).

19.2 Which Adaptive Landscape?

Since the Adaptive Landscape concept was first formulated, it has been presented in several different versions, and some of them are clearly logically inconsistent (Provine 1986) (Chapters 2–3). The first confusion stems from the independent axes in the landscape diagrams: do the axes refer to individual genotypic values, or population gene frequencies? In the first version of the Adaptive Landscape, the units on the axes occur in discrete classes, in the second version, the axes are treated as continuous variables. Provine (1986) and Pigliucci (2007) have argued that it is only the latter landscapes, which depict fitnesses as a function of gene frequencies, that are scientifically meaningful.

Although these criticisms have some value, it seems to us as if both types of landscapes could actually have some utility, albeit in different domains. For instance, Gavrilets (2004) frequently and creatively used fitness landscapes of individual genotypes to illustrate how the Bateson–Dobzhansky–Muller (BDM) model of hybrid fitness, genetic incompatibilities, and sexual conflict in mating pairs can be understood at the genomic level. Goodnight (Chapter 6) tries to resolve the problem of the discreteness of genotypes by pointing out that one could assume both that the horizontal axes represent the effects of a locus on an individual's phenotype, and that the axes are continuous. His argument is that the axes are the results of three-dimensional projections of landscapes of very high dimensionality (Chapter 6).

Gavrilets (2004) recently criticized the view put forward by Provine (1986) and others (Kaplan 2008; Pigliucci 2008) that it is only the fitness surface of gene frequencies that is meaningful and questioned the claim that the fitness surface of individual genotypes is meaningless. According to Gavrilets, this view is based on a misconception of Provine (1986; see also Pigliucci 2008 and Kaplan 2008 for similar views) it is wrong to claim that the two versions of fitness landscapes are "wholly incompatible" (Gavrilets 2004, pp. 32–33). Rather, Gavrilets argues:

> It is straightforward to find mathematically the average fitness of the population $\bar{w}$ if one knows the individual fitnesses. That is, the first version can be transformed in to the second version in a straightforward manner. However, knowing $\bar{w}$ indeed does not allow one to find individual fitnesses (Gavrilets 2004, p. 32).

Moreover, Gavrilets suggests that:

> Describing populations by listing genotype frequencies becomes equivalent to listing all genotypes present, which is exactly what is done in the first version of fitness landscapes. Finally, the process of speciation, during which a population splits into two different species, is impossible to describe in a framework where a population is the smallest unit (Gavrilets 2004, p. 33).

Although more certainly remains to be said on this issue, the landscape of discrete genotypes has certainly played an important role in illustrating and conceptualizing fitness epistasis—i.e. the interactions between genes and how they affect fitness (Wolf et al. 2000). Moreover, many evolutionary geneticists have frequently and successfully used the Adaptive Landscape of genotypes as a tool to understand gene interactions and their consequences (Brodie 2000; Phillips 2008).

By contrast, at the phenotypic level, some of the problems discussed above might actually disappear, as phenotypes are typically continuous even if their underlying genotypes are discrete (Arnold et al. 2001). For polygenic characters governed by many loci of small effect, which are typically the focus of study by evolutionary ecologists (Chapters 7–9), the axes in the fitness landscape are continuous variables (traits), which are governed by many discrete underlying genetic factors along many dimensions (loci). For these and other reasons, the controversy about discrete genotypes versus gene frequencies might become irrelevant at the phenotypic level (Simpson 1944; Schluter 2000; Arnold et al. 2001; Chapters 9, 12, 13, and 15). Moving up to the level of phenotypes might therefore reduce the dimensionality of the landscape considerably, although there are of course many other conceptual and practical problems in quantifying and describing phenotypes and creating phenotype ontologies (Houle 2010; Houle et al. 2010).

Another conceptual issue that has been discussed more recently is the unit of measurement on the vertical (fitness) axis. Fear and Price (1998) pointed out that the fitness surface of individuals will be steeper than the surface of mean fitness. This is because averaging fitness over many individuals "smoothes" the Adaptive Landscape, and reduces the height of both adaptive peaks and valleys. They recommended that the term "the Adaptive Landscape" should be reserved only to those fitness surfaces which show how population mean fitness changes as a function of mean trait values, and that the fitness surfaces of individuals should be clearly separated from the Adaptive Landscape. This redefinition should essentially make it impossible to estimate Adaptive Landscapes at the empirical level, and workers would then have to resort to quantifying individual fitness surfaces. Although some authors in this volume follow this strict definition (Chapters 3 and 9,), others are more liberal and take individual fitness surfaces as a proxy for the Adaptive Landscape (Chapters 7 and 13). This might be justified as the two types of surfaces are related to each other and qualitatively similar. Hence, they are not in totally separate domains.

19.3 Model or metaphor?

What is the Adaptive Landscape: a model, a metaphor or something else? Skipper and Dietrich (Chapter 2) argue that the Adaptive Landscape should best be viewed as a metaphor whose value is largely heuristic. Metaphors can be important, perhaps even necessary in science, even if they are not as solid as formal mathematical models. For instance, the metaphor of the "selfish gene," as coined by Richard Dawkins (Dawkins 1976) obviously had an enormous impact on the development of behavioral ecology and social evolution, and was a natural starting point to discuss the evolution of cooperation and the conflict between lower- and higher-level selection. By analogy, the Adaptive Landscape metaphor has been a natural starting point to think about various problems in population divergence and local adaptation (Chapters 5 and 6), evolutionary ecology (Chapter 7), and macroevolutionary dynamics (Chapters 13–15). These examples might be sufficient to argue that the Adaptive Landscape has been of great service in its heuristic value.

However, the Adaptive Landscape has also stimulated several theoretical approaches, which are more than mere metaphors. For instance, quantitative genetic models (Chapter 9), models of macroevolution (Chapter 13), metabolic networks (Chapter 17), and phenotype development (Chapter 18) are often mathematically explicit and yet have their conceptual basis in the Adaptive Landscape tradition. Perhaps the general idea of Adaptive Landscapes is not, and will never be, a formal mathematical model, but certain off-shoots and branches might definitely be more than only of metaphoric value. Moreover, as long as metaphors have *some* connection to formal models, they should be of value as a starting point, more formal and quantitative analyses.

To briefly return to the metaphor of the selfish gene and make another analogy: as long as one realizes that this metaphor focuses on the additive effects of genes and ignores epistasis and that there are explicit mathematical models that describe the dynamics, the metaphor of the selfish gene has some justification, even if it is not a complete model of evolution. Similarly, the Adaptive Landscape in the "Simpsonian" sense and at the level of phenotypic evolution is nowadays built on an explicit mathematical theory. Fitness landscapes can be rigorously described using a series of well-established formal equations (Arnold et al. 2001). In practice, the distinction between metaphors and formal models might not be as clear-cut as some philosophers argue (Kaplan 2008), and there may often be a gradual transition between them. Gavrilets (2004) argues that metaphors are similar to simple mathematical models that emphasize some features of the real world at the deliberate neglect of others. He further argues that fitness landscapes are a very useful metaphor for a number of important evolutionary processes (Gavrilets 2004). Finally, Arnold et al. (2001) argue that the Adaptive Landscape of phenotypes is clearly more than a metaphor, since there is an explicit mathematical theory available in evolutionary quantitative genetics. The Adaptive Landscape is therefore not only of heuristic value but is a quantitative concept, at least at the phenotypic level.

19.4 Achievements

In this chapter we have argued that the Adaptive Landscape has been a highly successful metaphor and scientific concept, judged by the many different research disciplines in evolutionary biology it has influenced, and the many researchers it has stimulated intellectually. Our opinion has not changed in that respect since we decided to assemble this volume, as we have also been inspired by the Adaptive Landscape in our own research programs in evolutionary ecology (Chapter 7). However, perhaps we were a bit surprised that the Adaptive Landscape still seems to be such a living idea which continues to stimulate research also in disciplines quite different from ours (Carneiro and Hartl 2010).

Some exciting and fast-moving fields where much work remains to be done within the paradigm of the Adaptive Landscape include protein evolution (Carneiro and Hartl 2010), metabolic network theory (Chapter 17), and the genotype–phenotype map (Chapter 18). We also need to develop new models of macroevolution based on not only evolution *on* the current Adaptive Landscape, but evolution *of* the landscape itself (Chapter 13). These are not trivial scientific problems, and solving them will require hard work, both conceptually and empirically. The Adaptive Landscape concept is no panacea for all problems in evolutionary biology. However, it seems increasingly clear to us, after editing this volume, that the Adaptive Landscape is not a dead scientific idea, but is, on the contrary, very much alive and active in the minds of many researchers. The proof of the pudding is in the eating, and the Adaptive Landscape has turned out to have a surprising adaptability, that has survived several criticisms in the past. We therefore surmise that one of the most recent declarations of "the end of the Adaptive Landscape metaphor" (Kaplan 2008) might have been slightly premature (see also Pigliucci 2008).

Although evolutionary-oriented molecular biologists increasingly direct their attentions to metabolic networks, "holey Adaptive Landscapes" and "ridges" in genotype space (Gavrilets 2004; Chapter 17), the Adaptive Landscape has traditionally had its mains proponents in the field of phenotypic evolution (Chapters 7–9; Arnold et al. 2001). Organismal biologists interested in phenotypes, their form and function, have traditionally been interested in genetic and phenotypic correlations among fitness-related characters, and they have asked questions about the possible genetic constraints on evolutionary change that are reflected in such correlations (Arnold 1992; Schluter 1996; Eroukhmanoff and Svensson 2011). The research field of phenotypic integration (Pigliucci and Preston 2004), for instance, focuses on such correlations and tries to understand their genetic and developmental basis and ecological significance. Such correlational patterns can of course be studied in their own right, and be analyzed without reference to the Adaptive Landscape at all, although phenotypic integration often becomes more interesting when it is projected onto an Adaptive Landscape (Merilä and Björklund 2004). Some evolutionary biologists would even go a step further and argue that the study of phenotypic integration risks becoming an phenomenological exercise which would be of little interest in itself and might be "lost in translation"
if it is not directly or indirectly coupled to an underlying Adaptive Landscape (Arnold 2005). The reason for this is that phenotypic and genetic correlations are expected to evolve by selection, and should ultimately become aligned with the directions of maximum fitness on the Adaptive Landscape (Cheverud 1984). Similar arguments might be put forward for the study of adaptive phenotypic plasticity, which might be enriched theoretically by coupling the development of alternative plastic phenotypes to underlying Adaptive Landscapes (Chapter 7).

For those of us who are rooted in the evolutionary ecology tradition of studying natural and sexual selection in the wild (Lande and Arnold 1983), the Adaptive Landscape has been a source of inspiration and starting point for many empirical investigations (see also Chapter 9). Thus, we are not so much interested in phenotypic integration patterns and genetic correlations *per se*, but rather their adaptive significance and their possible role as constraints on the evolutionary process.

19.5 Remaining conceptual problems: do we really need Adaptive Landscapes?

Obviously, we are somewhat positively biased in our view of the Adaptive Landscape, as we would never have initiated this volume if we did not think it had some value to the fields of ecology and evolutionary biology. That said, one could of course raise the critical question of whether the Adaptive Landscape has been useful at all to these scientific fields, or whether it has caused distraction towards unsolvable problems or even misled generations of biologists (Kaplan 2008; Pigliucci 2008)? These are of course important questions: would ecology and evolution be in a better state today if the Adaptive Landscape concept had never been formulated by Sewall Wright in 1932?

Some philosophers have argued that the Adaptive Landscape has largely caused confusion, even damage, and misled biologists to focus on non-existing problems such as "peak shifts" (Kaplan 2008; Pigliucci 2008). The arguments behind these criticisms is that in multidimensional landscapes, such as in genotype space, what might look like an adaptive valley in a three-dimensional graphical visualization, will in practice disappear when we consider additional genetic dimensions. This is because each organism is the product of thousands of interacting loci (Chapters 2 and 3) and multidimensionality in genotype space should therefore create "ridges" or regions of fitness neutrality, across which populations could travel between adaptive peaks (Gavrilets 2004). Such ridges in higher dimensional genotype space have also been denoted "extra-dimensional bypasses" (Gavrilets 2004), and they "bridge" the fitness valleys that are the foci in models of adaptive peak shifts (Price et al. 1993; Whitlock 1997; Price et al. 2003). The earlier-proposed solution to this problem (ridges and extra-dimensional bypasses) should thus suddenly make such models of adaptive peak shifts largely redundant. If true, then the problem of peak shifts ceases to be a problem at all, and turns out to be based on a misunderstanding. Have evolutionary biologists spent valuable time and effort trying to solve a problem that does not really exist, a problem that arose from a graphical illusion that only exists in a three-dimensional world?

There are several responses to this. First and foremost: although it is possible that the problem of peak shifts might be non-existent if we consider more multidimensional landscapes (Gavrilets 2004; see also Chapter 17), the Adaptive Landscape does not stand and fall with the problem of peak shifts. Nor does the Adaptive Landscape stand and fall with the validity of the shifting balance theory (SBT), that was so intimately connected with the landscape metaphor when it was first formulated by Sewall Wright in 1932 (Chapters 4–6). There is certainly more to the Adaptive Landscape than the problems of peak shifts and the validity of SBT, which has likewise been subject to considerable controversy and discussed extensively elsewhere (Coyne et al. 1997; Wade and Goodnight 1998). Even if "real" Adaptive Landscapes would be characterized by a single "global" peak, it would still be of biological interest to understand the rate of evolution towards this peak, how close populations are to adaptive peaks in natural populations (Arnold et al. 2001; Kingsolver et al. 2001; Estes and Arnold 2007) and the factors that displace populations from their adaptive peaks. Moreover, the highly successful research tradition of optimization analysis in behavioral ecology (as practiced in, e. g. in biomechanics and ecomorphology) is based on the explicit assumption that adaptive peaks (optima) do exist and that these optima can be reached by the process of gradual natural selection (Alerstam and Lindström 1990). Thus, even if peak shifts are not important, and even if Adaptive Landscapes do not exhibit the stability that Wright might have envisioned (Chapter 7) and even if SBT would turn out to be entirely wrong, the Adaptive Landscape would still have some utility in terms of inspiring empirical and theoretical research.

Few scientists do demand that a metaphor or concept necessarily explain *all* interesting problems to be useful, and this is also true for the Adaptive Landscape. The Adaptive Landscape metaphor has obviously played a more pronounced role in some research traditions than in others. For instance, the Adaptive Landscape holds a central position in the field of ecological speciation and the ecological theory of adaptive radiations (Schluter 2000), whereas it has probably been underutilized in studies of animal behavior (but see Chapter 7; Chapter 16) and problems of learning (Chapter 5). The Adaptive Landscape has traditionally and historically not played such as a central role in the fields of molecular evolution and development, although its importance seems to be on the rise also in these areas (Gavrilets 2004; Carneiro and Hartl 2010; Chapters 16 and 17).

Perhaps the current volume will promote a more pluralistic view and pragmatic attitude towards the Adaptive Landscape and its utility. Hopefully, some researchers will realize that there might never exist *the* Adaptive Landscape that will solve all problems or be useful for every circumstance. If so, we have achieved what we hoped for with this volume. The Adaptive Landscape that is useful for a researcher in evolutionary ecology (a phenotypic fitness surface) might be of little interest to a molecular biologist (who may think about a fitness surface of discrete genotypes) and vice versa. Rather than striving to find a philosophically entirely rigorous and completely logical definition of the Adaptive Landscape that everybody will agree upon, we might instead accept the difficulty or impossibility of ever finding such a definition and let several different concepts coexist and even compete with each other.

There are some interesting parallels between the Adaptive Landscape controversy and other great debates in evolutionary biology. For instance, empirically oriented life-history biologists find it very difficult to agree on a single fitness concept, but instead use "fitness" in an operational way and differently in different organismal groups (and this will probably always be the case, given the enormous empirical difficulties of quantifying basic fitness components in natural populations, let alone total fitness!). In spite of these inconsistencies among studies, we still think it fair to say that evolutionary biologists have made substantial progress in terms of our understanding fitness and life-history evolution, perhaps *because* (?) they did not first decide to "solve" the problem of which fitness concept to use, but rather went on to do their research directly. This is largely an empirical attitude that might cause dismay among some philosophers, but it has clearly worked as a research strategy in several areas of ecology and evolutionary biology.

Likewise, speciation biologists have not yet agreed upon *the* perfect species concept, but this has not prevented many of them from making impressive advances in the study of speciation. The Biological Species Concept (BSC; Mayr 1942) is certainly imperfect. For instance, BSC cannot be applied to asexually reproducing organisms and has many other problems. Still, BSC has had an enormous influence on empirical research, has certainly inspired many investigators and contributed to our knowledge about the nature of species and the process of speciation (Coyne and Orr 2004). Again, most biologists might be more interested in practical applications and tools to use in their everyday research, even if some philosophers find it problematic or impossible to do research under such circumstances and without rigorous concepts (Kaplan 2008). Finally, we quote Wagner in this volume (Chapter 17), who writes the following about the Adaptive Landscape:

> There may be philosophical quibbles about it (Provine 1986, discussed also in Chapter 2), but nothing speaks better to its success than its widespread usage almost a century after its conception. The various incarnations the Adaptive Landscape has taken in the hands of different researchers [e.g. Chapters 2, 9, and 19]...show that rather than abandoning the concept, we need to refine it for specific purposes.

It is in this spirit that we hope this volume will stimulate further research about Adaptive Landscapes, their utility, and limitations and to clarify remaining theoretical problems and empirical challenges.

19.6 Empirical challenges

Many of the conceptual and theoretical difficulties with Adaptive Landscapes have been discussed already in this chapter, and are certainly not trivial. However, from our perspective as evolutionary ecologists, we see empirical challenges which are at least as large as the theoretical ones. To study selection in natural population, estimate selection coefficients and visualize fitness surfaces, often requires very large sample sizes and considerable resources in terms of money and time (Chapter 7). Moreover, some of the most sophisticated statistical and analytical approaches that are currently available are currently not even possible to apply to field situations and outside laboratory systems of a few model organisms (Chapter 9). This might be the greatest limitation of the Adaptive Landscape concept for organismal biologists, ecologists, and students of phenotypic evolution. Obtaining such detailed data might either be too expensive or impossible to gather in the field. Nevertheless, it has been possible in some cases, and empirically estimated fitness surfaces are today quite commonly published in many ecological and evolutionary journals (such as *American Naturalist*, *Evolution*, or *Journal of Evolutionary Biology*). This research field has grown rapidly during the last two to three decades, following the publication of the statistical methods for estimating selection coefficients in natural populations in the early eighties (Lande and Arnold 1983; Endler 1986; Kingsolver et al. 2001), and somewhat later the development of visualization techniques of fitness surfaces, such as cubic spline regression and Generalized Additive Models (GAMs)(Schluter 1988; Schluter and Nychka 1994).

Today, the publication of selection gradients and selection differentials is often accompanied by visualization of fitness surfaces, which is a natural complement as these parameters describe the response surfaces, visualized as fitness surfaces (Phillips and Arnold 1989). With increasingly sophisticated statistical and analytical techniques rapidly becoming available to biologists (Chapter 9) it should now also be apparent that we are no longer limited to restricting fitness surfaces to three dimensions. On the contrary, powerful statistical tools are now available that can be applied to data on many traits and in many dimensions (Chapter 9). When Sewall Wright outlined his three-dimensional fitness surfaces in 1932, he did not of course have these statistical tools or computing power at his disposal, nor could he even have imagined that we would have such access.

19.7 Alternatives to the Adaptive Landscape

Scientific theories, concepts, and metaphors do not of course exist for their own sake, but they owe their justification to how useful they are to scientists who wish to solve concrete research problems. To cite the late great population geneticist J. B. S. Haldane: "No scientific theory is worth anything unless it enables us to predict something which is actually going on. Until that is done, theories are a mere game of words, and not such a good game as poetry" (Grant and Grant 1995). This strict principle should of course also hold for the Adaptive Landscape, it is not enough that it is captivating and beautiful, we should certainly require more from it. We have already concluded that the landscape will probably live on. We also think that is time to broaden the concept of the "Adaptive Landscape" to include not only the controversial three-dimensional plots that have been criticized by philosophers and biologists (Chapters 2 and 3), but to underscore that nowadays the concept has now a solid mathematical foundation, at least at the level of phenotypes (Arnold et al. 2001; Chapter 9).

As forcefully argued by Arnold and colleagues (Arnold et al. 2001), the utility of the Adaptive Landscape concept lies as much in the series of formal equations which describe the slope, curvature, and orientation of the fitness surface, as in the graphical illustrations of these surfaces. These parameters, referred to as selection differentials and selection gradients, have a solid mathematical and statistical foundation, which was laid down several decades ago (Lande and Arnold 1983). The parameters are regression coefficients for response surfaces, where the dependent variable (fitness) is regressed against one or several independent variables (traits) and hence have a quantitative and statistically rigorous foundation. The highly successful enterprise of studying natural and sexual selection in the field using regression methods (Lande and Arnold 1983) has resulted in an impressive amount of interesting information about the strength, direction and mode of selection in natural populations (Kingsolver et al. 2001; Kingsolver and Diamond 2011). Moreover, this highly active research area owes its existence directly to the "Simpsonian" phenotypic landscape (Simpson 1944), which was mathematically formalized by Russell Lande in the 1970s (Lande 1976). Therefore, this research field in evolutionary ecology which has taught us so much about selection in natural populations would probably not have existed without the Adaptive Landscape concept, as first suggested by Wright (1932) and later developed by Simpson (1944).

Statistically and analytically speaking, presenting statistics such as selection gradients (= slopes of fitness surfaces) is a complement to visualization of fitness surfaces (a form of Adaptive Landscape), not a replacement. Thus, contrary to Kaplan (2008), who has argued that the Adaptive Landscape should be abandoned and replaced by formal modeling, we do not see such a contradiction between visualization and formal modeling (i.e. parameter estimation). Instead, we think that visualization, formal modeling, and parameter estimation are intimately connected to each other. The Adaptive Landscape in its various versions (fitness surfaces) can thus be modeled as a response surface with fitness as a dependent variable, and the effects of underlying traits (independent variables) can be estimated and quantified using the statistics of selection differentials and selection gradients (Chapter 9). Visualization and parameter estimation are intimately and naturally connected to each other, and should therefore not be viewed as alternatives.

However, given that there are obviously some conceptual problems remaining with the Adaptive Landscape metaphor (Chapter 3), are there any viable alternative concepts for evolutionary biologists that would be more useful? Pigliucci suggests (Chapter 3) that the so-called "morphospaces" of organisms might be such a promising and more empirically tractable alternative to the Adaptive Landscape metaphor, based on his reading and interpretation by McGhee (2006). Clearly, morphospace analysis is an underutilized tool in evolutionary biology and among researchers of morphological evolution, but McGhee himself did not claim morphospace analysis would replace the Adaptive Landscape. Rather, he saw it as an extension and a natural complement to the Adaptive Landscape (McGhee 2006).

The Adaptive Landscape is intimately and historically connected with the Modern Synthesis (MS) in evolutionary biology, which started during the late 1930s, continued during the 1940s, and was more or less completed in the 1950s (Mayr and Provine 1998). Given this historical background, and given recent calls for an "Extended Evolutionary Synthesis" (EES)(Pigliucci 2007), one might wonder if the Adaptive Landscape metaphor will survive such a radical transformation of evolutionary biology that some envision will soon take place (Pigliucci and Müller 2010). However, we are not convinced that the MS is in as deep crisis as it is sometimes portrayed, nor that the EES will necessary replace the MS in the way that some think. Scientific syntheses, including MS, do not simply happen because we want them, or because it is declared that they must come. Rather, they grow organically, from below, and it may be several decades before we even realize that they have taken place and only then do they become the focus of study of historians of science and philosophers (Mayr and Provine 1998).

Other potential alternatives to the Adaptive Landscape are more modest and make less grand claims. For instance, adaptive dynamics (AD) is largely a phenotypic modeling tradition that has been developed somewhat in isolation from the mathematical population genetics tradition that is so intimately connected with the Adaptive Landscape (Chapter 14). Although AD has been criticized because it ignores important genetic details such as recombination constraints on sympatric speciation (Waxman and Gavrilets 2005a, 2005b), the traditionally asexual modeling tradition of AD has some clear logical connections to the Adaptive Landscape, in the focus on adaptive peaks and valleys and evolutionary branching that might result from these aspects of landscape topology (Chapter 14). Thus, from our perspective, we do not see AD as an alternative to the Adaptive Landscape, but rather as a specific modeling technique that has strong conceptual connections (albeit not always made explicit) to the Adaptive Landscape tradition, as outlined by Sewall Wright in 1932. The focus of AD is more on the ecological causes of fitness peaks and valleys, and how they arise than on the underlying genetics of the phenotypic traits under selection. However, these ecological issues have also been investigated more or less independently and empirically in the experimental evolutionary ecology tradition (Chapter 7) as well as in formal quantitative genetic models (Chapter 4). Thus, although AD is a valuable analytical tool that has been useful in terms of drawing attention to the important role of biotic interactions, density- and frequency-dependent (Chapter 14), it would be erroneous to claim that these ecological aspects of selection were ignored in the Adaptive Landscape tradition from Wright (1932) and onwards (see also Chapter 4 for a discussion about the different perspectives of Sewall Wright and Fisher about the evolutionary process).

19.8 Conclusion and the next 80 years

Clearly, the research presented in this volume bears testimony to an extremely powerful idea which, even if it has been a source of considerable confusion, and controversy, has nonetheless survived for eight decades and still holds a central position in evolutionary theory. There is absolutely no question in our mind that the Adaptive Landscape has inspired many biologists and research investigations, and will continue to do so in the future. Much of what we now take for granted in evolutionary biology would be difficult to envision without being aware of the historical influence of the idea of the Adaptive Landscape. If a scientific metaphor or concept such as the Adaptive Landscape should be judged by the number of studies and publications that has resulted from it, the Adaptive Landscape must surely be one of the most successful ideas in evolutionary biology, comparable in influence to (say) the BSC and the neutral theory of molecular evolution. There are probably tens of thousands of papers that have been published that explicitly or implicitly refer to the Adaptive Landscape or use it as a conceptual starting point.

What about the next 80 years? Will the Adaptive Landscape continue to hold such a central position in evolutionary biology, or will it soon be replaced by other metaphors or concepts? This is of course difficult to say, but at the moment we see no sign that any more powerful and useful alternatives are emerging. Even if the MS will eventually be replaced by an EES, as has recently been argued (Pigliucci and Müller 2010), the relationships between both phenotypes and fitness ("the phenotype–fitness map") and between genotype and phenotype ("the genotype–fitness map") are likely to be at the center of many biological investigations for many years to come. Both of these mappings efforts are of course easily and naturally studied and couched in the traditional framework and terminology of the Adaptive Landscape (Chapters 13 and 18). The Adaptive Landscape metaphor still seems to be evolving, and it is now making new incursions into previously neglected areas such as metabolic networks (Chapter 17). What we are still lacking is a macroevolutionary theory of how the Adaptive Landscape itself evolves, and not only microevolutionary dynamics on a given (static) landscape (Chapter 13). We still need understand also some classical problems such as evolutionary stasis (Chapter 15) and we anticipate that the Adaptive Landscape will play some role in solving this classical problem. We definitely need to understand more about agents of selection and the ecological factors influencing Adaptive Landscape topography (Chapters 4, 7, and 14). The Adaptive Landscape is of course interesting as a theoretical idea and in terms of its role in basic science, but it might also help us to understand more pressing applied problems in conservation biology, such as how fitness landscapes of native species will change following the establishment of invasive species (Chapter 12). Clearly, there are many other theoretical and empirical challenges ahead. We therefore anticipate that the coming eight decades will be as exciting as the past.

References

Alerstam, T. and Lindström, Å. (1990). Optimal bird migration: The relative importance of time, energy and safety. In E. D. Gwinner (ed.) Bird migration: the physiology and ecophysiology. Springer-Verlag, Berlin Heidelberg, pp. 331–351.

Arnold, S. J. (1992). Constraints on phenotypic evolution. American Naturalist, 140, S85–S107.

Arnold, S. J. (2005). The ultimate cause of phenotypic integration: lost in translation (book review). Evolution, 59, 2059–2061.

Arnold, S. J., Pfrender, M. E., and Jones, A. G. (2001). The Adaptive Landscape as a conceptual bridge between micro- and macroevolution. Genetica, 112–113, 9–32.

Brodie III, E. D. (2000). Why evolutionary genetics does not always add up. In J. B. Wolf, E. D. Brodie III, and M. J. Wade (eds.) Epistasis and the evolutionary process. Oxford University Press, New York, pp. 3–19.

Carneiro, M. and Hartl, D. L.. (2010). Colloquium papers: Adaptive landscapes and protein evolution. Proceedings of the National Academy of Sciences of the United States of America 107(Suppl 1), 1747–1751.

Cheverud, J. M. (1984). Quantitative genetics and developmental constraints on evolution by selection. Journal of theoretical Biology, 110, 155–171.

Coyne, J. A., Barton, N. H., and Turelli, M. (1997). Perspective: A critique of of Sewall Wright's shifting balance theory of evolution. Evolution, 51, 643–671.

Coyne, J. A. and Orr, H. A. (2004). Speciation. Sinauer Associates Inc., Sunderland, MA.

Dawkins, R. (1976). The Selfish Gene. Oxford University Press, Oxford.

Endler, J. A. (1986). Natural selection in the wild. Princeton University Press, Princeton.

Eroukhmanoff, F. and Svensson, E. I. (2011). Evolution and stability of the G-matrix during colonization of a novel environment. Journal of Evolutionary Biology, 24, 1363–1373.

Estes, S. and Arnold, S. J. (2007). Resolving the paradox of stasis: Models with stabilizing selection explain evolutionary divergence on all timescales. American Naturalist, 169, 227–244.

Fagerström, T. (1987). On theory, data and mathematics in ecology. Oikos, 50, 258–261.

Fear, K. and Price, T. (1998). The adaptive surface in ecology. Oikos, 82, 440–448.

Gavrilets, S. (2004). Fitness landscapes and the origin of species. Princeton University Press, Princeton, NJ.

Grant, P. R. and Grant, B. R. (1995). Predicting microevolutionary responses to directional selection on heritable variation. Evolution, 49, 241–251.

Houle, D. (2010). Numbering the hairs on our heads: The shared challenge and promise of phenomics. Proceedings of the National Academy of Sciences of the United States of America, 107, 1793–1799.

Houle, D., Govindaraju, D. R., and Omholt, S. (2010). Phenomics: the next challenge. Nature Reviews Genetics, 11, 855–866.

Kaplan, J. (2008). The end of the Adaptive Landscape metaphor? Biology & Philosophy, 23, 625–638.

Kingsolver, J. G. and Diamond, S. E. (2011). Phenotypic selection in natural populations: What limits directional selection? American Naturalist, 177, 346–357.

Kingsolver, J. G., Hoekstra, H. E., Hoekstra, J. M., Berrigan, D., Vignieri, S. N., Hill, C. E., et al. (2001). The strength of phenotypic selection in natural populations. American Naturalist, 157, 245–261.

Lande, R. (1976). Natural selection and random genetic drift in phenotypic evolution. Evolution, 30, 314–334.

Lande, R. and Arnold, S. J. (1983). The measurement of selection on correlated characters. Evolution, 37, 1210–1226.

Mayr, E. (1942). Systematics and the origin of species. Columbia University Press, New York.

Mayr, E. and Provine, W. B. (1998). The Evolutionary Synthesis. Perspectives on the Unification of Biology. Harvard University Press, Cambridge, MA.

McGhee, G. R. (2006). The Geometry of Evolution: Adaptive Landscapes and Theoretical Morphospaces. Cambridge University Press, Cambridge.

Merilä, J. and M. Björklund. (2004). Phenotypic integration as a constraint and adaptation. In Pigliucci, M. and K. Preston (eds.) Phenotypic integration: studying the ecology and evolution of complex phenotypes. Oxford University Press, Oxford, pp. 107–129.

Phillips, P. C. (2008). Epistasis—the essential role of gene interactions in the structure and evolution of genetic systems. Nature Review Genetics, 9, 855–867.

Phillips, P. C. and Arnold, S. J. (1989). Visualizing multivariate selection. Evolution, 43, 1209–1266.

Pigliucci, M. (2007). Do we need an extended evolutionary synthesis? Evolution, 61, 2743–2749.

Pigliucci, M. (2008). Sewall Wright's Adaptive Landscape: 1932 vs. 1988. Biology & Philosophy, 23, 591–603.

Pigliucci, M. and Müller, G. B. (2010). Evolution—The Extended Synthesis. The MIT Press, Cambridge, MA.

Pigliucci, M. and Preston, K. (2004). Phenotypic integration: studying the ecology and evolution of complex phenotypes. Oxford University Press, Oxford.

Price, T., Turelli, M., and Slatkin, M. (1993). Peak shifts produced by correlated response to selection. Evolution, 47, 280–290.

Price, T. D., Qvarnstrom, A., and Irwin, D. E. (2003). The role of phenotypic plasticity in driving genetic evolution. Proceedings of the Royal Society London Series B, 270, 1433–1440.

Provine, W. B. (1986). Sewall Wright and evolutionary biology. University of Chicago Press, Chicago, IL.

Schluter, D. (1988). Estimating the form of natural selection on a quantitative trait. Evolution, 42, 849–861.

Schluter, D. (1996). Adaptive radiation along genetic lines of least resistance. Evolution, 50, 1766–1774.

Schluter, D. (2000). The Ecology of Adaptive Radiation. Oxford University Press, Oxford.

Schluter, D. and Nychka, D. (1994). Exploring fitness surfaces. American Naturalist, 143, 597–616.

Simpson, G. G. (1944). Tempo and mode in evolution. Columbia University Press, New York.

Wade, M. J. and Goodnight, C. J. (1998). Perspective: The theories of Fisher and Wright in the context of metapopulations: when nature does many small experiments. Evolution, 52, 1537–1553.

Waxman, D. and Gavrilets, S. (2005a). 20 questions on adaptive dynamics. Journal of Evolutionary Biology, 18, 1139–1154.

Waxman, D. and Gavrilets, S. (2005b). Issues of terminology, gradient dynamics and the ease of sympatric speciation in adaptive dynamics. Journal of Evolutionary Biology, 18, 1214–1219.

Whitlock, M. C. (1997). Founder effects and peak shifts without genetic drift: Adaptive peak shifts occur easily when environments fluctuate slightly. Evolution, 51, 1044–1048.

Wolf, J. B., Brodie III, E. D., and Wade, M. J. (2000). Epistasis and the evolutionary process. Oxford University Press, New York.

Wright, S. (1932). The roles of mutation, inbreeding, crossbreeding and selection in evolution. Proceedings of the Sixth Annual Congress of Genetics 1, 356–366.

Author Index

Abrams, P. A. 239, 240
Adams, D. C. 211
Agrawal, A. A. 219
Agrawal, A. F. 249
Aguirre, W. E. 248
Alberch, P. 26, 36
Albert, R. 275
Alerstam, T. 303
Alfaro, M. E. 183
Allen, J. A. xvii, 94
Alonzo, S. H. 94
Alroy, J. 192
Altenberg, L. 151, 186, 219
Amundsen, T. 100, 106
Anderson, R. M. 95
Andersson, D. I. 176
Andersson, M. 92, 97, 110, 111, 114
Andrén, H. 193
Anfinsen, C. 12
Angilletta Jr., M. J. 100
Anholt, B. R. 131
Ankenman, B. 128–130, 134
Anstey, R. L. 247, 251
Antonovics, J. 194
Arendt, J. D. 191
Arif, S. 248
Armbruster, W. S. viii, xii, 150, 151, 154, 156, 157, 159, 161, 164, 214
Arnold, S. A. 219
Arnold, S. J. 26–29, 31, 36, 75, 89–92, 94, 111–115, 118, 119, 126–129, 131, 133, 134, 139, 140–142, 145, 146, 150, 152–154, 156, 162, 164, 181, 185, 195, 196, 205–207, 209, 210, 214–216, 220, 244, 251–253, 300–305
Arnqvist, G. 100, 120, 122, 131
Aronsson, M. 262, 268
Ashley, M. V. 193
Ashton, K. G. 209
Aspi, J. 76

Atkinson, C. T. 189
Ayala, F. J. 9, 93–95

Babajide, A. 276
Baguette, M. 173
Balogh, A. C. V. 261, 263–265, 267
Bambach, R. K. 246, 247
Barbaro, G. 186
Barrett, R. D. H. 173
Barrick, J. E. 171, 175
Barton, N. H. 9, 63, 78, 79, 126, 142, 145, 151, 239, 265
Bar-Yam, Y. 78
Basolo, A. L. 93
Bastolla, U. 276, 277
Bataillon, T. 175, 176
Bateman, A. J. xvii, 114–116, 123
Bates, D. 138
Baxter, S. W. 261, 264, 267
Beckie, H. J. 186
Behm, J. E. 189
Behrensmeyer, A. K. 246, 247, 248
Bell, G. 90, 126, 162, 164, 171, 173, 184, 195, 196, 208, 214, 218
Bell, M. A. ix, xii, 207, 219, 221, 243, 247, 248, 250, 252
Beltrán, M. 266
Benitez-Vieyra, S. 161
Benkman, C. W. 97, 113, 164, 182, 189
Bennett, A. B. 174
Bentsen, C. L. 118, 119, 131, 164
Berglund, A. 115
Bergstrom, C. T. 194
Berner, D. 216
Bharathan, G. 275
Billerbeck, J. M. 191
Birkhead, T. R. 120–122
Bischoff, R. J. 110
Bisgaard, S. 128–130, 134
Björklund, M. xvii, 302
Blackburn, T. M. 180
Blanckenhorn, W. U. 120
Blomberg, S. P. 209

Blows, M. W. 92, 100, 114, 117–119, 126–131, 136–144, 150, 152, 156, 164, 181, 186, 208, 209, 211, 219, 249
Boisduval, J. A. 266, 310
Bokma, F. 211
Bolker, B. M. 139
Bolnick, D. I. 95, 181, 193
Bolstad, G. H. viii, xii, 150, 157–159, 162, 164
Bonduriansky, R. 100
Bonneaud, C. 100
Bookstein, F. L. 164, 250
Borash, D. J. 95
Bork, P. 277
Both, C. 20, 28, 51, 80, 157, 175, 190, 191, 211, 231, 248, 274, 307
Bowerman, B. L. 137
Bowler, P. J. 244
Box, G. E. P. 128, 131, 133–136, 142, 169, 171, 172, 174, 175
Boyd, C. 188, 190
Boyd, L. H. 75
Bradshaw, A. D. 194, 195, 211
Bradshaw, H. D. 103
Bradshaw, W. E. 191
Brigandt, I. 26
Briggs, W. H. 67
Brisson, J. A. 84
Brodie, E. D. III, 89, 92, 93, 97, 103, 113, 118, 119, 127, 300
Brooks, R. 92, 114, 117, 118, 127, 129, 131–135, 142, 156
Brown, J. H. 189
Brown, J. L. 228, 239
Brown, J. S. 240
Brown, W. D. 122
Brown, W. L. 185
Bryant, E. H. 67
Buckley, L. B. 101
Buckling, A. 69, 169, 173
Bulmer, M. G. 126, 142
Burch, C. L. 177

309

AUTHOR INDEX

Bürger, R. 180, 185, 195, 227
Bush, G. L. 188, 190
Butler, M. A. 212
Butlin, R. K. 118

Caballero, A. 61, 66
Cahya, S. 134
Calsbeek, R. iii, vii, xi, xii, xiv, xv, xvii, 89, 94, 95, 97, 98, 100, 101, 164
Campbell, C. A. 93, 94
Campbell, D. R. 161, 164
Campbell, D. R. 191
Candolin, U. 101
Carlborg, Ö. 80
Carlson, S. M. 191
Carneiro, M. 301, 303
Carroll, S. B. 248, 274, 275
Carroll, S. P. 188, 190
Carson, H. L. 66
Carter, A. J. R. 219
Carter, G. S. 11
Caswell, H. 227
Causier, B. 275
Cerling, T. E. 251
Chaine, A. S. 98, 120
Chamberlain, J. A. 31–34
Champagnat, N. 228
Chan, Y. F. 219
Chao, L. 177
Chapman, T. 100
Charlesworth, B. 78
Charmantier, A. 190, 294
Cheetham, A. H. 250
Chen, L. B. 276
Cheng, C. C.-H. 276
Chenoweth, S. F. vii, xii, 92, 100, 116, 118, 126, 131, 137–139, 141–145, 219
Cheptou, P. O. 193
Cheverud, J. M. 65–67, 78, 79, 128, 219, 302
Chevin, L. M. 180
Chew, F. S. 190
Chippindale, A. K. 100
Chittka, L. 263
Chou, H-H. 177
Christiansen, F. B. 240
Ciliberti, S. 275, 276, 278, 279
Clark, R. 30, 267
Clarke, B. C. 94
Clarke, C. A. 259, 267
Cline, R. E. 272
Cody, M. L. 193
Colegrave, N. 69
Collar, D. C. 212
Colosimo, P. F. 216

Coltman, D. W. 191
Conrad, M. 271
Cooper, N. 211, 219
Cooper, T. F. viii, xii, 169, 171, 173
Cooper, V. S. 170, 174
Copley, R. R. 277
Copley, S. D. 272
Corl, A. 102, 103
Cornwallis, C. K. 100
Cott, H. B. 259
Cowperthwaite, M. C. 28
Cox, G. W. 190
Cox, R. M. xvii, 97, 98, 100, 126
Coyne, J. A. 3, 20–23, 30, 61, 64, 77–79, 126, 215, 260, 265, 303, 304
Craven, P. 131
Crawford, L. E. 262
Crespi, B. J. 97, 127, 150, 164, 193, 211
Cresswell, J. E. 164
Crick, R. E. 250
Crispo, E. 102

Damon, R. A. 68
Damuth, J. D. 75, 76, 157, 193
Dantas, G. 272
Darimont, C. T. 180, 191, 196
Darwin, C. R. 46, 59, 95, 101, 110, 115, 243, 246, 248, 253, 259
da Silva, J. 172
Daugulis, A. J. 272
Davidson, E. H. 275
Davison, A. C. 160
Dawkins, R. 301
Day, T. 96, 100
Dean, A. M. 173, 276
Del Castillo, E. 134
De León, L. F. 189
Dercole, F. 228, 233
de Visser, J. 292
DeWitt, T. 102
Diamond, S. E. 89, 91, 101, 305
DiBattista, J. D. 97, 191, 193
Dieckmann, U. 181, 185, 228, 230–234, 240, 241
Diekmann, O. 227, 228, 239
Dietrich, M. R. v, xii, 3, 9, 13, 16, 243, 301
Dingus, L. 246
Dobzhansky, T. xiv, 8–11, 19, 28, 43, 55, 62, 66, 244, 299
Dodson, M. M. 5
Doebeli, M. ix, xii, 95, 227, 231, 233–236, 238–241
Draper, N. R. 128, 137
Drury, D. W. 68, 69, 80, 83
Dudgeon, S. 216, 217
Dumont, B. L. 212

Duvall, D. 114, 115
Dvorak, J. 151
Dykhuizen, D. E. 173

Early, R. L. 59, 244
Eberhard, W. G. 101–103, 120, 122
Eckert, C. G. 94, 159
Edwards, A. W. F. 20
Edwards, O. R. 185
Einum, S. 95
Eldakar, O. T. 76
Eldredge, N. 11, 210, 216, 217, 246, 248, 250, 251, 253
Elena, S. F. 69, 169
Ellegren, H. 117
Ellstrand, N. C. 189
Emerson, B. C. 194
Emlen, S. T. 115
Endler, J. A. 111, 115, 180, 208, 304
Enfield, F. D. 66
Enquist, M. 261
Eroukhmanoff, F. xvii, 302
Erwin, D. H. 247, 248, 251, 254, 275
Eshel, I. 235
Estes, S. 29, 92, 126, 156, 162, 185, 195, 205, 206, 209, 210, 214–216, 251–253, 303
Evans, C. S. 111
Ewens, W. C. 62
Ewens, W. J. 44, 45, 47, 48
Eyre-Walker, A. 278

Fagerström, T. 299
Fahrig, L. 193
Fairbairn, D. J. 118
Falconer, D. S. 63, 78
Fear, K. K. 181, 182, 187
Fedorka, K. M. 100
Feist, A. M. 273
Feldgarden, M. 194
Feldman, M. W. 97
Felsenstein, J. 95, 144
Fenberg, P. B. 191
Fenster, C. B. 80
Ferea, T. L. 171
Ferenci, T. 171, 173
Ferrada, E. 277
Ferre-D'Amare, A. R. 90
Fisher, H. S. 119
Fisher, R. A. 5, 7, 9, 16, 19, 20, 41–57, 62, 63, 78, 90, 93, 95, 110, 113, 227, 228, 239, 259, 263, 306
Fitter, A. 189
Flatt, T. 292
Fleming, I. A. 115–118
Flury, B. 140

Fong, S. S. 171
Fontana, W. 21, 22, 24
Foote, M. 243
Forbes, A. A. 194
Ford, E. B. 42, 43, 53, 260
Forsgren, E. 100, 106
Forster, J. 273
Frank, S. A. vi, xii, 45–49, 53, 57, 74, 78, 83, 90, 239
Frankham, R. 220
Franks, D. W. 263, 267
Franks, S. J. 191
Freckleton, R. P. 209, 210
Freeman, S. 9
Frentiu, F. D. 145
Friberg, M. 97, 98
Friesen, M. L. 173
Fritts, T. H. 189
Futuyma, D. J. 9, 27, 205, 211, 212, 216, 217, 271

Gaggiotti, O. 77
Gaines, M. S. 218
Galen, C. 161
Gamberale-Stille, G. 262
Garant, D. 187, 193, 209, 294
Garcia, N. 67
Gavrilets, S. xv, 14, 20, 21, 23, 26–28, 30, 31, 36, 80, 81, 92, 100, 181, 218, 227, 249, 265, 271, 278, 285, 299–303, 306
Gelade, G. 262
Gelembiuk, G. W. 164, 219
Gerhardt, H. C. 92, 110, 131, 132
Geritz, S. A. H. 228, 237, 240
Geyer, C. J. 127, 130, 145, 164
Ghalambor, C. K. 190
Ghirlanda, S. 261
Gibson, G. 291
Gilbert, L. E. 266
Gilbert, S. F. 274
Gilchrist, G. W. 152, 154
Gillespie, J. H. 13, 14, 126, 205, 286
Gingerich, P. D. 209, 246, 251, 253
Gittleman, J. L. 209
Giurumescu, C. A. 276, 277, 279
Gleiser, G. 212
Golding, G. B. 276
Goldman, I. L. 67
Goldschmidt, R. B. 260, 265
Gomez, D. 212
Gomulkiewicz, R. 180, 184, 185, 195
Gonzalez, A. viii, xii, 92, 150, 180, 181, 184, 193, 195, 196
Goodman, M. 276

Goodnight, C. J. vi, xii, 62, 64–66, 68–70, 74, 75, 77–80, 82, 83, 215, 280, 300, 303
Gordon, S. P. 194
Gosden, T. P. vii, xii, 89, 98, 99–101, 120, 121, 126, 131, 162, 164, 208, 209, 309
Gould, F. 248
Gould, S. J. 210, 216, 217, 246, 250, 251, 253
Gowaty, P. A. 120
Grabowski, M. W. 219
Graham, C. H. 212
Grant, B. R. 91–93, 305
Grant, P. R. 91–93, 305
Green, P. J. 131
Greene, E. 118
Greenfield, M. D. 133
Grether, G. F. 111, 195
Groom, M. J. 181
Gross, M. R. 115–118
Gvozdik, L. 151
Gyllenberg, M. 238, 239

Habets, M. G. J. L. 173
Hadfield, J. D. 131, 139, 191
Hafner, M. 277, 279
Haglund, T. R. 248
Hahn, M. W. 220
Hairston Jr., N. G. 180
Haldane, J. B. S. 5, 7, 16, 63, 97, 150, 164, 305
Haley, C. 80
Hall, M. D. 122
Hallam, A. 5
Hamilton, T. 11
Hansen, T. F. viii, ix, xii, xv, 91, 141, 144, 145, 150, 151, 153, 154, 160–162, 184, 186, 205–207, 209–214, 216–219, 221, 251, 252
Hanski, I. 77, 194
Hardling, R. 100
Harmon, L. J. 209, 211, 216, 219
Hartl, D. L. 30, 301, 303
Harvey, P. H. 169
Head, M. L. 120
Heap, I. M. 194
Hedrick, P. W. 78, 126
Heino, M. 194
Heisler, I. L. 75, 76, 157
Hendry, A. P. viii, xii, 92, 150, 180, 181, 189, 191–193, 196, 209, 221, 248
Hereford, J. 91, 126, 127, 141, 154, 161, 164, 206–208, 211, 216
Herlihy, D. P. 120
Hermisson, J. 94, 220

Herre, E. A. 150
Herrera, C. M. 157
Herron, J. 9
Hesse, M. 18, 20
Hews, D. K. 95
Higgie, M. 219
Hill, J. K. 193
Hill, W. G. 60, 61, 63, 66
Hine, E. 98, 120, 136, 140, 143, 144
Hinkley, D. V. 160
Hipp, A. L. 212
Hoekstra, H. E. 91, 126, 145
Hoffman, A. 83, 211, 265, 266
Hohenlohe, P. A. 140, 145, 219
Holland, B. 100
Holland, J. H. 61
Holland, S. M. 246–248
Holt, R. D. 180, 184, 185, 195, 196, 218, 239
Holzapfel, C. M. 191
Houle, D. 126, 141, 142, 162, 181, 186, 205–209, 211, 217, 219, 221, 300
Huang, W. 276
Hübner, J. 266, 310
Hueber, S. D. 275
Huey, R. B. 93, 96, 100
Hughes, C. L. 275
Hughes, D. 176
Hughes, J. B. 180
Hunt, G. 209, 210, 219, 243, 247, 248, 250, 252, 253, 254, 310
Hunt, J. vii, xii, 111, 113, 115–119, 126, 133, 135, 142, 143
Hunter, J. S. 134, 135
Huttenlocher, J. 262
Huynen, M. 277

Ibanez-Escriche, N. 68
Irish, V. F. 275
Irwin, D. E. xvii
Ispolatov, Y. 238
Iwasa, Y. 114

Jablonski, D. 217
Jackson, J. B. C. 250
Jaeger, J. 275
Jain, S. K. 194
Janet, A. 3–5, 14
Janzen, F. J. 114, 127, 129, 164
Jasmin, J-N. 174
Jayakar, D. 164
Jensen, H. 152, 209
John, J. A. 137
Johnson, N. 60
Johnson, S. D. 161
Johnson, T. 142

Johnston, M. O. 164
Johnstone, R. A. 100
Jones, A. G. vii, xii, 110, 114, 115, 126
Jones, K. N. 188
Jones, L. E. 94
Joron, M. 260, 261, 264, 266, 267
Jukes, T. 12, 13

Kahn, S. E. 291
Kapan, D. D. 267
Kaplan, J. xv, 10, 18, 19, 20, 27, 29, 91, 300–302, 304, 305
Kassen, R. 93, 97, 173–176
Katz, A. J. 66
Kauffman, S. 21, 22, 271
Kaufman, T. C. 275
Kawecki, T. J. 126, 218
Keightley, P. D. 142, 278
Keitt, T. H. 196
Keller, L. F. 193
Kemp, T. S. 216, 218
Kerr, B. 94
Kettlewell, B. 194
Khan, A. I. 177
Kidwell, S. M. 246–248
Kikuchi, D. W. 267
Kimura, M. 13, 205
King, A. A. 212
King, J. 13
Kingsolver, J. G. H. 29, 89–91, 101, 111, 115–118, 126, 127, 145, 152, 154, 161, 191, 208, 304, 305
Kinnison, M. T. 180, 209, 248
Kirkpatrick, M. 61, 111, 112, 138, 208, 209, 239, 249
Kleina, L. 276
Klump, G. M. 110
Knust, R. 191
Kojima, K. 9
Kokko, H. 100
Kolm, N. 194
Kondrashov, A. 208
Kopp, M. 94
Korona, R. 94, 175
Kot, M. 231, 232
Kotler, B. P. 292
Kowalewski, M. 246, 247
Kozak, K. H. 101, 213
Krebs, J. R. 150
Krishnan, V. V. 95
Kruuk, L. E. B. 111
Kryazhimskiy, S. 175–177
Krzanowski, W. J. 140

Labandeira, C. 212
Labra, A. 212, 218

Lack, D. 181, 194
Laforsch, C. 102
Lahti, D. C. 190
Lande, R. 11, 12, 22, 26–29, 31, 36, 78, 91, 92, 97, 100, 111, 113, 114, 119, 126, 127, 129, 131, 133, 134, 139, 141–146, 150–154, 164, 181, 206, 207, 214, 215, 228, 239, 249, 250, 286, 287, 304, 305
Langbein, L. I. 157
Lank, D. B. 94
Laporte, L. F. 243, 246
Law, R. 228, 230, 232, 240
Law, W. 191
Layzer, D. 160
LeBoeuf, B. J. 98
Lee, C. E. 164, 219
Lee, M. C. 171, 175
Leimar, O. x, xii, 231, 259, 261, 263, 264, 267
Lemons, D. 275
Lenski, R. E. 69, 93, 169–175
Levin, S. 21, 22, 271
Levine, L. 9
Levins, R. 22, 77, 164
Levinton, J. S. 194, 254
Levitt, M. 278
Lewontin, R. C. 9, 10, 18, 20, 22, 283
Liang, Y. H. 276
Lichtman, A. J. 157
Lieberman, B. S. 216, 217
Lindström, Å. 303
Linksvayer, T. A. 61, 64
Lipman, D. 276
Littell, R. C. 137, 138
Lively, C. M. 94, 102
Lloyd, E. A. 18
Loewe, L. 60, 63
Lohmann, I. 275
Lopez-Fanjul, C. 64, 67
Losos, J. B. 97, 180, 194, 216
Lunzer, M. 177
Lynch, M. 17, 126, 128, 130, 150, 180, 195, 205, 206, 208, 214, 250
Lyon, B. E. 98, 120

Maad, J. 161
MacCarthy, T. 275, 276
MacDonald, S. E. 269
MacFadden, B. J. 213, 217, 246, 251
Mackay, T. 78
Maclaurin, J. 36
MacLean, R. C. 171
Majerus, M. E. N. 248
Mallet, J. L. B. x, xiii, 112, 187, 259, 260, 264, 266, 267

Manly, B. 129
Marchinko, K. B. 103
Marroig, G. 219
Marrow, P. 231, 233
Marsh, H. L. 262
Martins, E. P. xv, 144, 216, 218
May, R. M. 95
Maynard Smith, J. 12, 13, 94, 150, 216, 227, 235, 249
Mayr, E. xiv, 27, 62, 66, 89, 246, 304, 306
McCandlish, D. M. 90
McCleery, R. H. 150
McCollum, S. A. 102
McCoy, J. W. 3, 4, 5
McDonald, M. J. 173, 175
McGhee, G. R. 27, 30–36, 244, 252, 253, 306
McGinnis, W. 275
McGlothlin, J. W. xvii, 98
McGuigan, K. 92, 136, 138, 139, 144, 146, 209
McLachlan, G. J. 145
McNeilly, T. 194
McPhail, J. D. 248
McRae, B. H. 196
Mead, L. S. 113, 119
Meffert, L. M. 67
Merilä, J. 195, 302, 294
Merriam, G. 194
Metz, J. A. J. 227–229, 240
Meyer, J. R. 97
Meyer, K. 138
Meyers, L. A. 28
Miller, J. 276
Millien, V. viii, xiii, 180, 193
Milligan, B. G. 249
Mitchell, W. A. 250, 150
Mitchell-Olds, T. 29, 31, 36, 127, 129–131, 135, 144, 145, 164
Moland, E. 191
Moller, A. P. 142
Mooradd, J. A. 61
Moore, A. J. xvii
Moran, P. A. P. 20
Mueller, L. D. 95
Muir, W. M. 61, 62, 69, 70
Müller, G. B. xiv, 306
Mylius, S. D. 239

Naisbit, R. E. 119
Neff, B. D. 122
Neiman, M. 64
Nemec, V. 261
Newman, C. M. 249
Nicholson, A. J. 260, 265
Nijhout, H. F. 267, 291, 292

Niklas, K. J. 36
Nilsson, L. A. 161
Nosil, P. xv, 97, 101, 180, 181, 187, 193
Notley-McRobb, N. 171, 173
Nowak, M. 235
Nychka, D. 92, 114, 131, 145, 164, 304

O'Connell, R. T. 137
Ogata, H. 278
Olsen, E. M. 191–193
Olson, E. C. 212
Ord, T. J. 115
Oring, L. W. 115
Orr, H. A. 64, 67, 78, 126, 175, 180, 185, 195, 304
Orzack, S. H. 150, 221
Osorio, D. 263
Ostrowski, E. A. 171
Othmer, H. G. 275
Overton, J. M. 193

Pagel, M. D. 169
Palsson, B. O. 273
Palumbi, S. R. 180, 196
Papadopoulos, D. 175
Papageorgis, C. 264
Parchman, T. L. 189
Parker, G. A. 100, 120, 150
Parker, T. H. 209
Parmesan, C. 190, 191
Parsons, P. A. 212, 216
Parvinen, K. 238–240
Pavlicev, M. 130
Pavlovic, N. B. 240
Payseur, B. A. 212
Pearce, J. M. 262
Pearson, K. 11, 12, 150
Pélabon, C. viii, xiii, 150, 151, 160, 161, 221, 254
Pelosi, L. 171
Pérez-Barrales, R. viii, xiii, 150, 157, 159
Pergams, O. R. W. 193
Perini, L. 17, 18
Perrot, V. 176
Petry, D. 249
Pfennig, D. W. 191, 267
Pfrender, M. E. 290, 291
Phillimore, A. B. 190, 196
Phillips, B. L. 189
Phillips, P. C. xv, 64, 79, 92, 111, 113, 114, 118, 127, 128, 131, 134, 141, 144, 150, 151, 164, 300, 305
Pigliucci, M. v, xiii, xiv, xv, 18, 26, 31, 36, 37, 92, 102, 244, 246, 252, 254, 299, 300, 302, 306, 307
Pimm, S. L. 180

Pinheiro, J. 138
Pinto, N. 196
Pischedda, A. 100
Pitcher, T. E. 122
Pitnick, S. 100, 122
Pitt, J. N. 90
Pizzari, T. 120–122
Plutynski, A. 18, 20, 27
Polly, D. 219
Pomiankowski, A. 142
Pörtner, H. O. 191
Poulton, E. B. 94, 259, 260
Preston, K. 302
Preziosi, R. F. 118
Price, G. R. 44, 45
Price, T. D. 129, 181, 182, 187, 195, 206, 215, 221, 227, 235, 300, 303
Proulx, S. R. 151, 286
Provine, W. B. xiv, 3, 5, 7–9, 11, 18–20, 27, 42, 43, 59, 74, 75, 89, 95, 252, 279, 299, 300, 304, 306
Punnett, R. C. 259, 260, 262
Punzalan, D. 120, 131
Purvis, A. 211, 219

Qvarnstrom, A. 98

Rainey, P. B. 94, 173
Raman, K. 277
Ratterman, N. L. vii, xiii, 110, 114, 115
Raup, D. M. 27–29, 31–34
Rausher, M. D. 131, 164
Réale, D. 190
Reboud, X. 186
Rehmann, L. 272
Reidys, C. 276
Reimchen, T. E. 97
Relyea, R. A. 102
Rencher, A. C. 141
Rendell, L. 61
Rennell, D. 276
Rescorla, R. A. 261
Reynolds, R. J. 129, 130, 134
Reznick, D. N. 190, 194
Ricciardi, A. 189
Rice, S. H. x, xiii, 227, 283, 285–294
Rice, W. R. 100
Ridenhour, B. J. 103
Ridley, M. 20, 23
Rieseberg, L. H. 187
Rinaldi, S. 228
Ritchie, M. G. 119
Rodda, G. H. 189
Rodrigues, J. F. 273, 274
Roff, D. 150
Rogers, D. G. 159

Rokyta, D. R. 176
Rolff, J. 209
Rønning, B. 209
Rosenqvist, G. 98, 115
Rosenthal, G. G. 110
Rosenzweig, M. L. 95
Rost, B. 277
Roughgarden, J. 95
Roush, R. T. 186
Routman, E. 65, 66, 78, 79
Rowe, L. 100, 122, 142
Roy, K. 191
Rozen, D. E. 172, 173, 175
Rueffler, C. 181, 182, 185, 188
Rundle, H. D. vii, xiii, xv, 97, 98, 120, 126, 131, 140, 141, 180
Ruse, M. 3, 18, 19, 22
Ruxton, G. D. 212, 239, 259, 261, 263, 264, 267
Ryan, M. J. 94, 133

Sadler, P. M. 246, 247
Salick, J. 191
Samal, A. 274
Sanchez, L. 275
Sandholm, W. S. 227
Sanjuán, R. 177
Sax, D. F. 189
Sbordoni, V. 265
Scales, J. A. 212
Schamber, E. M. 61, 62, 69, 70
Scheiner, S. M. 102, 164
Schemske, D. W. 103, 127
Schierenbeck, K. A. 189
Schilling, C. H. 273
Schindel, D. E. 246, 247
Schlaepfer, M. A. 190
Schlichting, C. D. 37
Schluter, D. xv, 92, 95, 97, 101, 111, 114, 126, 131, 143, 144, 145, 150, 164, 174, 180, 181, 183, 185, 187, 194, 196, 208, 215, 216, 218, 219, 252, 300, 302–304
Schmidt, R. S. 262
Schmitt, J. 102
Schoustra, S. E. 93
Schrag, S. J. 176
Schtickzelle, N. 193
Schuett, G. W. 212
Schultz, E. T. 116
Schuster, P. 21, 276–279
Schwaegerle, K. E. 156, 214
Schwarz, D. 188, 189
Scoville, A. G. 290, 291
Searcoid, M. O. 273
Seehausen, O. 101, 119, 195
Serbezov, D. 111

Shannon, S. 150
Shapiro, M. D. 219, 252
Sharpe, D. M. T. 191
Shaw, K. L. 120
Shaw, R. G. 29, 31, 36, 127, 129–131, 145, 164
Sheldon, B. C. 294
Sheldon, P. R. 217
Sheldon, S. P. 117, 188
Sheppard, P. M. 259, 260, 263, 266, 267
Sherratt, T. N. 263, 267
Shimomura, K. 291, 292
Shine, R. 189
Shuster, S. M. 67, 94
Siepielski, A. M. xvii, 29, 36, 89–91, 93, 101, 103, 120, 126, 131, 162, 164, 181, 185, 214
Sigmund, K. 235
Sillén-Tullberg, B. 141
Silverman, B. W. 131
Simons, A. M. 164
Simonsen, A. K. 141
Simpson, G. G. xiv, 5, 8, 10–14, 19, 26–29, 31, 36, 62, 66, 89, 101, 111, 112, 114, 134, 150, 164, 180, 181, 185, 205, 206, 217, 220, 243–246, 248, 250, 252–254, 300, 305
Sinervo, B. xiv, 89, 93–96, 101, 102, 131, 132, 164
Singer, M. C. 190, 260, 264
Skipper, Jr., R. A. v, xiii, 3, 10, 13, 16, 18, 22, 243, 301
Slatkin, M. 45, 46, 53, 67, 90
Sletvold, N. 161
Smith, A. B. 247
Smith, D. C. 102
Smith, J. N. M. 131
Smith, T. B. 95, 101
Smocovitis, V. B. 9
Snook, R. R. 121
Sobel, J. M. 101
Sober, E. 150, 211
Socha, R. 261
Stadler, B. M. R. 21
Stadler, P. 21
Stamps, J. A. 95
Stanley, S. M. 251
Steadman, D. W. 189
Stebbins, G. L. 9, 19
Steiner, K. E. 161
Stenseth, N. C. 216
Stern, H. S. 114, 129, 164
Stevens, L. 67
Stewart, P. 5
Stinchcombe, J. R. 127, 131, 141, 156
Stockwell, C. A. 190, 196

Strauss, S. Y. 189
Stromberg, C. A. E. 213, 217
Sumedha 276, 277
Svensson, E. I. iii, vii, xi, xiii, xiv, xv, xvii, 1, 89, 95–102, 121, 126, 131, 132, 162, 164, 208, 209, 299, 302

Taborsky, M. 118
Takiguchi, M. 272
Taylor, E. B. 101, 181, 189, 193
Taylor, P. D. 194
Thery, M. 212
Thomas, C. D. vii, viii, xii, 13, 89, 150, 190, 193, 205
Thompson, d'A. W. 36
Thompson, J. N. 91, 100, 103, 216
Thornhill, R. 120
Thornton, J. 277
Tobler, M. 194
Todd, A. 276, 277
Tollrian, R. 102
Travisano, M. 94, 170–173
Treisman, A. M. 262
Treves, D. S. 173
Troje, N. F. 262
Tuda, M. 100
Tuljapurkar, S. 227
Turelli, M. 126, 142, 151
Turner, J. R. G. 10, 260, 264–267
Turner, P. E. 173
Tversky, A. 262

Uller, T. xvii, 100
Unckless, R. L. 180, 185, 195
Uyeda, J. C. 91, 102, 209, 210, 214, 221

Valone, T. J. 150
Vamosi, S. M. 97
van Alphen, J. J. M. 119
van Buskirk, J. 212
van der Meer, J. R. 272
van Homrigh, A. 119, 143
van Klinken, R. D. 185
van Tienderen, P. H. 154
van Tienderen, P. M. 207
van Valen, L. 31
Vasi, F. 93
Velicer, G. J. 171, 174
Verdu, M. 212
Vermeij, G. J. 216
Villaverde, A. 67, 83, 86
Vincent, T. L. 228, 239
Visser, M. E. 190, 292
Vitousek, P. M. 180
von Dassow, G. 275

Wade, M. J. vi, xiii, 62, 64–70, 75, 77–80, 83, 94, 156, 164, 215, 303
Wagner, A. x, xiii, 271–279, 304
Wagner, G. P. 84, 151, 186, 214, 219, 275, 292, 283
Wagner, P. J. 243
Wagner, W. E. 119
Wahba, G. 131
Wainwright, P. C. 249
Waller, D. M. 129, 193
Walsh, B. 126, 128, 130, 152, 156, 181, 186, 206, 208, 209, 249
Ward, J. M. 189
Warner, R. R. 94, 116
Waxman, D. 306
Webb, C. 238
Webster, G. L. 156
Weese, D. J. 97
Weinreich, D. M. 172, 177
West-Eberhard, M. J. 101–103
Whalon, M. E. 194
Wheeler, W. M. 5
White, M. J. D. 9, 10, 210
Whitlock, M. C. xv, 30, 64, 83, 111, 112, 170, 188, 303
Whittaker, R. H. 194
Wiens, J. J. 101, 212, 213, 216
Wilbur, W. 276
Wilczynski, W. 133
Willi, Y. 67, 209
Williams, G. C. 60, 63, 79, 205, 211, 217, 254
Williamson, M. 189
Willis, J. H. 64, 67, 78
Wilson, E. O. 185
Wolf, J. B. xv, 89, 102, 293, 300
Woods, R. J. 177
Wright, S. xiv, 3, 5–12, 14–23, 26, 27–32, 36, 41–45, 49, 50–56, 58–64, 66, 68, 74–78, 80, 83, 84, 89, 91, 103, 111, 112, 126, 150, 169, 181, 215, 227, 243, 244, 260, 264, 271, 283, 299, 302, 303, 305, 306
Wu, C. F. J. 129

Yang, X. 251
Yoder, J. B. 219
Yoshida, T. 94
Young, A. 193
Young, K. 97

Zar, J. H. 129
Zeng, Z. B. 139, 214
Zhang, J. 283
Zuk, M. 100

Subject Index

abiotic 93–5, 100, 102, 145, 218
accuracy 150–4, 160, 164
　accurate 18, 93, 154, 164, 266, 267
　inaccuracy 126, 150–2, 154–64
Adaptive Landscape 3, 5–23, 41, 42, 44, 47, 52, 54–6, 58, 59, 61, 62, 69, 70, 74–7, 81, 83, 89–97, 100–3, 110–14, 118, 119, 123, 150, 152, 154–6, 159, 162–4, 169, 170, 172, 173, 175–7, 180–9, 192, 195–7, 205–7, 210, 213–17, 220, 221, 243, 244, 246, 248–54, 260–4, 267, 271, 278, 279, 283, 285, 286, 288, 289, 294, 299–307
　adaptive zone 11, 111, 114, 205, 206, 213, 217, 218, 252
　surface 3–7, 9–11, 19–23, 31, 44, 54, 56, 59, 95, 96, 100, 111–15, 117–20, 126–31, 137–9, 141, 142, 145, 160–4, 170, 173, 177, 184, 185, 195, 196, 243, 244, 271, 284, 288, 290, 300, 303, 305
　see also fitness surfaces
additive 41, 42, 45, 48–50, 54, 58–61, 63–70, 82–4, 142–4, 206–8
　additively 49, 58, 66, 284, 286, 292–4
　additivity 45, 49, 50, 53, 60, 78, 80, 83
　nonadditive 49
Akaike (AIC) 250, 130, 138
Alcaligenes eutrophus 173
Allee effect 238
allele 9, 16, 26, 47, 48–51, 60–2, 82, 83, 252, 260, 271
　allelic 27, 61, 64, 82, 246
　allelomorph 59
amino acid 272, 276
ammonoid 31–3
analysis of variance (ANOVA) 64
　MANOVA 134
Anchitheriinae 29, 244, 251
Anolis sagrei 95, 97, 98, 212

anole 95
Anser indicus 276
apid 157
aposematic 259, 260, 264, 267
armaments 110, 122
artificial selection 59, 61, 66, 70, 110, 220, 252
assortative mating 32, 241
axes 6, 10, 19, 26, 29, 74–6, 81, 111, 142–4, 243–5, 300, 310

bacteria 66, 69, 94, 171, 173
　bacterial 97, 169, 175, 299
　bacteriophage 176
　bacterium 97, 173
Bateman gradient 114–16
behavior 61, 76, 94, 97, 102, 110, 111, 121, 186, 196, 250, 260, 303
　behavioral 94, 110, 111, 205, 207, 299, 301, 303
biotic 46, 60, 89, 93, 94, 96, 100, 101, 103, 218, 306
　abiotic 93–5, 100, 102, 145, 218
　xenobiotic 272
Biston betularia 97
blossom 157–9, 161–3
bottleneck 65
Brassica rapa 67
breeder 95
　breeding 54, 61, 65, 68, 69, 80, 83, 114–17, 206, 214, 310
　breeding value 65, 68
butterfly 193, 259, 261, 267

canonical analysis 114, 128–30, 134, 140, 142, 143
carrying capacity (K) 22, 67, 231–4, 236–8, 287, 289, 290
Cepea 96
chaotic 229, 240
Chicago 5, 9, 42
circuit 196, 272, 275, 276, 279
cladogenesis 180, 246

climate 93, 100, 180, 185, 188, 190, 191, 196, 197, 211–13
clone 170, 171, 173
　clonal 172, 173, 241
Coccinellidae 8
coevolution 97, 233, 264
　coevolutionary 103, 234
　see also evolution
collard lizards 84
competition 32, 42, 47, 49, 50, 93–8, 110, 115, 120, 122, 132, 173, 181, 183, 185, 218, 232, 236–8
complex systems 54
contextual analysis 75–6
continuously stable strategies (CSS) 170, 235, 240
converge 170, 210, 230, 231, 235, 237, 241, 264
　convergence 228–31, 233–7, 240, 241, 264
　convergent 216, 230, 233, 235–7, 240, 246
correlational selection, *see* natural selection
courtship 110
crypsis 259
Curculionidae 157
curvature 118, 127–9, 139, 150, 175, 214, 219, 236, 237, 284, 287, 289, 305
cuticular hydrocarbons 144
cytonuclear 65, 78

Dalechampia 152, 156, 157, 159–61
damselfly 120, 121
　calopterygid 97
Daphnia 290, 291
Darwin 46, 59, 95, 101, 110, 115, 243, 246, 248, 253, 259
　Darwinian 8, 9, 243, 248, 260
deme 44, 55, 58, 64–7, 69, 70, 78, 79, 81, 82
　demic 70

SUBJECT INDEX

demographic 78, 100, 186, 192, 196, 215
demography 84
density dependence 89, 95, 103, 182, 186
density dependent 96, 102, 287
determinant 230, 233, 292
determinism 286
development 82, 103, 151, 154, 181, 257, 275, 283–5, 288–94, 301–4
dimensionality 26, 31, 36, 74, 75, 81, 84, 210, 244, 249, 271, 277, 279
 dimensional 6, 16, 27, 28, 30, 32, 74, 75, 80, 261–3, 271–3, 275, 277, 279, 287, 290, 292
dimorphism 94, 100, 110, 126, 212, 237
 dimorphic 110, 115, 120
 sexual dimorphism
diploid 47, 48
dispersal 69, 184, 193, 194, 196, 239
disruptive selection, *see* natural selection
diversification 41, 63, 79, 100, 103, 180, 187, 190, 219, 235, 240, 241, 260, 266
DNA 12, 13, 272, 273, 276, 277
Dobzhansky–Muller 299
dominance 22, 43, 44, 49, 50, 54, 64, 65, 67, 68, 78–80, 82
drift 3, 8, 9, 13, 17, 21, 26, 28, 30–2, 41–5, 58–65, 68, 69, 77–84, 101, 118, 144, 169, 170–2, 185, 186, 211, 214, 215, 221, 240, 241, 243, 249, 250, 251, 253, 254, 260
Drosophila 8, 92, 119, 141, 142, 208, 209
dynamic 7, 8, 10, 89, 90, 92, 95, 96, 100, 101, 103, 114, 208, 228, 233, 244
 dynamical 16, 22, 23, 45, 52, 206, 220, 228, 230, 231, 234, 237, 239, 240
 dynamically 227, 239
 dynamics 5, 7–9, 11, 27, 41–3, 50–5, 84, 87, 93, 94, 110, 219, 220, 227–41, 306, 307

E. coli 93, 94, 170, 173, 175, 176, 273
effective population size (Ne) 29, 43, 63, 65, 67, 214, 215, 249, 251, 252, 279
eigenvector 128–30, 134, 136, 137, 139, 140, 288
 eigenfunction 136, 138, 139
 eigenstructure 129, 140

eigenvalues 128–30, 134, 139, 140, 145, 230, 233, 234
empirical 9, 10, 18, 20, 28, 31, 36, 58–66, 68–70, 89, 90, 95, 98, 100, 101, 103, 111–13, 116–19, 169, 176, 177, 261, 264, 267, 283, 300, 302–4, 306, 307
environment 4, 5, 7, 8, 11, 43, 45–50, 52, 53, 57, 58, 61, 62, 74, 75, 83, 84, 94, 102, 114, 115, 120, 152, 156, 162, 169–76, 184, 188, 196, 205, 216–18, 228, 229, 235, 240, 244, 273, 276, 290, 293, 294
enzyme 18, 272–4
 enzymatic 272, 277, 284
Eocene 11, 29, 244, 245
ephialtoid 265, 266, 310
Ephyris pendularia 94
epistasis 76, 79, 300, 172,
 epistacic 7, 16, 22, 41, 43, 48, 49, 52, 61, 64, 66, 68, 79, 80, 82, 83, 89, 220, 283, 292
 epistatically 170, 292
 interacting 65, 76, 82, 97, 100, 123, 170, 175, 188, 294
 interaction 7, 17, 18, 22, 51, 55, 58, 60, 62, 64, 70, 78–80, 82, 84, 89, 94, 102, 119, 136, 137, 156, 231, 272, 276, 292
Equidae 11, 28, 244, 245
 equids 31
evolution 5, 7–13, 26–8, 41–6, 58–64, 66, 69, 78–82, 84, 100–3, 110–12, 115, 117–19, 122, 123, 126, 133, 139, 170–7, 180, 181, 184–97, 205–11, 213–21, 227–9, 239–50, 252–4, 259–67, 283–92, 301–4
 evolutionarily 233–5, 237, 240
 macroevolution 12, 26, 27, 29, 89, 203, 205, 206, 215, 216, 219, 220, 244, 250, 252, 301
 macroevolutionary 11, 26, 27, 36, 83, 205, 206, 211, 213, 214, 216, 218, 220, 221, 299, 301, 307
 microevolution 26, 27, 126, 205, 206, 244, 246, 252
 microevolutionary 26, 28, 62, 87, 126, 144, 145, 205, 206, 219, 220, 243, 244, 253, 307
evolutionary stable strategy (ESS) 235
extinct 30, 32, 42, 229, 239
 extinction 11, 32, 35, 70, 78, 123, 172, 180, 184, 189, 192, 195, 215, 218, 232, 238, 239, 247, 251
extradimensional bypass 271

fecundity 59, 113, 191
fitness 3, 5, 6, 7, 9, 10, 17, 19, 20–2, 26–32, 36, 41, 42, 44–52, 54–70, 74–83, 89, 91–6, 100–3, 110–22, 126–31, 134, 135, 137–42, 145, 150–2, 154–64, 170–7, 180–97, 205–8, 214, 215, 227–41, 243–6, 249–52, 259, 260, 263, 264, 271, 278, 283–9, 292, 299–307
 fitnesses 20, 21, 48, 58, 76, 89, 94, 164, 299, 300
 multipeaked 42, 215, 267
 peak 17, 18, 29–33, 35, 36, 41–4, 74, 76–8, 81–4, 100–3, 113, 117, 132, 134, 170, 171, 173, 175, 176, 181–92, 194–7, 214–16, 244–6, 248–53, 259–64, 284, 285, 289, 302, 303
 see also valley
fitness surfaces 3, 26, 27, 29, 31, 36, 44, 54, 89, 91, 92, 95, 96, 100, 101, 112, 118, 126–8, 130, 131, 134, 135, 137–42, 145, 150, 154, 156, 160, 161, 163, 164, 170, 195, 300, 303, 304, 305
fixation 64, 66, 69, 79, 103, 172, 208, 240
fixed 17, 44, 46, 49, 68, 83, 102, 133, 136–9, 141, 155, 162, 164, 174, 205, 206, 220, 227, 291
fixed effect 136–8, 141
fossil record 11, 13, 205, 207–12, 214, 216, 243, 244, 246–54
 fossils 4, 246–8, 250, 251, 254
 geological record 243
F_{ST} 61, 64, 66–9
function 28, 33, 44, 116, 129, 134–6, 151–60, 162, 164, 214, 216, 220, 228, 229, 231, 232, 234–6, 238–40, 260–4, 274, 276, 277, 278, 283–9, 292–4
fundamental theorem 41, 42, 44–55, 63, 90, 227, 228, 239

Geospiza magnirostris 91, 183
Galapagos 91, 92, 183, 189, 190
game theory 227, 241
Gasterosteus aculeatus 95, 248, 250
Gaussian 145, 231, 238, 261, 262, 286
gene flow 61, 100, 181, 187, 193, 197, 216–18, 251
generalized cross-validation (GCV) 131, 134
generalized linear models 129, 135, 136
 link function 129, 135, 136

SUBJECT INDEX

genetic architecture 69, 89, 126, 219, 25211
genetic background 63–6, 68, 76, 82, 171, 172
genetic correlations 69, 100, 102, 208, 215, 219, 249, 252, 302
genomics 32, 90, 267, 300
　genome 13, 54, 59, 61, 65, 69, 151, 171, 175, 177, 205, 267, 272, 274, 292, 294
　genotype 19, 21, 22, 26–9, 36, 48, 51, 53, 55, 57, 58, 60, 61, 65, 74–6, 80, 89, 90, 102, 103, 153–4, 161, 164, 170, 172, 175–7, 205, 208, 260, 271–9, 283, 300, 302, 303, 307
　　genotypes 6, 9, 10, 21, 26–8, 32, 47–9, 51, 55, 59, 60, 62, 63, 75, 76, 80, 89, 94, 153–4, 172, 173, 177, 243, 244, 252, 271–9, 287, 292, 299, 300, 303
　　genotypic 6, 26–8, 31, 36, 60, 89, 90, 94, 131, 175, 177, 243, 244, 249, 279, 299
　　frequency 9, 10, 19, 26, 30–5, 43–5, 48, 52–7, 60, 61, 63, 65, 67, 70, 76, 79, 83, 89, 93, 94, 102, 103, 113, 131–4, 162, 170, 172, 181–3, 185, 188, 205, 227, 228, 236–40, 250, 259, 260, 264, 287
　　mapping 22, 26
　　see also phenotype
geometry 126–8, 131, 138, 142, 251, 283, 284, 287, 289
Geospiza fortis 91, 92, 183
G-matrix 140, 206, 209, 287, 290
Gobiusculus flavescens 100
goldschmidtian 267
gradualism 259
group selection 62, 63, 67, 76, 79

haplotypes 19
Heliconius 263, 266, 267
heritability 61, 64, 70, 75, 83, 118, 206, 207, 249, 251, 283, 290, 292–4
　heritable 50, 54, 60, 67, 75, 244, 248, 249, 252, 284, 293, 294
heterozygote 48
heuristic 10, 16, 18–20, 22, 23, 28, 29–32, 36, 95, 112, 115, 157, 165, 246, 267, 283, 301
hill climbing 41, 58, 59, 267
holey landscape 14, 20, 21, 23, 26, 27, 30, 36, 31, 80, 81, 92, 100, 181, 218, 227, 249, 265, 271, 278, 285, 299–303, 306
　Wright–Fisher 41, 50

Wrightian 21, 22, 43, 90, 205, 206, 264
　see also extradimensional bipass
homeobox 275
hypsodonty 212, 213, 218, 251
Hyracotheriinae 29, 244, 245, 251

Ichthyosaura alpestris 151
inbreeding 5, 16, 62, 64, 67, 78, 159, 162, 193, 299
industrial melanism 248
innovation
　innovations 271–6, 278
invasion 79, 228–36, 238, 239
Ischnura elegans 99, 120, 309
Ithomiinae 266

Jacobian matrix 230, 231, 233, 234

kernel 236–8

learning 30, 61, 260, 261, 303
least-squares 137, 127, 129, 130, 135, 145
linkage 45, 48, 53

macroevolution 12, 26, 27, 29, 89, 203, 205, 206, 215, 216, 219, 220, 244, 250, 252, 301
male–male competition
　male–male combat 111, 117
mass selection 21, 60, 61, 63, 77–9, 82
maternal effects 76, 180, 184, 187
maximum likelihood 129, 137, 138, 250
mean 45, 50, 53–5, 65, 66, 68, 82, 111, 113, 114, 141, 144, 145, 153, 156, 206, 210, 214, 235, 247, 249, 274, 277, 283, 286, 294
mean field 74, 78
Mendelian 5, 8, 16, 76, 244, 259
metabolism 272–4, 278
　metabolic 171, 195, 249, 271–4, 276–9, 301, 302, 307
　metabolite 173
metaphor 3, 5, 8, 17–20, 22, 23, 26–8, 30–2, 36, 54, 56, 58–61, 74, 75, 78, 81, 84, 95, 113, 169, 177, 196, 197, 206, 227, 243, 246, 251–3, 271, 301–3, 306, 307
microbe
　microbes 272, 278
　microbial 171, 176
microevolution 26, 27, 126, 205, 206, 244, 246, 252

microevolutionary 26, 28, 62, 87, 126, 144, 145, 205, 206, 219, 220, 243, 244, 253, 307
mimic 118, 259, 260, 262, 264, 265
mimicry 259–67, 299, 310
mimetic 263, 264, 266, 267, 310
model 259, 260, 262, 264, 265, 267
Müllerian 259–64, 266, 267
Miocene 11, 29, 217, 245, 246, 250
modern synthesis 26, 27, 62, 306
monomorphism 265
　monomorphic 228, 229, 231, 232, 235–8, 240, 241
Monte Carlo 129, 139, 274
Moraba scurra 9, 10
morphospaces 26–32, 31–7, 246
multipeaked 42, 215, 267
multiple regression 26, 29, 31, 36, 75, 76, 114, 126, 127, 129, 137, 138, 150
Musca domestica 67
mutationism 5
　mutationists 244

natural selection 12, 13, 26, 30–3, 36, 41–56, 58, 59, 61–3, 70, 89–93, 95–8, 101, 111, 114, 120, 123, 132, 141, 142, 144, 165, 170, 176, 177, 191, 205, 211, 214, 228, 243, 244, 246, 248, 250–3, 259, 303
　correlational 89, 102, 113, 114, 119, 127–9, 133, 138–40, 145, 156
　disruptive 91, 95, 96, 113, 118, 119, 121, 129, 164, 184, 185, 241, 284, 288
　frequency-dependent 9, 89, 93, 94, 102, 134, 227, 228, 236–8, 240, 259
　group 43, 62, 63, 67, 76, 79
　intensity 11, 17, 22, 23, 45, 46, 51, 70, 93, 98, 101, 115, 120, 132, 185, 194
　interdemic 21, 30, 58, 59, 62, 63, 64, 66, 69, 77, 82
　quadratic 118, 121, 127, 128, 138, 142, 150, 152, 153, 159, 215
　stabilizing 84, 91, 92, 113, 117–19, 121, 122, 127–9, 132–5, 142, 143, 152, 153, 156, 159, 162, 184, 185, 191, 192, 205, 209, 213, 215, 217, 219, 220, 221, 233, 236–8, 251–3, 277, 284, 285, 288, 292, 310, 261, 284
　see also artificial selection
nautilids 32, 35, 36
neo-Darwinian synthesis 8, 9, 248, 260

SUBJECT INDEX

networks 6, 12, 177, 271, 272, 274–7, 279, 299, 301, 302, 307
neutral theory 13, 205, 220, 307
niche 84, 97, 101, 212, 213, 216–18
novelty 81, 195, 283, 294

Oligocene 29, 244, 246, 249
one-dimensional 231, 234, 236, 237, 240, 261, 287
operational sex ratio 100
optimization 227, 303
  optimizing 30
  optimum 58, 59, 61, 64, 79, 102, 111, 117, 118, 145, 150–6, 159, 162–4, 185, 205, 212–18, 251, 252, 285, 292
ornaments 110, 113, 119, 122
Ozark Mountains 84

paleontology 8, 10, 11, 26, 32, 36, 37, 62, 243, 246, 248, 252, 253
  paleobiological 213, 220
panmixia 8, 52–4
Papilio dardanus 267
Papilio polytes 259
paradigm shift 90
parametric 129, 131, 160
  non-parametric 129, 131
partial regression coefficient 48, 127
Passerina amoena 118
peak shift 30, 31, 43, 44, 52, 84, 90, 102, 187, 215, 249, 250, 252, 261
peucedanoid 265, 266, 310
phase 17, 23, 44, 59, 60–7, 69, 70, 77–83, 122, 157, 159, 162, 193, 210, 218, 291, 292
phenotype 4, 21, 28, 29, 31, 36, 50, 61, 68, 74–6, 78, 79, 82, 83, 89, 103, 111, 112, 117, 122, 134, 144, 150, 152–6, 159, 161, 164, 171–3, 181, 185, 186, 196, 208, 219, 220, 227–32, 234, 236, 237, 238, 241, 249, 251, 260–4, 271–9, 283–94, 300, 301, 307
Phymata americana 120
Phyrnosoma mcalli 97
physiological epistasis 79
pipefish 115, 116
plant 61, 67, 157, 158, 180, 184, 190, 191
plasticity 3, 30, 98, 101–3, 180, 184, 187, 190, 196, 205, 289, 302
Plato
  Platonic 112
pleiotropy 76, 145, 171, 174, 214, 219, 220, 262

pleiotropic 175, 208, 219, 220, 262
plesiomorphic 251, 252
P-matrix 141
  see also G-matrix
Poecilia reticulata 118
poker 60
pollen 151, 154, 157–60
  pollinator 151, 154, 157, 159, 163
polygenic 76, 300
polymorphism 55, 94, 118, 173, 237, 240, 260, 265
  polymorphic 94, 103, 241, 259, 267
polyploidy 101, 187
population genetics 3, 5, 7, 16, 19–23, 26–8, 30, 42, 58, 74, 150, 152, 205, 215, 216, 220, 227, 253, 299, 306
population size 17, 22, 43, 53, 65, 76, 80–3, 169, 172, 183, 194, 195, 196, 214, 217, 230, 232, 249, 251
postcopulatory 111, 117, 120–3
precision 150, 151, 160, 161, 165
  imprecision 150–6, 159–64, 285
precopulatory 115, 117, 120–3
predation 89, 93, 94, 96–8, 102, 157, 159, 163, 194, 232, 259, 260, 291
  predator 97, 98, 102, 159, 212, 231–4, 260–4, 267, 275
predator–prey 97, 229, 231, 233, 261
projection pursuit 114, 131
protein 12, 13, 28–30, 36, 135, 271, 272, 276, 278, 279, 299, 301
Pseudomonas fluorescens 94, 97, 173
punctuated equilibria 31, 210, 216, 217, 246, 248, 249, 250, 251, 253

QTL 80, 292
quantitative genetics 27, 65, 67, 78, 83, 92, 126, 127, 131, 139, 141, 142, 145, 150, 206, 219, 228, 239, 251, 283, 301, 306
  see also population genetics

random effects 136–9, 163
random walk 13, 250, 251, 260, 274
rank 60, 130, 139, 287, 289, 290
recessive 65, 66, 252
recombination 21, 42, 45, 61, 63, 102, 212, 241, 262, 306
repeatability 169
  repeatable 97, 172
ribosome 176
ridge 4, 10, 152, 156, 185, 196, 263, 278, 285
RNA 12, 21, 28–30, 36, 271–3, 276, 277
rugged landscape 102, 170

salmon 115, 117
saltation 260, 262, 263, 265–7
  saltational 5, 259, 265, 267, 299
sedimentation 247
selection differential 206, 289
selection gradients 91, 114, 116, 126, 127–9, 130, 131, 133, 135–9, 141, 142, 145, 150, 159, 207, 208, 211, 214–16, 229–34, 236, 237, 240, 305
sex ratio 61, 100, 120, 227
sexual conflict 100, 103, 113, 116, 117, 122, 299
sexual selection 89, 97, 98, 100, 103, 110–23, 132–7, 141, 142, 144, 218, 302, 305
  female preference 98, 111, 118–20
  male–male competition 97, 98, 110, 111
  preferences 97, 98, 100, 110, 111, 113, 114, 117–20, 122, 123, 141, 213, 217
  runaway 113
  fisherian 21, 23, 53, 113, 263, 264
shifting balance (SBT) 3, 7, 8, 17, 19, 20, 22, 23, 30, 39, 41, 42, 44, 45, 54, 58, 59, 62, 63, 66, 69, 74, 76–8, 84, 111, 227, 260, 264, 267, 303
  see also peak shift
side-blotched lizards 102, 132
  Uta stansburiana 96, 102, 132
Simpsonian landscapes 89, 90, 205, 301, 305
slope 44, 75, 83, 113, 116, 236, 248, 260–3, 284, 285, 287–90, 292, 293
social 60, 61, 75, 76, 100, 111, 113, 115, 120, 301
speciation 19–21, 26, 28, 36, 59, 62, 78, 90, 100–2, 119, 123, 180, 194, 203, 210, 211, 218, 240, 241, 248, 278, 285, 299, 300, 303, 304, 306
  ecological speciation 28, 90, 100, 101, 180, 303
spermatophores 122
stability 4, 7, 92, 93, 97, 100, 173, 205, 228, 229, 230, 231, 233–5, 237, 240, 241, 303
stasis 90, 92, 126, 162, 205, 217, 219, 220, 249, 251, 307
statistical 9, 17, 27, 29, 31, 36, 46, 50, 51, 54, 61, 75, 79, 92, 93, 114, 127, 129, 130, 131, 136, 137, 140, 141, 143, 145, 150, 171, 210, 243, 250, 254, 304, 305
stigma 151, 154, 157, 159, 162

stochastic 31, 41, 42, 44, 91, 100, 211, 215, 216, 218, 220, 227, 228, 230, 239, 240, 241, 286
strategy 23, 60, 61, 118, 164, 176, 212, 235, 262, 304
stratigraphic 246–8, 250, 254
Syntomis 265, 266, 310

Taricha 97
Teleogryllus commodus 118, 122, 133
tensor 287–90
terpenoid 156
Tetrahymena thermophila 97
Thamnophis sirtalis 97
topography 6, 10, 13, 74, 75–7, 80, 81, 84, 96, 102, 163, 164, 184–6, 196, 197, 249, 307

trade-off 95, 103, 173, 174
Tribolium castaneum 66, 70, 80
  flour beetles 66, 67, 68, 69
Tribulus cistoides 91

USDA 59, 61

valley 13, 30, 41, 77, 82, 84, 101, 113, 176, 184–6, 215, 244, 249, 252, 253, 260–4, 267, 303
variance-covariance 130, 144, 150, 237
  covariance 65, 68, 126, 127, 132, 133, 136–40, 144, 154, 196, 206, 230, 247, 287, 290, 292, 293
  covariation 154, 293

covary 152
variance 44, 45, 47–52, 54, 60, 63–9, 78–80, 82–4, 93, 112, 123, 126, 127, 129, 130, 131, 135–45, 150–4, 161, 164, 181, 192, 197, 206–11, 214, 215, 220, 230, 233, 247, 249, 250, 251, 285, 287–9, 293, 294, 310
vector 4, 127, 130, 133, 136–40, 142, 152, 156, 229, 230, 284, 287–9, 293
vertebrate 211, 248

Wisconsin 5
wrasse 116

Zygaena 265, 266, 310